NONLINEAR SIGNAL and IMAGE PROCESSING

Theory, Methods, and Applications

THE ELECTRICAL ENGINEERING
AND APPLIED SIGNAL PROCESSING SERIES
Edited by Alexander Poularikas

The Advanced Signal Processing Handbook:
Theory and Implementation for Radar, Sonar,
and Medical Imaging Real-Time Systems
Stergios Stergiopoulos

The Transform and Data Compression Handbook
K.R. Rao and P.C. Yip

Handbook of Multisensor Data Fusion
David Hall and James Llinas

Handbook of Neural Network Signal Processing
Yu Hen Hu and Jenq-Neng Hwang

Handbook of Antennas in Wireless Communications
Lal Chand Godara

Noise Reduction in Speech Applications
Gillian M. Davis

Signal Processing Noise
Vyacheslav P. Tuzlukov

Digital Signal Processing with Examples in MATLAB®
Samuel Stearns

Applications in Time-Frequency Signal Processing
Antonia Papandreou-Suppappola

The Digital Color Imaging Handbook
Gaurav Sharma

Pattern Recognition in Speech and Language Processing
Wu Chou and Biing-Hwang Juang

Propagation Handbook for Wireless Communication System Design
Robert K. Crane

Nonlinear Signal and Image Processing: Theory, Methods, and Applications
Kenneth E. Barner and Gonzalo R. Arce

Forthcoming Titles

Smart Antennas
Lal Chand Godara

Soft Computing with MATLAB®
Ali Zilouchian

NONLINEAR SIGNAL and IMAGE PROCESSING

Theory, Methods, and Applications

Edited by
Kenneth E. Barner
Gonzalo R. Arce

CRC PRESS

Boca Raton London New York Washington, D.C.

Library of Congress Cataloging-in-Publication Data

Nonlinear signal and image processing / edited by Kenneth E. Barner, Gonzalo R. Arce.
    p. cm. — (Electrical engineering & applied signal processing)
    Includes bibliographical references and index.
    ISBN 0-8493-1427-5 (alk. paper)
    1. Signal processing—Digital techniques. 2. Image processing—Digital techniques. I.
Barner, Kenneth E. II. Arce, Gonzalo R. III. Electrical engineering and applied signal
processing series.

TK5102.9N66 2003
621.382′2—dc22
                                2003055587

Dedication

To the students and colleagues
who make working in the dynamic field of nonlinear signal processing so much fun
and whose efforts made this book possible.

Preface

Nonlinear signal processing methods continue to grow in popularity and use. This growth is due to one factor—performance. While it is true that linear methods continue to dominate in current practice, nonlinear methods are making steady progress in moving from theoretical explorations to practical implementations. Clearly, the advances in computing performance have accelerated this progress by making nonlinear methods more practical. Additionally, nonlinear theory continues to grow and yield a firm foundation upon which nonlinear methods can be developed, optimized, and analyzed. Nonlinear methods are thus being applied to address many of the most interesting and challenging signal processing problems of the day.

This book details recent advances in nonlinear theory and methods. A wide array of contemporary applications in which nonlinear methods are being applied to address challenging open problems are also presented. Although there is no single theory under which nonlinear methods are unified— these approaches are defined simply by what they are not, namely, linear— significant advances have been made in recent years in several branches of nonlinear theory. The first set of chapters in the book is therefore focused on recent advances in nonlinear signal processing theory. This set of chapters targets three critical areas of theory: (1) filter analysis, (2) nonlinear filter class design, and (3) signal analysis.

The filter analysis presented utilizes new nonlinear approaches to analyzing the performance of adaptive filters. By utilizing nonlinear analysis, adaptive filters can be characterized in terms of a state-space model that lends greater insight. Although no single theory unites the large number of nonlinear approaches reported in the literature, approaches founded on maximum likelihood principles define large filter classes that can address many challenging problems. Here, we present two such filter classes: fuzzy ordering theory-based filters and myriad filters. Fuzzy ordering unites the theories of rank ordering and fuzzy relations to yield a broad class of robust filters. Myriad filters are also a robust class of filters designed specifically, in this case, to address applications with stable distributions statistics. The signal analysis presented here is based on time-frequency distributions. Although time-frequency distributions have been studied for some time, the current results presented here show that the Wigner distribution can be represented as a dynamic equation, the solution of which is an ordinary or partial differential equation. The advantage of this approach is the insight into the nature of the solution provided by the analysis. Although this first set of chapters is focused primarily on theory, each includes numerous examples and applications illustrating the advantages of nonlinear approaches.

The remainder of the book focuses on the application of nonlinear methods to a wide array of contemporary applications. Nonlinear approaches, once used only in niche problems, are now becoming ubiquitous across the broad spectrum of applications that have signal processing components. Of the numerous important applications now dominating researcher efforts, we have chosen to focus on applications in communications and networking, imaging and video, and genomics.

Within the area of communications and networking, methods and results are presented for data traffic modeling, echo cancellation in mobile terminals, and blind and semiblind channel estimation. Imaging and video, natural applications for nonlinear approaches due to the nonstationarity of the signals and the nonlinearity of the human visual system, are addressed in several chapters. Specifically, current results on image and video enhancement, wavelet domain statistical image modeling and processing, image information organization and retrieval, and color image processing are presented. While imaging applications have, for some time, been addressed through nonlinear approaches, an increasingly important, but less mature, application certain to benefit from nonlinear approaches is the growing field of genomics. The application of nonlinear methods to select genomic applications is presented in a chapter on genetic regulatory networks.

Each of the chapters was contributed by leading researchers in the field of nonlinear signal processing. Without their hard work, a book covering such a wide array of applications simply would not be possible. Although the full depth of the contributions can only be appreciated by a thorough reading of the chapters, we attempt to summarize the main contributions in each chapter here so that researchers, practicing engineers, students, and any other readers of this book can go directly to the chapter of most relevance to their particular needs. The chapters can be read alone, but a fuller appreciation of current nonlinear methods, or at least those presented here, requires the reading of related chapters. The thorough reader is encouraged to explore these topics in greater depth. Further exploration can be started by examining the extensive reference list at the end of each chapter.

The set of chapters on nonlinear signal processing theory begins with Chapter 1, *Energy Conservation in Adaptive Filtering*, by Ali H. Sayed, Tareq Y. Alnaffouri, and Vitor H. Nascimento, which analyzes adaptive filter performance under a unified energy-conservation approach. This approach leads to a tractable analysis that provides not only information about stability and convergence behavior of the filter, but its steady-state performance as well. The next two chapters, *Fuzzy Methods in Nonlinear Signal Processing: Part I—Theory and Part II—Applications*, by Kenneth E. Barner, Yao Nie, and Yuzhong Shen, cover fuzzy extensions to rank-order and spatial-order based methods. In Part I, the importance of spatial order and rank order in filtering is derived from a maximum likelihood approach. Fuzzy extensions to the ordering concepts are derived that incorporate sample diversity; these extensions are applied, in Part II, to several signal, imaging, and communications problems.

Chapter 4, *Time-Frequency Wigner Distribution Approach to Differential Equations*, by Lorenzo Galleani and Leon Cohen, presents methods enabling one to cast the solution to an ordinary or partial differential equation, such as the Schrödinger equation, as a dynamical equation for the Wigner distribution. The advantage of this approach is that one gains considerable insight into the nature of the solution, and it leads to new analysis and approximations to the original equation. The set of chapters focused primarily on theory concludes with Chapter 5, *Weighted Myriad Filters*, by Gonzalo R. Arce, Juan G. Gonzalez, and Yinbo Li. This chapter considers the processing of stable processes, and focuses on the theory of M-estimation, which, for the special case of the Cauchy distribution, leads to the class of weighted myriad filters. Fast implementations, filter design, and optimization procedures are presented along with imaging and equalization applications.

The remainder of the book concentrates on critical contemporary applications that are being increasingly addressed through nonlinear methods. This look at applications begins with a set of three chapters addressing important problems in communications. The first chapter in this set, Chapter 6, *Data Traffic Modeling—A Signal Processing Perspective*, by Athina P. Petropulu and Xueshi Yang, concentrates on statistical analysis and modeling of broadband heterogeneous data-network traffic. This chapter develops mathematical tools to characterize the self-similar and impulsive nature of data traffic. The presented statistical characterizations provide insights into the physical understanding of data traffic.

Giovanni L. Sicuranza, Alberto Carini, and Andrea Fermo address the important problem of acoustic echo cancellation in Chapter 7, *Nonlinear Adaptive Filters for Acoustic Echo Cancellation in Mobile Terminals*. The increasing use of mobile communications terminals has made echo cancellation a critical component in the development of high-quality mobile communications services. This chapter shows that effective echo cancellation is achieved with polynomial, or Volterra, filtering. The authors present computationally efficient implementations and develop the optimization and tracking algorithms necessary for Volterra echo cancellers to be introduced into practical handset devices. The final communications-focused contribution, Chapter 8, *Blind and Semiblind Channel Estimation*, by Visa Koivunen, Mihai Enescu, and Marius Sirbu, develops nonlinear channel equalization methods for multiple-input multiple-output (MIMO) systems. Channel equalization is critical to improving the spectral efficiency of communications systems and the authors cover blind and semiblind methods for the widely used GSM, DS-CDMA, and OFDM wireless systems.

Perhaps the area within which nonlinear methods found their earliest acceptance and enjoy the greatest dominance is the field of image processing.

Indeed, nonlinear methods dominate in nearly all aspects of image-based applications. The nonstationarity of images, the importance of visual cues such as edges, and the nonlinearity of the human visual system all contribute to the success of nonlinear methods in imaging applications. In recognition of the importance of nonlinear methods in imaging applications, the next four chapters focus on various aspects of the image processing field.

The set of chapters on imaging applications begins with the problem of image enhancement, addressed by Richard R. Schultz and Robert L. Stevenson in Chapter 9, *Bayesian Image and Video Enhancement Using a Non-Gaussian Prior*. The enhancements addressed here cover the key problems of image magnification, removal of block-DCT compression artifacts, and the superresolution enhancement of digital video. The statistical modeling and processing of images are addressed in Chapter 10, *Statistical Image Modeling and Processing Using Wavelet Domain Hidden Markov Models*, by Guoliang Fan and Xiang-Gen Xia. This chapter shows that the problems of image denoising and segmentation, as well as texture analysis and synthesis, can be effectively addressed through wavelet domain HMMs.

Self-organization is an important concept in nonlinear methods that, while applicable to many applications, can be used to address several fundamental imaging applications. In Chapter 11, *Self-Organizing Maps and Their Applications in Image Processing, Information Organization, and Retrieval*, by Constantine Kotropoulos and Ioannis Pitas, the theory, learning algorithms, and analysis of SOMs are presented along with their application in image quantization, segmentation, and document organization and retrieval. The final chapter on imaging applications, Chapter 12, *Nonlinear Techniques for Color Image Processing*, by Bogdan Smolka, Konstantinos N. Plataniotis, and Anastasios N. Venetsanopoulos, addresses the particular challenges and opportunities that arise when processing color images. A new digital paths approach, which utilizes the connection between image pixels rather than traditional filtering window structures, is presented for the filtering of color images. This approach is an extension of adaptive noise reduction filtering and anisotropic diffusion techniques and is shown to have advantages over traditional methods.

We end the book with a chapter on genetic regulatory networks, which are an important and rapidly developing research area in computational genomics. Authors Ilya Shmulevich and Edward R. Dougherty of Chapter 13, *Genetic Regulatory Networks: A Nonlinear Signal Processing Perspective*, present Boolean network approaches to the modeling and analysis of genetic regulatory networks. This chapter shows that the nonlinear signal processing theory originally developed to address filtering problems can play an important role in the modeling and inference of genetic networks as well as the analysis of gene expression data.

We wish to thank all of the researchers that contributed to this work, as a book on the broad topics of nonlinear signal processing theory, methods, and applications could not have been compiled without their expertise and considerable efforts. The presented work is, of course, but a fraction of the dynamic and ever-expanding body of work on nonlinear signal processing. We hope readers find the selected topics representative, interesting, and informative. For those interested in a deeper investigation of the presented topics, the extensive bibliography at the end of each chapter serves as an excellent link to current references.

As a final note, we would like to point out that much of the presented work had its origins in the 2001 Nonlinear Signal and Image Processing (NSIP) workshop held in Baltimore's Inner Harbor, which we had the honor of co-chairing. It has been a privilege to work with CRC Press and the contributing authors since the NSIP workshop on the concept, development, and final completion of this book. We also welcome feedback from readers, who can contact us at barner@ece.udel.edu and arce@ece.udel.edu.

Kenneth E. Barner
Gonzalo R. Arce

Editors

Kenneth E. Barner received a B.S.E.E. degree (*magna cum laude*) from Lehigh University, Bethlehem, PA, in 1987 and M.S.E.E. and Ph.D. degrees from the University of Delaware, Newark, in 1989 and 1992, respectively. For his dissertation "Permutation Filters: A Group Theoretic Class of Non-Linear Filters," Dr. Barner received the Allan P. Colburn Prize in Mathematical Sciences and Engineering for the most outstanding doctoral dissertation in the engineering and mathematical disciplines.

Dr. Barner was the duPont Teaching Fellow and a visiting lecturer at the University of Delaware in 1991 and 1992, respectively. From 1993 to 1997 he was an assistant research professor in the Department of Electrical and Computer Engineering at the University of Delaware and a research engineer at the duPont Hospital for Children. He is currently an associate professor in the Department of Electrical and Computer Engineering at the University of Delaware. Dr. Barner is the recipient of a 1999 NSF Career award. He was the co-chair of the 2001 IEEE EURASIP Nonlinear Signal and Image Processing (NSIP) Workshop and a guest editor for a special issue of the *EURASIP Journal of Applied Signal Processing on Nonlinear Signal and Image Processing*. Dr. Barner is a member of the Nonlinear Signal and Image Processing Board and is a senior member of the IEEE. He is also serving as an associate editor of the *IEEE Transactions on Signal Processing*, the IEEE *Signal Processing Magazine*, and the *IEEE Transaction on Neural Systems and Rehabilitation Engineering*. Dr. Barner is also a member of the Editorial Board of the *EURASIP Journal of Applied Signal Processing*. His research interests include signal and image processing, robust signal processing, nonlinear systems, communications, haptic and tactile methods, and universal access.

Gonzalo R. Arce received a Ph.D. degree from Purdue University, West Lafayette, IN, in 1982. Since 1982 he has been with the faculty of the Department of Electrical and Computer Engineering at the University of Delaware where he is the Charles Black Evans Professor and department chairman. Funded broadly by federal agencies and industry, his research interests include statistical and nonlinear signal processing, multimedia security, electronic imaging, and signal processing for communications and networks. Dr. Arce received the NSF Research Initiation Award. He is a Fellow of the IEEE for his contributions on nonlinear signal processing and its applications. Dr. Arce was the co-chair of the 2001 EUSIPCO/IEEE Workshop on

Nonlinear Signal and Image Processing (NSIP'01), co-chair of the 1991 SPIE's Symposium on Nonlinear Electronic Imaging, and the co-chair of the 2002 and 2003 SPIE ITCOM conferences. Dr. Arce has served as associate editor for the *IEEE Transactions for Signal Processing*, senior editor of the *Applied Signal Processing Journal*, guest editor for the *IEEE Transactions on Image Processing*, and guest editor for *Optics Express*. He is co-author of the textbook *Digital Halftoning* (Marcel Dekker, 2001). Dr. Arce is a frequent consultant to industry in the areas of image printing and digital video and he holds five U.S. patents.

Contributors

Tareq Y. Al-Naffouri Department of Electrical Engineering, Stanford University, Stanford, California

Gonzalo R. Arce Department of Electrical and Computer Engineering, University of Delaware, Newark, Delaware

Kenneth E. Barner Department of Electrical and Computer Engineering, University of Delaware, Newark, Delaware

Alberto Carini Department of Electrical Electronic and Computer Engineering, University of Trieste, Trieste, Italy

Leon Cohen Department of Physics, Hunter College, City University of New York, New York, New York

Edward R. Dougherty Department of Electrical Engineering, Texas A&M University, College Station, Texas

Mihai Enescu Department of Electrical and Communications Engineering, Helsinki University of Technology, Helsinki, Finland

Guoliang Fan School of Electrical and Computer Engineering, Oklahoma State University, Stillwater, Oklahoma

Andrea Fermo Department of Electrical Electronic and Computer Engineering, University of Trieste, Trieste, Italy

Lorenzo Galleani Dipartimento di Elettronica, Politecnico di Torino, Torino, Italy

Juan G. Gonzalez NameTech, LLC, Weston, Florida

Visa Koivunen Department of Electrical and Communications Engineering, Helsinki University of Technology, Helsinki, Finland

Constantine Kotropoulos   Department of Informatics, Aristotle University of Thessaloniki, Thessaloniki, Greece

Yinbo Li   Department of Electrical and Computer Engineering, University of Delaware, Newark, Delaware

Vitor H. Nascimento   Department of Electrical Engineering, University of Sao Paulo, Sao Paulo, Brazil

Yao Nie   Department of Electrical and Computer Engineering, University of Delaware, Newark, Delaware

Athina P. Petropulu   Department of Electrical and Computer Engineering, Drexel University, Philadelphia, Pennsylvania

Ioannis Pitas   Department of Informatics, Aristotle University of Thessaloniki, Thessaloniki, Greece

Konstantinos N. Plataniotis   The Edward S. Rogers Sr. Department of Electrical and Computer Engineering, University of Toronto, Toronto, Ontario, Canada

Ali H. Sayed   Department of Electrical Engineering, University of California, Los Angeles, California

Richard R. Schultz   Department of Electrical Engineering, University of North Dakota, Grand Forks, North Dakota

Yuzhong Shen   Department of Electrical and Computer Engineering, University of Delaware, Newark, Delaware

Ilya Shmulevich   Department of Pathology, The University of Texas M.D. Anderson Cancer Center, Houston, Texas

Giovanni L. Sicuranza   Department of Electrical Electronic and Computer Engineering, University of Trieste, Trieste, Italy

Marius Sirbu   Department of Electrical and Communications Engineering, Helsinki University of Technology, Helsinki, Finland

Bogdan Smolka   Department of Automatic Control, Silesian University of Technology, Gliwice, Poland

Robert L. Stevenson   Department of Electrical Engineering, University of Notre Dame, Notre Dame, Indiana

Anastasios N. Venetsanopoulos   Faculty of Applied Science and Engineering, University of Toronto, Toronto, Ontario, Canada

Xiang-Gen Xia   Department of Electrical and Computer Engineering, University of Delaware, Newark, Delaware

Xueshi Yang   Seagate Research, Seagate Technology, Pittsburgh, Pennsylvania

Contents

1

Energy Conservation in Adaptive Filtering

Ali H. Sayed, Tareq Y. Al-Naffouri, and Vitor H. Nascimento

CONTENTS

1.1 Introduction

The study of the steady-state and transient performance of adaptive filters is a challenging task because of the nonlinear and stochastic nature of their update equations (e.g., References 1 to 4). The purpose of this chapter is to provide an overview of an energy-conservation approach to study the performance of adaptive filters in a unified manner.[4] The approach is based on showing that certain *a priori* and *a posteriori* errors maintain an energy balance for all time instants.[5–7] When examined under expectation, this energy balance leads to a variance relation that characterizes the dynamics of an adaptive filter.[10–14] An advantage of the energy framework is that it allows us to push the algebraic manipulations of variables to a limit, and to eliminate unnecessary cross-terms before appealing to expectations. This is a useful step because it is usually easier to handle random variables algebraically than under expectations, especially for higher-order moments. A second advantage of the energy arguments is that they can be pursued without restricting the distribution of the input data. To illustrate this point, we have opted not to restrict the regression data to Gaussian or white in most of the discussions below. Instead, all results are derived for arbitrary input distributions. Of course, by specializing the results to particular distributions, some known results from the literature can be recovered as special cases of the general framework.

As for most adaptive filter analysis, progress is difficult without relying on simplifying assumptions. In the initial part of our presentation, we derive exact energy-conservation and variance relations that hold for a large class of adaptive filters without any approximations. Subsequent discussions will call upon simplifying assumptions to make the analysis more tractable. The assumptions tend to be reasonable for small step sizes and long filters.

1.2   The Data Model

Consider reference data $\{\mathbf{d}(i)\}$ and regression data $\{\mathbf{u}_i\}$, assumed related via the linear regression model

$$\mathbf{d}(i) = \mathbf{u}_i w^o + \mathbf{v}(i) \tag{1.1}$$

for some $M \times 1$ unknown column vector w^o that we wish to estimate. Here $\mathbf{u}_i$ is a regressor, taken as a row vector, and $\mathbf{v}(i)$ is measurement noise. Observe that we are using boldface letters to denote random quantities, which will be our convention throughout this chapter. Also, all vectors in our presentation are column vectors except for the regressor $\mathbf{u}_i$. In this way, the inner product between $\mathbf{u}_i$ and w^o is written simply as $\mathbf{u}_i w^o$ without the need for transposition symbols.

In Equation 1.1, $\{\mathbf{d}(i), \mathbf{u}_i, \mathbf{v}(i)\}$ are random variables that satisfy the following conditions:

a. $\{\mathbf{v}(i)\}$ is zero-mean, independent and identically distributed with variance $E\,\mathbf{v}^2(i) = \sigma_v^2$.
b. $\mathbf{v}(i)$ is independent of $\mathbf{u}_j$ for all i, j.   (1.2)
c. The regressor $\mathbf{u}_i$ is zero-mean and has covariance matrix $E\,\mathbf{u}_i^T \mathbf{u}_i = R_u > 0$.

In the first part of the chapter we focus on data-normalized adaptive filters for generating estimates for w^o, specifically, on updates of the form

$$\mathbf{w}_i = \mathbf{w}_{i-1} + \mu \frac{\mathbf{u}_i^T}{g[\mathbf{u}_i]} \mathbf{e}(i), \quad i \geq 0, \tag{1.3}$$

where

$$\mathbf{e}(i) = \mathbf{d}(i) - \mathbf{u}_i \mathbf{w}_{i-1} \tag{1.4}$$

is the estimation error at iteration i, and $g[\mathbf{u}_i] > 0$ is some function of $\mathbf{u}_i$. Typical choices for g are

$$g[\mathbf{u}] = 1 \quad \text{(LMS)}, \qquad g[\mathbf{u}] = \|\mathbf{u}\|^2 \quad \text{(NLMS)}, \qquad g[\mathbf{u}] = \epsilon + \|\mathbf{u}\|^2 \quad (\epsilon\text{-NLMS}).$$

The initial condition $\mathbf{w}_{-1}$ of Equation 1.3 is assumed to be independent of all $\{\mathbf{d}(j), \mathbf{u}_j, \mathbf{v}(j)\}$. Later in the chapter we study adaptive filters with error nonlinearities in their update equations — see Equation 1.52.

Our purpose is to examine the transient and steady-state performance of such data-normalized filters in a unified manner (i.e., uniformly for all g). The first step in this regard is to establish an energy-conservation relation that holds for a large class of adaptive filters, and then use it as the basis of all subsequent analysis.

1.3   Energy-Conservation Relation

Let $\tilde{\mathbf{w}}_i = w^o - \mathbf{w}_i$ denote the weight-error vector at iteration i, and let Σ denote some $M \times M$ positive-definite matrix. Define further the weighted *a priori* and *a posteriori* errors:

$$e_a^\Sigma(i) \stackrel{\Delta}{=} \mathbf{u}_i \Sigma \tilde{\mathbf{w}}_{i-1}, \qquad e_p^\Sigma(i) \stackrel{\Delta}{=} \mathbf{u}_i \Sigma \tilde{\mathbf{w}}_i. \tag{1.5}$$

When $\Sigma = I$, we recover the standard definitions

$$e_a(i) = \mathbf{u}_i \tilde{\mathbf{w}}_{i-1}, \qquad e_p(i) = \mathbf{u}_i \tilde{\mathbf{w}}_i. \tag{1.6}$$

The freedom in selecting Σ will be seen to be useful in characterizing several aspects of the dynamic behavior of an adaptive filter. For now, we shall treat Σ as an arbitrary weighting matrix.

It turns out that the errors $\{\tilde{\mathbf{w}}_i, \tilde{\mathbf{w}}_{i-1}, e_a^\Sigma(i), e_p^\Sigma(i)\}$ satisfy a fundamental energy-conservation relation. To arrive at the relation, we subtract w^o from both sides of Equation 1.3 to obtain

$$\tilde{\mathbf{w}}_i = \tilde{\mathbf{w}}_{i-1} - \mu \frac{\mathbf{u}_i^T}{g[\mathbf{u}_i]} \mathbf{e}(i), \tag{1.7}$$

and then multiply Equation 1.7 by $\mathbf{u}_i \Sigma$ from the left to conclude that

$$e_p^\Sigma(i) = e_a^\Sigma(i) - \mu \frac{\|\mathbf{u}_i\|_\Sigma^2}{g[\mathbf{u}_i]} \mathbf{e}(i), \tag{1.8}$$

where the notation $\|\mathbf{u}_i\|_\Sigma^2$ denotes the squared weighted Euclidean norm of $\mathbf{u}_i$, specifically,

$$\|\mathbf{u}_i\|_\Sigma^2 = \mathbf{u}_i \Sigma \mathbf{u}_i^T.$$

Relation 1.8 can be used to express $\mathbf{e}(i)/g[\mathbf{u}_i]$ in terms of $\{\mathbf{e}_p^\Sigma(i), \mathbf{e}_a^\Sigma(i)\}$ and to eliminate this term from Equation 1.7. Doing so leads to the equality

$$\|\mathbf{u}_i\|_\Sigma^2 \cdot \tilde{\mathbf{w}}_i + \mathbf{u}_i^T \mathbf{e}_a^\Sigma(i) = \|\mathbf{u}_i\|_\Sigma^2 \cdot \tilde{\mathbf{w}}_{i-1} + \mathbf{u}_i^T \mathbf{e}_p^\Sigma(i). \tag{1.9}$$

By equating the weighted Euclidean norms of both sides of this equation, we arrive, after a straightforward calculation, at the relation:

$$\|\mathbf{u}_i\|_\Sigma^2 \cdot \|\tilde{\mathbf{w}}_i\|_\Sigma^2 + \left(\mathbf{e}_a^\Sigma(i)\right)^2 = \|\mathbf{u}_i\|_\Sigma^2 \cdot \|\tilde{\mathbf{w}}_{i-1}\|_\Sigma^2 + \left(\mathbf{e}_p^\Sigma(i)\right)^2. \tag{1.10}$$

This energy relation is an exact result that shows how the energies of the weight-error vectors at two successive time instants are related to the energies of the *a priori* and *a posteriori* estimation errors.* In addition, it follows from $\mathbf{e}(i) = \mathbf{u}_i \tilde{\mathbf{w}}_{i-1} + \mathbf{v}(i)$, and from Equation 1.7, that the weight-error vector satisfies

$$\tilde{\mathbf{w}}_i = \left(I - \mu \frac{\mathbf{u}_i^T \mathbf{u}_i}{g[\mathbf{u}_i]}\right)\tilde{\mathbf{w}}_{i-1} - \mu \frac{\mathbf{u}_i^T}{g[\mathbf{u}_i]}\mathbf{v}(i). \tag{1.11}$$

1.4   Weighted Variance Relation

The result (Equation 1.10) with $\Sigma = I$ was developed in Reference 5 and subsequently used in a series of works to study the robustness of adaptive filters (e.g., References 6 through 9). It was later used in References 10 through 12 to study the steady-state and tracking performance of adaptive filters. The incorporation of a weighting matrix Σ in References 13 and 14 turns out to be useful for transient (convergence and stability) analysis.

In transient analysis we are interested in characterizing the time evolution of the quantity $\mathrm{E}\,\|\tilde{\mathbf{w}}_i\|_\Sigma^2$, for some Σ of interest (usually, $\Sigma = I$ or $\Sigma = R_u$). To arrive at this evolution, we use Equation 1.8 to replace $\mathbf{e}_p^\Sigma(i)$ in Equation 1.10

---

* Later in Section 1.13 we provide an interpretation of the energy relation (Equation 1.10) in terms of Snell's law for light propagation.

in terms of $\mathbf{e}_a^\Sigma(i)$ and $\mathbf{e}(i)$. This step yields, after expanding and grouping terms,

$$
\|\mathbf{u}_i\|_\Sigma^2 \cdot \|\tilde{\mathbf{w}}_i\|_\Sigma^2 = \|\mathbf{u}_i\|_\Sigma^2 \cdot \|\tilde{\mathbf{w}}_{i-1}\|_\Sigma^2 + \frac{\mu^2(\|\mathbf{u}_i\|_\Sigma^2)^2}{g^2[\mathbf{u}_i]} \|\tilde{\mathbf{w}}_{i-1}\|_{\mathbf{u}_i^T\mathbf{u}_i}^2
$$
$$
+ \frac{\mu^2(\|\mathbf{u}_i\|_\Sigma^2)^2}{g^2[\mathbf{u}_i]} \mathbf{v}^2(i) - \frac{\mu\|\mathbf{u}_i\|_\Sigma^2}{g[\mathbf{u}_i]} \|\tilde{\mathbf{w}}_{i-1}\|_{\Sigma\mathbf{u}_i^T\mathbf{u}_i+\mathbf{u}_i^T\mathbf{u}_i\Sigma}^2
$$
$$
+ 2\mu^2 \frac{(\|\mathbf{u}_i\|_\Sigma^2)^2}{g^2[\mathbf{u}_i]} \mathbf{v}(i)\mathbf{e}_a(i) - 2\mu \frac{\|\mathbf{u}_i\|_\Sigma^2}{g[\mathbf{u}_i]} \mathbf{v}(i)\mathbf{e}_a^\Sigma(i). \qquad (1.12)
$$

Assuming the event $\|\mathbf{u}_i\|_\Sigma^2 = 0$ occurs with zero probability, we can eliminate $\|\mathbf{u}_i\|_\Sigma^2$ from both sides of Equation 1.12 and take expectations to arrive at

$$
\mathrm{E}\|\tilde{\mathbf{w}}_i\|_\Sigma^2 = \mathrm{E}\left(\|\tilde{\mathbf{w}}_{i-1}\|_{\Sigma'}^2\right) + \mu^2\sigma_v^2\mathrm{E}\left(\frac{\|\mathbf{u}_i\|_\Sigma^2}{g^2[\mathbf{u}_i]}\right), \qquad (1.13)
$$

where the weighting matrix Σ' is defined by

$$
\Sigma' = \Sigma - \frac{\mu}{g[\mathbf{u}_i]}\Sigma\mathbf{u}_i^T\mathbf{u}_i - \frac{\mu}{g[\mathbf{u}_i]}\mathbf{u}_i^T\mathbf{u}_i\Sigma + \frac{\mu^2\|\mathbf{u}_i\|_\Sigma^2}{g^2[\mathbf{u}_i]}\mathbf{u}_i^T\mathbf{u}_i. \qquad (1.14)
$$

Observe that Σ' is a random matrix due to its dependence on the data (and, hence, the use of the boldface notation for it). The matrix Σ, on the other hand, is not random.

1.4.1 Independent Regressors

Relations 1.11, 1.13, and 1.14 characterize the dynamic behavior of data-normalized adaptive filters for generic input distributions; they are all exact relations. Still, Recursion 1.13 is difficult to propagate as it requires the evaluation of the expectation

$$
\mathrm{E}\left(\|\tilde{\mathbf{w}}_{i-1}\|_{\Sigma'}^2\right) = \mathrm{E}\left(\tilde{\mathbf{w}}_{i-1}^T\Sigma'\tilde{\mathbf{w}}_{i-1}\right).
$$

The difficulty is due to the fact that Σ' is a random matrix that depends on $\mathbf{u}_i$, and $\tilde{\mathbf{w}}_{i-1}$ is dependent on prior regressors as well. To progress further in the analysis, we assume that

the $\{\mathbf{u}_i\}$ are independent and identically distributed, $\qquad (1.15)$

which allows us to deal with Σ' independently from $\tilde{\mathbf{w}}_{i-1}$. This so-called independence assumption is commonly used in the literature. Although rarely applicable, it gives good results for small step sizes.

Under Assumption 1.15, it is easy to verify that $\tilde{\mathbf{w}}_{i-1}$ becomes independent of Σ' and, consequently, that

$$
\mathrm{E}\left[\|\tilde{\mathbf{w}}_{i-1}\|_{\Sigma'}^2\right] = \mathrm{E}\left[\|\tilde{\mathbf{w}}_{i-1}\|_{\mathrm{E}[\Sigma']}^2\right],
$$

with the weighting matrix Σ' replaced by its mean, which we shall denote by Σ'. In this way, the variance Recursion 1.13 becomes

$$\mathrm{E}\,\|\tilde{\mathbf{w}}_i\|_\Sigma^2 = \mathrm{E}\,\|\tilde{\mathbf{w}}_{i-1}\|_{\Sigma'}^2 + \mu^2\sigma_v^2\mathrm{E}\left(\frac{\|\mathbf{u}_i\|_\Sigma^2}{g^2[\mathbf{u}_i]}\right), \tag{1.16}$$

with deterministic weighting matrices $\{\Sigma, \Sigma'\}$ and where, by evaluating the expectation of Equation 1.14,

$$\Sigma' = \Sigma - \mu\Sigma\mathrm{E}\left(\frac{\mathbf{u}_i^T\mathbf{u}_i}{g[\mathbf{u}_i]}\right) - \mu\mathrm{E}\left(\frac{\mathbf{u}_i^T\mathbf{u}_i}{g[\mathbf{u}_i]}\right)\Sigma + \mu^2\mathrm{E}\left(\frac{\|\mathbf{u}_i\|_\Sigma^2}{g^2[\mathbf{u}_i]}\mathbf{u}_i^T\mathbf{u}_i\right). \tag{1.17}$$

Observe that the expression for Σ' is data dependent *only*.

Finally, taking expectations of both sides of Equation 1.11, and using Equation 1.15, we find that

$$\mathrm{E}\,\tilde{\mathbf{w}}_i = \left(I - \mu\mathrm{E}\left(\frac{\mathbf{u}_i^T\mathbf{u}_i}{g[\mathbf{u}_i]}\right)\right)\cdot\mathrm{E}\,\tilde{\mathbf{w}}_{i-1}. \tag{1.18}$$

Expressions 1.16 through 1.18 show that studying the transient behavior of a data-normalized adaptive filter in effect requires evaluating the three multivariate moments:

$$\mathrm{E}\left(\frac{\|\mathbf{u}_i\|_\Sigma^2}{g^2[\mathbf{u}_i]}\right), \quad \mathrm{E}\left(\frac{\mathbf{u}_i^T\mathbf{u}_i}{g[\mathbf{u}_i]}\right), \quad \text{and} \quad \mathrm{E}\left(\frac{\|\mathbf{u}_i\|_\Sigma^2}{g^2[\mathbf{u}_i]}\mathbf{u}_i^T\mathbf{u}_i\right),$$

which are functions of $\mathbf{u}_i$ only. In terms of these moments, Relations 1.16 through 1.18 can now be used to characterize the dynamic behavior of adaptive filters under independence Assumption 1.15. We start with the mean-square (transient) behavior.

1.5   Mean-Square Behavior

Let σ denote the $M^2 \times 1$ column vector that is obtained by stacking the columns of Σ on top of each other, written as $\sigma = \mathrm{vec}(\Sigma)$. Likewise, let $\sigma' = \mathrm{vec}(\Sigma')$. We shall also use the $\mathrm{vec}^{-1}(\cdot)$ notation and write $\Sigma = \mathrm{vec}^{-1}(\sigma)$ to recover Σ from σ. Similarly, $\Sigma' = \mathrm{vec}^{-1}(\sigma')$.

Then using the Kronecker product notation,[16] and the following property, for arbitrary matrices $\{X, Y, Z\}$ of compatible dimensions,

$$\mathrm{vec}(XYZ) = (Z^T \otimes X)\mathrm{vec}(Y).$$

We can easily verify that Relation 1.17 for Σ' transforms into the linear vector relation

$$\sigma' = F\sigma,$$

where F is $M^2 \times M^2$ and given by

$$F = I - \mu A + \mu^2 B, \tag{1.19}$$

in terms of the symmetric matrices $\{A, B\}$,

$$A = (P \otimes I_M) + (I_M \otimes P),$$

$$B = E\left(\frac{\mathbf{u}_i^T \mathbf{u}_i \otimes \mathbf{u}_i^T \mathbf{u}_i}{g^2[\mathbf{u}_i]}\right), \tag{1.20}$$

$$P = E\left(\frac{\mathbf{u}_i^T \mathbf{u}_i}{g[\mathbf{u}_i].}\right).$$

Actually, A is positive-definite (because P is) and B is nonnegative-definite. Using the column notation σ, and the relation $\sigma' = F\sigma$, we can write Equations 1.16 through 1.17 as

$$E\|\tilde{\mathbf{w}}_i\|^2_{\text{vec}^{-1}(\sigma)} = E\|\tilde{\mathbf{w}}_{i-1}\|^2_{\text{vec}^{-1}(F\sigma)} + \mu^2\sigma_v^2 E\left(\frac{\|\mathbf{u}_i\|^2_\sigma}{g^2[\mathbf{u}_i]}\right),$$

which we shall rewrite more succinctly, by dropping the $\text{vec}^{-1}(\cdot)$ notation and keeping the weighting vectors, as

$$E\|\tilde{\mathbf{w}}_i\|^2_\sigma = E\|\tilde{\mathbf{w}}_{i-1}\|^2_{F\sigma} + \mu^2\sigma_v^2 E\left(\frac{\|\mathbf{u}_i\|^2_\sigma}{g^2[\mathbf{u}_i]}\right). \tag{1.21}$$

Now, as mentioned earlier, in transient analysis we are interested in the evolution of $E\|\tilde{\mathbf{w}}_i\|^2$ and $E\|\tilde{\mathbf{w}}_i\|^2_{R_u}$; the former quantity is the filter mean-square deviation while the second quantity relates to the filter mean-square error (or learning) curve because

$$E\mathbf{e}^2(i) = E\mathbf{e}_a^2(i) + \sigma_v^2 = E\|\tilde{\mathbf{w}}_{i-1}\|^2_{R_u} + \sigma_v^2.$$

The quantities $\{E\|\tilde{\mathbf{w}}_i\|^2, E\|\tilde{\mathbf{w}}_i\|^2_{R_u}\}$ are in turn special cases of $E\|\tilde{\mathbf{w}}_i\|^2_\Sigma$ obtained by choosing $\Sigma = I$ or $\Sigma = R_u$. Therefore, in the sequel, we focus on studying the evolution of $E\|\tilde{\mathbf{w}}_i\|^2_\Sigma$ for arbitrary Σ.

From Equation 1.21 we see that to evaluate $E\|\tilde{\mathbf{w}}_i\|^2_\sigma$, we need $E\|\tilde{\mathbf{w}}_i\|^2_{F\sigma}$ with weighting vector $F\sigma$. This term can be deduced from Equation 1.21 by writing it for $\sigma \leftarrow F\sigma$, i.e.,

$$E\|\tilde{\mathbf{w}}_i\|^2_{F\sigma} = E\|\tilde{\mathbf{w}}_{i-1}\|^2_{F^2\sigma} + \mu^2\sigma_v^2 E\left(\frac{\|\mathbf{u}_i\|^2_{F\sigma}}{g^2[\mathbf{u}_i]}\right),$$

with the weighted term $E\|\tilde{\mathbf{w}}_i\|^2_{F^2\sigma}$. This term can in turn be deduced from Equation 1.21 by writing it for $\sigma \leftarrow F^2\sigma$. Continuing in this fashion, for successive powers of F, we arrive at

$$E \|\tilde{\mathbf{w}}_i\|^2_{F^{M^2-1}\sigma} = E \|\tilde{\mathbf{w}}_{i-1}\|^2_{F^{M^2}\sigma} + \mu^2 \sigma_v^2 E \left(\frac{\|\mathbf{u}_i\|^2_{F^{M^2-1}\sigma}}{g^2[\mathbf{u}_i]} \right)$$

in terms of the M^2-power of F (recall that F is $M^2 \times M^2$).

Fortunately, this procedure terminates. To see this, let $p(x) = \det(xI - F)$ denote the characteristic polynomial of F, say,

$$p(x) = x^{M^2} + p_{M^2-1}x^{M^2-1} + p_{M^2-2}x^{M^2-2} + \cdots + p_1 x + p_0,$$

with coefficients $\{p_i\}$. Then, since $p(F) = 0$ in view of the Cayley–Hamilton theorem,[16] we have

$$E \|\mathbf{w}_i\|^2_{F^{M^2}\sigma} = \sum_{k=0}^{M^2-1} -p_k E \|\mathbf{w}_i\|^2_{F^k \sigma}.$$

Putting these results together, we conclude that the transient (mean-square) behavior of the filter (Equation 1.3) is described by an M^2-dimensional state-space model of the form:

$$\mathcal{W}_i = \mathcal{F}\mathcal{W}_{i-1} + \mu^2 \sigma_v^2 \mathcal{Y}, \tag{1.22}$$

where the $M^2 \times 1$ vectors $\{\mathcal{W}_i, \mathcal{Y}\}$ are defined by

$$\mathcal{W}_i = \begin{bmatrix} E \|\tilde{\mathbf{w}}_i\|^2_\sigma \\ E \|\tilde{\mathbf{w}}_i\|^2_{F\sigma} \\ \vdots \\ E \|\tilde{\mathbf{w}}_i\|^2_{F^{M^2-2}\sigma} \\ E \|\tilde{\mathbf{w}}_i\|^2_{F^{M^2-1}\sigma} \end{bmatrix}, \quad \mathcal{Y} = \begin{bmatrix} E \left(\|\mathbf{u}_i\|^2_\sigma / g^2[\mathbf{u}_i] \right) \\ E \left(\|\mathbf{u}_i\|^2_{F\sigma} / g^2[\mathbf{u}_i] \right) \\ \vdots \\ E \left(\|\mathbf{u}_i\|^2_{F^{M^2-2}\sigma} / g^2[\mathbf{u}_i] \right) \\ E \left(\|\mathbf{u}_i\|^2_{F^{M^2-1}\sigma} / g^2[\mathbf{u}_i] \right) \end{bmatrix}, \tag{1.23}$$

and the $M^2 \times M^2$ coefficient matrix $\mathcal{F}$ is given by

$$\mathcal{F} = \begin{bmatrix} 0 & 1 & & & & \\ 0 & 0 & 1 & & & \\ 0 & 0 & 0 & 1 & & \\ \vdots & & & & & \\ 0 & 0 & 0 & & & 1 \\ -p_0 & -p_1 & -p_2 & \cdots & & -p_{M^2-1} \end{bmatrix}. \tag{1.24}$$

The entries of $\mathcal{Y}$ can be written more compactly as

$$\mathcal{Y} = \text{col} \left\{ \text{Tr}(Q \, \text{vec}^{-1}(F^k \sigma)), \quad k = 0, 1, \ldots, M^2 - 1 \right\},$$

where

$$Q = \mathrm{E} \left(\frac{\mathbf{u}_i^T \mathbf{u}_i}{g^2[\mathbf{u}_i]} \right), \tag{1.25}$$

and the notation $\mathrm{vec}^{-1}(F^k \sigma)$ recovers the weighting matrix that corresponds to the vector $F^k \sigma$.

When $\Sigma = I$, the evolution of the top entry of $\mathcal{W}_i$ in Equation 1.22 describes the mean-square deviation of the filter, i.e., $\mathrm{E} \|\tilde{\mathbf{w}}_i\|^2$. If, on the other hand, Σ is chosen as $\Sigma = R_u$, the evolution of the top entry of $\mathcal{W}_i$ describes the excess mean-square error (or learning curve) of the filter, i.e., $\mathrm{E} \|\tilde{\mathbf{w}}_i\|_{R_u}^2 = \mathrm{E} e_a^2(i)$.

The learning curve can also be characterized more explicitly as follows. Let $r = \mathrm{vec}(R_u)$ and choose $\sigma = r$. Iterating Equation 1.21 we find that

$$\mathrm{E} \|\tilde{\mathbf{w}}_i\|_r^2 = \|\tilde{\mathbf{w}}_{-1}\|_{F^{i+1}r}^2 + \mu^2 \sigma_v^2 \mathrm{E} \left[\frac{\|\mathbf{u}_i\|_{(I+F+\cdots+F^i)r}^2}{g^2[\mathbf{u}_i]} \right],$$

that is,

$$\mathrm{E} \|\tilde{\mathbf{w}}_i\|_r^2 = \|\tilde{\mathbf{w}}_{-1}\|_{a_i}^2 + \mu^2 \sigma_v^2 b(i),$$

where the vector a_i and the scalar $b(i)$ satisfy the recursions

$$a_i = F a_{i-1}, \quad a_{-1} = r,$$

$$b(i) - b(i-1) + \mathrm{E} \left[\frac{\|\mathbf{u}_i\|_{a_{i-1}}^2}{g^2[\mathbf{u}_i]} \right], \quad b(-1) = 0.$$

Usually $\mathbf{w}_{-1} = 0$ so that $\tilde{\mathbf{w}}_{-1} = w^o$. Using the definitions for $\{a_i, b(i)\}$, it is easy to verify that

$$\mathrm{E} e_a^2(i) = \mathrm{E} e_a^2(i-1) + \|w^o\|_{F^{i-1}(F-I)r}^2 + \mu^2 \sigma_v^2 \mathrm{Tr}(Q\mathrm{vec}^{-1}(F^{i+1}r)), \tag{1.26}$$

which describes the learning curve of data-normalized adaptive filters as in Equation 1.3. Further discussions on the learning behavior of adaptive filters can be found in Reference 17.

1.6 Mean-Square Stability

Recursion 1.22 shows that the adaptive filter will be mean-square stable if, and only if, the matrix $\mathcal{F}$ is a stable matrix; i.e., all its eigenvalues lie inside the unit circle. But since $\mathcal{F}$ has the form of a companion matrix, its eigenvalues coincide with the roots of $p(x)$, which in turn coincide with the eigenvalues of F. Therefore, the mean-square stability of the adaptive filter requires the matrix F in Equation 1.19 to be a stable matrix.

Now it can be verified that matrices F of the form of Equation 1.19, for arbitrary $\{A > 0, B \geq 0\}$, are stable for all values of μ in the range:

$$0 < \mu < \min \left\{ \frac{1}{\lambda_{\max}(A^{-1}B)}, \frac{1}{\max\{\lambda(H) \in \mathbb{R}^+\}} \right\}, \qquad (1.27)$$

where the second condition is in terms of the largest positive real eigenvalue of the block matrix,

$$H = \begin{bmatrix} A/2 & -B/2 \\ I_{M^2} & 0 \end{bmatrix},$$

when it exists. Because H is not symmetric, its eigenvalues may not be positive or even real. If H does not have any real positive eigenvalue, then the upper bound on μ is determined by $1/\lambda_{\max}(A^{-1}B)$ alone.*

Likewise, the mean-stability of the filter, as dictated by Equation 1.18, requires the eigenvalues of $(I - \mu P)$ to lie inside the unit circle or, equivalently,

$$\mu < 2/\lambda_{\max}(P). \qquad (1.28)$$

Combining Equations 1.27 and 1.28 we conclude that the filter is stable in the mean and mean-square senses for step-sizes in the range

$$\mu < \min \left\{ \frac{2}{\lambda_{\max}(P)}, \frac{1}{\lambda_{\max}(A^{-1}B)}, \frac{1}{\max\{\lambda(H) \in \mathbb{R}^+\}} \right\}. \qquad (1.29)$$

1.7   Steady-State Performance

Steady-state performance results can also be deduced from Equation 1.21. Assuming the filter is operating in steady state, Recursion 1.21 gives in the limit

$$\lim_{i \to \infty} \mathrm{E} \|\tilde{\mathbf{w}}_i\|^2_{(I-F)\sigma} = \mu^2 \sigma_v^2 \mathrm{E} \left[\frac{\|\mathbf{u}_i\|^2_\sigma}{g^2[\mathbf{u}_i]} \right].$$

This expression allows us to evaluate the steady-state value of $\mathrm{E} \|\tilde{\mathbf{w}}_i\|^2_S$ for any weighting matrix S, by choosing σ such that

$$(I - F)\sigma = \mathrm{vec}(S),$$

i.e.,

$$\sigma = (I - F)^{-1}\mathrm{vec}(S).$$

---

*The condition involving $\lambda_{\max}(A^{-1}B)$ in Equation 1.27 guarantees that all eigenvalues of F are less than 1, while the condition involving H ensures that all eigenvalues of F are larger than -1.

In particular, the filter excess mean-square error, defined by

$$\text{EMSE} = \lim_{i \to \infty} \text{E}\, \text{e}_a^2(i)$$

corresponds to the choice $S = R_u$ since, by virtue of the independence Assumption 1.15, $\text{E}\, \text{e}_a^2(i) = \text{E}\, \|\tilde{\mathbf{w}}_{i-1}\|_{R_u}^2$. In other words, we should select σ as

$$\sigma_{\text{emse}} = (I - F)^{-1}\text{vec}(R_u).$$

On the other hand, the filter mean-square deviation, defined as

$$\text{MSD} = \lim_{i \to \infty} \text{E}\, \|\tilde{\mathbf{w}}_i\|^2$$

is obtained by setting $S = I$, i.e.,

$$\sigma_{\text{msd}} = (I - F)^{-1}\text{vec}(I).$$

Let $\{\Sigma_{\text{emse}}, \Sigma_{\text{msd}}\}$ denote the weighting matrices that correspond to the vectors $\{\sigma_{\text{emse}}, \sigma_{\text{msd}}\}$, i.e.,

$$\Sigma_{\text{emse}} = \text{vec}^{-1}(\sigma_{\text{emse}}), \qquad \Sigma_{\text{msd}} = \text{vec}^{-1}(\sigma_{\text{msd}}).$$

Then we are led to the following expressions for the filter performance:

$$\text{EMSE} = \mu^2 \sigma_v^2 \text{Tr}(Q\Sigma_{\text{emse}}),$$
$$\text{MSD} = \mu^2 \sigma_v^2 \text{Tr}(Q\Sigma_{\text{msd}}). \tag{1.30}$$

Alternatively, we can also write

$$\text{EMSE} = \mu^2 \sigma_v^2 \text{vec}^T(Q)\sigma_{\text{emse}} = \mu^2 \sigma_v^2 \text{vec}^T(Q)(I - F)^{-1}\text{vec}(R_u),$$
$$\text{MSD} = \mu^2 \sigma_v^2 \text{vec}^T(Q)\sigma_{\text{msd}} = \mu^2 \sigma_v^2 \text{vec}^T(Q)(I - F)^{-1}\text{vec}(I). \tag{1.31}$$

While these steady-state results are obtained here as a consequence of variance Relation 1.21, which relies on independence Assumption 1.15, it turns out that steady-state results can also be deduced in an alternative manner that does not rely on using the independence condition. This alternative derivation starts from Equation 1.10 and uses the fact that $\text{E}\, \|\tilde{\mathbf{w}}_i\|^2 = \text{E}\, \|\tilde{\mathbf{w}}_{i-1}\|^2$ in steady state to derive expressions for the filter EMSE; the details are spelled out in References 11 and 12.

1.8   Small-Step-Size Approximation

Returning to the expression of F in Equation 1.19, and to the performance results (Equation 1.30), we see that they are defined in terms of moment matrices $\{A, B, P, Q\}$. These moments are generally not easy to evaluate for arbitrary input distributions and data nonlinearities g. This fact explains why it is common in the literature to resort to Gaussian or whiteness assumptions on the regression data.

In our development so far, all results concerning filter transient performance, stability, and steady-state performance (e.g., Equations 1.22, 1.26, 1.29, and 1.30) have been derived without restricting the distribution of the regression data to being Gaussian or white. To simplify the analysis, we keep the input distribution generic and appeal instead to approximations pertaining to the step-size value, to the filter length, and also to a fourth-order moment approximation. In this section, we discuss small-step-size approximation.

To begin with, even though we may not have available explicit values for the moments $\{A, B, P, Q\}$ in general, we can still assert the following. If the distribution of the regression data is such that the matrix B is finite, then there always exists a small enough step size for which F (and, hence, the filter) is stable. To see this, observe first that the eigenvalues of $I - \mu A$ are given by

$$\{1 - \mu[\lambda_k(P) + \lambda_j(P)]\}$$

for all combinations $1 \leq j, k \leq M$ of the eigenvalues of P. Now if B is bounded, then the maximum eigenvalue of F is bounded by

$$\lambda_{\max}(F) \leq 1 - 2\mu\lambda_{\min}(P) + \mu^2\beta$$

for some finite positive scalar β (e.g., $\beta = \lambda_{\max}(B)$). The upper bound on $\lambda_{\max}(F)$ is a quadratic function of μ, and it is easy to verify that the values of this function are less than 1 for step sizes in the range $(0, 2\lambda_{\min}(P)/\beta)$. Because $\lambda_{\min}(P)/\beta$ is positive, we conclude that there should exist a small enough μ such that F is stable and, consequently, the filter is mean-square stable.

Now for such small step sizes, we may ignore the quadratic term in μ that appears in Equation 1.17, and approximate the variance relation (1.16 to 1.17) by

$$\begin{aligned}
\mathrm{E}\,\|\tilde{\mathbf{w}}_i\|_{\Sigma}^2 &= \mathrm{E}\,\|\tilde{\mathbf{w}}_{i-1}\|_{\Sigma'}^2 + \mu^2\sigma_v^2\mathrm{E}\left(\frac{\|\mathbf{u}_i\|_{\Sigma}^2}{g^2[\mathbf{u}_i]}\right), \\
\Sigma' &= \Sigma - \mu\Sigma P - \mu P\Sigma,
\end{aligned} \tag{1.32}$$

or, equivalently, using the weighting vector notation, by

$$\begin{aligned}
\mathrm{E}\,\|\tilde{\mathbf{w}}_i\|_{\sigma}^2 &= \mathrm{E}\,\|\tilde{\mathbf{w}}_{i-1}\|_{F\sigma}^2 + \mu^2\sigma_v^2\mathrm{E}\left(\frac{\|\mathbf{u}_i\|_{\sigma}^2}{g^2[\mathbf{u}_i]}\right), \\
F &= I - \mu A,
\end{aligned}$$

where $P = \mathrm{E}\,(\mathbf{u}_i^T \mathbf{u}_i / g[\mathbf{u}_i])$. Moreover, because $I - F = \mu A$, we can also approximate the EMSE and MSD performances (Equation 1.30) of the filter by

$$\mathrm{EMSE} \approx \mu \sigma_v^2 \mathrm{Tr}(Q \Sigma_{\mathrm{emse}}),$$

$$\mathrm{MSD} \approx \mu \sigma_v^2 \mathrm{Tr}(Q \Sigma_{\mathrm{msd}}),$$

(1.33)

where now $\{\Sigma_{\mathrm{emse}}, \Sigma_{\mathrm{msd}}\}$ denote the weighting matrices that correspond to the vectors

$$\sigma_{\mathrm{emse}} = A^{-1} \mathrm{vec}(R_u), \qquad \sigma_{\mathrm{msd}} = A^{-1} \mathrm{vec}(I).$$

That is, $\{\Sigma_{\mathrm{emse}}, \Sigma_{\mathrm{msd}}\}$ are the unique solutions of the Lyapunov equations:

$$P \Sigma_{\mathrm{msd}} + \Sigma_{\mathrm{msd}} P = I \quad \text{and} \quad P \Sigma_{\mathrm{emse}} + \Sigma_{\mathrm{emse}} P = R_u.$$

It is easy to verify that $\Sigma_{\mathrm{msd}} = P^{-1}/2$ so that the MSD expression can be written more explicitly as

$$\mathrm{MSD} \approx \frac{\mu \sigma_v^2}{2} \mathrm{Tr}(Q P^{-1}). \tag{1.34}$$

For example, in the special case of LMS, for which $g[\mathbf{u}] = 1$ and $P = R_u = Q$, the above expressions give for small step sizes:

$$\mathrm{EMSE} \approx \frac{\mu \sigma_v^2 \mathrm{Tr}(R_u)}{2}, \qquad \mathrm{MSD} \approx \frac{\mu \sigma_v^2 M}{2} \quad \text{(LMS)}. \tag{1.35}$$

Using the simplified variance relation (Equation 1.32), we can also describe the dynamic behavior of the mean-square deviation of the filter by means of an M-dimensional state-space model, as opposed to the M^2-dimensional model (Equation 1.22). To see this, let $P = U \Delta U^T$ denote the eigen-decomposition of $P > 0$, and introduce the transformed quantities:

$$\overline{\mathbf{w}}_i = U^T \tilde{\mathbf{w}}_i, \quad \overline{\mathbf{u}}_i = \mathbf{u}_i U, \quad \overline{\Sigma} = U^T \Sigma U, \quad \overline{Q} = U^T Q U.$$

Then the variance relation (Equation 1.32) can be equivalently rewritten as*

$$\mathrm{E}\,\|\overline{\mathbf{w}}_i\|_{\overline{\Sigma}}^2 = \mathrm{E}\,\|\overline{\mathbf{w}}_{i-1}\|_{\overline{\Sigma}'}^2 + \mu^2 \sigma_v^2 \mathrm{E}\left(\frac{\|\overline{\mathbf{u}}_i\|_{\overline{\Sigma}}^2}{g^2[\overline{\mathbf{u}}_i]}\right),$$

$$\overline{\Sigma}' = \overline{\Sigma} - \mu \overline{\Sigma} \Delta - \mu \Delta \overline{\Sigma}. \tag{1.36}$$

The expression for $\overline{\Sigma}'$ shows that it will be diagonal as long as $\overline{\Sigma}$ is diagonal. Therefore, because we are free to choose Σ (and, consequently, $\overline{\Sigma}$), we can

---

*Usually, $g[\cdot]$ is invariant under orthogonal transformations, i.e., $g[\mathbf{u}_i] = g[\overline{\mathbf{u}}_i]$. This is the case for LMS, NLMS, and ϵ-NLMS.

assume that $\overline{\Sigma}'$ is diagonal. In this way, $\{\overline{\Sigma}, \overline{\Sigma}'\}$ will be fully characterized by their diagonal entries. Thus let $\{\overline{\sigma}, \overline{\sigma}'\}$ denote $M \times 1$ vectors that collect the diagonal entries of $\{\overline{\Sigma}, \overline{\Sigma}'\}$, i.e.,

$$\overline{\sigma} = \mathrm{diag}(\overline{\Sigma}), \quad \overline{\sigma}' = \mathrm{diag}(\overline{\Sigma}').$$

Then from Equation 1.36 we find that

$$\overline{\sigma}' = \overline{F}\overline{\sigma},$$

where $\overline{F}$ is the $M \times M$ matrix

$$\overline{F} = I - \mu\overline{A}, \quad \overline{A} = 2\Delta.$$

Repeating the arguments that led to Equation 1.22 we can then establish that, for sufficiently small step sizes, the evolution of $\mathrm{E}\,\|\overline{\mathbf{w}}_i\|_{\overline{\sigma}}^2$ is described by the following M-dimensional state-space model:

$$\overline{\mathcal{W}}_i = \overline{\mathcal{F}}\,\overline{\mathcal{W}}_{i-1} + \mu^2\sigma_v^2\overline{\mathcal{Y}}, \tag{1.37}$$

where the $M \times 1$ vectors $\{\overline{\mathcal{W}}_i, \overline{\mathcal{Y}}\}$ are defined by

$$\overline{\mathcal{W}}_i = \begin{bmatrix} \mathrm{E}\,\|\overline{\mathbf{w}}_i\|_{\overline{\sigma}}^2 \\ \mathrm{E}\,\|\overline{\mathbf{w}}_i\|_{\overline{F}\overline{\sigma}}^2 \\ \vdots \\ \mathrm{E}\,\|\overline{\mathbf{w}}_i\|_{\overline{F}^{M-2}\overline{\sigma}}^2 \\ \mathrm{E}\,\|\overline{\mathbf{w}}_i\|_{\overline{F}^{M-1}\overline{\sigma}}^2 \end{bmatrix}, \quad \overline{\mathcal{Y}} = \begin{bmatrix} \mathrm{E}\left(\|\overline{\mathbf{u}}_i\|_{\overline{\sigma}}^2/g^2[\overline{\mathbf{u}}_i]\right) \\ \mathrm{E}\left(\|\overline{\mathbf{u}}_i\|_{\overline{F}\overline{\sigma}}^2/g^2[\overline{\mathbf{u}}_i]\right) \\ \vdots \\ \mathrm{E}\left(\|\overline{\mathbf{u}}_i\|_{\overline{F}^{M-2}\overline{\sigma}}^2/g^2[\overline{\mathbf{u}}_i]\right) \\ \mathrm{E}\left(\|\overline{\mathbf{u}}_i\|_{\overline{F}^{M-1}\overline{\sigma}}^2/g^2[\overline{\mathbf{u}}_i]\right) \end{bmatrix}, \tag{1.38}$$

and the $M \times M$ coefficient matrix $\overline{\mathcal{F}}$ is given by

$$\overline{\mathcal{F}} = \begin{bmatrix} 0 & 1 & & & \\ 0 & 0 & 1 & & \\ 0 & 0 & 0 & 1 & \\ \vdots & & & & \\ 0 & 0 & 0 & & 1 \\ -\overline{p}_0 & -\overline{p}_1 & -\overline{p}_2 & \cdots & -\overline{p}_{M-1} \end{bmatrix}, \tag{1.39}$$

where the $\{\overline{p}_i\}$ are the coefficients of the characteristic polynomial of $\overline{F}$. If we select $\overline{\sigma} = \mathrm{vec}(I)$, then

$$\|\overline{\mathbf{w}}_i\|_{\overline{\sigma}}^2 = \|\overline{\mathbf{w}}_i\|^2 = \|U^T\tilde{\mathbf{w}}_i\|^2 = \|\tilde{\mathbf{w}}_i\|^2$$

because U is orthogonal. In this case, the top entry of $\overline{\mathcal{W}}_i$ will describe the evolution of the filter MSD.

When P and R_u have identical eigenvectors, e.g., as in LMS for which $g[\mathbf{u}] = 1$ and $P = R_u$, then the evolution of the learning curve of the filter can also be read from Equation 1.37. To see this, let λ be the column vector consisting of the eigenvalues of R_u. Choosing $\overline{\sigma} = \lambda$ gives

$$\|\overline{\mathbf{w}}_i\|_{\overline{\sigma}}^2 = \|\overline{\mathbf{w}}_i\|_\lambda^2 = \overline{\mathbf{w}}_i^T \Lambda \overline{\mathbf{w}}_i = \tilde{\mathbf{w}}_i^T R_u \tilde{\mathbf{w}}_i = \|\tilde{\mathbf{w}}_i\|_{R_u}^2,$$

so that the EMSE behavior of the filter can be read from the top entry of the resulting state-vector $\overline{\mathcal{W}}_i$.

1.9 Applications to Selected Filters

We now illustrate the application of the results of the earlier sections, as well as some extensions of these results, to selected adaptive filters.

1.9.1 The NLMS Algorithm

Our first example derives performance results for NLMS by showing how to relate it to LMS. In NLMS, $g[\mathbf{u}] = \|\mathbf{u}\|^2$, and the filter recursion takes the form

$$\mathbf{w}_i = \mathbf{w}_{i-1} + \mu \frac{\mathbf{u}_i^T}{\|\mathbf{u}_i\|^2}[\mathbf{d}(i) - \mathbf{u}_i \mathbf{w}_{i-1}].$$

Introduce the transformed variables:

$$\check{\mathbf{u}}_i = \frac{\mathbf{u}_i}{\|\mathbf{u}_i\|}, \quad \check{\mathbf{d}}(i) = \frac{\mathbf{d}(i)}{\|\mathbf{u}_i\|}, \quad \check{\mathbf{v}}(i) = \frac{\mathbf{v}(i)}{\|\mathbf{u}_i\|}. \tag{1.40}$$

Then the NLMS recursion can be rewritten as

$$\mathbf{w}_i = \mathbf{w}_{i-1} + \mu \check{\mathbf{u}}_i^T \check{e}(i)$$

with

$$\check{e}(i) = \check{\mathbf{d}}(i) - \check{\mathbf{u}}_i \mathbf{w}_{i-1}.$$

In other words, we find that NLMS can be regarded as an LMS filter with respect to the variables $\{\check{\mathbf{d}}(i), \check{\mathbf{u}}_i\}$. Moreover, these variables satisfy a model similar to that of $\{\mathbf{d}(i), \mathbf{u}_i\}$, as given by Equations 1.1 and 1.2, specifically

$$\check{\mathbf{d}}(i) = \check{\mathbf{u}}_i w^o + \check{\mathbf{v}}(i),$$

where:

1. The sequence $\{\check{\mathbf{v}}(i)\}$ is iid with variance

$$E \check{\mathbf{v}}^2(i) = \check{\sigma}_v^2 = \sigma_v^2 E \left(\frac{1}{\|\mathbf{u}_i\|^2} \right).$$

2. The sequence $\mathbf{v}(i)$ is independent of $\mathbf{u}_j$ for all $i \neq j$.
3. The covariance matrix of $\check{\mathbf{u}}_i$ is

$$\check{R}_u = \mathrm{E}\,\check{\mathbf{u}}_i^T \check{\mathbf{u}}_i = \mathrm{E}\left(\frac{\mathbf{u}_i^T \mathbf{u}_i}{\|\mathbf{u}_i\|^2}\right) > 0.$$

4. The random variables $\{\check{\mathbf{v}}(i), \check{\mathbf{u}}_i\}$ are zero mean.

These conditions allow us to repeat the previous derivation of the variance and mean relations (Equations 1.16 through 1.18) using the transformed variables (Equation 1.40). In this way, the performance of NLMS can be deduced from that of LMS. In particular, from Equation 1.35 we obtain for NLMS:

$$\mathrm{MSD} \approx \frac{\mu\check{\sigma}_v^2 M}{2} = \frac{\mu\sigma_v^2 M}{2}\mathrm{E}\left(\frac{1}{\|\mathbf{u}_i\|^2}\right) \tag{1.41}$$

and

$$\lim_{i \to \infty} \mathrm{E}\,\check{\mathbf{e}}_a^2(i) \approx \frac{\mu\check{\sigma}_v^2 \mathrm{Tr}(\check{R}_u)}{2} = \frac{\mu\check{\sigma}_v^2}{2},$$

because $\mathrm{Tr}(\check{R}_u) = 1$, and where $\check{\mathbf{e}}_a(i) = \check{\mathbf{d}}(i) - \check{\mathbf{u}}_i \mathbf{w}_{i-1}$. However, the filter EMSE relates to the limiting value of $\mathrm{E}\,\mathbf{e}_a^2(i)$ and not $\mathrm{E}\,\check{\mathbf{e}}_a^2(i)$. To find this limiting value, we first note from the definitions of $\mathbf{e}_a(i)$ and $\check{\mathbf{e}}_a(i)$ that

$$\frac{1}{\|\mathbf{u}_i\|^2} \cdot \mathbf{e}_a^2(i) = \check{\mathbf{e}}_a^2(i).$$

Then if we introduce the steady-state separation assumption*

$$\mathrm{E}\left(\frac{1}{\|\mathbf{u}_i\|^2} \cdot \mathbf{e}_a^2(i)\right) \approx \frac{\mathrm{E}\,\mathbf{e}_a^2(i)}{\mathrm{E}\,\|\mathbf{u}_i\|^2} \quad \text{as } i \longrightarrow \infty,$$

so that

$$\lim_{i \to \infty} \mathrm{E}\,\mathbf{e}_a^2(i) = \mathrm{Tr}(R_u) \cdot \left(\lim_{i \to \infty} \mathrm{E}\,\check{\mathbf{e}}_a^2(i)\right),$$

we obtain

$$\mathrm{EMSE} = \frac{\mu\sigma_v^2 \mathrm{Tr}(R_u)}{2}\mathrm{E}\left(\frac{1}{\|\mathbf{u}_i\|^2}\right). \tag{1.42}$$

An alternative method to evaluate the steady-state (as well as transient) performance of NLMS is to treat it as a special case of the results developed in Section 1.8 by setting $g(\mathbf{u}) = \|\mathbf{u}\|^2$. In this case, the variance relation (Equation 1.36) would become

$$\begin{cases} \mathrm{E}\,\|\overline{\mathbf{w}}_i\|_{\overline{\Sigma}}^2 = \mathrm{E}\,\|\overline{\mathbf{w}}_{i-1}\|_{\overline{\Sigma}'}^2 + \mu^2\sigma_v^2\mathrm{E}\left[\frac{\|\overline{\mathbf{u}}_i\|_{\overline{\Sigma}}^2}{\|\check{\mathbf{u}}_i\|^4}\right], \\ \overline{\Sigma}' = \overline{\Sigma} - \mu\overline{\Sigma}\Delta - \mu\Delta\overline{\Sigma}. \end{cases}$$

---

*The assumption is reasonable for longer filters.

Moreover, the EMSE and MSD expressions (1.33 and 1.34) would give

$$\text{MSD} = \frac{\mu \sigma_v^2 \text{Tr}(QP^{-1})}{2},$$

$$\text{EMSE} = \mu \sigma_v^2 \text{Tr}(Q\Sigma_{\text{emse}}),$$

(1.43)

where now

$$P = \text{E}\left(\frac{\mathbf{u}_i^T \mathbf{u}_i}{\|\mathbf{u}_i\|^2}\right), \quad Q = \text{E}\left(\frac{\mathbf{u}_i^T \mathbf{u}_i}{\|\mathbf{u}_i\|^4}\right)$$

and Σ_{emse} is the unique solution of $P\Sigma_{\text{emse}} + \Sigma_{\text{emse}}P = R_u$. Expressions 1.44 are alternatives to (1.41) and (1.42).

1.9.2 The RLS Algorithm

Our second example pertains to the recursive least-squares algorithm:

$$\mathbf{w}_i = \mathbf{w}_{i-1} + \mathbf{P}_i \, \mathbf{u}_i^T [\mathbf{d}(i) - \mathbf{u}_i \mathbf{w}_{i-1}], \quad i \geq 0$$

(1.44)

$$\mathbf{P}_i = \alpha^{-1}\left[\mathbf{P}_{i-1} - \frac{\alpha^{-1}\mathbf{P}_{i-1}\mathbf{u}_i^T\mathbf{u}_i\mathbf{P}_{i-1}}{1 + \alpha^{-1}\mathbf{u}_i\mathbf{P}_{i-1}\mathbf{u}_i^T}\right],$$

(1.45)

where the data $\{\mathbf{d}(i), \mathbf{u}_i\}$ satisfy Equations 1.1 and 1.2, and the regressors satisfy independence Assumption 1.15. In the above, $0 \ll \alpha \leq 1$ is a forgetting factor and $\mathbf{P}_{-1} = \epsilon^{-1}I$ for a small positive ϵ.

Compared with the LMS-type recursion (Equation 1.3), the RLS update includes the matrix factor $\mathbf{P}_i$ multiplying $\mathbf{u}_i^T$ from the left. Moreover, $\mathbf{P}_i$ is a function of both current and prior regressors. Still, the energy-conservation approach of Sections 1.3 and 1.4 can be extended to deal with this more general case. In particular, it is straightforward to verify that Equation 1.10 is now replaced by

$$\|\mathbf{u}_i\|_{\mathbf{P}_i\Sigma\mathbf{P}_i}^2 \cdot \|\tilde{\mathbf{w}}_i\|_\Sigma^2 + \left(e_a^{\mathbf{P}_i\Sigma}(i)\right)^2 = \|\mathbf{u}_i\|_{\mathbf{P}_i\Sigma\mathbf{P}_i}^2 \cdot \|\tilde{\mathbf{w}}_{i-1}\|_\Sigma^2 + \left(e_p^{\mathbf{P}_i\Sigma}(i)\right)^2.$$

(1.46)

Under expectation, Equation 1.46 leads to

$$\text{E}\|\tilde{\mathbf{w}}_i\|_\Sigma^2 = \text{E}\|\tilde{\mathbf{w}}_{i-1}\|_{\Sigma'}^2 + \sigma_v^2 \text{E}\|\mathbf{u}_i\|_{\mathbf{P}_i\Sigma\mathbf{P}_i}^2,$$

$$\Sigma' = \Sigma - \Sigma\text{E}(\mathbf{P}_i\mathbf{u}_i^T\mathbf{u}_i) - \text{E}(\mathbf{u}_i^T\mathbf{u}_i\mathbf{P}_i)\Sigma + \text{E}\left[\|\mathbf{u}_i\|_{\mathbf{P}_i\Sigma\mathbf{P}_i}^2\mathbf{u}_i^T\mathbf{u}_i\right].$$

(1.47)

However, the presence of the matrix $\mathbf{P}_i$ makes the subsequent analysis rather challenging; this is because $\mathbf{P}_i$ is dependent not only on $\mathbf{u}_i$ but also on all prior regressors $\{\mathbf{u}_j, \, j \leq i\}$.

To make the analysis more tractable, whenever necessary, we approximate and replace the random variable $\mathbf{P}_i$ in steady-state by its respective mean value.* Now because

$$\mathbf{P}_i^{-1} = \alpha^{i+1}\epsilon I + \sum_{j=0}^{i}\alpha^{i-j}\mathbf{u}_j^*\mathbf{u}_j,$$

we find that, as $i \to \infty$, and because $\alpha < 1$,

$$\lim_{i\to\infty} \mathrm{E}\left(\mathbf{P}_i^{-1}\right) = \frac{R_u}{1-\alpha} \overset{\Delta}{=} P^{-1}.$$

That is, the mean value of $\mathbf{P}_i^{-1}$ tends to $R_u/(1-\alpha)$. In comparison, the evaluation of the limiting mean value of $\mathbf{P}_i$ is generally more difficult. For this reason, we content ourselves with the approximation

$$\mathrm{E}\,\mathbf{P}_i \approx \left[\mathrm{E}\,\mathbf{P}_i^{-1}\right]^{-1} = (1-\alpha)R_u^{-1} = P, \quad \text{as } i \to \infty.$$

This is an approximation, of course, because even though $\mathbf{P}_i$ and $\mathbf{P}_i^{-1}$ are the inverses of one another, it does not hold that their means will have the same inverse relation.[†]

Replacing $\mathbf{P}_i$ by $P = (1-\alpha)R_u^{-1}$, we find that variance Relation 1.48 becomes

$$\mathrm{E}\,\|\tilde{\mathbf{w}}_i\|_{\Sigma}^2 = \mathrm{E}\,\|\tilde{\mathbf{w}}_{i-1}\|_{\Sigma'}^2 + \sigma_v^2(1-\alpha)^2\mathrm{E}\,\|\mathbf{u}_i\|_{R_u^{-1}\Sigma R_u^{-1}}^2,$$

$$\Sigma' = \Sigma - 2(1-\alpha)\Sigma + (1-\alpha)^2\mathrm{E}\left[\|\mathbf{u}_i\|_{R_u^{-1}\Sigma R_u^{-1}}^2\mathbf{u}_i^T\mathbf{u}_i\right].$$

Introduce the eigen-decomposition $R_u = U\Lambda U^T$, and define the transformed variables

$$\overline{\mathbf{w}}_i \overset{\Delta}{=} U^T\tilde{\mathbf{w}}_i, \quad \overline{\mathbf{u}}_i \overset{\Delta}{=} \mathbf{u}_i U, \quad \overline{\Sigma} \overset{\Delta}{=} U^T\Sigma U.$$

Assume further, for the sake of illustration, that the regressors $\{\mathbf{u}_i\}$ are Gaussian. Then

$$\mathrm{E}\left[\|\overline{\mathbf{u}}_i\|_{\Lambda^{-1}\overline{\Sigma}\Lambda^{-1}}^2\overline{\mathbf{u}}_i^T\overline{\mathbf{u}}_i\right] = 2\Lambda\mathrm{Tr}(\Lambda^{-1}\overline{\Sigma}) + \overline{\Sigma}$$

and the variance relation becomes

$$\mathrm{E}\,\|\overline{\mathbf{w}}_i\|_{\overline{\Sigma}}^2 = \mathrm{E}\,\|\overline{\mathbf{w}}_{i-1}\|_{\overline{\Sigma}'}^2 + \sigma_v^2(1-\alpha)^2\mathrm{E}\,\|\overline{\mathbf{u}}_i\|_{\Lambda^{-1}\overline{\Sigma}\Lambda^{-1}}^2,$$

$$\overline{\Sigma}' = \alpha^2\overline{\Sigma} + 2(1-\alpha)^2\Lambda\mathrm{Tr}(\Lambda^{-1}\overline{\Sigma}).$$

It follows that $\overline{\Sigma}'$ will be diagonal if $\overline{\Sigma}$ is. If we further introduce the M-dimensional column vectors

$$\lambda = \mathrm{diag}\{\Lambda\}, \quad a = \mathrm{diag}\{\Lambda^{-1}\}, \quad \overline{\sigma} = \mathrm{diag}\{\overline{\Sigma}\},$$

---

* This approximation essentially amounts to an ergodicity assumption on the regressors.
† It turns out that the approximation is reasonable for Gaussian regressors.

then the above recursion for $\overline{\Sigma}'$ is equivalent to

$$\overline{\sigma}' = \overline{F}\overline{\sigma} \quad \text{where} \quad \overline{F} = \alpha^2 I + 2(1-\alpha)^2 \lambda a^T.$$

Let $\overline{\Sigma}_{\text{msd}}$ denote the weighting matrix that corresponds to the vector

$$\overline{\sigma}_{\text{msd}} = (I - \overline{F})^{-1}\text{diag}(I).$$

Let also $\overline{\Sigma}_{\text{emse}}$ denote the weighting matrix that corresponds to the vector

$$\overline{\sigma}_{\text{emse}} = (I - \overline{F})^{-1}\lambda.$$

Then, because

$$\text{MSD} = \sigma_v^2 (1-\alpha)^2 \text{E} \|\overline{\mathbf{u}}_i\|^2_{\Lambda^{-1}\overline{\Sigma}_{\text{msd}}\Lambda^{-1}},$$

$$\text{EMSE} = \sigma_v^2 (1-\alpha)^2 \text{E} \|\overline{\mathbf{u}}_i\|^2_{\Lambda^{-1}\overline{\Sigma}_{\text{emse}}\Lambda^{-1}},$$

we can verify after some algebra that

$$\text{MSD} = \frac{\sigma_v^2 \sum_{k=1}^{M}(1/\lambda_k)}{\frac{1+\alpha}{1-\alpha} - 2M},$$

$$\text{EMSE} = \frac{\sigma_v^2 M}{\frac{1+\alpha}{1-\alpha} - 2M}.$$

(1.48)

1.9.3 Leaky-LMS

Our third example extends the energy-conservation and variance relations of Sections 1.3 and 1.4 to leaky-LMS updates of the form:

$$\mathbf{w}_i = (1 - \mu\alpha)\mathbf{w}_{i-1} + \mu\mathbf{u}_i^T \mathbf{e}(i), \quad i \geq 0$$

$$\mathbf{e}(i) = \mathbf{d}(i) - \mathbf{u}_i\mathbf{w}_{i-1},$$

where α is a positive scalar. The data $\{\mathbf{d}(i), \mathbf{u}_i\}$ are still assumed to satisfy Equations 1.1 and 1.2, with the regressors satisfying independence Assumption 1.15.

Repeating the arguments of Sections 1.3 and 1.4, it is straightforward to verify that the variance and mean relations (Equations 1.16 through 1.18) extend to the following (see Reference 18):

$$\text{E} \|\tilde{\mathbf{w}}_i\|^2_\Sigma = \text{E} \|\tilde{\mathbf{w}}_{i-1}\|^2_{\Sigma'} + \mu^2 \sigma_v^2 \text{E} \|\mathbf{u}_i\|^2_\Sigma + 2\alpha\mu \, (w^o)^T \Sigma J \text{E} \tilde{\mathbf{w}}_{i-1} + \alpha^2\mu^2 \|w^o\|^2_\Sigma,$$

$$\Sigma' = \Sigma - \mu(\text{E}\,\mathbf{U}_i)\Sigma - \mu\Sigma(\text{E}\,\mathbf{U}_i) + \mu^2\text{E}\,(\mathbf{U}_i\Sigma\mathbf{U}_i),$$

$$\text{E}\,\tilde{\mathbf{w}}_i = J\,\text{E}\,\tilde{\mathbf{w}}_{i-1} + \alpha\mu w^o.$$

where

$$U_i = \alpha I + \mathbf{u}_i^T \mathbf{u}_i, \qquad J = \mathrm{E}\,(I - \mu U_i) = (1 - \alpha\mu)I - \mu R_u.$$

Frequently, $w_{-1} = 0$, so that $\mathrm{E}\,\tilde{w}_{-1} = w^o$. We will make this assumption to simplify the analysis, although it is not necessary for stability or steady-state results.

Therefore, by iterating the recursion for $\mathrm{E}\,\tilde{w}_i$ we can verify that

$$\mathrm{E}\,\tilde{\mathbf{w}}_{i-1} = C_i w^o, \quad i \geq 0,$$

where

$$C_i = J^i + \alpha\mu(I + J + \cdots + J^{i-1}).$$

It then follows that the term below, which appears in the recursion for $\mathrm{E}\,\|\tilde{\mathbf{w}}_i\|_\Sigma^2$, can be expressed in terms of $\|w^o\|^2$ as

$$2\alpha\mu\,(w^o)^T \Sigma J\,\mathrm{E}\,\tilde{\mathbf{w}}_{i-1} = \alpha\mu\|w^o\|_{\Sigma J C_i + C_i J \Sigma}^2.$$

Now repeating the arguments of Section 1.5 we can verify that the transient behavior of the leaky filter is characterized by the following state-space model:

$$\mathcal{W}_i = \mathcal{F}\mathcal{W}_{i-1} + \mu \mathcal{Y}_i,$$

where $\mathcal{W}_i$ is the M^2-dimensional vector

$$\mathcal{W}_i \triangleq \begin{bmatrix} \mathrm{E}\,\|\tilde{\mathbf{w}}_i\|_\sigma^2 \\ \mathrm{E}\,\|\tilde{\mathbf{w}}_i\|_{F\sigma}^2 \\ \mathrm{E}\,\|\tilde{\mathbf{w}}_i\|_{F^2\sigma}^2 \\ \vdots \\ \mathrm{E}\,\|\tilde{\mathbf{w}}_i\|_{F^{M^2-1}\sigma}^2 \end{bmatrix},$$

and $\mathcal{F}$ is the $M^2 \times M^2$ companion matrix

$$\mathcal{F} = \begin{bmatrix} 0 & 1 & & & & \\ 0 & 0 & 1 & & & \\ 0 & 0 & 0 & 1 & & \\ \vdots & & & & & \\ 0 & 0 & 0 & & 1 \\ -p_0 & -p_1 & -p_2 & \cdots & -p_{M^2-1} \end{bmatrix},$$

with

$$p(x) \triangleq \det(xI - F) = x^{M^2} + \sum_{k=0}^{M^2-1} p_k x^k$$

denoting the characteristic polynomial of the matrix

$$F = I - \mu A + \mu^2 B,$$

where

$$A = (E\,\mathbf{U}_i \otimes I) + (I \otimes E\,\mathbf{U}_i),$$
$$B = E\,(\mathbf{U}_i \otimes \mathbf{U}_i).$$

Moreover,

$$\mathcal{y}_i \;=\; \mu \sigma_v^2 \begin{bmatrix} E\,|\mathbf{u}_i\|_\sigma^2 \\ E\,\|\mathbf{u}_i\|_{F\sigma}^2 \\ E\,\|\mathbf{u}_i\|_{F^2\sigma}^2 \\ \vdots \\ E\,\|\mathbf{u}_i\|_{F^{M^2-1}\sigma}^2 \end{bmatrix} + \alpha \begin{bmatrix} \|w^o\|_{(\alpha\mu I + S_i)\sigma}^2 \\ \|w^o\|_{(\alpha\mu I + S_i)F\sigma}^2 \\ \|w^o\|_{(\alpha\mu I + S_i)F^2\sigma}^2 \\ \vdots \\ \|w^o\|_{(\alpha\mu I + S_i)F^{M^2-1}\sigma}^2 \end{bmatrix},$$

where S_i is the $M^2 \times M^2$ matrix

$$S_i \triangleq (J\,C_i \otimes I_M) + (I_M \otimes C_i J).$$

It follows that the filter is stable in the mean and mean-square senses for step sizes in the range

$$\mu < \min \left\{ \frac{2}{\alpha + \lambda_{\max}(R_u)},\; \frac{1}{\lambda_{\max}(A^{-1}B)},\; \frac{1}{\max\{\lambda(H) \in \mathbb{R}^+\}} \right\},$$

where

$$H = \begin{bmatrix} A/2 & -B/2 \\ I & 0 \end{bmatrix}.$$

It also follows that in steady state,

$$\lim_{i \to \infty} E\,\tilde{\mathbf{w}}_i = \alpha(\alpha I + R_u)^{-1} w^o,$$

$$\text{MSD} = \mu^2 \sigma_v^2 E\left(\|\mathbf{u}_i\|_{(I-F)^{-1}\text{vec}(I)}^2 \right) + \alpha^2 \mu^2 \|w^o\|_{T(I-F)^{-1}\text{vec}(I)}^2, \quad (1.49)$$

$$\text{EMSE} = \mu^2 \sigma_v^2 E\left(\|\mathbf{u}_i\|_{(I-F)^{-1}\text{vec}(R_u)}^2 \right) + \alpha^2 \mu^2 \|w^o\|_{T(I-F)^{-1}\text{vec}(R_u)}^2,$$

where T is the $M^2 \times M^2$ matrix

$$T = I + ((I - J)^{-1}J \otimes I) + (I \otimes (I - J)^{-1}J).$$

1.10   Fourth-Order Moment Approximation

Instead of the small-step-size approximation of Section 1.8, we can choose to approximate the fourth-order moment that appears in the expression for Σ' in Equation 1.17 as

$$\mathrm{E}\left(\frac{\|\mathbf{u}_i\|_\Sigma^2}{g^2[\mathbf{u}_i]}\mathbf{u}_i^T\mathbf{u}_i\right) \approx \mathrm{E}\left(\frac{\mathbf{u}_i^T\mathbf{u}_i}{g[\mathbf{u}_i]}\right)\cdot\mathrm{E}\left(\frac{\|\mathbf{u}_i\|_\Sigma^2}{g[\mathbf{u}_i]}\right) = P\mathrm{Tr}(\Sigma P),$$

where $P = \mathrm{E}\left(\mathbf{u}_i^T\mathbf{u}_i/g[\mathbf{u}_i]\right)$. In this way, Expression 1.17 for Σ' would become

$$\Sigma' = \Sigma - \mu\Sigma P - \mu P\Sigma + \mu^2 P\mathrm{Tr}(P\Sigma), \tag{1.50}$$

which is fully characterized in terms of the single moment P. If we now let $P = U\Delta U^T$ denote the eigen-decomposition of $P > 0$, and introduce the transformed quantities:

$$\overline{\mathbf{w}}_i = U^T\tilde{\mathbf{w}}_i, \quad \overline{\mathbf{u}}_i = \mathbf{u}_i U, \quad \overline{\Sigma} = U^T\Sigma U.$$

Then variance Relations 1.16 and 1.50 can be equivalently rewritten as

$$\mathrm{E}\|\overline{\mathbf{w}}_i\|_{\overline{\Sigma}}^2 = \mathrm{E}\|\overline{\mathbf{w}}_{i-1}\|_{\overline{\Sigma}'}^2 + \mu^2\sigma_v^2\mathrm{E}\left(\frac{\|\overline{\mathbf{u}}_i\|_\Sigma^2}{g^2[\overline{\mathbf{u}}_i]}\right),$$

$$\overline{\Sigma}' = \overline{\Sigma} - \mu\overline{\Sigma}\Delta - \mu\Delta\overline{\Sigma} + \mu^2\Delta\mathrm{Tr}(\overline{\Sigma}\Delta). \tag{1.51}$$

The expression for $\overline{\Sigma}'$ shows that it will be diagonal as long as $\overline{\Sigma}$ is diagonal. Thus let again

$$\overline{\sigma} = \mathrm{diag}(\overline{\Sigma}), \quad \overline{\sigma}' = \mathrm{diag}(\overline{\Sigma}').$$

Then from Equation 1.51 we find that

$$\overline{\sigma}' = \overline{F}\overline{\sigma},$$

where $\overline{F}$ is $M \times M$ and given by

$$\overline{F} = I - \mu\overline{A} + \mu^2, \overline{B}, \quad \overline{A} = 2\Delta, \quad \overline{B} = \mu^2\delta\delta^T,$$

where $\delta = \mathrm{diag}(\Delta)$. Repeating the arguments that led to Equation 1.22 we can establish that, under the assumed fourth-order moment approximation, the evolution of $\mathrm{E}\|\overline{\mathbf{w}}_i\|_{\overline{\sigma}}^2$ is described by an M-dimensional state-space model similar to Equation 1.37.

1.11   Long Filter Approximation

In addition to the small-step-size and fourth-order moment approximations of Sections 1.8 and 1.10, we can also resort to a long filter approximation and derive simplified transient and steady-state performance results for data-normalized filters of form 1.3. We postpone this discussion until Section 1.12.5, whereby the simplified results will be obtained as a special case of the theory we develop below for adaptive filters with error nonlinearities.

1.12   Adaptive Filters with Error Nonlinearities

The analysis in the earlier sections has focused on data-normalized adaptive filters of the form 1.3. We now extend the energy-based arguments to filters with error nonlinearities in their update equations. This class of filters is usually more challenging to study. For this reason, we resort to a long filter assumption in order to make the analysis more tractable, as we explain in the following.

Thus consider filter updates of the form

$$\mathbf{w}_i = \mathbf{w}_{i-1} + \mu \mathbf{u}_i^T f[\mathbf{e}(i)], \quad i \geq 0 \tag{1.52}$$

where

$$\mathbf{e}(i) = \mathbf{d}(i) - \mathbf{u}_i \mathbf{w}_{i-1} \tag{1.53}$$

is the estimation error at iteration i, and f is some function of $\mathbf{e}(i)$. Typical choices for f are

$$f[\mathbf{e}] = \mathbf{e} \ \text{(LMS)}, \quad f[\mathbf{e}] = \text{sign}(\mathbf{e}) \ \text{(sign-LMS)}, \quad f[\mathbf{e}] = \mathbf{e}^3 \ \text{(LMF)}.$$

The initial condition $\mathbf{w}_{-1}$ of Equation 1.52 is assumed to be independent of all $\{\mathbf{d}(j), \mathbf{u}_j, \mathbf{v}(j)\}$.

The same argument that was employed in Section 1.3 can be repeated here to verify that the energy relation (Equation 1.10) still holds. Indeed, subtracting w^o from both sides of Equation 1.52 we obtain

$$\tilde{\mathbf{w}}_i = \tilde{\mathbf{w}}_{i-1} - \mu \mathbf{u}_i^T f[\mathbf{e}(i)], \tag{1.54}$$

and multiplying Equation 1.54 by $\mathbf{u}_i \Sigma$ from the left we find that

$$\mathbf{e}_p^\Sigma(i) = \mathbf{e}_a^\Sigma(i) - \mu \|\mathbf{u}_i\|_\Sigma^2 f[\mathbf{e}(i)]. \tag{1.55}$$

Relation 1.55 can be used to express $f[\mathbf{e}(i)]$ in terms of $\{\mathbf{e}_p^\Sigma(i), \mathbf{e}_a^\Sigma(i)\}$ and to eliminate it from Equation 1.54. Doing so leads to the equality

$$\|\mathbf{u}_i\|_\Sigma^2 \cdot \tilde{\mathbf{w}}_i + \mathbf{u}_i^T \mathbf{e}_a^\Sigma(i) = \|\mathbf{u}_i\|_\Sigma^2 \cdot \tilde{\mathbf{w}}_{i-1} + \mathbf{u}_i^T \mathbf{e}_p^\Sigma(i), \tag{1.56}$$

and by equating the weighted Euclidean norms of both sides of this equation we arrive again at Equation 1.10, which is repeated here for ease of reference,

$$\|\mathbf{u}_i\|_\Sigma^2 \cdot \|\tilde{\mathbf{w}}_i\|_\Sigma^2 + \left(e_a^\Sigma(i)\right)^2 = \|\mathbf{u}_i\|_\Sigma^2 \cdot \|\tilde{\mathbf{w}}_{i-1}\|_\Sigma^2 + \left(e_p^\Sigma(i)\right)^2. \tag{1.57}$$

1.12.1   Variance Relation for Error Nonlinearities

Now recall that in transient analysis we are interested in characterizing the time evolution of the quantity $\mathrm{E} \|\tilde{\mathbf{w}}_i\|_\Sigma^2$, for some Σ of interest (usually, $\Sigma = I$ or $\Sigma = R_u$). To characterize this evolution, we replace $e_p^\Sigma(i)$ in Equation 1.57 by its expression (Equation 1.55) in terms of $e_a^\Sigma(i)$ and $\mathbf{e}(i)$ to obtain

$$\|\mathbf{u}_i\|_\Sigma^2 \cdot \|\tilde{\mathbf{w}}_i\|_\Sigma^2 = \|\mathbf{u}_i\|_\Sigma^2 \cdot \|\tilde{\mathbf{w}}_{i-1}\|_\Sigma^2 + \mu^2 \left(\|\mathbf{u}_i\|_\Sigma^2\right)^2 f^2[\mathbf{e}(i)]$$
$$- 2\mu \|\mathbf{u}_i\|_\Sigma^2 e_a^\Sigma(i) f[\mathbf{e}(i)].$$

Assuming the event $\|\mathbf{u}_i\|_\Sigma^2 = 0$ occurs with zero probability, we can eliminate $\|\mathbf{u}_i\|_\Sigma^2$ from both sides and take expectations to arrive at

$$\mathrm{E} \|\tilde{\mathbf{w}}_i\|_\Sigma^2 = \mathrm{E} \|\tilde{\mathbf{w}}_{i-1}\|_\Sigma^2 - 2\mu \mathrm{E} \left(e_a^\Sigma(i) f[\mathbf{e}(i)]\right) + \mu^2 \mathrm{E} \left(\|\mathbf{u}_i\|_\Sigma^2 f^2[\mathbf{e}(i)]\right), \tag{1.58}$$

which is the equivalent of Equation 1.13 for filters with error nonlinearities. Observe, however, that the weighting matrices for $\mathrm{E} \|\tilde{\mathbf{w}}_i\|_\Sigma^2$ and $\mathrm{E} \|\tilde{\mathbf{w}}_{i-1}\|_\Sigma^2$ are still identical because we did not substitute $\{e_a(i), \mathbf{e}(i)\}$ by their expressions in terms of $\tilde{\mathbf{w}}_{i-1}$. The reason we did not do so here is because of the nonlinear error function f. Instead, to proceed, we show how to evaluate the expectations

$$\mathrm{E} \left(e_a^\Sigma(i) f[\mathbf{e}(i)]\right) \quad \text{and} \quad \mathrm{E} \left(\|\mathbf{u}_i\|_\Sigma^2 f^2[\mathbf{e}(i)]\right). \tag{1.59}$$

These expectations are generally hard to compute because of f. To facilitate their evaluation, we assume that the filter is long enough to justify, by central limit theorem arguments, that

$$e_a(i) \quad \text{and} \quad e_a^\Sigma(i) \quad \text{are jointly Gaussian random variables.} \tag{1.60}$$

1.12.1.1   *Evaluation of* $\mathrm{E}(e_a^\Sigma f[\mathbf{e}])$

Using Statement 1.60 we can evaluate the first expectation, $\mathrm{E} \left(e_a^\Sigma(i) f[\mathbf{e}(i)]\right)$, by appealing to Price's theorem.[15] The theorem states that if $\mathbf{x}$ and $\mathbf{y}$ are jointly Gaussian random variables that are independent from a third random variable $\mathbf{z}$, then

$$\mathrm{E}\,\mathbf{x}k(\mathbf{y} + \mathbf{z}) = \frac{\mathrm{E}\,\mathbf{xy}}{\mathrm{E}\,\mathbf{y}^2} \mathrm{E}\,\mathbf{y}k(\mathbf{y} + \mathbf{z}),$$

TABLE 1.1

Expressions for h_G and h_U for Some Error
Nonlinearities

Algorithm	Error Nonlinearity	$\{h_G, h_U\}$
LMS	$f[e] = e$	$h_G = 1$ $h_U = E\,e_a^2(i) + \sigma_v^2$
sign-LMS	$f[e] = \text{sign}[e]$	$h_G = \sqrt{\dfrac{2}{\pi}}\dfrac{1}{\sqrt{E\,e_a^2(i) + \sigma_v^2}}$ $h_U = 1$
LMF	$f[e] = e^3$	$h_G = 3\left(E\,e_a^2(i) + \sigma_v^2\right)$ $h_U = 15\left(E\,e_a^2(i) + \sigma_v^2\right)^3$

Note: In the least-mean-fourth (LMF) case, we assume Gaussian
noise for simplicity.

where $k(\cdot)$ is some function of $\mathbf{y} + \mathbf{z}$. Using this result, together with the
equality $\mathbf{e}(i) = \mathbf{e}_a(i) + \mathbf{v}(i)$, we obtain

$$E\,\mathbf{e}_a^\Sigma(i)f[\mathbf{e}(i)] = E\,\mathbf{e}_a^\Sigma(i)\mathbf{e}_a(i)\frac{E\,\mathbf{e}_a(i)f[\mathbf{e}(i)]}{E\,\mathbf{e}_a^2(i)} \triangleq \left(E\,\mathbf{e}_a^\Sigma(i)\mathbf{e}_a(i)\right)\cdot h_G,$$

where the function h_G is defined by

$$h_G \triangleq \frac{E\,\mathbf{e}_a(i)f[\mathbf{e}(i)]}{E\,\mathbf{e}_a^2(i)}. \tag{1.61}$$

Clearly, because $\mathbf{e}_a(i)$ is Gaussian, the expectation $E\,\mathbf{e}_a(i)f[\mathbf{e}(i)]$ depends on
$\mathbf{e}_a(i)$ only through its second moment, $E\,\mathbf{e}_a^2(i)$. This means that h_G itself is only
a function of $E\,\mathbf{e}_a^2(i)$. The function $h_G[\cdot]$ can be evaluated for different choices
of the error nonlinearity $f[\cdot]$, as shown in Table 1.1.

1.12.1.2 Evaluation of $E(\|\mathbf{u}_i\|_\Sigma^2 f^2[\mathbf{e}])$

To evaluate the second expectation, $E\left(\|\mathbf{u}_i\|_\Sigma^2 f^2[\mathbf{e}(i)]\right)$, we resort to a separation
assumption; i.e., we assume that the filter is long enough so that

$$\|\mathbf{u}_i\|_\Sigma^2 \quad \text{and} \quad f^2[\mathbf{e}(i)] \quad \text{are uncorrelated.} \tag{1.62}$$

This assumption allows us to write

$$E\left(\|\mathbf{u}_i\|_\Sigma^2 f^2[\mathbf{e}(i)]\right) = \left(E\,\|\mathbf{u}_i\|_\Sigma^2\right)\cdot\left(E\,f^2[\mathbf{e}(i)]\right) \triangleq \left(E\,\|\mathbf{u}_i\|_\Sigma^2\right)\cdot h_U,$$

where the function h_U is defined by

$$h_U \triangleq E\,f^2[\mathbf{e}(i)]. \tag{1.63}$$

Again, since $\mathbf{e}_a(i)$ is Gaussian and independent of the noise, the function h_U is
a function of $E\,\mathbf{e}_a^2(i)$ only. The function h_U can also be evaluated for different
error nonlinearities, as shown in Table 1.1.

1.12.2   Independent Regressors

Using the definitions of h_U and h_G, we can rewrite the variance relation (Equation 1.58) more compactly as

$$\mathrm{E}\,\|\tilde{\mathbf{w}}_i\|_{\Sigma}^2 = \mathrm{E}\,\|\tilde{\mathbf{w}}_{i-1}\|_{\Sigma}^2 - 2\mu h_G \mathrm{E}\left(\mathbf{e}_a^{\Sigma}(i)\mathbf{e}_a(i)\right) + \mu^2 h_U \mathrm{Tr}(R_u\Sigma). \qquad (1.64)$$

As it stands, this relation is still difficult to propagate because it requires the evaluation of $\mathrm{E}\,\mathbf{e}_a^{\Sigma}(i)\mathbf{e}_a(i)$, and this expectation is not trivial in general. This is because of possible dependencies among the successive regressors $\{\mathbf{u}_i\}$. However, if we again resort to independence Assumption 1.15, then it is easy to verify that

$$\mathrm{E}\,\mathbf{e}_a^{\Sigma}(i)\mathbf{e}_a(i) = \mathrm{E}\,\|\tilde{\mathbf{w}}_{i-1}\|_{\Sigma R_u}^2,$$

so that Equation 1.64 becomes

$$\mathrm{E}\,\|\tilde{\mathbf{w}}_i\|_{\Sigma}^2 = \mathrm{E}\,\|\tilde{\mathbf{w}}_{i-1}\|_{\Sigma}^2 - 2\mu h_G \mathrm{E}\,\|\tilde{\mathbf{w}}_{i-1}\|_{\Sigma R_u}^2 + \mu^2 h_U \mathrm{Tr}(R_u\Sigma). \qquad (1.65)$$

We now illustrate the application of this result by considering two cases separately. We start with the simpler case of white input data followed by correlated data.

1.12.3   White Regression Data

Assume first that $R_u = \sigma_u^2 I$ and select $\Sigma = I$. Then Equation 1.65 becomes

$$\mathrm{E}\,\|\tilde{\mathbf{w}}_i\|^2 = \mathrm{E}\,\|\tilde{\mathbf{w}}_{i-1}\|^2 - 2\mu h_G \sigma_u^2 \mathrm{E}\,\|\tilde{\mathbf{w}}_{i-1}\|^2 + \mu^2 M \sigma_u^2 h_U. \qquad (1.66)$$

Note that all terms on the right-hand side are dependent on $\mathrm{E}\,\|\tilde{\mathbf{w}}_{i-1}\|^2$ only; this is because h_G and h_U are functions of $\mathrm{E}\,\mathbf{e}_a^2(i)$ and, for white input data, $\mathrm{E}\,\mathbf{e}_a^2(i) = \sigma_u^2 \mathrm{E}\,\|\tilde{\mathbf{w}}_{i-1}\|^2$. We therefore find that Recursion 1.66 characterizes the evolution of $\mathrm{E}\,\|\tilde{\mathbf{w}}_i\|^2$. Two special cases help demonstrate this fact.

1.12.3.1   Transient Behavior of LMS

When $f[\mathbf{e}] = \mathbf{e}$ we obtain the LMS algorithm,

$$\mathbf{w}_i = \mathbf{w}_{i-1} + \mu \mathbf{u}_i^T \mathbf{e}(i) \qquad (1.67)$$

Using the following expressions from Table 1.1,

$$h_U = \sigma_u^2 \mathrm{E}\,\|\tilde{\mathbf{w}}_{i-1}\|^2 + \sigma_v^2, \quad h_G = 1.$$

we obtain

$$\mathrm{E}\,\|\tilde{\mathbf{w}}_i\|^2 = \left(1 - 2\mu\sigma_u^2 + \mu^2\sigma_u^4 M\right)\mathrm{E}\,\|\tilde{\mathbf{w}}_{i-1}\|^2 + \mu^2 M \sigma_u^2 \sigma_v^2 \qquad (1.68)$$

which is a linear recursion in $\mathrm{E}\,\|\tilde{\mathbf{w}}_i\|^2$; it characterizes the transient behavior of LMS for white input data.

1.12.3.2   Transient Behavior of sign-LMS

When $f[\mathbf{e}] = \text{sign}(\mathbf{e})$ we obtain the sign-LMS algorithm,

$$\mathbf{w}_i = \mathbf{w}_{i-1} + \mu \mathbf{u}_i^T \text{sign}[\mathbf{e}(i)]. \tag{1.69}$$

Using the following expressions from Table 1.1,

$$h_U = 1, \quad h_G = \sqrt{\frac{2}{\pi}} \frac{1}{\sqrt{\sigma_u^2 \mathrm{E}\|\tilde{\mathbf{w}}_{i-1}\|^2 + \sigma_v^2}},$$

we obtain

$$\mathrm{E}\|\tilde{\mathbf{w}}_i\|^2 = \left(1 - \sqrt{\frac{8}{\pi}} \frac{\mu \sigma_u^2}{\sqrt{\sigma_u^2 \mathrm{E}\|\tilde{\mathbf{w}}_{i-1}\|^2 + \sigma_v^2}}\right) \mathrm{E}\|\tilde{\mathbf{w}}_{i-1}\|^2 + \mu^2 M \sigma_u^2, \tag{1.70}$$

which is now a nonlinear recursion in $\mathrm{E}\|\tilde{\mathbf{w}}_i\|^2$; it characterizes the transient behavior of sign-LMS for white input data.

1.12.3.3   Transient Behavior of LMF

When $f[\mathbf{e}] = \mathbf{e}^3$ we obtain the LMF algorithm,

$$\mathbf{w}_i = \mathbf{w}_{i-1} + \mu \mathbf{u}_i^T \mathbf{e}^3(i). \tag{1.71}$$

Using the following expressions from Table 1.1,

$$h_G = 3\left(\mathrm{E}|\mathbf{e}_a(i)|^2 + \sigma_v^2\right), \quad h_U = 15\left(\mathrm{E}|\mathbf{e}_a(i)|^2 + \sigma_v^2\right)^3,$$

we obtain

$$\mathrm{E}\|\tilde{\mathbf{w}}_i\|^2 = f\mathrm{E}\|\tilde{\mathbf{w}}_{i-1}\|^2 + 15\mu^2 M \sigma_u^2 \sigma_v^6, \tag{1.72}$$

where

$$
\begin{aligned}
f = {}& \left[1 + \mu \sigma_u^2 \sigma_v^2 \left(45\mu M \sigma_u^2 \sigma_v^2 - 2\right) + \mu \sigma_u^4 \left(45\mu M \sigma_u^2 \sigma_v^2 - 2\right)\right] \mathrm{E}\|\tilde{\mathbf{w}}_{i-1}\|^2 \\
& + 15\mu^2 \sigma_u^8 M (\mathrm{E}\|\tilde{\mathbf{w}}_{i-1}\|^2)^2,
\end{aligned}
$$

which is a nonlinear recursion in $\mathrm{E}\|\tilde{\mathbf{w}}_i\|^2$; it characterizes the transient behavior of LMF for white input data.

1.12.4   Correlated Regression Data

When the input data is correlated, different weighting matrices will appear on both sides of the variance relation (Equation 1.65). Indeed, writing Equation 1.65 for $\Sigma = I$ yields

$$\mathrm{E}\|\tilde{\mathbf{w}}_i\|^2 = \mathrm{E}\|\tilde{\mathbf{w}}_{i-1}\|^2 - 2\mu h_G \mathrm{E}\|\tilde{\mathbf{w}}_{i-1}\|_{R_u}^2 + \mu^2 \mathrm{Tr}(R_u) \cdot h_U,$$

with the weighted term $\mathrm{E}\,\|\tilde{\mathbf{w}}_{i-1}\|_{R_u}^2$. This term can be deduced from Equation 1.65 by writing it for $\Sigma = R_u$, which leads to

$$\mathrm{E}\,\|\tilde{\mathbf{w}}_i\|_{R_u}^2 = \mathrm{E}\,\|\tilde{\mathbf{w}}_{i-1}\|_{R_u}^2 - 2\mu h_G \mathrm{E}\,\|\tilde{\mathbf{w}}_{i-1}\|_{R_u^2}^2 + \mu^2 h_U \mathrm{Tr}\left(R_u^2\right),$$

with the weighted term $\mathrm{E}\,\|\tilde{\mathbf{w}}_i\|_{R_u^2}^2$. This term can in turn be deduced from Equation 1.65 by writing it for $\Sigma = R_u^2$. Continuing in this fashion, for successive powers of R_u, we arrive at

$$\mathrm{E}\,\|\tilde{\mathbf{w}}_i\|_{R_u^{M-1}}^2 = \mathrm{E}\,\|\tilde{\mathbf{w}}_{i-1}\|_{R_u^{M-1}}^2 - 2\mu h_G \mathrm{E}\,\|\tilde{\mathbf{w}}_{i-1}\|_{R_u^M}^2 + \mu^2 h_U \mathrm{Tr}\left(R_u^M\right).$$

As before, this procedure terminates. To see this, let $p(x) = \det(xI - R_u)$ denote the characteristic polynomial of R_u, say,

$$p(x) = x^M + p_{M-1}x^{M-1} + p_{M-2}x^{M-2} + \cdots + p_1 x + p_0.$$

Then, since $p(R_u) = 0$ in view of the Cayley–Hamilton theorem, we have

$$\mathrm{E}\,\|\tilde{\mathbf{w}}_i\|_{R^M}^2 = -p_0\mathrm{E}\,\|\tilde{\mathbf{w}}_i\|^2 - p_1\mathrm{E}\,\|\tilde{\mathbf{w}}_i\|_{R_u}^2 - \cdots - p_{M-1}\mathrm{E}\,\|\tilde{\mathbf{w}}_i\|_{R_u^{M-1}}^2.$$

This result indicates that the weighted term $\mathrm{E}\,\|\tilde{\mathbf{w}}_i\|_{R^M}^2$ is fully determined by the prior weighted terms.

Putting these results together, we find that the transient behavior of Filter 1.52 is now described by a nonlinear M-dimensional state-space model of the form

$$\mathcal{W}_i = \mathcal{F}\mathcal{W}_{i-1} + \mu^2 h_U \mathcal{Y}, \tag{1.73}$$

where the $M \times 1$ vectors $\{\mathcal{W}_i, \mathcal{Y}\}$ are defined by

$$\mathcal{W}_i \triangleq \begin{bmatrix} \mathrm{E}\,\|\tilde{\mathbf{w}}_i\|^2 \\ \mathrm{E}\,\|\tilde{\mathbf{w}}_i\|_{R_u}^2 \\ \vdots \\ \mathrm{E}\,\|\tilde{\mathbf{w}}_i\|_{R_u^{M-2}}^2 \\ \mathrm{E}\,\|\tilde{\mathbf{w}}_i\|_{R_u^{M-1}}^2 \end{bmatrix}, \quad \mathcal{Y} \triangleq \begin{bmatrix} \mathrm{Tr}(R_u) \\ \mathrm{Tr}(R_u^2) \\ \vdots \\ \mathrm{Tr}(R_u^{M-1}) \\ \mathrm{Tr}(R_u^M) \end{bmatrix}, \tag{1.74}$$

and the $M \times M$ coefficient matrix $\mathcal{F}$ is given by

$$\mathcal{F} \triangleq \begin{bmatrix} 1 & -2\mu h_G & & & & \\ 0 & 1 & -2\mu h_G & & & \\ 0 & 0 & 1 & -2\mu h_G & & \\ \vdots & & & & & \\ 0 & 0 & & & 1 & -2\mu h_G \\ 2\mu p_0 h_G & 2\mu p_1 h_G & \cdots & & 2\mu p_{M-2}h_G & 1+2\mu p_{M-1}h_G \end{bmatrix}.$$

The evolution of the top entry of $\mathcal{W}_i$ describes the mean-square deviation of the filter, $E \|\tilde{\mathbf{w}}_i\|^2$, while the evolution of the second entry of $\mathcal{W}_i$ relates the learning behavior of the filter because

$$E\,\mathbf{e}^2(i) = E\,\mathbf{e}_a^2(i) + \sigma_v^2 \;=\; E\,\|\tilde{\mathbf{w}}_{i-1}\|_{R_u}^2 + \sigma_v^2.$$

1.12.5   Long Filter Approximation

The earlier results on filters with error nonlinearities can be used to provide an alternative simplified analysis of adaptive filters with data nonlinearities as in Equation 1.3. We did this in Sections 1.8 and 1.10 by resorting to simplifications that resulted from the small-step-size and fourth-order moment approximations.

Indeed, starting from Equation 1.10, substituting $\mathbf{e}_p^{\Sigma}(i)$ in terms of $\{\mathbf{e}_a^{\Sigma}(i),\,\mathbf{e}(i)\}$ from (Equation 1.8), and taking expectations, we arrive at the variance relation

$$E\,\|\tilde{\mathbf{w}}_i\|_{\Sigma}^2 = E\,\|\tilde{\mathbf{w}}_{i-1}\|_{\Sigma}^2 \;-\; 2\mu E\left(\frac{\mathbf{e}_a^{\Sigma}(i)\mathbf{e}(i)}{g[\mathbf{u}_i]}\right) \;+\; \mu^2 E\left(\frac{\|\mathbf{u}_i\|_{\Sigma}^2\,\mathbf{e}^2(i)}{g^2[\mathbf{u}_i]}\right). \qquad (1.75)$$

This relation is equivalent to Equation 1.13, except that in Equation 1.13 we proceeded further and expressed the terms $\mathbf{e}_a^{\Sigma}(i)\mathbf{e}(i)$ and $\mathbf{e}^2(i)$ as weighted norms of $\tilde{\mathbf{w}}_{i-1}$. Relation 1.75 has the same form as variance relation 1.58 used for filters with error nonlinearities. Observe in particular that the function $\mathbf{e}/g[\mathbf{u}]$ in data-normalized filters plays the role of $f[\mathbf{e}]$ in nonlinear error filters.

Now by following the arguments of Section 1.12.1, and under the following assumptions:

$\mathbf{e}_a(i)$ and $\mathbf{e}_a^{\Sigma}(i)$ are jointly Gaussian random variables,

$\|\mathbf{u}_i\|_{\Sigma}^2$ and $g[\mathbf{u}_i]$ are independent of $\mathbf{e}(i)$, and $\qquad\qquad(1.76)$

the regressors $\mathbf{u}_i$ are independent and identically distributed,

we can evaluate the expectations

$$E\left(\frac{\mathbf{e}_a^{\Sigma}(i)\mathbf{e}(i)}{g[\mathbf{u}_i]}\right) \qquad \text{and} \qquad E\left(\frac{\|\mathbf{u}_i\|_{\Sigma}^2\,\mathbf{e}^2(i)}{g^2[\mathbf{u}_i]}\right),$$

and conclude that variance Relation 1.75 reduces to

$$E\,\|\tilde{\mathbf{w}}_i\|_{\Sigma}^2 = E\,\|\tilde{\mathbf{w}}_{i-1}\|_{\Sigma}^2 \;-\; 2\mu h_G E\left(\mathbf{e}_a^{\Sigma}(i)\mathbf{e}_a(i)\right) \;+\; \mu^2 E\left(\frac{\|\mathbf{u}_i\|_{\Sigma}^2}{g^2[\mathbf{u}_i]}\right)\left(E\,\mathbf{e}_a^2(i) + \sigma_v^2\right),$$

where now

$$h_G \triangleq \frac{\mathrm{E}\left(e_a^2(i)/g[\mathbf{u}_i]\right)}{\mathrm{E}\,e_a^2(i)} = \mathrm{E}\left(\frac{1}{g[\mathbf{u}_i]}\right) \tag{1.77}$$

in view of the independence assumptions in Listing 1.76.

If we again use $\mathrm{E}\,e_a^2(i) = \mathrm{E}\,\|\tilde{\mathbf{w}}_{i-1}\|_{R_u}$, then we arrive at

$$\mathrm{E}\,\|\tilde{\mathbf{w}}_i\|_\Sigma^2 = \mathrm{E}\,\|\tilde{\mathbf{w}}_{i-1}\|_\Sigma^2 - \mu h_G \mathrm{E}\,\|\tilde{\mathbf{w}}_{i-1}\|_{\Sigma R_u + R_u \Sigma}^2 + \mu^2 \mathrm{E}\left(\frac{\|\mathbf{u}_i\|_\Sigma^2}{g^2[\mathbf{u}_i]}\right)$$
$$\left(\mathrm{E}\,\|\tilde{\mathbf{w}}_{i-1}\|_{R_u}^2 + \sigma_v^2\right), \tag{1.78}$$

which is the extension of Equation 1.65 to data-normalized filters. We now illustrate the application of this result to the transient analysis of some data-normalized adaptive filters.

1.12.5.1 *White Regression Data*

Assume first that $R_u = \sigma_u^2 I$ and select $\Sigma = I$. Then Equation 1.78 becomes

$$\mathrm{E}\,\|\tilde{\mathbf{w}}_i\|^2 = \left(1 - 2\mu\sigma_u^2 h_G + \mu^2\sigma_u^2 \mathrm{E}\left(\frac{\|\mathbf{u}_i\|^2}{g^2[\mathbf{u}_i]}\right)\right) \mathrm{E}\,\|\tilde{\mathbf{w}}_{i-1}\|^2 + \mu^2\sigma_v^2 \mathrm{E}\left(\frac{\|\mathbf{u}_i\|^2}{g^2[\mathbf{u}_i]}\right). \tag{1.79}$$

For the special case of LMS, when $g[\mathbf{u}] = 1$, h_G in Equation 1.77 becomes $h_G = 1$ and Equation 1.79 reduces to

$$\mathrm{E}\,\|\tilde{\mathbf{w}}_i\|^2 = \left(1 - 2\mu\sigma_u^2 + \mu^2\sigma_u^4 M\right) \mathrm{E}\,\|\tilde{\mathbf{w}}_{i-1}\|^2 + \mu^2 M \sigma_u^2 \sigma_v^2. \tag{1.80}$$

This is the same recursion we obtained before for LMS when trained with white input data.

For the special case of NLMS, $g[\mathbf{u}] = \|\mathbf{u}\|^2$, and Relation 1.79 reduces to

$$\|\tilde{\mathbf{w}}_i\|^2 = \left(1 - 2\mu\sigma_u^2 \mathrm{E}\left(\frac{1}{\|\mathbf{u}_i\|^2}\right) + \mu^2\sigma_u^2 \mathrm{E}\left(\frac{1}{\|\mathbf{u}_i\|^2}\right)\right) \mathrm{E}\,\|\tilde{\mathbf{w}}_{i-1}\|^2$$
$$+ \mu^2\sigma_v^2 \mathrm{E}\left(\frac{1}{\|\mathbf{u}_i\|^2}\right). \tag{1.81}$$

1.12.5.2 *Correlated Regression Data*

When the input data are correlated, different weighting matrices will appear on both sides of variance Relation 1.78. Indeed, writing 1.78 for $\Sigma = I$ yields

$$\mathrm{E}\,\|\tilde{\mathbf{w}}_i\|^2 = \mathrm{E}\,\|\tilde{\mathbf{w}}_{i-1}\|^2 - 2\mu h_G \mathrm{E}\,\|\tilde{\mathbf{w}}_{i-1}\|_{R_u}^2 + \mu^2 \mathrm{E}\left(\frac{\|\mathbf{u}_i\|^2}{g^2[\mathbf{u}_i]}\right)\left(\mathrm{E}\,\|\tilde{\mathbf{w}}_{i-1}\|_{R_u}^2 + \sigma_v^2\right)$$

with the weighted term $\mathrm{E}\,\|\tilde{\mathbf{w}}_{i-1}\|_{R_u}$. This term can be deduced from Relation 1.78 by writing it for $\Sigma = R_u$, which leads to

$$\mathrm{E}\,\|\tilde{\mathbf{w}}_i\|_{R_u}^2 = \mathrm{E}\,\|\tilde{\mathbf{w}}_{i-1}\|_{R_u}^2 - 2\mu h_G \mathrm{E}\,\|\tilde{\mathbf{w}}_{i-1}\|_{R_u^2}^2 + \mu^2 \mathrm{E}\left(\frac{\|\mathbf{u}_i\|_{R_u}^2}{g^2[\mathbf{u}_i]}\right)\left(\mathrm{E}\,\|\tilde{\mathbf{w}}_{i-1}\|_{R_u}^2 + \sigma_v^2\right)$$

with the weighted term $\mathrm{E}\,\|\tilde{\mathbf{w}}_i\|_{R_u^2}^2$ and so forth. The procedure terminates and leads to the following state-space model:

$$\mathcal{W}_i = \left(\mathcal{F} + \mu^2 \mathcal{Y} e_2^T\right)\mathcal{W}_{i-1} + \mu^2 \sigma_v^2 \mathcal{Y}, \tag{1.82}$$

where the $M \times 1$ vectors $\{\mathcal{W}_i, \mathcal{Y}\}$ are defined by

$$\mathcal{W}_i \triangleq \begin{bmatrix} \mathrm{E}\,\|\tilde{\mathbf{w}}_i\|^2 \\ \mathrm{E}\,\|\tilde{\mathbf{w}}_i\|_{R_u}^2 \\ \vdots \\ \mathrm{E}\,\|\tilde{\mathbf{w}}_i\|_{R_u^{M-2}}^2 \\ \mathrm{E}\,\|\tilde{\mathbf{w}}_i\|_{R_u^{M-1}}^2 \end{bmatrix}, \qquad \mathcal{Y} \triangleq \begin{bmatrix} \mathrm{E}\left(\|\mathbf{u}_i\|^2/g^2[\mathbf{u}_i]\right) \\ \mathrm{E}\left(\|\mathbf{u}_i\|_{R_u}^2/g^2[\mathbf{u}_i]\right) \\ \vdots \\ \mathrm{E}\left(\|\mathbf{u}_i\|_{R_u^{M-2}}^2/g^2[\mathbf{u}_i]\right) \\ \mathrm{E}\left(\|\mathbf{u}_i\|_{R_u^{M-1}}^2/g^2[\mathbf{u}_i]\right) \end{bmatrix}, \tag{1.83}$$

the $M \times M$ matrix $\mathcal{F}$ is given by

$$\mathcal{F} \triangleq \begin{bmatrix} 1 & -2\mu h_G & & & & \\ 0 & 1 & -2\mu h_G & & & \\ 0 & 0 & 1 & -2\mu h_G & & \\ \vdots & & & & & \\ 0 & 0 & & & 1 & -2\mu h_G \\ 2\mu p_0 h_G & 2\mu p_1 h_G & \cdots & & 2\mu p_{M-2} h_G & 1 + 2\mu p_{M-1} h_G \end{bmatrix}$$

and

$$e_2 = \mathrm{col}\{0, 1, 0, \ldots, 0\}.$$

Also,

$$h_G = \mathrm{E}\left(\frac{1}{g[\mathbf{u}_i]}\right).$$

The evolution of the top entry of $\mathcal{W}_i$ describes the mean-square deviation of the filter, $\mathrm{E}\,\|\tilde{\mathbf{w}}_i\|^2$, while the evolution of the second entry of $\mathcal{W}_i$ relates to the learning behavior of the filter. The model (Equation 1.82) is an alternative to Equation 1.22 for adaptive filters with data nonlinearities; it is based on assumptions in Listing 1.76.

1.12.5.3 Steady-State Performance

The variance Relation 1.78 can also be used to approximate the steady-state performance of data-normalized adaptive filters. Writing it for $\Sigma = I$,

$$\mathrm{E}\,\|\tilde{\mathbf{w}}_i\|^2 = \mathrm{E}\,\|\tilde{\mathbf{w}}_{i-1}\|^2 - 2\mu h_G \mathrm{E}\,\|\tilde{\mathbf{w}}_{i-1}\|_{R_u}^2 + \mu^2 \mathrm{E}\left(\frac{\|\mathbf{u}_i\|^2}{g^2[\mathbf{u}_i]}\right)\left(\mathrm{E}\,\|\tilde{\mathbf{w}}_{i-1}\|_{R_u}^2 + \sigma_v^2\right) \tag{1.84}$$

and setting, in steady state,

$$\lim_{i \to \infty} \mathrm{E} \, \|\tilde{w}_i\|^2 \; = \; \lim_{i \to \infty} \mathrm{E} \, \|\tilde{w}_{i-1}\|^2,$$

we obtain

$$0 = -2\mu \mathrm{E} \left(\frac{1}{g[\mathbf{u}_i]} \right) \mathrm{EMSE} + \mu^2 \mathrm{E} \left(\frac{\|\mathbf{u}_i\|^2}{g^2[\mathbf{u}_i]} \right) (\mathrm{EMSE} + \sigma_v^2),$$

so that the excess mean-square error, $\mathrm{E} \, \mathbf{e}_a^2(\infty)$, is given by

$$\mathrm{EMSE} = \frac{\mu \sigma_v^2 \mathrm{Tr}(Q)}{2\mathrm{E} \left(1/g[\mathbf{u}_i] \right) - \mu \mathrm{Tr}(Q)}, \tag{1.85}$$

where $Q = \mathrm{E} \left(\mathbf{u}_i^T \mathbf{u}_i / g^2[\mathbf{u}_i] \right)$. For LMS we have $g[\mathbf{u}] = 1$ and $Q = R_u$, and the above expression reduces to

$$\mathrm{EMSE} = \frac{\mu \sigma_v^2 \mathrm{Tr}(R_u)}{2 - \mu \mathrm{Tr}(R_u)} \qquad \text{(LMS)}.$$

For NLMS we have $g[\mathbf{u}] = \|\mathbf{u}\|^2$ and $Q = \mathrm{E} \left(\mathbf{u}_i^T \mathbf{u}_i / \|\mathbf{u}_i\|^4 \right)$, so that

$$\mathrm{EMSE} \approx \frac{\mu \sigma_v^2}{2 - \mu} \qquad \text{(NLMS)}.$$

1.12.5.4 Stability

Recursion 1.84 can be rearranged as

$$\mathrm{E} \, \|\tilde{w}_i\|^2 = \mathrm{E} \, \|\tilde{w}_{i-1}\|^2 + \mu \left(\mu \mathrm{Tr}(Q) - 2h_G \right) \mathrm{E} \, \|\tilde{w}_{i-1}\|_{R_u}^2 + \mu^2 \sigma_v^2 \mathrm{Tr}(Q).$$

It is now easy to see that $\mathrm{E} \, \|\tilde{w}_i\|^2$ converges for step sizes satisfying

$$\mu \mathrm{Tr}(Q) - 2h_G < 0,$$

or, equivalently,

$$0 < \mu < \frac{2h_G}{\mathrm{Tr}(Q)} \; = \; 2\mathrm{E} \left(\frac{1}{g[\mathbf{u}_i]} \right) \frac{1}{\mathrm{Tr}(Q)}.$$

For LMS, this simplified analysis results in the condition $\mu < 2/\mathrm{Tr}(R_u)$. For NLMS, $\mathrm{Tr}(Q) = E(1/\|\mathbf{u}_i\|^2)$ and the condition on μ becomes $\mu < 2$.

1.13 An Interpretation of the Energy Relation

We end our discussions in this chapter by making a connection between the energy relation (Equation 1.10) and Snell's law of optics. We reconsider Equation 1.10 and assume first that $\Sigma = I$ so that

$$\|\mathbf{u}_i\|^2 \cdot \|\tilde{w}_i\|^2 + \mathbf{e}_a^2(i) = \|\mathbf{u}_i\|^2 \cdot \|\tilde{w}_{i-1}\|^2 + \mathbf{e}_p^2(i). \tag{1.86}$$

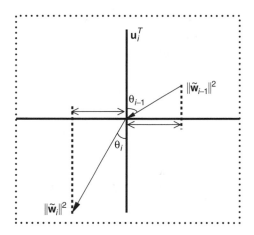

FIGURE 1.1
An interpretation of the energy-conservation relation (Equation 1.10) by means of an analogy with Snell's law of optics.

Let θ_i denote the acute angle between the column vectors $\{\tilde{\mathbf{w}}_i, \mathbf{u}_i^T\}$. Likewise, let θ_{i-1} denote the acute angle between $\{\tilde{\mathbf{w}}_{i-1}, \mathbf{u}_i^T\}$. Then

$$\mathbf{e}_a^2(i) = \|\mathbf{u}_i\|^2 \cdot \|\tilde{\mathbf{w}}_{i-1}\|^2 \cdot \cos^2(\theta_{i-1}), \quad \text{and} \quad \mathbf{e}_p^2(i) = \|\mathbf{u}_i\|^2 \cdot \|\tilde{\mathbf{w}}_i\|^2 \cdot \cos^2(\theta_i).$$

Substituting into Equation 1.86 and collecting terms we find that it reduces to

$$\|\tilde{\mathbf{w}}_{i-1}\|^2 \sin^2(\theta_{i-1}) = \|\tilde{\mathbf{w}}_i\|^2 \sin^2(\theta_i). \tag{1.87}$$

Equality 1.87 resembles a famous result in optics, known as Snell's law, which relates the refraction indices of two media with the sines of the incident and refracted rays of light, i.e.,

$$\eta_1 \sin\theta_1 = \eta_2 \sin\theta_2,$$

where θ_1 and θ_2 are the angles of incidence and refraction, respectively; both angles are measured relative to the direction that is orthogonal to the surface separating both media. This analogy suggests that we can relate the operation of an adaptive filter, at each iteration, to that of a fictitious ray traveling from one medium to another. The magnitudes $\|\tilde{\mathbf{w}}_{i-1}\|$ and $\|\tilde{\mathbf{w}}_i\|$ play the role of refraction indices of the media, while $\{\theta_{i-1}, \theta_i\}$ play the role of the incidence and refraction angles of the ray. Alternatively, we can interpret the result (Equation 1.87) as shown in Figure 1.1. An incident vector of norm $\|\tilde{\mathbf{w}}_{i-1}\|$ impinges on the separation layer at an angle θ_{i-1} with respect to $\mathbf{u}_i^T$, while a refracted vector of norm $\|\tilde{\mathbf{w}}_i\|$ leaves the layer at an angle θ_i, also with respect to $\mathbf{u}_i^T$. Relation 1.87 then amounts to saying that the projections of these vectors along the horizontal direction should have equal norms.

More generally, when a positive-definite weighting matrix Σ is present in Equation 1.10, we let $\{\theta_i, \theta_{i-1}\}$ denote acute angles whose squared cosines are given by

$$
\cos_\Sigma^2(\theta_{i-1}) \triangleq \frac{\left(\mathbf{e}_a^\Sigma(i)\right)^2}{\|\tilde{\mathbf{w}}_{i-1}\|_\Sigma^2 \cdot \|\mathbf{u}_i\|_\Sigma^2}, \quad \cos_\Sigma^2(\theta_i) \triangleq \frac{\left(\mathbf{e}_p^\Sigma(i)\right)^2}{\|\tilde{\mathbf{w}}_i\|_\Sigma^2 \cdot \|\mathbf{u}_i\|_\Sigma^2}. \tag{1.88}
$$

The subscript Σ in $\cos_\Sigma(\cdot)$ indicates that a weighting matrix Σ is used in computing it. With this notation, it is straightforward to verify that energy relation 1.57 becomes

$$
\|\tilde{\mathbf{w}}_{i-1}\|_\Sigma^2 \sin_\Sigma^2(\theta_{i-1}) = \|\tilde{\mathbf{w}}_i\|_\Sigma^2 \sin_\Sigma^2(\theta_i), \tag{1.89}
$$

which is a natural extension of Equation 1.87.

1.14   Concluding Remarks

This chapter describes an energy-conservation approach to studying the performance of adaptive filters. By studying the energy balance at each iteration, the dynamic behavior of an adaptive filter can be characterized in terms of a variance relation (e.g., Equations 1.16, 1.64, and 1.65) and, subsequently, in terms of a state-space model (e.g., Equations 1.22 and 1.73). The approach does not restrict the input data to Gaussian or white distributions. In addition to providing information about the stability and convergence behavior of the filter, the energy-conservation arguments also help characterize the steady-state performance of the filter. Although the analysis in this chapter has relied on independence Assumption 1.15, steady-state results can be obtained without relying on this assumption (see, e.g., References 11 and 12).

Acknowledgments

This work was supported in part by the National Science Foundation under grants ECS-9820765 and CCR-0208573. The work of T.Y. Al-Naffouri was also supported by a fellowship from King Fahd University of Petroleum & Minerals, Dhahran, Saudi Arabia.

References

1. S. Haykin, *Adaptive Filter Theory*, Englewood Cliffs, NJ: Prentice-Hall, 1996.
2. B. Widrow and S.D. Stearns, *Adaptive Signal Processing*, Englewood Cliffs, NJ: Prentice-Hall, 1985.
3. O. Macchi, *Adaptive Processing: The LMS Approach with Applications in Transmission*, New York: Wiley, 1995.
4. A.H. Sayed, *Fundamentals of Adaptive Filtering*, Wiley, New York, 2003.
5. A.H. Sayed and M. Rupp, A time-domain feedback analysis of adaptive algorithms via the small gain theorem, *Proc. SPIE*, 2563, 458–469, 1995.
6. A.H. Sayed and M. Rupp, Robustness issues in adaptive filtering, in *DSP Handbook*, Boca Raton, FL: CRC Press, 1998, chap. 20.
7. M. Rupp and A.H. Sayed, A time-domain feedback analysis of filtered-error adaptive gradient algorithms, *IEEE Trans. Signal Process.*, 44(6), 1428–1439, 1996.
8. A.H. Sayed and M. Rupp, An l_2-stable feedback structure for nonlinear adaptive filtering and identification, *Automatica*, 33(1), 13–30, 1997.
9. M. Rupp and A.H. Sayed, On the convergence of blind adaptive equalizers for constant-modulus signals, *IEEE Trans. Commun.*, 48(5), 795–803, 2000.
10. J. Mai and A.H. Sayed, A feedback approach to the steady-state performance of fractionally-spaced blind adaptive equalizers, *IEEE Trans. Signal Process.*, 48(1), 80–91, 2000.
11. N.R. Yousef and A.H. Sayed, A unified approach to the steady-state and tracking analyses of adaptive filters, *IEEE Trans. Signal Process.*, 49(2), 314–324, 2001.
12. N.R. Yousef and A.H. Sayed, Ability of adaptive filters to track carrier offsets and random channel nonstationarities, *IEEE Trans. Signal Process.*, 50(7), 1533–1544, 2002.
13. T.Y. Al-Naffouri and A.H. Sayed, Transient analysis of data-normalized adaptive filters, *IEEE Trans. Signal Process.*, 51(3), 639–652, 2003.
14. T.Y. Al-Naffouri and A.H. Sayed, Transient analysis of adaptive filters with error nonlinearities, *IEEE Trans. Signal Process.*, 51(3), 653–663, 2003.
15. R. Price, A useful theorem for nonlinear devices having Gaussian inputs, *IEEE Trans. Info. Theor.* 4, 69–72, 1958.
16. R.A. Horn and C.R. Johnson, *Matrix Analysis*, New York: Cambridge University Press, 1985.
17. V.H. Nascimento and A.H. Sayed, On the learning mechanism of adaptive filters, *IEEE Trans. Signal Process.*, 48(6), 1609–1625, 2000.
18. A.H. Sayed and T.Y. Al-Naffouri, Mean-square analysis of normalized leaky adaptive filters, *Proc. ICASSP*, 6, 3873–3876, 2001.

2

Fuzzy Methods in Nonlinear Signal Processing: Part I—Theory

Kenneth E. Barner

CONTENTS

2.1 Introduction

Nonlinear signal processing methods, unlike linear methods, lack a unified and universal set of tools for analysis and design. Hundreds of nonlinear signal processing algorithms have been proposed. Most of the proposed methods, although well tailored for a given application, are not generally applicable. Although nonlinear signal processing is a dynamic, rapidly growing field, a large class of nonlinear signal processing algorithms can be studied with fundamentals that are well formulated. In this chapter, we approach the filtering problem from a maximum likelihood (ML) approach. It is shown that the ML optimization leads directly to the class of linear filters for signals with Gaussian statistics, and to the class of nonlinear weighted median filters for signals with double exponential, or Laplacian, distributions.

The ML development that leads to the class of weighted median filters shows that this class of filters operates jointly on spatial* and rank (SR) order information. The joint utilization of SR information has proved advantageous for two primary reasons: (1) spatial ordering can be used to exploit

---

*The ordering is spatial in two-dimensional signals cases, such as images, and temporal for one-dimensional time sequences.

0-8493-1427-5/04/$0.00+$1.50
© 2004 by CRC Press LLC

correlations between neighboring samples and (2) rank order can be used to isolate outliers and ensure robust behavior. Although the exploitation of SR ordering information in nonlinear filtering algorithms has yielded good results, traditional ordering information is based on a crisp relationship. Such crisp relations yield no information on important quantities such as sample spread or diversity.

The simple relaxation of the ordering relation from a crisp (binary) operator to a more general affinity (real-valued) operator leads to the concept of fuzzy SR orderings. Thus, fuzzy SR orderings not only relate spatial and rank orderings, but also contain information on sample spread (affinity). Powerful fuzzy nonlinear filtering algorithms can be realized by embedding fuzzy SR ordering information into the filter structure. Such filters can be simply realized as (1) generalizations of existing nonlinear filters that employ fuzzy, rather than crisp, ordering relations; (2) generalizations of linear filters that embed fuzzy SR ordering information into the traditional weighted sum filter structure; or (3) new filter structures specifically designed to exploit fuzzy SR information.

Fuzzy SR methods in signal processing is a broad topic covered here in two chapters. This first chapter theoretically motivates the use of fuzzy SR ordering information, details the theory of fuzzy ordering and fuzzy order statistics, and develops several classes of fuzzy nonlinear filters based on these concepts. Specifically, the classes of affine filters and fuzzy weighted median filters are developed. In the next chapter these methods are applied to a wide range of signal processing and communications problems.

The topics covered in this chapter are organized as follows: Section 2.2 begins with a theoretical discussion of ML estimation and formally develops the concept of SR ordering. The crisp ordering relation is relaxed in Section 2.3, which develops the concepts of sample affinity and the resulting fuzzy SR ordering. Fuzzy filter generalizations are developed in Section 2.4, where we focus on the fuzzy weighted median and affine filter classes. In the case of affine filters, the discussion is limited to the two important median affine and center affine filter subclasses. Extensions to multivariate data are covered in Section 2.5 and conclusions are drawn in Section 2.6. The next chapter presents the results of applying the filters developed here to several important signal processing and communications problems, including robust frequency selective filtering, Synthetic Aperture Radar image filtering, time-frequency domain filtering, multiresolution signal representations, surface smoothing, image smoothing, zooming, and deblocking, and multiuser detection.

2.2   Maximum Likelihood Estimation and Spatial–Rank Ordering

2.2.1   ML Estimation

To motivate the development of theoretically sound signal processing methods, consider first the modeling of observation samples. In all but trivial cases, nondeterministic methods must be used. As most signals have random components, probability-based models form a powerful set of modeling methods. Accordingly, signal processing methods have deep roots in statistical estimation theory.

Consider a set of N observation samples. In most applications, the observation samples are captured by a moving window centered at some position $\mathbf{n}$, where we consider the general case of a vector index to account for multidimensional signals. Such samples will be denoted as $\mathbf{x}[\mathbf{n}] = [x_1[\mathbf{n}], x_2[\mathbf{n}], \ldots, x_N[\mathbf{n}]]^T$. For notational convenience, we drop the index $\mathbf{n}$, unless necessary for clarity.

Assume now that we model these samples as independent and identically distributed (i.i.d.). Each observation sample is then characterized by the common probability density function (pdf) $f_\beta(x)$, where β is the mean, or location, of the distribution. Often β is information-carrying and unknown, and thus must be estimated. The ML estimate of the location is achieved by maximizing, with respect to β, the probability of observing $x_1, x_2, \ldots, x_N$. For i.i.d. samples, this results in

$$\hat{\beta} = \arg\max_\beta \prod_{i=1}^{N} f_\beta(x_i). \tag{2.1}$$

Thus, the value of β that maximizes the product of the pdfs constitutes the ML estimate.

The degree to which the ML estimate accurately represents the location is, to a large extent, dependent on how accurately the model distribution represents the true distribution of the observation process. To allow for a wide range of sample distributions, the commonly assumed Gaussian distribution can be generalized by allowing the exponential rate of tail decay to be a free parameter. This results in the *generalized Gaussian* density function:

$$f_\beta(x) = ce^{-(|x-\beta|/\sigma)^p}, \tag{2.2}$$

where p governs the rate of tail decay, $c = p/(2\sigma\Gamma(1/p))$, and $\Gamma(\cdot)$ is the Gamma function. This includes the standard Gaussian distribution as a special case ($p = 2$). For $p < 2$, the tails decay slower than in the Gaussian case, resulting in a heavier-tailed distribution. Of particular interest is the case $p = 1$, which yields the double exponential, or Laplacian, distribution,

$$f_\beta(x) = \frac{1}{2\sigma}e^{-|x-\beta|/\sigma}. \tag{2.3}$$

The ML criteria can be applied to optimally estimate the location of a set of N samples distributed according to the generalized Gaussian distribution, yielding

$$\hat{\beta} = \arg\max_\beta \prod_{i=1}^{N} ce^{-(|x-\beta|/\sigma)^p} = \arg\min_\beta \sum_{i=1}^{N} |x_i - \beta|^p. \tag{2.4}$$

Determining the ML estimate is thus equivalent to minimizing

$$G_p(\beta) = \sum_{i=1}^{N} |x_i - \beta|^p \tag{2.5}$$

with respect to β. For the Gaussian case ($p = 2$), this reduces to the sample mean, or average:

$$\hat{\beta} = \arg\min_{\beta} G_2(\beta) = \frac{1}{N} \sum_{i=1}^{N} x_i. \tag{2.6}$$

A much more robust estimator is realized if the underlying sample distribution is taken to be the heavy-tailed Laplacian distribution ($p = 1$). In this case, the ML estimator of location is given by the value β that minimizes the sum of absolute deviations,

$$G_1(\beta) = \sum_{i=1}^{N} |x_i - \beta|, \tag{2.7}$$

which can easily be shown to be the sample median:

$$\hat{\beta} = \arg\min_{\beta} G_1(\beta) = \text{MED}[x_1, x_2, \ldots, x_N]. \tag{2.8}$$

The sample mean and median thus play analogous roles in location estimation: While the mean is associated with the Gaussian distribution, the median is related to the Laplacian distribution, which has heavier tails and provides a better model for many signals, such as images, as well as those contaminated with impulsive outliers.[1–4]

Although the median is a robust estimator that possesses many optimality properties, the performance of the median filter is limited by the fact that it is spatially blind. That is, all observation samples are treated equally regardless of their location within the observation window. This limitation is a direct result of the i.i.d. assumption made in the filter development. A much richer class of filters is realized if this assumption is relaxed to the case of independent, but not identically distributed, samples.

Consider the generalized Gaussian distribution case where the observation samples have a common location parameter β, but where each x_i has a (possibly) unique scale parameter σ_i. Incorporating the unique scale parameters into the ML criteria yields a location estimate given by the value of β minimizing

$$G_p(\beta) = \sum_{i=1}^{N} \frac{1}{\sigma_i^p} |x_i - \beta|^p. \tag{2.9}$$

In the special case of the standard Gaussian distribution ($p = 2$), the ML estimate reduces to the normalized weighted average

$$\hat{\beta} = \arg\min_{\beta} \sum_{i=1}^{N} \frac{1}{\sigma_i^2}(x_i - \beta)^2 = \frac{\sum_{i=1}^{N} w_i \cdot x_i}{\sum_{i=1}^{N} w_i}, \qquad (2.10)$$

where $w_i = 1/\sigma_i^2 > 0$. In the heavier-tailed Laplacian distribution special case ($p = 1$), the ML estimate reduces to the weighted median (WM), originally introduced more than 100 years ago by Edgemore,[5] and defined as

$$\hat{\beta} = \arg\min_{\beta} \sum_{i=1}^{N} \frac{1}{\sigma_i}|x_i - \beta| = \text{MED}[w_1 \diamond x_1, w_2 \diamond x_2, \dots, w_N \diamond x_N], \qquad (2.11)$$

where $w_i = 1/\sigma_i > 0$ and $\diamond$ is the replication operator defined as

$$w_i \diamond x_i = \overbrace{x_i, x_i, \dots, x_i}^{w_i \text{ times}}.$$

A small yet very important, special example of the weighted median filter (as well as the weighted sum filter) is the identity operator. Assuming the samples constitute an ordered (temporal or spatial) set from an observed process, let $\delta_c = (N+1)/2$ be the index of the center observation sample. Then it is easy to see that

$$x_{\delta_c} = \text{MED}[w_1 \diamond x_1, w_2 \diamond x_2, \dots, w_N \diamond x_N], \qquad (2.12)$$

for $w_{\delta_c} = 1$ and $w_i = 0$ for $i \neq \delta_c$. Thus the weighted median has two important special cases: (1) the standard median filter, which operates strictly on rank order information, and (2) the identity filter, which operates strictly on spatial order.

To illustrate the importance of these cases, and the corresponding orderings on which they are based, consider the filtering of a moving average (MA) process corrupted by Laplacian noise. Figure 2.1 shows the correlation between the desired MA process and the filter outputs, for the identity and median cases, as a function of the signal-to-noise ratio (SNR) in the corrupted observation. The figure shows that for high SNRs, the identity filter output (central observation sample) has the highest correlation with the desired output, while for low SNRs the median has the highest correlation. Thus, this simple example illustrates the importance of spatial order in high-SNR cases and rank order in low-SNR cases.

The general formulation of the weighted median filter attempts to exploit both spatial ordering, through repetition of samples, and rank ordering, through median selection. The filter is thus able to exploit spatial correlations among neighboring samples and limit the influence of outliers. A more formal consideration of spatial and rank ordering can be obtained by considering the full ordering relations between observed samples.

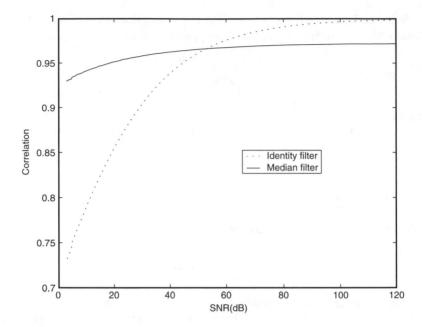

FIGURE 2.1
Correlation coefficient between an MA process and the identity and (window size 5) median filter outputs as a function of SNR in the Laplacian noise-corrupted observation.

2.2.2   Spatial–Rank Ordering

To formally relate the spatial ordering and rank ordering of samples in a signal processing application, consider again the typical case in which an observation window passes over an observation sequence in a predefined scanning pattern. At each location $\mathbf{n}$ the observation window covers N samples, which can be indexed according to their spatial location and written in vector form:

$$\mathbf{x}_\ell[\mathbf{n}] = [x_1[\mathbf{n}], x_2[\mathbf{n}], \ldots, x_N[\mathbf{n}]]^T. \tag{2.13}$$

The subscript ℓ has now been added to explicitly indicate that the samples are indexed according to their natural spatial order within the observation signal or image. A second natural ordering of the observed samples is rank order, which yields the order statistics of the observation samples,

$$x_{(1)}[\mathbf{n}] \leq x_{(2)}[\mathbf{n}] \leq \cdots \leq x_{(N)}[\mathbf{n}]. \tag{2.14}$$

Writing the order statistics in vector form yields the rank order observation vector:

$$\mathbf{x}_L[\mathbf{n}] = [x_{(1)}[\mathbf{n}], x_{(2)}[\mathbf{n}], \ldots, x_{(N)}[\mathbf{n}]]^T. \tag{2.15}$$

Again, the spatial location of the observation window is only to be shown explicitly when needed for clarity. Thus, we write the spatial order and rank order observation vectors as simply x_ℓ and x_L.

A crisp, or binary, relation between the samples of two sets A and B can be denoted by a crisp membership function $\mu_C(a, b): A \times B \mapsto \{0, 1\}$, $a \in A, b \in B$. Note that both the spatial and rank-ordered samples constitute the same set, $X = \{x_1, \ldots, x_N\} = \{x_{(1)}, \ldots, x_{(N)}\}$. Thus, to relate the spatial and rank orderings of the samples in X, we can define the SR matrix

$$
\mathbf{R} = \begin{bmatrix} R_{1,(1)} & \cdots & R_{1,(N)} \\ \vdots & \ddots & \vdots \\ R_{N,(1)} & \cdots & R_{N,(N)} \end{bmatrix}, \tag{2.16}
$$

where $R_{i,(j)} = \mu_C(x_i, x_{(j)})$ and

$$
\mu_C(x_i, x_{(j)}) = \begin{cases} 1 & \text{if } x_i \text{ has rank } j \ (x_i \leftrightarrow x_{(j)}) \\ 0 & \text{otherwise} \end{cases}. \tag{2.17}
$$

This crisp relation produces a binary relation matrix, i.e., $R_{i,(j)} \in \{0, 1\}$.

The matrix $\mathbf{R}$ contains the full joint SR information of the observation set X. Thus, $\mathbf{R}$ can be used as a transformation between the two orderings and to extract the marginal vectors x_ℓ and x_L. The transformations yielding the rank and spatial order indexes are given by

$$
\mathbf{r} = \mathbf{R}[1 : N] \quad \text{and} \quad \mathbf{s} = \mathbf{R}^T[1 : N], \tag{2.18}
$$

where $[1 : N] = [1, 2, \ldots, N]^T$, and $\mathbf{s} = [s_1, s_2, \ldots, s_N]^T$ and $\mathbf{r} = [r_1, r_2, \ldots, r_N]^T$ are the spatial and rank order index vectors, respectively, i.e., $x_{s_j} \leftrightarrow x_{(j)}$ and $x_i \leftrightarrow x_{(r_i)}$ for $i, j = 1, 2, \ldots, N$. Similarly, the spatial and rank ordered samples are related by

$$
x_\ell = \mathbf{R}x_L \quad \text{and} \quad x_L = \mathbf{R}^T x_\ell. \tag{2.19}
$$

As an illustrative example, suppose a three-sample observation window is used and a particular spatial order observation vector is given by $x_\ell = [10, 1, 2]^T$. This results in the SR matrix

$$
\mathbf{R} = \begin{bmatrix} 0 & 0 & 1 \\ 1 & 0 & 0 \\ 0 & 1 & 0 \end{bmatrix}, \tag{2.20}
$$

from which we can obtain the spatial and rank order indexes $\mathbf{s} = \mathbf{R}^T[1, 2, 3]^T = [2, 3, 1]^T$, $\mathbf{r} = \mathbf{R}[1, 2, 3]^T = [3, 1, 2]^T$, and the spatial and rank order samples $x_\ell = \mathbf{R}[1, 2, 10]^T = [10, 1, 2]^T$, $x_L = \mathbf{R}^T[10, 1, 2]^T = [1, 2, 10]^T$.

The SR matrix fully relates the spatial and rank orderings of the observed samples. Thus, the structure of the SR matrix captures spatial correlations of the data (spatial order information) and indicates which samples are likely to be outliers and which samples are likely to be reliable (rank order information). To illustrate the structure of the SR matrix for typical signals, and to show how this structure changes with the underlying signal characteristics, the statistics of the SR matrix are examined for two types of signals.

Consider first a one-dimensional statistical sequence consisting of an MA process. This process is generated by simple finite impulse response (FIR) filtering of white Gaussian noise. In the case of white noise samples (no filtering), all spatial–rank order combinations are equally likely and the expected SR matrix is uniform,

$$
E[\mathbf{R}] =
\begin{bmatrix}
\frac{1}{N} & \cdots & \frac{1}{N} \\
\vdots & \ddots & \vdots \\
\frac{1}{N} & \cdots & \frac{1}{N}
\end{bmatrix},
\tag{2.21}
$$

where $E[\cdot]$ denotes the expectation operator. Figure 2.2a shows $E[\mathbf{R}]$ for the white noise case when the window size is $N = 15$. As expected, all spatial order pairs are equally likely. Applying a lowpass FIR filter with a cutoff frequency of $\omega_c = 0.33$ to the noise to generate an MA process yields a time sequence with a greater concentration of low frequency power. This greater concentration of low frequency power gives the time series increasingly sinusoidal structure, which is reflected in the resulting SR matrix, as shown in Figure 2.2b. As the figure shows, extreme samples are most likely to be located in the first or last observation window location and monotonic observations are more likely than other observations. Decreasing the cutoff of the FIR filter to $\omega_c = 0.25$ and $\omega_c = 0.20$ increases the sinusoidal nature of the time domain sequence and increases the structure of the corresponding SR matrices, as illustrated in Figure 2.2c and d.

Similar results hold for images. Consider the original and a contaminated Gaussian noise corrupted version of the well-known image Lenna, which is utilized in Section 3.2 and shown in Figure 3.18. The expected SR matrices for an $N = 15$ one-dimensional observation window passing over these images are shown in Figure 2.3. As expected, the SR matrix for the original image has a structure that corresponds to the underlying image. The heavily corrupted noisy image, however, has lost much of the underlying structure as all SR order pairs are nearly equally likely. Thus, just as additive noise tends to decorrelate samples and flatten the power spectral density (PSD) of a signal, it also tends to flatten the expected SR matrix. Appropriate filtering can help restore the SR structure to that of the underlying signal.

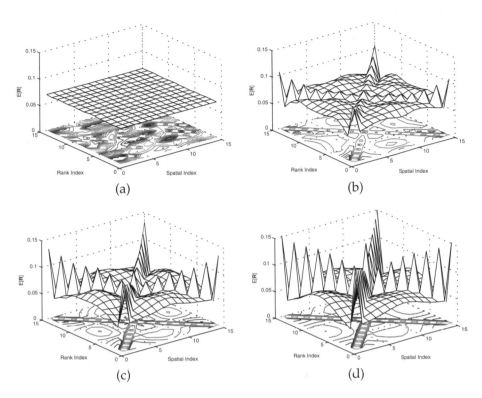

FIGURE 2.2
Expected SR matrices for lowpass FIR filtered white noise (MA) processes. The lowpass filter cutoff frequency in the examples are (a) $\omega_c = 1.0$ (no filtering), (b) $\omega_c = 0.33$, (c) $\omega_c = 0.25$, and (d) $\omega_c = 0.20$.

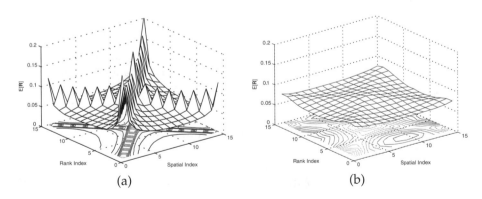

FIGURE 2.3
Expected SR matrix for the image (a) Lenna and (b) Lenna corrupted with contaminated Gaussian noise.

2.3   Sample Affinity and Fuzzy Spatial–Rank Ordering

2.3.1   Sample Affinity

The SR orderings of samples provide valuable information characterizing the observed samples. Strict use of traditional ordering, however, can often lead to suboptimal performance. For example, strict use of rank ordering can be misleading as it is often assumed that extreme rank samples should be discarded. Maximum and minimum samples, however, are not necessarily outliers and may, in fact, be information-bearing samples that should not be merely discarded. Thus, traditional ordering methods discard such important measures as sample value, spread, and diversity.

More general relations between samples can be realized by adopting a fuzzy set theory approach. Consider a real-valued fuzzy membership function

$$\mu_R(a, b) : A \times B \mapsto [0, 1] | a \in A, b \in B \qquad (2.22)$$

that describes the relations between the samples in the sets A and B. The relation function $\mu_R(a, b)$ can be any shape that reflects the most relevant information between samples for the problem at hand. Thus, a wide range of fuzzy membership functions can be defined. In fact, the crisp relation function $\mu_C(a, b)$ can be viewed as a special case of the fuzzy membership function. For the ordering and filter generalizations considered here, we base the membership on the concept of affinity and accordingly impose the following conditions on the membership function:

1. $\lim_{|a-b| \to 0} \mu_R(a, b) = 1$,
2. $\lim_{|a-b| \to \infty} \mu_R(a, b) = 0$,
3. $\mu_R(a_1, b_1) \geq \mu_R(a_2, b_2)$ for $|a_1 - b_1| \leq |a_2 - b_2|$.

The intuitive justification for these conditions is that two identical samples should have high affinity, or relation 1, while infinitely distant samples should have low affinity, or relation 0. Additionally, the affinity between samples should increase as the distance between them decreases. Many membership functions, such as rectangular and triangular membership functions, satisfy the constraints. Here we have utilized the commonly used Gaussian membership function:

$$\mu_G(a, b) = e^{-(a-b)^2/2\sigma^2}, \qquad (2.23)$$

where $\sigma > 0$ controls the spread of the membership functions.

To illustrate the effect of the Gaussian membership function, consider again the example $\mathbf{x}_\ell = [10, 1, 2]^T$. The relation between $x_3 = x_{(2)}$ and each of the observation samples is illustrated in Figure 2.4 for membership functions

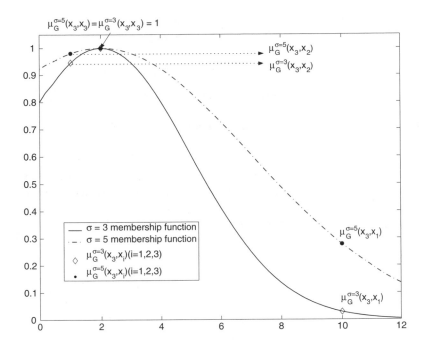

FIGURE 2.4
The affinity-based relation between $x_3 = 2$ and the elements of $\mathbf{x}_\ell = [10, 1, 2]^T$ for membership function spread parameters $\sigma = 3$ and $\sigma = 5$. (From K.E. Barner et al., *EURASIP J. Appl. Signal Process. Spec. Iss. Nonlinear Signal Image Process.*, Dec. 2001. © 2001 Hindawi. With permission.)

with spread $\sigma = 2$ and $\sigma = 5$. The figure shows that the similarly valued samples x_2 and x_3 have a strong relation in both cases, but that the relation between the more distant x_1 and x_3 samples is significantly reduced as σ is decreased.

2.3.2   Fuzzy Spatial–Rank Ordering

The affinity-based relations between all observation samples can be represented in a fuzzy SR matrix as a straightforward generalization of the crisp SR matrix.[4,6–8] Simply replacing the crisp relation with the more general fuzzy relation yields

$$\tilde{\mathbf{R}} = \begin{bmatrix} \tilde{R}_{1,(1)} & \cdots & \tilde{R}_{1,(N)} \\ \vdots & \ddots & \vdots \\ \tilde{R}_{N,(1)} & \cdots & \tilde{R}_{N,(N)} \end{bmatrix}, \tag{2.24}$$

where $\tilde{R}_{i,(j)} = \mu_{\tilde{R}}(x_i, x_{(j)})$. This fuzzy relation produces a real-valued relation matrix, i.e., $\tilde{R}_{i,(j)} \in [0, 1]$. In this context $\mu_{\tilde{R}}(x_i, x_{(j)})$, and equivalently $\tilde{R}_{i,(j)}$, denotes the degree to which x_i and $x_{(j)}$ are related.

The real-domain fuzzy rank and spatial vectors $\tilde{r}$ and $\tilde{s}$ are now defined in an analogous manner to their crisp counterparts:

$$\tilde{r} = \tilde{R}[1:N] \qquad \text{and} \qquad \tilde{s} = \tilde{R}^T[1:N]. \tag{2.25}$$

Similarly, the crisp and fuzzy spatial and rank order vectors are given by $\tilde{x}_\ell = \tilde{R}x_L$ and $\tilde{x}_L = \tilde{R}^T x_\ell$.

A careful inspection of the fuzzy terms indicates that the ranges of values have been increased beyond that of their crisp counterpart. Specifically, $\tilde{x}_i, \tilde{x}_{(j)} \in [x_{(1)}, \sum_{l=1}^N x_l]$, and $\tilde{r}_i, \tilde{s}_i \in [1, \sum_{l=1}^N l]$. To yield more intuitive values, these terms can be restricted to the same range as their crisp counterpart by normalizing the rows or columns of the fuzzy SR matrix. Because the rows and columns correspond to space and rank, respectively, we designate $\tilde{R}_\ell$ and $\tilde{R}_L$ to be the row (spatial) and column (rank) normalized fuzzy SR matrices.

The normalized fuzzy rank and spatial index vectors are now given by $\tilde{r} = \tilde{R}_\ell[1:N]$ and $\tilde{s} = \tilde{R}_L^T[1:N]$. Similarly, $\tilde{x}_\ell = \tilde{R}_\ell x_L$ and $\tilde{x}_L = \tilde{R}_L^T x_\ell$. Given this normalization, the appropriate bounds hold $\tilde{r}_i, \tilde{s}_{(j)} \in [1, N]$ and $\tilde{x}_i, \tilde{x}_{(j)} \in [x_{(1)}, x_{(N)}]$. Carrying out the matrix expressions for a single term yields the following expressions for $\tilde{r}_i$ and $\tilde{x}_{(j)}$:

$$\tilde{r}_i = \frac{\sum_{j=1}^N j \tilde{R}_{i,(j)}}{\sum_{j=1}^N \tilde{R}_{i,(j)}}, \tag{2.26}$$

and

$$\tilde{x}_{(j)} = \frac{\sum_{i=1}^N x_i \tilde{R}_{i,(j)}}{\sum_{i=1}^N \tilde{R}_{i,(j)}}. \tag{2.27}$$

Thus $\tilde{r}_i$ is a normalized weighted sum of the integers $1, 2, \ldots, N$ and $\tilde{x}_{(j)}$ is a normalized weighted sum of the samples $x_1, x_2, \ldots, x_N$. The weights in each case are the affinity relations between samples.

To illustrate the value in utilizing fuzzy relations, consider again the example $x_\ell = [10, 1, 2]^T$ and Gaussian membership function ($\sigma = 3$). The fuzzy SR matrix in this case is

$$\tilde{R} = \begin{bmatrix} 0.0111 & 0.0286 & 1.0000 \\ 1.0000 & 0.9460 & 0.0111 \\ 0.9460 & 1.0000 & 0.0286 \end{bmatrix}, \tag{2.28}$$

and the normalized fuzzy SR matrices are

$$\tilde{R}_\ell = \begin{bmatrix} 0.0107 & 0.0275 & 0.9618 \\ 0.5110 & 0.4834 & 0.0057 \\ 0.4791 & 0.5065 & 0.0145 \end{bmatrix} \tag{2.29}$$

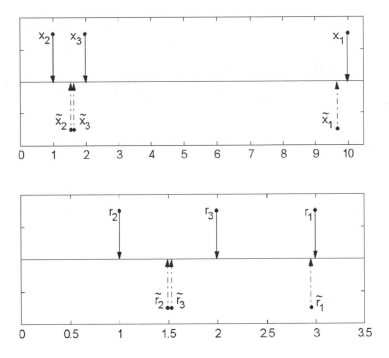

FIGURE 2.5
Comparison of crisp and fuzzy (a) samples and (b) ranks for the observation $\mathbf{x}_\ell = [10, 1, 2]^T$. (From K.E. Barner et al., *EURASIP J. Appl. Signal Process. Spec. Iss. Nonlinear Signal Image Process.*, Dec. 2001. © 2001 Hindawi. With permission.)

and

$$\tilde{\mathbf{R}}_L = \begin{bmatrix} 0.0057 & 0.0145 & 0.9618 \\ 0.5110 & 0.4791 & 0.0107 \\ 0.4834 & 0.5065 & 0.0275 \end{bmatrix}. \tag{2.30}$$

The resulting fuzzy rank vector is $\tilde{\mathbf{r}} = [2.9512, 1.4947, 1.5354]^T$. The resulting fuzzy space- and rank-ordered sample vectors are $\tilde{\mathbf{x}}_\ell = [9.6840, 1.5344, 1.6367]^T$ and $\tilde{\mathbf{x}}_L = [1.5344, 1.6367, 9.6840]^T$, respectively. The comparison of $\tilde{\mathbf{r}}$ and $\tilde{\mathbf{x}}_\ell$ with their crisp counterparts is illustrated in Figure 2.5. As the figure shows, $\tilde{x}_1 \approx x_1$ and $\tilde{r}_1 \approx r_1$. This is a result of the fact that $x_1 \gg x_2, x_3$ in relation to the spread of the membership function. Conversely, since x_2 and x_3 are similarly valued, and thus highly related, $\tilde{r}_2 \approx \tilde{r}_3 \approx (r_2 + r_3)/2$ and $\tilde{x}_2 \approx \tilde{x}_3 \approx (x_2 + x_3)/2$. Thus, fuzzy ranking and order statistics reflect not only the ordering of samples but also their spread. These concepts and properties are considered further in the following section.

2.3.3 Properties of Fuzzy Spatial–Rank Ordering

The fuzzy SR matrix, indices, and samples possess powerful properties that are useful in relating sample orderings and values. Several of the most important properties are highlighted below. A fuller discussion of properties is given in References 4, 6 through 8, and 10.

2.3.3.1 Superordination Property

The fuzzy SR matrix is a superset of the crisp SR matrix, containing information on the spread of the samples in addition to the crisp SR information.

This property arises from the fact that a fuzzy relation can be mapped to its crisp counterpart through thresholding: $R_{i,(j)} = T_1(\tilde{R}_{i,(j)})$, where $T_\delta(a)$ is the thresholding operation that yields 1 if $a \geq \delta$ and 0 otherwise. Extending this element-wise to the fuzzy SR matrix yields $\mathbf{R} = T_1(\tilde{\mathbf{R}})$, showing that the fuzzy SR matrix can be reduced to it crisp counterpart by thresholding.* In case of a row (column) normalized fuzzy SR matrix, elements are thresholded with δ set to the maximum row (column) element.

2.3.3.2 Reduction Property

As the membership function spread decreases, fuzzy relations, ranks, and order statistics reduce to their crisp counterparts. Conversely, as the membership function spread increases to infinity, all fuzzy relations become equal, fuzzy ranks converge to the median, and fuzzy order statistics converge to the sample mean.

This property describes the limiting behavior, with respect to membership function spread, of the fuzzy SR matrix. It is easy to see $\lim_{\sigma \to 0} \tilde{\mathbf{R}} = \mathbf{R}$. Conversely, $\sigma \to \infty$ indicates infinite spread of the membership function and the relation between all samples is equal, i.e., $\tilde{R}_{i,(j)} = 1$ for $i, j = 1, 2, \ldots, N$. For row (column) normalized fuzzy matrix, $\lim_{\sigma \to \infty} \tilde{R}_{i,(j)} = 1/N$. From this, it is easy to see that $\lim_{\sigma \to 0} \tilde{\mathbf{r}} = \mathbf{r}$ and $\lim_{\sigma \to \infty} r_i = (N+1)/2$. Also, $\lim_{\sigma \to 0} \tilde{x}_{(j)} = x_{(j)}$ and $\lim_{\sigma \to \infty} \tilde{x}_{(j)} = \bar{x} = 1/N \sum_l^N x_l, i, j = 1, 2, \ldots, N$.

2.3.3.3 Spread-Sensitivity Property

The intersample spacing is captured in the fuzzy SR matrix.

If the membership function covers the range $[0, x_{(N)} - x_{(1)}]$, and is continuous and continuously decreasing (as a function of $|x_i - x_j|$), then the sample spread can be exactly determined from $\tilde{\mathbf{R}}$. That is, $|x_i - x_j| = \mu_R^{-1}(\tilde{R}_{i,(r_j)})$. Thus, the observation vector can be recovered from $\tilde{\mathbf{R}}$ to within a constant. It is easy to see that the observation vector can be also be recovered from the normalized SR matrix.

This property explains why the fuzzy SR matrix captures sample spread information. To illustrate, consider two strictly increasing sequences, one linear, the other exponential. While their corresponding crisp SR matrices are diagonal identity matrices, concealing the sample spacing, the fuzzy SR matrices reflect the sample spread. This is illustrated in Figure 2.6 for a sequence

---

* This assumes all observation values are unique. For the case of equally valued samples, stable sorting is utilized to assign each sample a unique crisp rank index and this rank can be extracted from the fuzzy SR matrix by simply preserving the original time order of equally valued samples.

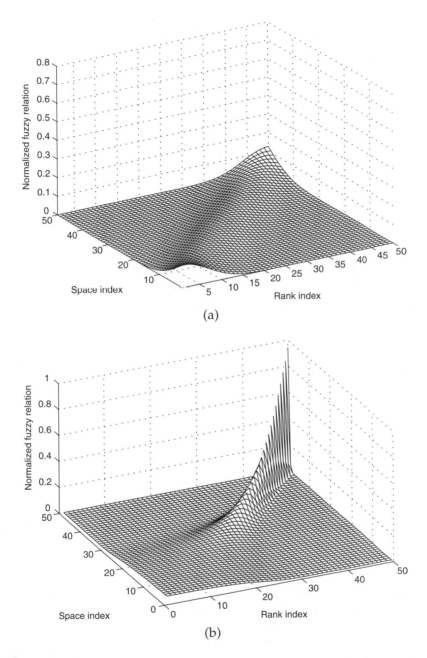

FIGURE 2.6
The row normalized fuzzy SR matrices for (a) linearly increasing and (b) exponentially increasing sequences. (From K.E. Barner et al., *EURASIP J. Appl. Signal Process. Spec. Iss. Nonlinear Signal Image Process.*, Dec. 2001. © 2001 Hindawi. With permission.)

of length $N = 20$. From the figure it is clear that the greater separation between the later exponential samples is captured in the fuzzy SR matrix.

2.3.3.4   Element Invariant Property

The fuzzy samples and the fuzzy order statistics constitute the same set and the SR relation is invariant to the fuzzy ordering.

Analytically, this property means if x_i is the jth order statistic, i.e., $x_i = x_{(j)}$, then $\tilde{x}_i = \tilde{x}_{(j)}$.[9] This property is a direct consequence of the definitions of the fuzzy samples and the fuzzy order statistics. As a result, a one-to-one mapping can be established between $\tilde{X} = \{\tilde{x}_1, \tilde{x}_2, \ldots, \tilde{x}_N\} = \{\tilde{x}_{(1)}, \tilde{x}_{(2)}, \ldots, \tilde{x}_{(N)}\}$ and $X = \{x_1, x_2, \ldots, x_N\} = \{x_{(1)}, x_{(2)}, \ldots, x_{(N)}\}$. For each element $x \in X$, its image in $\tilde{X}$ is the weighted average of all the elements of X, where the weight of each element is its membership function value with respect to x. Moreover, the element and its image have the same spatial and rank indexes in each set.

2.3.3.5   Order Invariant Property

The fuzzy order statistics have the same rank orders as their crisp counterparts if the membership function obeys certain conditions, which are satisfied by Gaussian, triangular, and uniform membership functions.

This property guarantees that the rank of a fuzzy order statistic's value is consistent with its rank index, i.e., $\tilde{x}_{(1)} \leq \tilde{x}_{(2)} \leq \ldots \tilde{x}_{(N)}$.[9] With this property, the fuzzy samples and the crisp samples have the same internal rank order relations. This property holds if and only if the membership function $\mu_{\tilde{R}}(\cdot, \cdot)$ is such that

$$C(x, t, \Delta t) = \frac{\mu_{\tilde{R}}(x, t + \Delta t)}{\mu_{\tilde{R}}(x, t)}$$

is a monotonically nondecreasing function of $x \in \Re$, for $\forall\, t, \Delta t \in \Re, \Delta t \geq 0$. This constraint controls the decay of the membership function and is satisfied by many common membership functions, such as the Gaussian, triangular, and uniform functions.

2.3.3.6   Distribution of Fuzzy Order Statistics

Since the fuzzy order statistics are formed as weighted averages of crisp samples, the distributions of fuzzy order statistics contain hybrid characteristics of both the crisp order statistics and the local means. The contribution of each component is jointly controlled by the membership function spread and the local distribution of the crisp samples. Thus, the distribution of a fuzzy order statistic can be approximated by a linear combination of the distribution of the corresponding crisp order statistic and the local mean,[10]

$$f_{\tilde{x}_{(j)}}(x) = (1 - p(x))^\lambda f_{x_{(j)}}(x) + p(x) f_{\Sigma}(x), \tag{2.31}$$

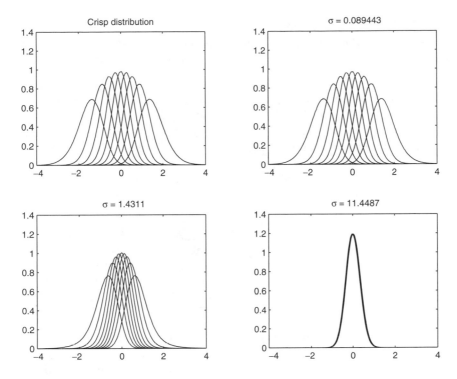

FIGURE 2.7

The distribution of the crisp and fuzzy order statistics of i.i.d. zero mean Gaussian samples with variance 4 for the $N = 9$ case. (From K.E. Barner et al., *EURASIP J. Appl. Signal Process. Spec. Iss. Nonlinear Signal Image Process.*, Dec. 2001. © 2001 Hindawi. With permission.)

where $f_{\tilde{x}_{(j)}}(\cdot)$ and $f_{x_{(j)}}(\cdot)$ are the distribution of the jth fuzzy order statistic and crisp order statistic, respectively. In this expression, $f_{\Sigma}(x)$ denotes the distribution of the local average of the crisp samples within a neighborhood around x, where the size of the neighborhood is controlled by the membership function spread, and $\lambda \geq 0$ is a constant related to the total number of the samples. The parameter $p(\cdot) \in (0, 1)$, which controls the mixing, is a function of both x and the membership function spread. For fixed x, larger spread leads to higher value of $p(x)$ and the fuzzy order statistics behave more like the local mean.

As an example of the role played by the membership function spread, consider the case of i.i.d. zero mean, unit variance Gaussian observation samples with $N = 9$. The distribution of $\tilde{x}_{(j)}$ is shown in Figure 2.7 for several membership function spread values. The figure illustrates that as σ grows and increased averaging is introduced, the variances of the fuzzy order statistics are reduced and their means are shifted to the center of the original distributions. Thus, the membership function spread controls the migration of $\tilde{x}_{(j)}$ from the crisp order statistic $x_{(j)}$ to the sample mean $\bar{x}$.

2.3.3.7   Clustering Property

Samples from a finite number of populations yield fuzzy samples (fuzzy ranks) that are averages of the observation samples (crisp ranks) within the individual populations.

Let the samples be from L populations with $\mu(x_i, x_k) = 1$ for $i, k \in I_l$, $\mu(x_i, x_k) = 0$ for $i \in I_l, k \in I_m, l \neq m$ and $\cup_{l=1}^{L} I_l = \{1, 2, \ldots, N\}$. Also, let p_i be the population index of x_i, i.e., $x_i \in I_{p_i}$. In this case, the fuzzy relation between samples in different populations is 0 and the relation between samples in the same population I_l is, in the normalized case, $1/\|I_l\|$, where $\|I_l\|$ denotes the number of samples in I_l. This results in the fuzzy time and rank samples

$$\tilde{\mathbf{x}}_i = \tilde{\mathbf{x}}_{r_i} = \frac{1}{\|I_{p_i}\|} \sum_{k \in I_{p_i}} x_k \qquad (2.32)$$

and

$$\tilde{\mathbf{r}}_i = \frac{1}{\|I_{p_i}\|} \sum_{k \in I_{p_i}} r_k. \qquad (2.33)$$

Note that the fuzzy rank indices are obtained by averaging crisp ranks, which, in the case of equally valued samples, form an integer subsequence, $r_{\min}, r_{\min} + 1, \ldots, r_{\max}$, where $r_{\min}$ and $r_{\max}$ are the minimum and maximum rank of samples in a given population. Averaging such a subsequence results in a value that is always a multiple of $\frac{1}{2}$. Thus, fuzzy ranks are typically such that $\tilde{r}_i \in \{1, 1\frac{1}{2}, 2, 2\frac{1}{2}, \ldots, N\}$.

Relaxing the membership relations between samples in the above property, we see that the fuzzy relations introduce averaging among similarly valued samples, where similarity is determined by the membership function shape and spread.

2.4   Fuzzy Filter Definitions

Having established the concepts of affinity and fuzzy SR orderings, these concepts can now be adopted into filtering structures. One method for accomplishing the inclusion is through the simple modification of established filtering algorithms. Thus, existing algorithms that have proved useful can be updated to include affinity or fuzzy ordering information. The additional degrees of freedom introduced by fuzzy methods, and thus the consideration of sample spread by the filtering algorithm, lead to improved performance.

The fuzzy generalization of two broad classes is considered here. First, the general class of affine filters is established.[11] Affine filters are realized as a simple extension of the broad class of weighted sum filters in which affinity, or sample spread, weighting is introduced in the weighted sum output. The second class of filters established is the fuzzy generalization of weighted median filters.[7,8] Fuzzy weighted median filters are realized by simply inserting fuzzy samples into the standard weighted median filter formulation. As is demonstrated in Section 3.2, the affinity and fuzzy generalizations lead to improved performance.

2.4.1 Affine Filters

Affine filters are realized by including an affinity weighting, to a specified reference point, in a standard weighted sum filter. Two important subclasses of affine filters are the median affine and center affine filter classes. The median affine filter class is established by utilizing the median sample as the reference point and introducing the affinity weighting into the standard weighted sum filter. The center affine filter class is realized by setting the central observation sample as the reference point and introducing the affinity weighting into the weighted sum of order statistics filter.

2.4.1.1 Median Affine Filters

The standard linear FIR filter has been successfully applied to many problems. Indeed, by formulating the filter output as a weighted sum of spatially, or temporally, ordered samples, important filter characteristics are realized, such as frequency selectivity. Linear FIR filters, however, perform poorly in the presence of outliers. This performance can be improved if a simple valid reference point can be established and the validity of each sample measured in reference to the reference point. This is the motivation behind the median affine filter.

The median operator is robust, and can thus serve to establish a valid reference point for many signal statistic cases. Thus, the standard weighted sum FIR filter given in Equation 2.9 can be made more robust by weighting each sample according to its affinity to the median reference point. Utilizing again $\delta = (N+1)/2$ as the central index, the median affine filter is defined as

$$\text{MAFF}[\mathbf{x}] = \frac{\sum_{i=1}^{N} w_i \tilde{R}_{i,(\delta)} x_i}{\sum_{i=1}^{N} |w_i| \tilde{R}_{i,(\delta)}}, \tag{2.34}$$

where the w_i are the filter weights and $\tilde{R}_{i,(\delta)}$ is the affinity of the ith observation with respect to the median reference point, $\text{MED}(\mathbf{x}) = x_{(\delta)}$.

The filter structure in Equation 2.34 weights each observation twice: first, according to its reliability, and second, according to its natural order, Figure 2.8. Median affine estimates are therefore based on observations that are both reliable and favorable due to their natural order. Observations that fail to meet either criterion have only a limited influence on the estimate.

By varying the dispersion of the affinity (or membership) function, certain properties of the median affine filter can be stressed: a large value of σ

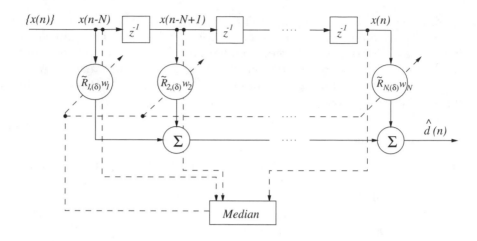

FIGURE 2.8
Structure of the (unnormalized) median affine filter.

emphasizes the linear properties of the filter, while a small value puts more weight on its median properties. Of special interest are the limiting cases. For $\sigma \to \infty$, the affinity function is constant on its entire domain. The estimator, therefore, weights all observations strictly according to their natural order, i.e.,

$$\lim_{\sigma \to \infty} \text{MAFF}[\mathbf{x}] = \frac{\sum_{i=1}^{N} w_i x_i}{\sum_{i=1}^{N} w_i} \tag{2.35}$$

and the median affine estimator reduces to a normalized linear filter. In contrast, for $\sigma \to 0$ the affinity function shrinks to an impulse at $\text{MED}(\mathbf{x})$. Thus, the constant weights w_i are disregarded and the estimate is equal to the median, i.e.,

$$\lim_{\sigma \to 0} \text{MAFF}[\mathbf{x}] = \text{MED}(\mathbf{x}). \tag{2.36}$$

In addition to the limiting cases, the median affine filter includes important filters, such as the MTM filter,[12] as subclasses.

The median affine filter also possesses several desirable properties,[11] including: (1) data-adaptiveness, (2) translation invariance, (3) the ability to suppress impulses, and (4) the ability to preserve signal trends and discontinuities. These properties lend understanding to the filtering process and help explain the improved performance achieved through the incorporation of sample affinity into the filter structure.

2.4.1.2   Center Affine Filters

A second important subclass of affine filters is the class of center affine filters. While the median serves as an acceptable reference point for certain applications, the observation sample located (spatially) in the center of the observation window is particularly important in other applications. Thus, the class of center affine filters is based on the concept of affinity to the central observation sample.

Center affine filters are related to median affine filters through a simple change in ordering. That is, rather than a generalization of the linear filter based on affinity to the median, the center affine filter is a generalization of the order-statistic weighted sum (L) filter based on affinity to the central spatial sample. Accordingly, the output of the center affine filter is given by

$$\text{CAFF}[\mathbf{x}] = \frac{\sum_{i=1}^{N} w_{(i)} \tilde{R}_{\delta,(i)} x_{(i)}}{\sum_{i=1}^{N} |w_{(i)}| \tilde{R}_{\delta,(i)}}, \tag{2.37}$$

where the $w_{(i)}$ are the filter coefficients and $\tilde{R}_{i,(\delta)}$ is the affinity of the ith order statistic, $x_{(i)}$, with respect to the central (spatial) sample reference point, x_δ.

The center affine filter weights each order-statistic according to its affinity to the central observation sample and according to its rank. The filter output, therefore, is based mainly on those order statistics that are simultaneously close to the central observation sample and preferable due to their rank order. Note that, as opposed to the median affine filters, here the temporal weights $\tilde{R}_{\delta,(i)}$ are time varying and the rank order weights $w_{(i)}$ are constant. Like the median affine filter, the center affine filter with nonnegative weights reduces to its basic structures at the limits of the dispersion parameter σ:

$$\lim_{\sigma \to 0} \text{CAFF}[\mathbf{x}] = x_\delta \quad \text{and} \quad \lim_{\sigma \to \infty} \text{CAFF}[\mathbf{x}] = \frac{\sum_{i=1}^{N} w_{(i)} x_{(i)}}{\sum_{i=1}^{N} |w_{(i)}|}. \tag{2.38}$$

Thus, the center affine filter reduces to the identity filter and the L-filter, with coefficients $w_{(i)}$, for $\sigma \to 0$ and $\sigma \to \infty$, respectively.

2.4.1.3 Optimization

To appropriately tune the performance of an affine filter, the filter coefficients can be optimized. Two approaches to optimizing the parameters are presented. First, a simple suboptimal procedure that addresses only the affinity spread function is presented. A more comprehensive optimization procedure that addresses both the filter weights and affinity spread function is then presented. This optimization is based on a stochastic adaptive procedure. In both cases, the presented methods address the optimization of the median affine filter. The methods can be applied to the design of center affine filters by simply interchanging corresponding quantities.

A simple and intuitive design procedure can be derived from the fact that the median affine filter behaves like a linear filter for $\sigma \to \infty$. Setting σ to a large initial value allows the use of the multitude of linear filter design methods to find the w_i coefficients of the median affine filter. Holding the w_i coefficients constant, the filter performance can, in general, be improved by gradually reducing the value of σ until a desired level of robustness is achieved. During the actual filtering process σ is fixed. Since this process strengthens the median-like properties, while weakening the influence of the FIR filter weights, this procedure is referred to as the *medianization* of a linear FIR filter.

The median affine filter can also be adaptively optimized under the mean square error (MSE) criteria in an approach that has been applied to related filter structures, such as radial basis functions.[13,14] Consider first the optimization of σ for a fixed set of filter coefficients. To simplify the notation, let $\gamma = \sigma^2$. Then, under the MSE criteria, the cost function to be minimized is

$$J(\gamma) \triangleq E[e^2] = E[(d - \hat{d})^2], \tag{2.39}$$

where $e = d - \hat{d}$ is the filtering error, and $E[\cdot]$ stands for the statistical expectation operator. The optimization problem can be stated as the minimization of $J(\gamma)$, where γ is restricted to nonnegative, real-valued numbers. Because of the nonlinearity of the median affine estimate, this is a nonlinear optimization problem.

Although finding a closed form solution is intractable, an iterative LMS-type approach can be adopted.[11,15] Under this approach, γ is indexed and updated according to

$$\gamma(n+1) = \gamma(n) - \mu_\gamma \frac{\partial J}{\partial \gamma}(n), \tag{2.40}$$

where μ_γ is the appropriately chosen step size. For the case of positive valued weights, differentiating $J(\gamma)$ with respect to γ, substituting in the above, and performing some simplification yields the update

$$\gamma(n+1) = \gamma(n) + \mu_\gamma \frac{(d(n) - \hat{d}(n))}{\gamma^2(n)}$$

$$\times \left(\sum_{i=1}^{N} w_i \tilde{R}_{i,(\delta)}(x_i(n) - \hat{d}(n))(x_i(n) - x_{(\delta)}(n))^2 \right). \tag{2.41}$$

At each iteration the positive constraint can be enforced by a simple maximum operator, $\gamma(n+1) = \max\{\gamma(n+1), 0\}$.

Adopting a similar approach for the optimization of the filter weights, in which $J(\gamma)$ is differentiated with respect to the filter weights, yields the update

$$w_i(n+1) = w_i(n) - \mu_w \frac{\partial J}{\partial w_i}(n) \tag{2.42}$$

for $i = 1, 2, \ldots, N$, where μ_w is the step size and

$$\frac{\partial J}{\partial w_i}(n) = -(d(n) - \hat{d}(n)) \left[\tilde{R}_{i,(\delta)}(n) \sum_{k=1}^{N} w_k \tilde{R}_{k,(\delta)}(n)(x_i(n) - x_k(n)) \right]. \tag{2.43}$$

Both iteration updates are computationally simple and can be modified to address the case of negative filter weights. Additionally, they can be applied in an alternating procedure to yield a globally optimal set of affine filter parameters.

2.4.2 Fuzzy Weighted Median Filters

2.4.2.1 Fuzzy Median Filters

The median filter is widely used due to its effectiveness in preserving signal structures, such as edges and monotonic regions, while smoothing noise. In many practical applications, however, it is advantageous to introduce some weighted averaging of samples, particularly among similarly valued samples. In such cases the fuzzy median can be utilized. The fuzzy median filter is realized by simply replacing the crisp order statistics with their fuzzy counterparts,[6–8] and is thus defined as

$$\text{FMED}[\mathbf{x}] = \tilde{x}_{(\delta)} = \frac{\sum_{i=1}^{N} x_i \, \tilde{R}_{i,(\delta)}}{\sum_{i=1}^{N} \tilde{R}_{i,(\delta)}}, \qquad (2.44)$$

where $\tilde{R}_{i,(\delta)}$ is the fuzzy relation between x_i and $x_{(\delta)}$. Note that this is simply a special case of the median affine filter in which all spatial order samples are giving equal weight, $w_i = 1$ for $i = 1, 2, \ldots, N$.

To illustrate the advantages of using fuzzy techniques in a median structure, consider the filtering of an ideal step edge corrupted by additive Gaussian noise. Figure 2.9 shows the ensemble average of the outputs of crisp and fuzzy median filters with a window size of $N = 9$ operating on the Gaussian noise-corrupted step signal. It can be seen that while the crisp median filter smoothes the additive noise, it also introduces significant edge smoothing. The fuzzy median, in contrast, preserves the edge transition while performing more effective noise smoothing.

This improved performance can be attributed to the fact that the fuzzy median is an unbiased estimator at all window locations spanning the edge transition (as well as in strictly uniform regions).[7] The crisp median, however, is biased at all window locations that span the edge transition and only unbiased in strictly uniform regions. It can also be shown that the variance of the fuzzy median estimator is less than its crisp counterpart. In non-ideal cases, such that as depicted in Figure 2.9, some smoothing does occur due to imperfect relation between samples on a common sides of the edge transition and some nonzero relation among samples on opposite side of the transitions.

As a second example, consider the filtering of a chirp image. Figure 2.10 shows the original, Gaussian noise-corrupted, median-filtered, and fuzzy median-filtered chirp images. The results show that, as in the one-dimensional case, the fuzzy extension to the median filter improves the edge preservation and noise-smoothing capabilities.

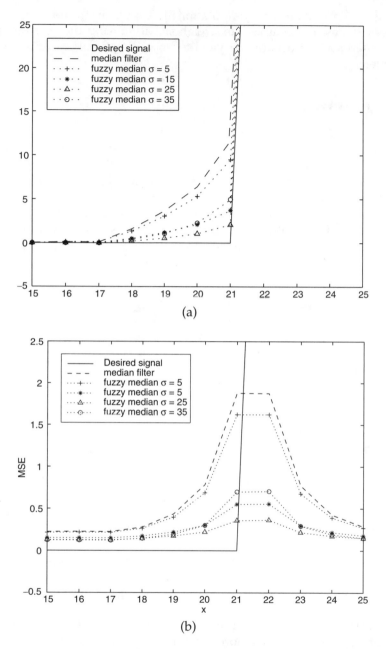

FIGURE 2.9
(a) Step signal of height 100, corrupted by Gaussian noise of variance 10, filtered with median filter and fuzzy median filter and (b) the resulting mean square filtering error. (From K.E. Barner et al., *EURASIP J. Appl. Signal Process. Spec. Iss. Nonlinear Signal Image Process.*, Dec. 2001. © 2003 IEEE. With permission.)

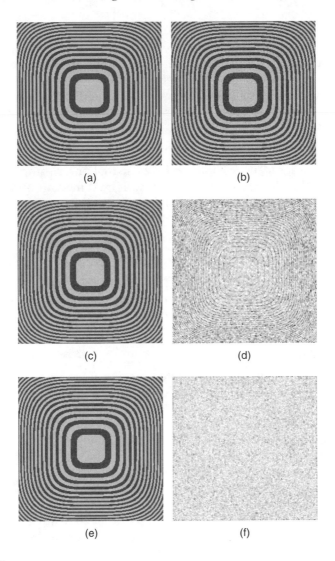

FIGURE 2.10
Filtered chirp images and the resulting scaled error images: (a–b) Original image and the image corrupted by additive Gaussian noise with variance 100, (c–d) median (RMSE = 7.0, MAE = 5.3), and (e–f) fuzzy median (RMSE = 4.1, MAE = 3.1) filter. Each filter utilized a 3 × 3 observation window. (From K.E. Barner et al., *EURASIP J. Appl. Signal Process. Spec. Iss. Nonlinear Signal Image Process.*, Dec. 2001. © 2001 Hindawi. With permission.)

2.4.2.2 Fuzzy Weighted Median Filters

Recall that the median filter can be generalized by weighting each of the spatially ordered samples prior to sorting and median selection. This results in the expression given in Equation 2.11, and repeated here for convenience:

$$\text{WMED}[\mathbf{x}] = \text{MED}[w_1 \diamond x_1, w_2 \diamond x_2, \ldots, w_N \diamond x_N], \qquad (2.45)$$

where $\diamond$ is the replication operator. Note that although the discussion presented in Section 2.2 restricted the replication weights to positive integer values, the filter can easily incorporate real-valued (positive and negative) weights utilizing the procedures in Reference 16.

The weighted median filter can be generalized to the fuzzy weighted median filter by simply replacing the crisp observation samples with their fuzzy counterparts,

$$\text{FWMED}[\mathbf{x}] = \text{MED}[w_1 \diamond \tilde{x}_1, w_2 \diamond \tilde{x}_2, \ldots, w_N \diamond \tilde{x}_N]. \tag{2.46}$$

As the fuzzy order statistics and the fuzzy spatial order samples constitute the same set, this definition of the fuzzy weighted median is equivalent to first weighting the (crisp) spatial order samples and then selecting the fuzzy median from this expanded set, i.e.,

$$\text{FWMED}[\mathbf{x}] = \text{MED}[w_1 \diamond \tilde{x}_1, w_2 \diamond \tilde{x}_2, \ldots, w_N \diamond \tilde{x}_N] \tag{2.47}$$

$$= \text{FMED}[w_1 \diamond x_1, w_2 \diamond x_2, \ldots, w_N \diamond x_N]. \tag{2.48}$$

Under both equivalent definitions, the fuzzy weighted median is able to exploit spatial correlations, through spatial weighting; limit the influence of outliers, through ranking and median selection; and introduce selected weighted averaging of similarly valued samples, through the use of fuzzy samples.

2.4.2.3 Optimization

The optimization of the fuzzy weighted median filter follows an approach similar to that presented for the class of affine filters. In the fuzzy weighted median filter case, however, the error criteria chosen is the mean absolute error (MAE) criteria. This is because the MAE criteria arise naturally out of the ML development under the Laplacian assumption. Additionally, median-type filters are typically applied to signals with heavy-tailed distributions. Utilizing the MSE for such signals tends to overemphasize the influence of outliers in the optimization procedure, and thus lower-power error criteria are typically employed.

Consider first the optimization of the fuzzy order statistics, from which the optimization of the fuzzy median follows directly. Under the MAE criteria, the cost to be minimized is $J(\gamma) = E(|d - \hat{d}|)$, where the optimization of the jth order statistic is achieved by setting $\hat{d} = \tilde{x}_{(j)}$. Differentiating this cost criteria with respect to γ, substituting into a stochastic gradient-based algorithm, and replacing the expectation operator with instantaneous estimates yields

$$\gamma(n+1) = \gamma(n) - \mu_\gamma \frac{\partial J(\gamma)}{\partial \gamma}(n) \tag{2.49}$$

$$= \gamma(n) + \mu_\gamma \frac{\text{sgn}(d(n) - \tilde{x}_j(n))}{\gamma^2(n)}$$

$$\times \frac{\sum_{i=1}^{N} \tilde{R}_{i,(j)}(x_i(n) - \tilde{x}_j(n))(x_i(n) - x_{(j)}(n))^2}{\sum_{i=1}^{N} \tilde{R}_{i,(j)}}, \tag{2.50}$$

where μ_γ is the step size and, as before, a positivity constraint is placed on γ.

The optimization of the fuzzy median follows directly from the above by simply setting $j = \delta = (N+1)/2$. Thus, in the fuzzy median case there is only a single parameter to optimize. In the case of an FWM filter, the spatial weights and membership function spread parameter must be optimized. The optimization of the WM filter spatial weights is similarly carried out under the MAE criteria utilizing a gradient-based algorithm.[17,18] Such an optimization yields the following weight update expression:

$$w_j(n+1) = w_j(n) + \mu_\omega e(n)\mathrm{sgn}(w_j(n))\mathrm{sgn}[\mathrm{sgn}(w_j(n))x_j(n) - \hat{d}(n)]. \quad (2.51)$$

As in the previous case, this optimization can be combined with the iterative membership function optimization expression and applied in an alternating fashion. Although this does not guarantee globally optimal results, the procedure is simple, computationally efficient, and has yielded good results.

2.5 Extensions to Multivariate Data

The fuzzy relations and resulting filtering algorithms can be readily extended to the multivariate data case. Indeed, multivariate signals arise naturally in many applications, such as color image processing, velocity estimation, three-dimensional surface moving, etc. Such signals can be processed in a component-wise fashion, although this approach fails to exploit correlations between components. Therefore, the more appropriate approach is to operate directly on the multivariate data, taking advantage of the natural correlations between signal components.

The fuzzy SR relations, upon which all of the discussed methods are founded, are based on two relations: (1) spatial–rank ordering and (2) fuzzy affinity relations between observation samples. The spatial, or temporal, ordering of the samples is defined naturally by the observation window configuration and movement. While no universally accepted concept of rank ordering exists for multivariate data, numerous ordering approaches have been defined, e.g., sum distance ordering. Thus, to extend the fuzzy SR relations concepts and filtering extensions to multivariate data, one need only select the appropriate rank ordering procedure and define a fuzzy relation between vector-valued samples.

Since fuzzy relations are a function of the distance between points, care must be taken in defining the appropriate multivariate distance metric $D(\mathbf{a}, \mathbf{b})$ between two vectors, $\mathbf{a}, \mathbf{b} \in \Re^M$. The selection of the vector difference metric is application dependent. For example, if the directions that $\mathbf{a}, \mathbf{b} \in \Re^M$ represent are the main features of concern, then the angle between $\mathbf{a}$ and $\mathbf{b}$ is a good difference metric.[19] Conversely, if the distance between vectors is the feature of concern, then the L_p norm[20] is the appropriate metric. Numerous other metrics, of course, are also possible, depending on the problem at hand.

Following selection of the distance metric $D(\cdot, \cdot)$, a vector-based membership function can be defined, for example, based on the Gaussian function,

$$\mu_G(\mathbf{a}, \mathbf{b}) = e^{-D(\mathbf{a},\mathbf{b})^2/2\sigma^2}, \tag{2.52}$$

where σ, as before, is the spread parameter. The multivariate median affine and center affine filters are now simply defined as

$$\text{MAFF}[\mathbf{x}] = \frac{\sum_{i=1}^{N} w_i \tilde{R}_{i,(\delta)}\mathbf{x}_i}{\sum_{i=1}^{N} |w_i|\tilde{R}_{i,(\delta)}} \tag{2.53}$$

and

$$\text{CAFF}[\mathbf{x}] = \frac{\sum_{i=1}^{N} w_{(i)} \tilde{R}_{\delta,(i)}\mathbf{x}_{(i)}}{\sum_{i=1}^{N} |w_{(i)}|\tilde{R}_{\delta,(i)}}. \tag{2.54}$$

The fuzzy vector median and fuzzy vector weighted median filters are formed as similar multivariate extensions:

$$\text{FMED}[\mathbf{x}] = \tilde{\mathbf{x}}_{(\delta)} = \frac{\sum_{i=1}^{N} \mathbf{x}_i \tilde{R}_{i,(\delta)}}{\sum_{i=1}^{N} \tilde{R}_{i,(\delta)}} \tag{2.55}$$

and

$$\text{FWMED}[\mathbf{x}] = \text{MED}[w_1 \diamond \tilde{\mathbf{x}}_1, w_2 \diamond \tilde{\mathbf{x}}_2, \dots, w_N \diamond \tilde{\mathbf{x}}_N] \tag{2.56}$$

$$= \text{FMED}[w_1 \diamond \mathbf{x}_1, w_2 \diamond \mathbf{x}_2, \dots, w_N \diamond \mathbf{x}_N]. \tag{2.57}$$

To illustrate the advantages of using fuzzy techniques in a multivariate application, we reconsider the edge filtering example in Figure 2.9 and extend it to vector data. Figure 2.11a shows a two-dimensional directional signal with an abrupt edge transition. Gaussian noise-corrupted observations of this

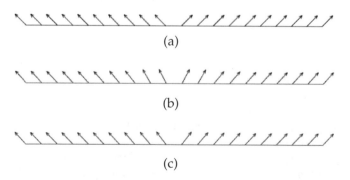

(a)

(b)

(c)

FIGURE 2.11
Filtering of directional signal corrupted by additive Gaussian noise: (a) original signal, (b) vector median, and (c) fuzzy vector median ensemble outputs.

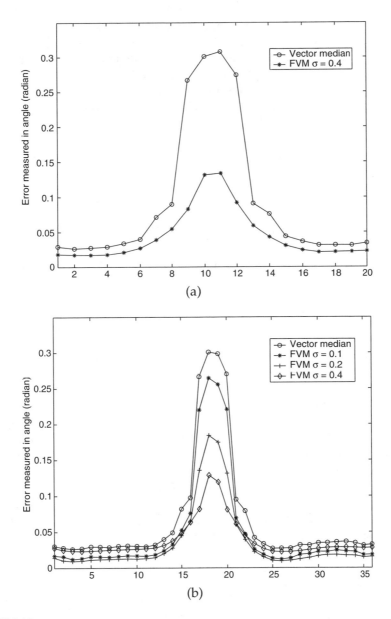

FIGURE 2.12
Filtering MSE values for the directional signal in Figure 2.11: (a) vector median and fuzzy vector median with $\sigma = 0.4$ and (b) vector median and fuzzy vector median with varying spread parameters.

signal were filtered by vector median and fuzzy vector median filters, with ensemble average outputs shown in Figure 2.11b and c. It can be seen that, while the vector median filter effectively smoothes the noise, it also introduces significant edge smoothing. The fuzzy vector median, in contrast, preserves the edge transition while performing more effective noise smoothing. The average filtering error is shown in Figure 2.12 for both filters and various membership function spread parameters. These results indicate that, as in the scalar case, the fuzzy filter extensions exhibit superior noise smoothing *and* edge preservation.

2.6 Conclusions

This chapter has developed several broad classes of nonlinear filters from a maximum likelihood perspective. It is shown that for heavy-tailed Laplacian distributions the ML development leads directly to the well-known median and weighted median filters. This development shows the fundamental importance of considering both spatial order and rank order of samples. The SR concept was formally established and then extended to the fuzzy case in which sample affinity, or spread (diversity), is considered. The concepts of sample affinity and affinity-based ordering lead to several classes of fuzzy filters. The fuzzy filters can, in general, be constructed as generalizations of existing filters that incorporate fuzzy measures or as new filtering paradigms based on fuzzy concepts. This chapter covers two broad classes of fuzzy filters: affine filters, which are generalizations of weighted sum filters that incorporate sample spread, and fuzzy weighted median filters, which generalize the standard class of weighted median filters by incorporating fuzzy samples. For both broad filter classes, optimization procedures are developed and extensions to multivariate data are addressed.

The following chapter builds on the theory developed here and shows that the incorporation of affinity measures into filtering algorithms yields additional degrees of freedom, enabling sample spread to be considered in the decision process. This consideration of sample spread leads to improved performance, which is demonstrated in a wide array of signal processing and communications problems. These include robust frequency selective filtering; Synthetic Aperture Radar image filtering; time-frequency plain filtering; multiresolution signal representations; surface smoothing; image smoothing, zooming, and deblocking; and multiuser detection.

References

1. G.R. Arce, Y.T. Kim, and K.E. Barner, Order-statistic filtering and smoothing of time series: Part 1, in *Order Statistics and Their Applications*, C.R. Rao and

N. Balakrishnan, Eds., Vol. 16 of *Handbook of Statistics*, Amsterdam, the Netherlands: Elsevier Science, 1998, 525–554.

2. K.E. Barner and G.R. Arce, Order–statistic filtering and smoothing of time series: Part 2, in *Order Statistics and Their Applications*, C.R. Rao and N. Balakrishnan, Eds., Vol. 16 of *Handbook of Statistics*, Amsterdam, the Netherlands: Elsevier Science, 1998, 555–602.

3. G.R. Arce and K.E. Barner, Nonlinear signals and systems, in *Encyclopedia of Electrical and Electronics Engineering*, J.G. Webster, Ed., New York: John Wiley & Sons, 1998, 612–630.

4. K.E. Barner and R.C. Hardie, Spatial–rank order selection filters, in *Nonlinear Image Processing*, S.K. Mitra and G. Sicuranza, Eds., San Diego, CA: Academic Press, 2001.

5. F.Y. Edgeworth, A new method of reducing observations relating to several quantities, *Philos. Mag. (Fifth Ser.)*, Vol. 24, 1887.

6. K.E. Barner, A. Flaig, and G.R. Arce, Fuzzy time–rank relations and order statistics, *IEEE Signal Process. Lett.*, 5, 252–255, 1998.

7. K.E. Barner, Y. Nie, and W. An, Fuzzy ordering theory and its use in filter generalizations, *EURASIP J. Appl. Signal Process. Spec. Iss. Nonlinear Signal Image Process.*, Dec. 2001.

8. A. Flaig, K.E. Barner, and G.R. Arce, Fuzzy ranking: theory and applications, *Signal Process. Spec. Iss. Fuzzy Process.*, 80, 1017–1036, 2000.

9. Y. Nie and K.E. Barner, The output distribution of fuzzy weighted median filters, paper presented at the 45th IEEE Midwest Symposium on Circuit and System (MWSCAS02), August, Tulsa, OK, 2002.

10. Y. Nie and K.E. Barner, The fuzzy transformation and its application in image processing, *IEEE Trans. Signal Process.*, Apr. 2003, Submitted for publication.

11. A. Flaig, G.R. Arce, and K.E. Barner, Affine order statistic filters: a data-adaptive filtering framework for nonstationary signals, *IEEE Trans. Signal Process.*, 46, 2101–2112, 1998.

12. Y.H. Lee and S.A. Kassam, Generalized median filtering and related nonlinear filtering techniques, *IEEE Trans. Acoust. Speech Signal Process.*, 33, 672–683, 1985.

13. I. Cha and S.A. Kassam, Channel equalization using adaptive complex radial basis function networks, *IEEE J. Select Areas Commun.*, 13, 964–975, 1995.

14. I. Cha and S.A. Kassam, RBFN restoration of nonlinearly degraded images, *IEEE Trans. Image Process.*, 5, 122–131, 1996.

15. S. Haykin, *Adaptive Filter Theory*, Englewood Cliffs, NJ: Prentice-Hall, 1991.

16. G.R. Arce, A generalized weighted median filter structure admitting negative weights, *IEEE Trans. Signal Process.*, 46, 1998.

17. G.R. Arce and J. Paredes, Image enhancement with weighted medians, in *Nonlinear Image Processing*, S. Mitra and G. Sicuranza, Eds., San Diego, CA: Academic Press, 2000, 27–67.

18. G.R. Arce and J.L. Paredes, Recursive weighted median filters admitting negative weights and their optimization, *IEEE Trans. Signal Process.*, 48, 768–779, 2000.

19. P.E. Trahanias and A.N. Venetsanopoulos, Vector directional filters—a new class of multichannel image processing filters, *IEEE Trans. Image Process.*, October, 528–534, 1993.

20. J. Astola, J. Haavisto, and Y. Neuvo, Vector median filters, in *Proc. IEEE*, April, 678–689, 1990.

3

Fuzzy Methods in Nonlinear Signal Processing: Part II—Applications

Kenneth E. Barner, Yao Nie, and Yuzhong Shen

CONTENTS

3.1 Introduction

This chapter builds on the theory of fuzzy methods in nonlinear signal processing developed in Part I of this two-chapter set. The theory developed in Part I shows that for the heavy-tailed Laplacian distribution, the maximum likelihood estimation criteria yield the well-known median and weighted median filters. Moreover, this development shows the importance of spatial order and rank order in estimator, or filter, development. This importance leads to the formalization of spatial and rank (SR) order theory and the more general fuzzy SR order theory. The fuzzy generalization is particularly important in that it enables sample spread, or affinity, to be included in the SR information.

Through use of the more general fuzzy ordering relation, well-established filtering paradigms can be generalized to include sample affinity, or spread, in the filtering process. The generalizations presented in the theory development focus on two broad classes of fuzzy filters: affine filters, which are generalizations of weighted sum filters that incorporate sample spread, and fuzzy weighted median filters, which generalize the standard class of weighted median filters by incorporating fuzzy samples. In both cases, there are well-established optimization procedures and generalizations to multivariate data.

The focus of this chapter is the application of the developed filters to a wide array of signal processing and communications problems. Indeed, because the affinity and fuzzy concepts are easily embedded in existing methods, nearly all problems can be successfully addressed through affinity- or fuzzy-based methods. To give an appreciation of the performance gains that can be achieved through more general methods, results are presented for fuzzy theory-based filters applied to the problems of robust frequency selective filtering, Synthetic Aperture Radar image filtering, time-frequency domain filtering, multiresolution signal representations, surface smoothing, image smoothing, zooming, and deblocking, and multiuser detection. In each case, we compare the performance of the fuzzy generalizations to their crisp counterparts as well as other methods reported in the literature designed to address the problem at hand. The presented results show that the fuzzy generalizations yield improved performance over existing methods. These results, coupled with the wide applicability of the fuzzy methods, indicate that fuzzy SR ordering and the resulting filter generalizations constitute important tools for modern signal processing applications.

This chapter is organized as follows. Applications of affine filters are covered in Section 3.2. There, the median affine filter is applied to the problems of robust frequency-selective filtering and signal multiresolution representations. Also in this section, the center affine filter is applied to the problems of inverse Synthetic Aperture Radar and time-frequency domain filtering, as well as image deblocking. Section 3.3 covers problems best addressed by fuzzy ordering and fuzzy weighted median filtering methods. Specifically, fuzzy weighted median type methods are applied to the problems of image (gray scale and color) and surface smoothing, image zooming, and noisy image sharpening. Also, the fuzzy ranks are used to address the problem of multiuser detection. Last, conclusions are drawn in Section 3.4.

3.2   Affine Filter Applications

Affine filters have a broad spectrum of potential applications due to their wide range of filter characteristics and their flexibility. The different structures that the filters can take on lend themselves to different problems. The first problem considered is robust frequency selective filtering, in which temporal (spatial) order of samples must be exploited to gain frequency selectivity, while some rank ordering should be considered to ensure robust behavior. This problem is effectively addressed through median affine filtering. These same characteristics allow median affine filters to be effectively employed in multiresolution signal decomposition problems, which are considered next. The following three problems focus on cases where the central observation sample plays a crucial role, and are thus effectively addressed through center affine filtering. Specifically, we apply center affine filtering to inverse Synthetic Aperture Radar image filtering, image deblocking, and time-frequency cross-term filtering.

3.2.1 Robust Frequency-Selective Filtering

Filters that are jointly robust and frequency-selective are of great interest. Although several robust lowpass filters exist, the design of robust bandpass and highpass filters remains a challenging task. The robust median affine filters can be easily designed to yield robust, frequency-selective behavior. Comparisons to the frequency-selective finite impulse response–weighted order statistic (FIR-WOS)[1] and $L\ell$ filters[2–4] show that median affine bandpass filters can achieve better performance with significantly fewer coefficients.

The performance of the compared filters is illustrated using a 2048-sample quad-chirp (sinusoidal waveform with quadratically increasing frequency) additively contaminated by α-stable noise ($\alpha = 1.2$) to simulate an impulsive environment,[5] Figure 3.1. The corresponding mean square error (MSE) and mean absolute error (MAE) are given in Table 3.1. As expected, the impulsive noise is detrimental to the performance of the linear filter. The FIR-WOS filter is able to trim the impulses, albeit at the cost of varying attenuation in the pass-band. Because of the employed weighted order statistic structure, however, this filter is not able to stop very low frequencies, requiring additional measures to complete the desired bandpass operation. The $L\ell$-filter is clearly more robust, but suffers from the blind rejection of the extreme order statistics. This blind rejection results in artifacts over most of the frequency spectrum. This effect is even stronger for higher signal-to-noise ratios (SNR). The median affine filter preserves the desired frequency band well, while strongly attenuating the impulsive noise. This superior performance is reflected in an MSE that is roughly half of that achieved by the $L\ell$-filter. By comparing the linear estimate to that of the median affine filter, it can be seen that the latter behaves exactly like the linear filter whenever no impulses are present. Finally, note that the FIR-WOS and $L\ell$-filters utilize $2N + 1$ and N^2 coefficients, respectively. The median affine filter, in contrast, requires only $N + 1$ coefficients.

3.2.2 Multiresolution Signal Representations

The previous example is a simple illustration that shows the median affine filter is frequency selective and robust. As such, it yields improved results over traditional linear filters. Here we generalize the widely used multiresolution structure, which is based on linear filtering, to a nonlinear decomposition based on median and median affine filters.

Multiresolution representations have been widely accepted as important analysis and processing tools that have been applied to a wide array of problems. The most widely used multiresolution procedures are based on wavelets, which enables the decomposition of a signal into the sum of a lower

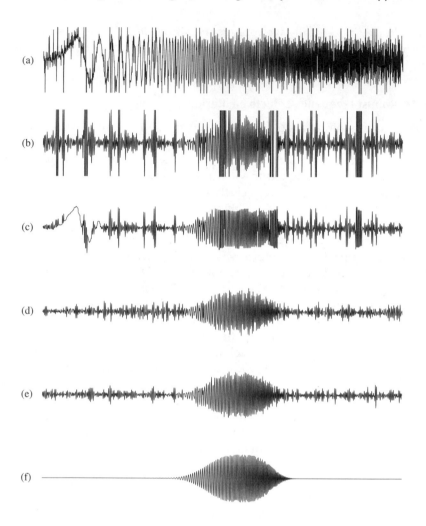

FIGURE 3.1

(a) The noise-corrupted chirp and the result of the (b) linear FIR bandpass, (c) FIR-WOS, (d) $L\ell$, and (e) median affine filters. The desired output is (f) and MSE and MAE results are reported in Table 3.1. (From A. Flaig et al., *IEEE Trans. Signal Process.*, 46, 2101–2112, Aug. 1998. © 1998 IEEE. With permission.)

resolution (or coarse) signal plus a detail signal. Each coarse approximation can, in turn, be decomposed further, yielding a coarser signal and a detail signal at that resolution.[6,7]

It has been shown[8] that the calculations of the wavelet decomposition can be accomplished using the quadrature mirror filters (QMF), Figure 3.2. Within this decomposition let $\text{Wav}_{\text{low}}[\cdot]$ and $\text{Wav}_{\text{high}}[\cdot]$ to be the one-dimensional (1D) lowpass and highpass filtering operations on each of the rows of a

TABLE 3.1

Comparison of Filter Performance on
Noise-Corrupted Chirp Shown in Figure 3.1

Filter	MSE	MAE
Identity	1.07×10^6	23.72
FIR	1.72×10^5	33.05
FIR-WOS	0.0989	0.2028
$L\ell$	0.0205	0.1085
MAFF	**0.0128**	**0.0862**

Note: Bold indicates best performance.

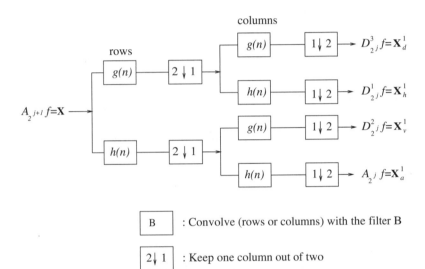

FIGURE 3.2
The quadrature mirror filter multiresolution wavelet decomposition structure. (Reproduced from
M. Asghar and K.E. Barner, *IEEE Trans. Visualization Comput. Graphics*, 7, 76–93, Mar. 2001. ©
2001 IEEE. With permission.)

two-dimensional (2D) data set. If we take $\mathbf{X}$ to be a 2D data matrix, then
$\mathbf{X}$ can be decomposed into its approximate and detail coefficients as follows:

$$\mathbf{X}_a^1 = \left((\downarrow 2)\text{Wav}_{\text{low}}[((\downarrow 2)\text{Wav}_{\text{low}}[\mathbf{X}])^{\text{T}}] \right)^{\text{T}}, \tag{3.1}$$

$$\mathbf{X}_h^1 = \left((\downarrow 2)\text{Wav}_{\text{low}}[((\downarrow 2)\text{Wav}_{\text{high}}[\mathbf{X}])^{\text{T}}] \right)^{\text{T}}, \tag{3.2}$$

$$\mathbf{X}_v^1 = \left((\downarrow 2)\text{Wav}_{\text{high}}[((\downarrow 2)\text{Wav}_{\text{low}}[\mathbf{X}])^{\text{T}}] \right)^{\text{T}}, \tag{3.3}$$

$$\mathbf{X}_d^1 = \left((\downarrow 2)\text{Wav}_{\text{high}}[((\downarrow 2)\text{Wav}_{\text{high}}[\mathbf{X}])^{\text{T}}] \right)^{\text{T}}. \tag{3.4}$$

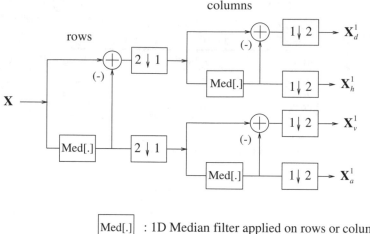

Med[.]   : 1D Median filter applied on rows or columns

$2\!\downarrow 1$: Keep one column out of two

$1\!\downarrow 2$: Keep one row out of two

FIGURE 3.3

The nonlinear filter multiresolution structure used to generate the nonlinear decomposition. (From M. Asghar and K.E. Barner, *IEEE Trans. Visualization Comput. Graphics, 7,* 76–93, Mar. 2001. © 2001 IEEE. With permission.)

In these expressions X_a^1 is the first-level approximation of X, and $X_h^1, X_v^1,$ and X_d^1 correspond to horizontal, vertical, and diagonal details, respectively. It is also possible to perfectly reconstruct the original signal from the approximation and detail coefficients.

The multiresolution structure can be relaxed to allow decompositions based on nonlinear filters. As a simple modification, a 1D median filter can be used in the multiresolution structure shown in Figure 3.2. In this case, the lowpass filter $h(n)$ is replaced by the median filter. Since there is no highpass median filter, some modification of the multiresolution structure is required. Instead of highpass filtering the data, the detail signal can be obtained by subtracting the median filtered data from the original data before decimation (Figure 3.3). In this case, the level-one decomposed signal and detail signals are given by

$$X_a^1 = \big((\downarrow 2)\, \mathrm{MED}[((\downarrow 2)\, \mathrm{MED}[X]^T)^T]\big)^T, \tag{3.5}$$

$$X_v^1 = \big((\downarrow 2)((\downarrow 2)\, \mathrm{MED}[X]^T - \mathrm{MED}[(\downarrow 2)\, \mathrm{MED}[X]^T])\big)^T, \tag{3.6}$$

$$X_h^1 = \big((\downarrow 2)\, \mathrm{MED}[(\downarrow 2)(X - \mathrm{MED}[X])^T]\big)^T, \tag{3.7}$$

$$X_d^1 = \big((\downarrow 2)((\downarrow 2)(X - \mathrm{MED}[X])^T - \mathrm{MED}[(\downarrow 2)(X - \mathrm{MED}[X])^T])\big)^T. \tag{3.8}$$

Perfect reconstruction is possible from the approximation and detail components only if the data is not decimated after the filtering operation. Decimation, however, is advantageous in many applications as it reduces the size of the data set.

Linear multiresolution techniques tend to smooth the edges in signals while nonlinear techniques, such as median filters, are known to preserve edges. A median filter, however, completely removes details with spatial span smaller than half of the window size. Conversely, linear filters do not take into account the spread of samples, and thus spread the effect of a pulse to its neighboring samples. The median affine filter provides a bridge between these two extremes and provides additional flexibility to the user in that a single parameter is required to specify the degree of filter nonlinearity. This flexibility allows the filter to behave as a linear filter, as a median filter, or as a hybrid filter combining the properties of the linear and nonlinear filters. Employing the median affine filter in the same structure yields

$$\mathbf{X}_a^1 = \left(({\downarrow}\, 2)\, \mathrm{MAFF}[(({\downarrow}\, 2)\, \mathrm{MAFF}[\mathbf{X}])^{\mathrm{T}}]\right)^{\mathrm{T}}, \tag{3.9}$$

$$\mathbf{X}_v^1 = \left(({\downarrow}\, 2)(({\downarrow}\, 2)\, \mathrm{MAFF}[\mathbf{X}]^{\mathrm{T}} - \mathrm{MAFF}[({\downarrow}\, 2)\, \mathrm{MAFF}[\mathbf{X}]^{\mathrm{T}}])\right)^{\mathrm{T}}, \tag{3.10}$$

$$\mathbf{X}_h^1 = \left(({\downarrow}\, 2)\, \mathrm{MAFF}[({\downarrow}\, 2)(\mathbf{X} - \mathrm{MAFF}[\mathbf{X}])^{\mathrm{T}}]\right)^{\mathrm{T}}, \tag{3.11}$$

$$\mathbf{X}_d^1 = \left(({\downarrow}\, 2)(({\downarrow}\, 2)(\mathbf{X} - \mathrm{MAFF}[\mathbf{X}])^{\mathrm{T}} - \mathrm{MAFF}[({\downarrow}\, 2)(\mathbf{X} - \mathrm{MAFF}[\mathbf{X}])^{\mathrm{T}}])\right)^{\mathrm{T}}. \tag{3.12}$$

The digital elevation model (DEM) data used to illustrate the performance of the multiresolution techniques are from the Lake Charles data set. The actual size of this data set is 1200×1200. However, for illustration purposes only certain sections of the data set are rendered at one time. A size 150×150 portion of the data set is shown in Figure 3.4. The approximations of the Lake Charles data set produced by the level-one biorthogonal wavelet, median filter, median affine filter, and scale invariant median affine filter[9] decompositions are shown in Figure 3.5. As the figure shows, the original data set has a number of sharp edges that are smeared in the wavelet decomposed surface. In comparison, the median-based decomposition contains edges that preserve sharp transitions. The hybrid behavior of the median affine filter produces a decomposition that smoothes fine details while preserving edge integrity.

The differences in the behaviors of the multiresolution techniques are more marked in the level-two decomposed approximations, which are shown in Figure 3.6. The figures show that the median-based multiresolution decomposition perfectly preserves large-scale edge features while completely removing small details. At the other extreme, the wavelet decomposition significantly smoothes fine details as well as large-scale edge features. The median affine decomposition can be varied between these limiting cases through control over the spread parameter γ. This hybrid response allows edges to be well

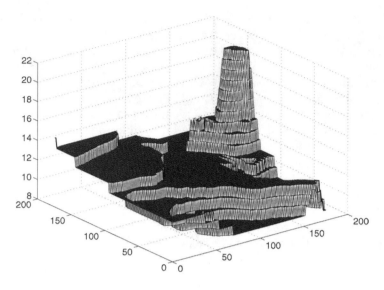

FIGURE 3.4
Original Lake Charles (150×150) DEM data set. (From M. Asghar and K.E. Barner, *IEEE Trans. Visualization Comput. Graphics, 7,* 76–93, Mar. 2001. © 2001 IEEE. With permission.)

preserved in the decomposition while simultaneously smoothing fine details, as the figures show.

In determining the behavior of a decomposition method, it is instructive to examine the detail signals, $\mathbf{X}_h^i$, $\mathbf{X}_v^i$, and $\mathbf{X}_d^i$. Utilizing the multiresolution structure in Figure 3.3, the level-one detail signals of the Lake Charles decomposition are shown in Figure 3.7. The figure shows that the detail signals in the linear decomposition contain significant power due to the edge smoothing of the wavelet transform. In contrast, the edge-preserving properties of the median filter are evident in the identically zero detail coefficients along most edges in the median decomposition. The detail signals show that the median affine decomposition has better edge preservation character than the linear decomposition, while also smoothing details in a more desirable fashion than the median.

As a final example, consider the case in which the observed data set is corrupted by independent, additive noise. Specifically, let 10% of the samples in the Lake Charles data set be corrupted by zero mean, unit variance Gaussian noise. The additive noise can either be attributed to the sensing, transmission, and storage of the data, or can be interpreted as small-scale details. In either case, the noise terms should be significantly attenuated in the decomposition approximations. Figure 3.8 shows a realization of the corrupted Lake Charles data set and the resulting level-one approximations for the wavelet, median, and scale invariant median affine decompositions. As the figure shows, the wavelet decomposition spreads the noise terms across neighboring samples

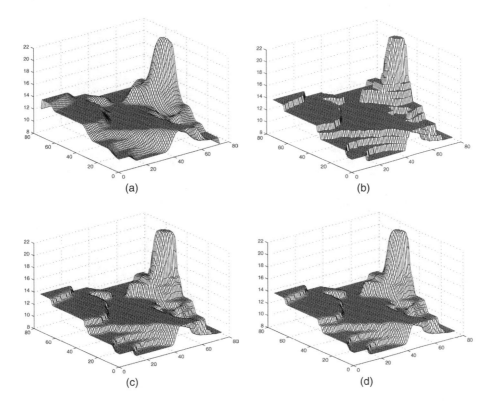

FIGURE 3.5
Level-one decomposition of Lake Charles DEM using (a) biorthogonal wavelet, (b) median, (c) median affine, and (d) scale invariant median affine methods. (From M. Asghar and K.E. Barner, *IEEE Trans. Visualization Comput. Graphics*, 7, 76–93, Mar. 2001. © 2001 IEEE. With permission.)

while only modestly reducing the magnitude. The median decomposition, in contrast, completely eliminates the noise terms. The noise is nearly eliminated in the median affine decomposition, resulting in an approximation with a slight texture indicating the presence of the noise/details in the observed data set. Moreover, the large-scale edges are well preserved in the median affine decomposition, yielding an approximation with the most desirable characteristics.

These examples illustrate that the QMF decomposition structure can be easily extended to incorporate nonlinear filters. Furthermore, the median affine-based decomposition has advantages over linear- and median filter-based decompositions of clean and noise-corrupted data sets.[9,10] In particular, the median affine-based decomposition is controlled by a single parameter that governs the level of detail smoothing, resulting in a flexible multiresolution decomposition that is effective in smoothing fine detail while simultaneously preserving large-scale features and sharp edge transitions.

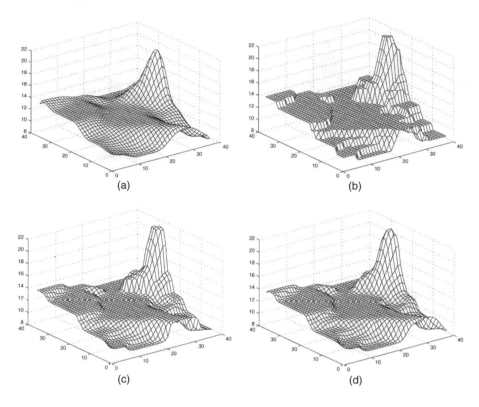

FIGURE 3.6
Level-two decomposition of Lake Charles DEM using (a) biorthogonal wavelet, (b) median, (c) median affine, and (d) scale invariant median affine methods. (From M. Asghar and K.E. Barner, *IEEE Trans. Visualization Comput. Graphics*, 7, 76–93, Mar. 2001. © 2001 IEEE. With permission.)

3.2.3   Inverse Synthetic Aperture Radar Image Filtering

The previous examples covered several illustrative cases in which affinity to the median plays a critical role. We now consider several examples in which affinity to the central observation sample plays an important role and leads to improved performance. We start by considering the filtering of Inverse Synthetic Aperture Radar (ISAR) images. ISAR images have attracted increasing interest in the context of target classification as a result of their high resolution.[11] Such images emerge from the mapping of the reflectivity density function of the target onto the range-Doppler plane. Difficulties in target identification arise from the fact that radar backscatters from the target are typically embedded in heavy clutter noise. For proper target classification, it is therefore necessary to remove the noise without altering the target backscatters. From a signal processing point of view, ISAR images constitute nonstationary 2D signals. The target backscatters are represented as pulselike features in non-Gaussian noise.

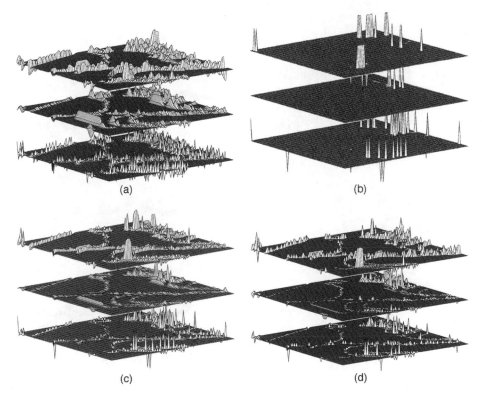

(a) (b)

(c) (d)

FIGURE 3.7
Level-one horizontal (top), vertical (middle), and diagonal (bottom) detail signals for the Lake Charles data set. Decomposition method: (a) biorthogonal wavelet, (b) median, (c) median affine, and (d) scale invariant median affine. (From M. Asghar and K.E. Barner, *IEEE Trans. Visualization Comput. Graphics*, 7, 76–93, Mar. 2001. © 2001 IEEE. With permission.)

We compare the performance of a center affine filter with that of a WOS-filter and an $L\ell$-filter on the ISAR image* depicted in Figure 3.9a. This image is a 128×128, 8 bits/pixel intensity image of a B-727. A pair of synthetic images, Figure 3.9b, is used to optimize each of the filters. Figure 3.10 shows the $L\ell$, WOS, and center affine filtering outputs and errors. An examination of the figure shows that the WOS-filter eliminates the noise well, but blurs details of the plane. The $L\ell$-filter preserves the plane much better, but is not very effective in removing the clutter noise. The center affine filter removes the background noise to a large extent while preserving the plane and all its details. The superior performance of the center affine filter can be explained by its affinity-based sample preference. While operating in the background noise, all observations are close to the center sample. Thus, the center

---

*Data provided by Victor C. Chen, Airborne Radar, Radar Division, Naval Research Laboratory, Washington, D.C. 20375.

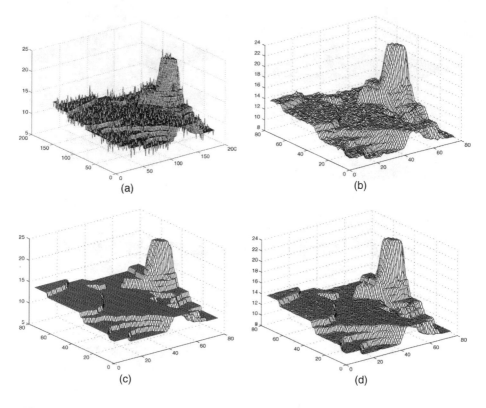

FIGURE 3.8
Level-one approximations of the (a) Lake Charles data set with 10% of samples corrupted by Gaussian noise. Decomposition methods: (b) biorthogonal wavelet, (c) median, and (d) scale invariant median affine. (From M. Asghar and K.E. Barner, *IEEE Trans. Visualization Comput. Graphics*, 7, 76–93, Mar. 2001. © 2001 IEEE. With permission.)

affine filter behaves like an L-filter, smoothing the clutter noise. When encountering backscatter from the plane, the center affine filter considers only those pixels with intensity similar to the center sample, thus preserving the plane details.

3.2.4   Image Deblocking

Another important imaging problem in which the central observation plays a key role is image deblocking. Blocking artifacts are common forms of interference in images that have been compressed, or otherwise processed, in a block-wise fashion. Such artifacts are recognized shortcomings of discrete cosine transform (DCT) based compression algorithms, such as the widely used JPEG method. In DCT coded images, two types of blocking artifacts are prominent: (1) grid noise in monotone areas and (2) ringing artifacts around strong edges. Grid noise is due to the independent application of the DCT

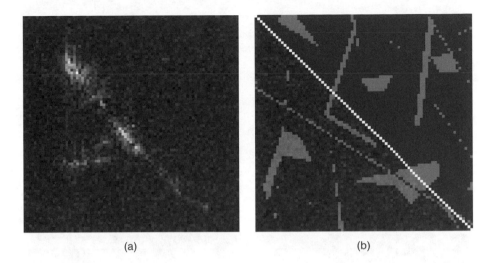

(a)   (b)

FIGURE 3.9
(a) ISAR image of B-727 and (b) portion of synthetic original and noisy training images used for optimization.

to adjacent blocks, while the ringing artifacts are attributable to Gibbs phenomenon and caused by the lossy quantization of frequency coefficients.

Many spatial and transform domain deblocking techniques have been developed.[12] However, most techniques either (1) have high computational complexity or (2) are limited to addressing certain types of artifacts. It is important to note that blocking artifacts manifest as relatively weak noisy edges. The center affine filter can thus be used to smooth these artifacts while simultaneously sharpening the stronger true edges. Additionally, the complexity of the center affine filter is only slightly greater than that of the linear filter and significantly less than most deblocking algorithms.

To evaluate the ability of the center affine filter to reduce blocking artifacts, we apply the method to JPEG images with various quality factors (QF). The QF maps to a scaling parameter of the standard quantization table to control the compression ratio. The range of QF is from 1 to 100, where 100 indicates no quantization. Note that different compression ratios lead to different noise, or artifact, levels. The center affine filter membership function and spread parameter can be optimized for a specified compression level. Here we use a Gaussian membership function and uniform spatial weights for all cases, but allow the spread parameter to vary with the compression level.

We compare the center affine method with that proposed in Reference 13, which we denote as the "2Mod" method. This provides a fair comparison as both methods have similar computational complexity and the 2Mod method is adopted in the MPEG-4 standard. The size of all test images is 256 × 256. For each QF, we use Lenna as the training image to optimize the spread parameter

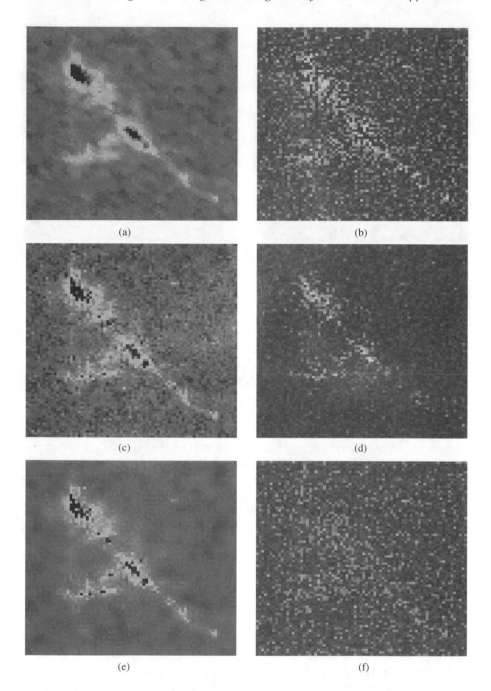

FIGURE 3.10

Filter output and difference image for filters operating on the ISAR image in Figure 3.9a. WOS filter (a) output and (b) difference image, $L\ell$-filter (c) output and (d) difference image, and center affine filter (e) output and (f) difference image. (From A. Flaig et al., *IEEE Trans. Signal Process.*, 46, 2101–2112, Aug. 1998. © 1998 IEEE. With permission.)

TABLE 3.2

Comparison of Deblocking Results

Image (QF)	Bitrate (bpp)	Input (dB)	CAFF (dB)	2Mod (dB)
Lenna(30)	0.69	30.33	**30.84**	30.55
Lenna(20)	0.54	29.19	**29.72**	29.44
Lenna(10)	0.37	27.23	**27.95**	27.72
Lenna(7)	0.31	26.12	**26.99**	26.77
Pepper(30)	0.73	30.75	**31.42**	31.02
Pepper(20)	0.58	29.51	**30.36**	29.88
Pepper(10)	0.39	27.36	**28.51**	27.97
Pepper(7)	0.32	26.09	**27.32**	26.87
Clock(30)	0.52	32.99	**33.81**	33.32
Clock(20)	0.43	31.53	**32.42**	31.92
Clock(10)	0.32	28.75	**29.71**	29.37
Clock(7)	0.27	27.36	**28.27**	28.10

Note: The PSNR values are listed and bold indicates the best performance.

and utilize the resulting spread parameter to deblock the other images. The peak SNR (PSNR) performance for each method and QF are listed in Table 3.2. In all cases, the center affine method produces consistent PSNR improvement over the 2Mod method. This improvement can be clearly seen in the resulting images, which are shown in Figure 3.11 for the Lenna and Pepper cases. Note that both the grid noise, e.g., in the hat surface, and ringing artifacts and staircase noise, e.g., along the hat edges, are successfully removed by the center affine method. Moreover, this method preserves the original image edges. The 2Mod method, in contrast, does not remove the ringing artifacts. These points are clearly seen in the image enlargements.

3.2.5 Time-Frequency Cross-Term Filtering

The central observation sample also plays an important role in time-frequency plane filtering. Specifically, we consider here cross-term filtering of the Wigner distribution (WD).[14] Motivating this application is the fact that quadratic time-frequency representations (TFRs) are powerful tools for the analysis of signals.[15] Among quadratic TFRs, the WD, which for continuous time is given by

$$WD_x(t, f) = \int_\tau x\left(t + \frac{\tau}{2}\right) x^*\left(t - \frac{\tau}{2}\right) e^{-j2\pi f\tau} d\tau, \tag{3.13}$$

satisfies a number of desirable mathematical properties and features optimal time-frequency concentration.[16–20] By using the WD, subtle signal features may be detected, especially for signals having short length and high time-frequency variation. However, despite the desirable properties of the WD, its use in practical applications has often been limited by the presence of

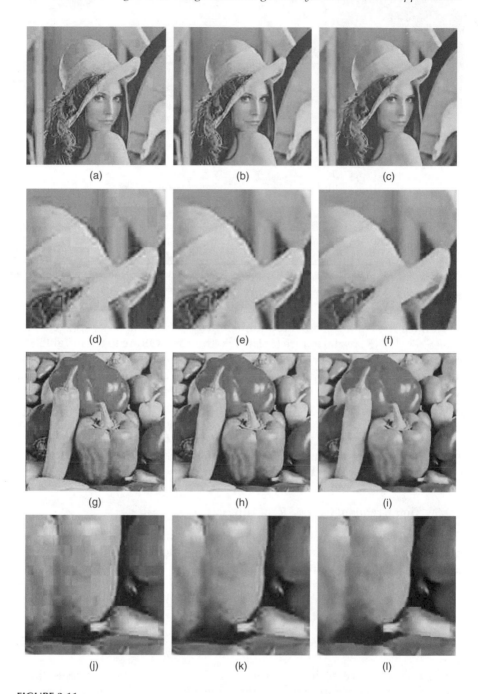

FIGURE 3.11
JPEG deblocking results: From left to right, JPEG image and 2Mod and center affine filter outputs.
(a to c): Lenna image; (d to f): Lenna image enlargements; (g to i): peppers image; (j to l): peppers
image enlargements.

cross terms. For instance, the WD of the sum of two signals $x(t) + y(t)$

$$WD_{x+y}(t, f) = WD_x(t, f) + 2\Re(WD_{x,y}(t, f)) + WD_y(t, f) \tag{3.14}$$

has a cross-term $2\Re(WD_{x,y}(t, f))$ in addition to the two autocomponents, where the cross WD is defined as

$$WD_{x,y}(t, f) = \int_{-\infty}^{+\infty} x\left(t + \frac{\tau}{2}\right) y^*\left(t - \frac{\tau}{2}\right) e^{-j2\pi f \tau} d\tau. \tag{3.15}$$

The cross or interference terms are often a serious problem in practical applications, especially if a WD outcome is to be visually analyzed by a human analyst. Cross terms lie between two autocomponents and are oscillatory, with their frequencies increasing with increasing distance in time-frequency between the two autocomponents.[14,19,21] For real-valued bandpass signals, the WD of the analytic signal is generally used because removal of the negative frequency components also eliminates cross terms between positive and negative frequency components. Although helpful in eliminating cross terms, cross terms between multiple components in the analytic signal still make interpretation difficult. Because cross terms have oscillations of relatively high frequency, they can be attenuated by means of a smoothing operation that corresponds to the convolution of the WD with a 2D smoothing kernel. For most methods, the smoothing tends to produce the following effects:

1. A (desired) partial attenuation of the interference terms

2. An (undesired) broadening of signal terms, i.e., a loss of time-frequency concentration

3. A (sometimes undesired) loss of some of the mathematical properties of the WD

The interference terms (ITs) of the WD are a consequence of the WD bilinear (or quadratic) structure. That is, they occur in the case of a multicomponent signal and can be identified mathematically with quadratic cross terms.[15,16,19] According to the quadratic superposition law, the WD of an N-component signal consists of N signal terms and $[N(N-1)]/2$ ITs. Each signal component generates a signal term, and each pair of signal components generates an IT. While the number of signal terms grows linearly with the number N of signal components, the number of ITs grows quadratically with N.[19] This is shown in Figure 3.12.

The goal of filtering in the WD plane is thus to reduce, or eliminate, the cross terms while preserving the autoterms. To achieve this result, a center affine filter can be applied to the WD. To take advantage of the structure of the filtering problem, the following modifications can be made to the standard center affine filtering operation:[14]

1. The absolute values of the samples, $|W_{m_1,m_2}|$, rather than their actual values, are utilized to calculate the respective affinities. Accordingly,

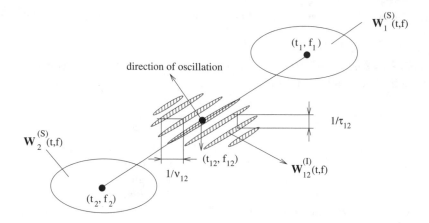

FIGURE 3.12

Interference geometry of Wigner distribution. (From G.R. Arce and S.R. Hasan, *IEEE Trans. Signal Process.*, 48, 2321–2331, Aug. 2000. © 2000 IEEE. With permission.)

the reference point of the affinity function is set as the absolute value of the center-sample, $|W_c|$.

2. The affinity spread parameter γ is made proportional to the local variance at the observation window location. Thus, a higher value of γ is obtained, in general, for an observation window centered in a cross-term region than for a window centered in an autoterm region.

While modification 1 may not affect the affinities of samples in the autoterm region (as most of the samples are already positive), it drastically changes the affinities of samples in the cross-term region. Specifically, both positive and negative samples have higher affinities, resulting in more samples contributing to the estimate and trending the filter to have linear lowpass characteristics. Condition 2 further ensures that more samples (including both positive and negative samples) receive higher affinities in the cross-term region while assigning very low affinities to most of the samples in the autoterm region. This scenario is depicted in Figure 3.13.

To evaluate the cross-term reduction, we consider the special case of a signal composed of the sum of time-frequency-shifted signals $x(t) = c_1 x_1(t) + c_2 x_2(t)$ where $x_i(t) = x_0(t - t_i)e^{j2\pi f_i t}$, for $i = 1, 2$. Consider the case where $x_0(t)$ is a Gaussian atom. Interest in a Gaussian-shaped atom derives from the fact that it has good locality (energy concentration) in the joint $t - f$ domain, and its WD is nonnegative.[22] Figure 3.14 demonstrates the center affine filtered WD of the signal composed of two Gaussian atoms.

To demonstrate the relative performance of the center affine filtered WD, a comparison of the experimental results provided by several time-frequency representations is shown. Consider first a 128-point test signal comprised of

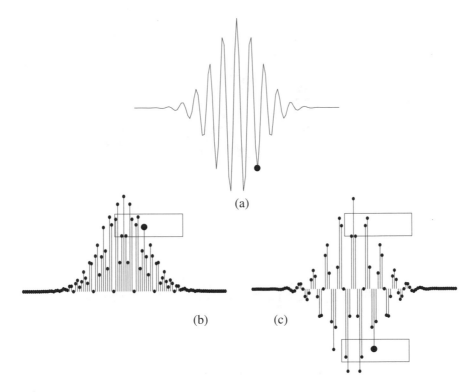

(a)

(b)   (c)

FIGURE 3.13
Nonlinear filtering mechanism of a sample in a cross-term component: (a) center sample point (dot) to be filtered in a cross term; (b) samples within square window represent samples affine to the center sample (absolute values); (c) filtering of original samples according to their affinities (samples within the two square windows enter in the filtering operation). (From G.R. Arce and S.R. Hasan, *IEEE Trans. Signal Process.*, 48, 2321–2331, Aug. 2000. © 2000 IEEE. With permission.)

one parabolic chirp, one sinusoidal pulse, and two parallel Gaussian pulses:

$$x(n) = r_{1,64}(n)\left(e^{-\frac{j\pi n}{8}} + e^{-\frac{j3\pi}{8}(n+0.5n^2+0.33n^3)}\right) + r_{75,124}(n)e^{-\frac{(n-99)^2}{100}}\left(e^{\frac{j\pi n}{4}} + e^{\frac{j3\pi}{4}}\right)$$

(3.16)

with $r_{a,b}(n)$ the gating function

$$r_{a,b}(n) = \begin{cases} 1, & a \leq n \leq b \\ 0, & \text{otherwise.} \end{cases}$$

(3.17)

Consider the results first reported in Reference 14. Figure 3.15a shows the WD of the signal having autocomponents well localized but numerous high-amplitude oscillating cross terms. Figure 3.15b shows the pseudo smoothed WD (using a 13-point Gaussian time smoothing window and 31-point Gaussian frequency smoothing window) of the test signal, which reduces the cross terms by both frequency and time direction smoothing.

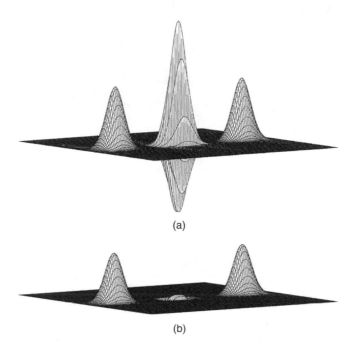

FIGURE 3.14

(a) Discrete WD of a signal composed of two Gaussian atoms and (b) center affine filtered distribution.

The interpretation of this signal is much easier, but the component localization becomes coarser. Figure 3.15c shows the Choi–Williams distribution of the given signal with kernel width $\sigma = 1$. Again, most of the cross terms are minimized at the cost of reduced localization. Obvious problems are also visible, especially when the signal components overlap in time and frequency. Figure 3.15d shows the results of the Baraniuk–Jones method-1.[23] This method fails to track the smoothly varying parabolic chirp and, as a result, that component appears as two connected linear chirps. Figure 3.15e shows the representation given in Reference 24. In this scheme a time-adaptive radial Gaussian kernel is used. The results are adequate, although some loss in autocomponent localization occurs. In addition, the computation cost is high as the kernel is computed in a local window sliding over the signal. Figure 3.15f shows the center affine filtered TFR. An almost complete reduction in cross terms is attained without losing the resolution and localization provided by the WD.

As a second example, consider a signal consisting of three Gaussian atoms: $B(t - T) + B(t) + B(t + T)$. The two outer atoms generate cross terms overlapping with the center atom. As illustrated in Figure 3.16a, the local variance of the WD around the center Gaussian atom is approximately equal to the local variance of the cross-term-only regions. To illustrate the interference term canceling ability of the center affine filter with respect to the standard

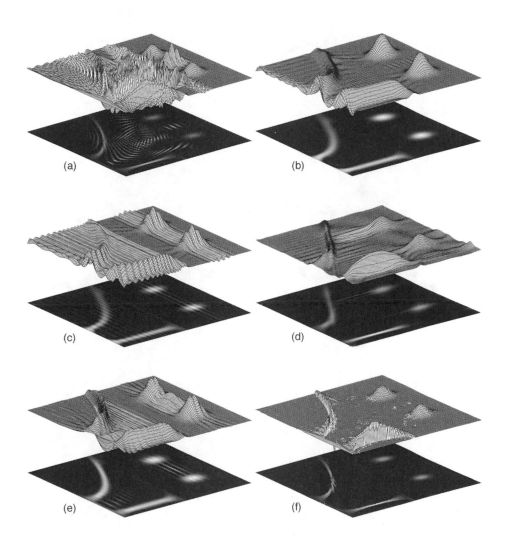

FIGURE 3.15

Time-frequency representation of the signal given by Equation 3.16 using (a) WD, (b) pseudo smoothed WD, (c) Choi-Williams distribution with spread factor $\sigma = 1$, (d) Baraniuk-Jones distribution (Method 1), (e) Baraniuk–Jones (method 2), and (f) center affine filtering. (From G.R. Arce and S.R. Hasan, *IEEE Trans. Signal Process.*, 48, 2321–2331, Aug. 2000. © 2000 IEEE. With permission.)

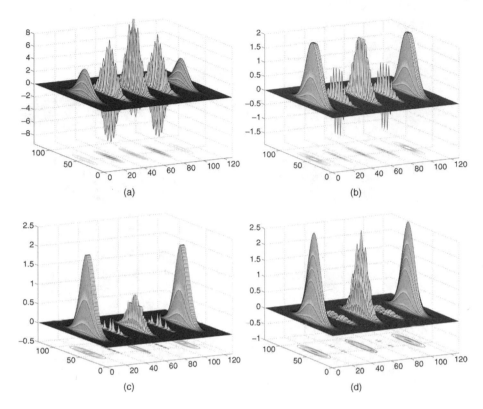

FIGURE 3.16

(a) WD of three Gaussian atoms and the (b) a 7 × 7 median, (c) 9 × 9 median, and (d) center affine filter outputs. (From G.R. Arce and S.R. Hasan, *IEEE Trans. Signal Process.*, 48, 2321–2331, Aug. 2000. © 2000 IEEE. With permission.)

median filter, we apply each filter to the original WD. The results for the 7 × 7 median, 9 × 9 median, and center affine filters are shown in Figure 3.16b to d. As the figure shows, the result of applying a median filter to the WD is not adequate because, although the cross terms are somewhat reduced, significant distortion is introduced into the autoterm components. The center affine filter, in contrast, significantly reduces the cross terms while preserving the autoterms.

As a final practical example, consider the processing of protein sequence data.[25] The amino acid sequence of protein may be considered as a 20-symbol alphabet sequence, or it may be considered a sequence of numerical values reflecting various physiochemical aspects of the amino acids, such as hydrophobicity, bulkiness, or electron–ion interaction potential (EIIP).[26] Analysis of the numerical representation of sequences often identifies characteristic patterns that may be too weak to be detected as patterns of symbols.[27] It has been shown[27,28] that proteins of a given family have a common characteristic frequency component related to their function. However, because frequency analysis alone contains no spatial information, there is no indication regarding which residues contribute to the frequency components. Through the use of the WD, information involving secondary structure and biologically active sites can be retained.

This result is illustrated using the fibroblast growth factors class of proteins.[25] Fibroblast growth factors constitute a family of proteins that affect the growth, migration, differentiation, and survival of certain cells. The amino acid sequence of a basic fibroblast growth factor (FGF) for humans is 155 amino acids (or residues) long. The amino acid representation of the sequence can be converted from an alphabetic string to a numerical signal by using the EIIP for each amino acid in the sequence.

The Resonant Recognition Model (RRM)[27] is a physicomathematical model that analyzes the interaction of a protein and its target using digital signal processing methods. One application of this model involves prediction of a protein's biologically active sites. In this technique, a Fourier transform is applied to the numerical protein sequence and a characteristic peak frequency is determined for a particular protein's function. What is lacking in this method is the ability to reliably identify the individual amino acids that contribute to that peak frequency. The time-frequency representation (TFR) of the amino acid sequence, however, preserves both the frequency information along with the spatial relationships. The discrete WD TFR of the human basic FGF is shown in Figure 3.17. As illustrated in the previous examples, cross terms make interpretation of this representation difficult. After application of the center affine filter, Figure 3.17b, cross terms are minimized and primary terms are retained. The retained terms correspond to the activation sites and characteristic frequency component. Indeed, experiments have shown that the potential cell attachment sites of FGF are between residues 46–48 and residues 88–90. The characteristic frequency has been shown in the literature as 0.4512,[29] which is in agreement with the presented results.[25]

3.3 Fuzzy Ordering and Fuzzy Median Filter Applications

Fuzzy ordering and fuzzy order statistic-based methods can also be applied to a wide range of problems. Here we consider several representative applications in which the fuzzy weighted median and fuzzy ordering show advantages over their traditional crisp counterparts. Specifically, the fuzzy weighted median filter is applied to image smoothing (color and gray scale), surface smoothing, image zooming, and noisy image sharpening. In the image-smoothing case, the use of fuzzy samples in the weighted median structure leads to improved performance, especially when background noise with a Gaussian distribution is considered. In such cases, the fuzzy weighted

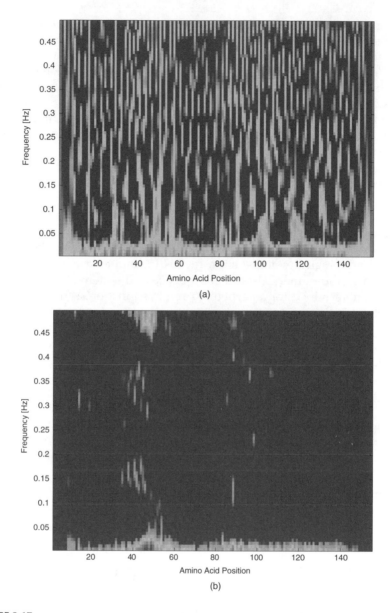

FIGURE 3.17
(a) WD TFR of basic FGF human and (b) the center affine-filtered representation indicating the activation sites.

median preserves image details, such as edges, while removing outlying samples and appropriately smoothing background noise. Similar results hold for surface smoothing, which is becoming an increasingly important problem with the growing utilization of three-dimensional (3D) models in engineering, medical, and entertainment applications. The fuzzy weighted median also shows improved results over traditional linear and crisp weighted median filters in image zooming, or interpolation. Interpolations based on fuzzy methods show improved edge representations while avoiding blocking artifacts in uniform regions. Presented results also indicate that fuzzy methods have similar advantages in the sharpening of noisy images. Last, fuzzy ranking is utilized in a multiuser detection scheme. The detection results show that the fuzzy ranking concept yields improved results compared to crisp ranking schemes, which do not consider the spread of samples.

3.3.1   Image Smoothing

Noise smoothing is an important component in many applications. This is particularly true in image processing, where sensor irregularities, atmosphere interference, or transmission/storage errors often introduce noise into the image capture, transmission, or display processes. Additionally, images are nonstationary signals within which important visual cues are represented by sharp transitions, such as edges. Consequently, linear filters do not yield satisfactory results in image noise filtering applications. The development of nonlinear image-smoothing filters has thus become an active area of research.

The median filter and its generalizations have, perhaps, gathered the most attention in addressing noise smoothing in images.[1,30-37] The weighted median filter has been widely researched as an effective filter for removing noise, especially noise with heavy-tailed distributions. The median and weighted median filters have not, however, shown good performance in processing Gaussian noise-contaminated images. The median and weighted median filter structures can be effectively utilized to process a wide range of corrupting noise processes if the more general fuzzy realizations are utilized.

The performance gain realized through the fuzzy generalizations is evaluated here for the fuzzy median and fuzzy weighted median filter cases.[38] The filters are compared to their crisp counterparts in a noise-smoothing application where we consider two cases: an image corrupted by varying levels of (1) impulsive noise and (2) contaminated Gaussian noise. The presented results utilize the images Lenna and Albert, Figure 3.18. Corrupted versions of the Lenna image are used as the input to each of the filters considered. Optimization is performed utilizing Lenna and noise with the same statistics as the observation image. The robustness of the algorithms, with respect to changing image statistics, is demonstrated by applying filters optimized on Albert to the image Lenna. In all cases, a 5×5 observation window is utilized.

Consider first the case of Salt and Pepper impulsive noise, where the probability of noise contamination is p and varies from 0.05 to 0.45. Figure 3.19a shows the MAE for each of the filters operating on the impulse-corrupted versions of Lenna as a function of the noise contamination probability. As the figure shows, the fuzzy and crisp weighted median filters have nearly identical

(a)   (b)

FIGURE 3.18
Two 512 × 512 8-bit/pixel gray scale images: (a) Lenna and (b) Albert.

performance. This similarity in performance is a result of the optimization that yields a membership spread parameter of $\lambda = 0.01$, which is very small compared to the range of signal values, [0, 255]. Thus, in the impulsive noise case, the optimal weighted median filter utilizes crisp order statistics as the output. This is an intuitive result because in the impulsive case, samples are either unaltered or bear no information. Thus, if the filter is centered on a sample that is not corrupted, it is best to simply perform the identity operation and weighted averaging has no advantage. On the other hand, if the filter is centered on a corrupted sample, then this sample contains no information about the original signal, and the best output selection is a crisp order statistics that can preserve local structure. Thus, there is no advantage to averaging samples and the optimization procedure correctly yields a crisp filter. Similar results are obtained in the median case.

Consider next the performance of the filters operating in a contaminated Gaussian noise environment. In this case, the observation image is corrupted by $\Phi(10, 100, \epsilon)$ contaminated Gaussian noise, where ϵ indicates mixing proportion of two Gaussian processes with variance 10 and 100. The MAE performance of each of the filters operating in the contaminated Gaussian environment is shown in Figure 3.19b as a function of the contamination parameter, for ϵ in the range 0 to 0.45. In this case there is a clear separation in the performance of the crisp and fuzzy filters. Selected weighted averaging among similarly valued samples is advantageous in this case because there is always background Gaussian noise, regardless of the mixture parameter value. The figure shows a fairly consistent performance advantage for the fuzzy filters across the entire range of contamination. The performance gain is primarily due to two factors: the fuzzy versions of the filters (1) smooth the background

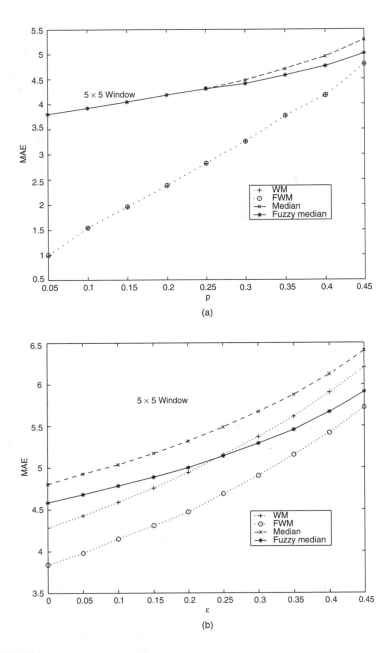

FIGURE 3.19

MAE for the median, fuzzy median, weighted median, and fuzzy weighted median filters operating on the image Lenna corrupted by (a) impulsive noise with impulse probability p and (b) $\Phi(10, 100, \epsilon)$-contaminated Gaussian noise.

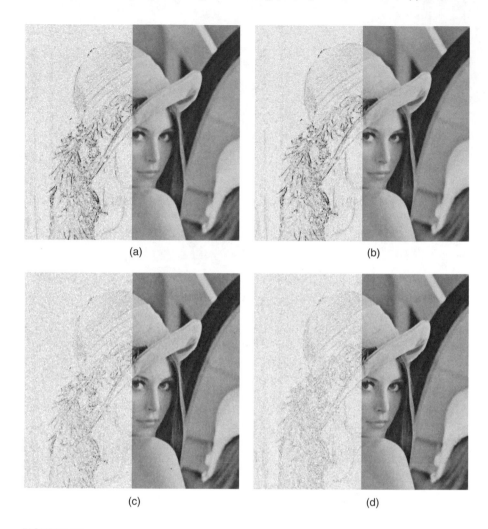

FIGURE 3.20
Filtered image and the scaled difference for Lenna processed by the (a) median (MAE $= 4.8028$), (b) fuzzy median (MAE $= 4.5732$), (c) weighted median (MAE $= 4.2729$), and (d) fuzzy weighted median (MAE $= 3.8259$) filters.

noise more effectively and (2) preserve edges better in the presence of additive noise than their crisp counterparts.

A subjective evaluation of the filter outputs also illustrates the superior performance of the fuzzy filter generalizations. Figure 3.20 shows the output images for each of the filters in the $\Phi(10, 100, 0)$ case. The right half of each image in the figure shows the output of a particular filter while the left half is an error image. As the images show, the fuzzy filter generalizations yield superior noise smoothing performance while simultaneously preserving more image structure than the crisp filter realizations.

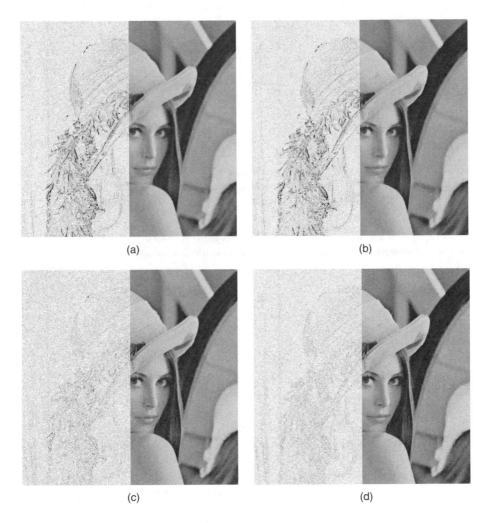

FIGURE 3.21
Filtered image and the scaled difference for Lenna processed by the (a) median (MAE = 4.8028), (b) fuzzy median (MAE = 4.5930), (c) weighted median (MAE = 4.7503), and (d) fuzzy weighted median (MAE = 4.0469) filters. Filter optimizations are performed on the image Albert.

To evaluate the robustness of the filters to changing image statistics, we consider the filtering of the image Lenna by filters optimized on the image Albert. Figure 3.21 shows the results of this operation for the same noise case as considered in Figure 3.20. A comparison of the two figures shows that little loss in performance is realized by optimizing on a different image. In fact, although the resulting MAE is slightly higher in this case, the results for the Albert optimized filters appear visually superior. This is most likely due to the fact that Albert contains a significant number of edges and, consequently, the optimization yields filters that are designed to place a high priority on edge preservation. When applied to the Lenna image, this results in outputs with superior edge preservation compared to the Lenna optimized filters, which place a somewhat reduced priority to edge preservation due to the fairly uniform nature of the Lenna image.

3.3.2   Color Image Smoothing

Image smoothing is also an important problem in color image processing. Indeed, color images are now ubiquitous from professional to recreational and home use. Color images can be represented in various color spaces, such as CIE XYZ, CIE LUV, CIE LAB, RGB, and CMYK, among others. The RGB color space is, by far, the most broadly adopted hardware representation due to its correspondence to the tristimulus theory. In the RGB color space, each pixel is represented by three components, corresponding to red, green, and blue, respectively. Thus a color image in RGB space is a multivariate signal in $\Re^3$.

Numerous generalizations or extensions of median-type filters have been developed to address noise smoothing in color images. These methods include the vector median[39] using the L_2 ($\text{MED}_v(L_2)$) and L_1 ($\text{MED}_v(L_1)$) norms, vector directional filters (VDF),[40] and directional distance filters (DDF).[41] These methods can be compared to the fuzzy vector median (FMED_v) as well as marginal processing of each color component by the MED and FMED filters.[42] As an illustration of a color imaging application, we consider the filtering of images corrupted by component-independent and component-correlated contaminated Gaussian noise, $N(20, 300, 0.1)$. For each filter, a 5×5 observation window is utilized. Results are presented for the widely utilized balloon image. The component-wise MSE and MAE (CMSE and CMAE) and normalized MSE (NMSE) quantitative error measures are evaluated and processed images are presented for subjective viewing.

The performance of the compared methods under the various error criteria is reported in Table 3.3 for the independent and correlated noise cases. Bold entries indicate the best performance. The tabulated results indicate that the fuzzy methods yield the best performance. The results also indicate that marginal processing is often sufficient for component-independent noise, but that multivariate approaches produce the best results for correlated corruption. This is an intuitive result as the multivariate approaches can exploit correlations in both the underlying signal and the additive noise. The tabulated results can be confirmed through a visual inspection of the images, which are shown in Figure 3.22 and Figure 3.23 for the independent and correlated noise cases, respectively. Inspection of the images shows that the affinity-based fuzzy approaches, in comparison to the other median-based methods, yield smoother uniform regions *and* improved detail retention. Additional results can be found in Reference 42.

TABLE 3.3

Color Image Denoising Results for Balloon Image Corrupted
by Component Independent and Correlated Gaussian Noise

Method	CMSE	NMSE	CMAE
Independent			
Noisy	1760	0.1083	24
MED	51	0.0032	5.3
$MED_v(L_2)$	94	0.0058	7.5
$MED_v(L_1)$	91	0.0056	7.4
VDF	91	0.0056	7.1
DDF	131	0.0081	8.7
FMED	**41**	**0.0026**	**4.7**
$FMED_v$	46	0.0029	5.1
Correlated			
Noisy	1765	0.1086	24
MED	51	0.0031	5.3
$MED_v(L_2)$	52	0.0032	5.4
$MED_v(L_1)$	53	0.0033	5.4
VDF	73	0.0045	6.4
DDF	67	0.0041	6.1
FMED	41	0.0026	**4.7**
$FMED_v$	**40**	**0.0025**	**4.7**

Note: The MED and FMED filters are applied marginally to each component and all other methods operate directly on the multivariate data. Bold indicates best performance.

3.3.3 Surface Smoothing

Multivariate processing is also integral to surface smoothing. The importance of this problem is growing with advances in computing power and 3D acquisition technology as there has been increasing deployment of 3D models in engineering, medical, and entertainment applications. These 3D models are usually stored and rendered as surfaces represented by polygonal, primarily triangular, meshes. It is important to note that in all stages of the 3D model construction process noise is inevitably introduced, due, for example, to measurement errors, sampling resolution limitations, algorithmic errors, etc. This is also true for 3D medical images reconstructed from computed tomography (CT) or magnetic resonance imaging (MRI) volumetric data. Surface smoothing, or denoising, adjusts vertex positions so that the overall surface becomes smoother while keeping mesh connectivity, or topology, unchanged. Surface smoothing is an active area of research.

To consider the surface smoothing problem, the triangular surface representation must first be defined. A triangular mesh is characterized by its topology and geometry. The topology is specified by a set of vertices V, a set of edges E, and a set of faces or triangles F. The symbols V, E, and F are also used to denote the cardinality of corresponding sets. An individual vertex is denoted as $v = \{i\}$, while $e = \{i, j\}$ and $f = \{i, j, k\}$ are used to denote an

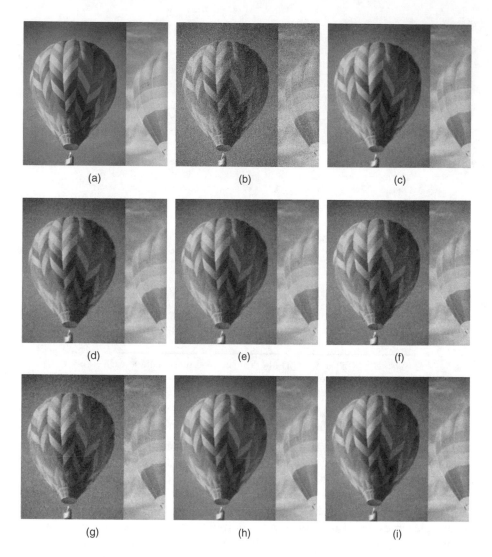

FIGURE 3.22
Filtering of balloon corrupted by component-independent contaminated Gaussian noise: (a) original, (b) noisy, (c) MED, (d) $MED_v(L_2)$, (e) $MED_v(L_1)$, (f) VDF, (g) DDF, (h) FMED, and (i) $FMED_v$ filtered images. (Original color images are depicted here in black and white.)

individual edge and face, respectively. In addition, vertices, edges, and faces are often represented by their corresponding indices. The geometry of a mesh is specified by all the vertex positions, denoted as $\mathbf{x}_i \in \Re^3$, $1 \leq i \leq V$. The vertex positions of a noisy triangular mesh are altered by surface smoothing algorithms while the topology is kept unchanged.

If a mesh is corrupted by noise, then the face normals are similarly corrupted.[43,44] The noisy normals can be effectively smoothed through fuzzy

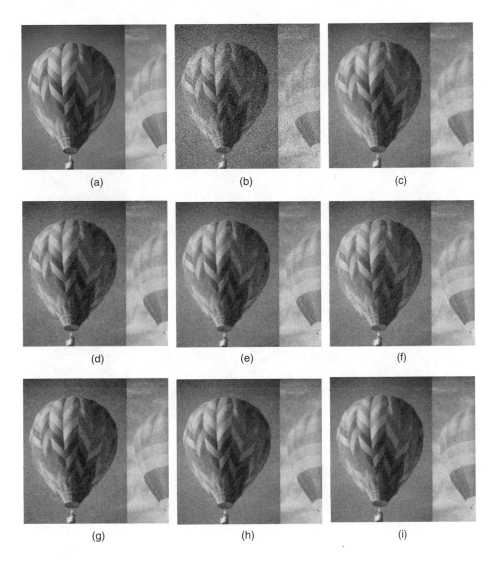

FIGURE 3.23
Filtering of balloon corrupted by component-correlated contaminated Gaussian noise: (a) original, (b) noisy, (c) MED, (d) $MED_v(L_2)$, (e) $MED_v(L_1)$, (f) VDF, (g) DDF, (h) FMED, and (i) $FMED_v$ filtered images. (Original color images are depicted here in black and white.)

vector median filtering.[43] The $FMED_v$ filter window, for each face i of the mesh, includes itself and its neighborhood, which is the set of faces that have a common edge with face i. Because the information in a normal vector is represented by the vector orientation (angle) rather than length, we adopt the angle between vectors[40] as the vector difference metric used in the fuzzy relation operator.

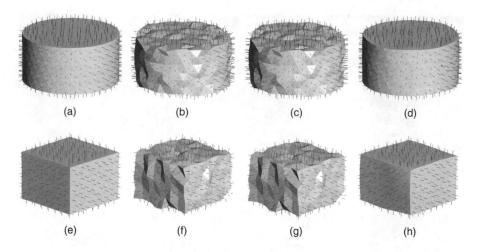

FIGURE 3.24
Surface smoothing of cylinder ($V = 365$, $F = 726$) and cube ($V = 244$, $F = 484$): (a) and (e) original objects; (b) and (f) noise-corrupted objects; (c) and (g) following FMED$_v$ face normal smoothing; (f) and (h) following vertex position update.

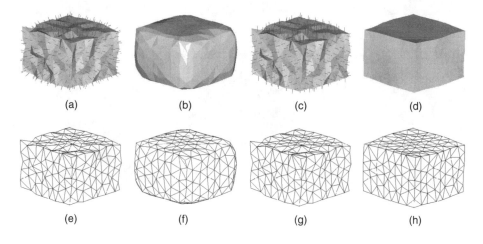

FIGURE 3.25
Smoothing of cube: (a) Taubin's normal smoothing; (b) vertex update operating on Taubin's result; (c) FMED$_v$ normal smoothing; (d) vertex update operating on FMED$_v$ result. The triangular meshes for a to d are shown in e to h.

To illustrate the surface normal smoothing process, Figure 3.24 shows experiments on two synthetic objects: a cylinder and a cube. Figure 3.24a and e show the original objects, whose vertex positions are then corrupted by additive Gaussian noise, Figure 3.24b and f. Note that both the vertex positions and face normals (represented by stubs on the surface) contain noise. The noisy face normals are smoothed by 50 iterations of fuzzy vector median

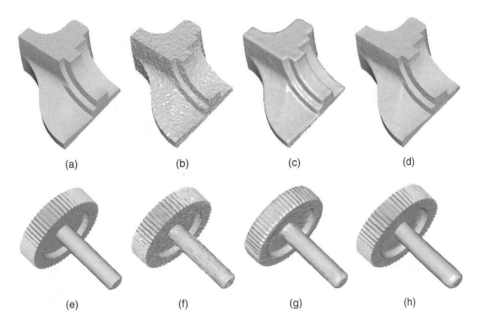

FIGURE 3.26
Smoothing of fan disk and gear: (a) and (e) original surfaces; (b) and (f) noisy surfaces; (c) and (g) results of Taubin's linear anisotropic filtering method; (d) and (h) FMED$_v$-based results.

filtering ($\sigma = 0.4$) and shown in Figure 3.24c and g. Note that the surface normals processed by the FMED$_v$ are very close to the original surface normals. Finally, the reconstructed surfaces, obtained using the vertex update procedure discussed next, are shown in Figure 3.24d and h. These results demonstrate that the FMED$_v$ filter is effective at removing noise from surface normals, while preserving desired features, such as crisp edges.

The vertex positions can be reconstructed based on the smoothed surface normals using the least-square error (LSE) method proposed in Reference 44 and adopted in Reference 43. Under this methodology, the vertex positions are updated as[44]

$$\mathbf{x}_i' \leftarrow \mathbf{x}_i + \lambda \sum_{j \in i^*} \sum_{f \in F_{ij}} \mathbf{n}_f \, \mathbf{n}_f^t (\mathbf{x}_j - \mathbf{x}_i), \tag{3.18}$$

where $\mathbf{x}_i$ and $\mathbf{x}_i'$ are the current and updated vertex position vectors, respectively, and $\mathbf{n}_f$ is the surface normal of face f. Also, i^* denotes the neighborhood of vertex i, i.e., the set of vertices that are connected to vertex i by an edge, and F_{ij} denotes the set of faces that contains the edge $\{i, j\}$. For comparison, Figure 3.25 and Figure 3.26 show the surface smoothing results using Taubin's[44] and the FMED$_v$-based methods operating on a cube and fan disk. Taubin's method diffuses edges and corners. Furthermore, this method significantly reduces the size of triangles at edges and corners. The FMED$_v$ method, in contrast, not only removes most of the noise, but also preserves features

$$
\begin{bmatrix} \bullet & \bullet & \bullet \\ \bullet & \bullet & \bullet \\ \bullet & \bullet & \bullet \end{bmatrix} \Rightarrow \begin{bmatrix} \bullet & \square & \bullet & \square & \bullet & \square \\ \triangle & \circ & \triangle & \circ & \triangle & \circ \\ \bullet & \square & \bullet & \square & \bullet & \square \\ \triangle & \circ & \triangle & \circ & \triangle & \circ \\ \bullet & \square & \bullet & \square & \bullet & \square \\ \triangle & \circ & \triangle & \circ & \triangle & \circ \end{bmatrix}
$$

FIGURE 3.27
Polyphase interpolation. Left: Original image; original pixels denoted by "•." Right: Zoomed image; interpolated pixels denoted by "△," "□," and "○."

such as edges and corners. Moreover, the triangle size is maintained within these features.

3.3.4   Image Zooming

The resizing of signals is an important problem in many applications. This is particularly important for image data, as many forms of information are represented in images. Moreover, the rapid growth of communications systems, such as the Internet, has made the exchange and display of image-based information widely available. Within such systems, images are often stored or transmitted at one size and displayed at another. As an example, images are often transmitted at a reduced pixel resolution to save storage or bandwidth and then zoomed, or interpolated, to a higher pixel density for display. Such operations have traditionally been performed utilizing linear operators. Linear operations, however, do not yield desirable results, especially for nonstationary signals such as images. Improved zooming results can be realized through the utilization of nonlinear interpolators. In particular, we show that the fuzzy weighted median filter produces superior zooming compared to linear and traditional weighted median filters.

Numerous structures exist for zooming, but most are based on first inserting zero-valued pixels between existing pixels and then interpolating the appropriate values of the inserted pixels from the surrounding samples. As an example, consider an algorithm that zooms an image by a factor of two. In this case, an empty array is constructed with twice the number of rows and columns as the original image. The original pixels are inserted into the array in alternating rows and columns, as indicated in Figure 3.27. In the figure, pixel values from the original image are designated by a "•." The "new" pixels to be interpolated are indicated by the symbols "△," "□," and "○." The unique symbol given to each pixel results from the differing set of neighbors for each case. Because there are four cases, including the original image pixels, this approach is referred to as polyphase interpolation.

In the bilinear approach, the average of the two original image pixels neighboring a "△" sample is used to set the pixel value of these samples. A similar

two-neighbor average is used to set the "□" pixel values. Last, the four original samples neighboring each "∘" pixel are averaged to set the values of these
remaining samples.

A similar polyphase approach utilizing weighted median, or fuzzy
weighted median, filters can be developed as a straightforward extension of
the linear approach.[38,45] In this case, the algorithm first interpolates "∘" pixels
as the weighted median of the original pixels at its four corners. As each of
these samples is equally distant from the sample to be estimated, they are
considered equally reliable and each is assigned a weight of 1. The remaining
pixels are determined by taking the weighted median of the four neighboring
samples for each case. More specifically, for each "□" pixel the two original
pixels to its left and right are assigned weight 1, while the two interpolated
"∘" pixels above and below are assigned weight 0.5. This lower weighting
reflects the fact that the interpolated samples are less reliable than the original pixels. Similarly, the original pixels above and below each "△" pixel are
assigned weight 1 and the interpolated pixels to the right and left are assigned
weight 0.5. The choice of 0.5 for the weights reflects the reliability of the estimated samples. It should be noted that, due to the structure of the weighted
median and the number of samples utilized, any assignment of weight values
between zero and one for the estimated samples yields identical results.

A fuzzy weighted median filter can be implemented in the above-mentioned
algorithm simply by replacing the pixels used for interpolation by their fuzzy
counterparts. Thus, the interpolation process can be illustrated as follows.
Consider an image represented by an array of pixel values denoted as $\{a_{i,j}\}$.
Then the fuzzy weighted median polyphase interpolation performs the following mapping:

$$
\begin{bmatrix} a_{1,1} & a_{1,2} & a_{1,3} \\ a_{2,1} & a_{2,2} & a_{2,3} \\ a_{3,1} & a_{3,2} & a_{3,3} \end{bmatrix} \Rightarrow
\begin{bmatrix}
x_{1,1}^{\bullet} & x_{1,1}^{\square} & x_{1,2}^{\bullet} & x_{1,2}^{\square} & x_{1,3}^{\bullet} & x_{1,3}^{\square} \\
x_{1,1}^{\triangle} & x_{1,1}^{\circ} & x_{1,2}^{\triangle} & x_{1,2}^{\circ} & x_{1,3}^{\triangle} & x_{1,3}^{\circ} \\
x_{2,1}^{\bullet} & x_{2,1}^{\square} & x_{2,2}^{\bullet} & x_{2,2}^{\square} & x_{1,3}^{\bullet} & x_{2,3}^{\square} \\
x_{2,1}^{\triangle} & x_{2,1}^{\circ} & x_{2,2}^{\triangle} & x_{2,2}^{\circ} & x_{1,3}^{\triangle} & x_{2,3}^{\circ} \\
x_{3,1}^{\bullet} & x_{3,1}^{\square} & x_{3,2}^{\bullet} & x_{3,2}^{\square} & x_{3,3}^{\bullet} & x_{3,3}^{\square} \\
x_{3,1}^{\triangle} & x_{3,1}^{\circ} & x_{3,2}^{\triangle} & x_{3,2}^{\circ} & x_{3,3}^{\triangle} & x_{3,3}^{\circ}
\end{bmatrix}, \tag{3.19}
$$

where $a_{i,j}$ is the value of the pixel in the ith row and jth column of the original
image. The pixels in the interpolated image are determined as

$$ x_{i,j}^{\bullet} = a_{i,j}, \tag{3.20} $$

$$ x_{i,j}^{\circ} = \mathrm{MED}\left(\tilde{a}_{i,j}^{\circ}, \tilde{a}_{i+1,j}^{\circ}, \tilde{a}_{i,j+1}^{\circ}, \tilde{a}_{i+1,j+1}^{\circ}\right), \tag{3.21} $$

$$ x_{i,j}^{\square} = \mathrm{MED}\left(\tilde{a}_{i,j}^{\square}, \tilde{a}_{i,j+1}^{\square}, 0.5 \diamond \tilde{x}_{i-1,j}^{\square}, 0.5 \diamond \tilde{x}_{i+1,j}^{\square}\right), \tag{3.22} $$

$$ x_{i,j}^{\triangle} = \mathrm{MED}\left(\tilde{a}_{i,j}^{\triangle}, \tilde{a}_{i+1,j}^{\triangle}, 0.5 \diamond \tilde{x}_{i,j-1}^{\triangle}, 0.5 \diamond \tilde{x}_{i,j+1}^{\triangle}\right). \tag{3.23} $$

In the above the samples $\tilde{a}_{i,j}^{\circ}$, $\tilde{a}_{i,j}^{\square}$, and $\tilde{a}_{i,j}^{\triangle}$ are the fuzzy samples based on observation windows centered at locations $\circ$, $\square$, and $\triangle$, respectively. Note that for the same pixel location these samples can take on different values, i.e., $\tilde{a}_{i,j}^{\circ}$, $\tilde{a}_{i,j}^{\square}$, and $\tilde{a}_{i,j}^{\triangle}$ may not be equal to each other. This is because the observation window is centered at a different location in each case and each fuzzy sample is set according to the relationship among the samples in the given window.

To illustrate the zooming performance of the bilinear, weighted median, and fuzzy weighted median methods, consider the zooming of the image area outlined in Figure 3.28a. The results of zooming this area by a factor of two utilizing each method is shown in Figure 3.28a to d. As the images show, the linear method introduces high-frequency distortions along the edges. This is clearly seen by examining the white leg of the chair. The weighted median reduces this distortion somewhat, but introduces blocking artifacts, which can be seen in smooth areas such as facial regions. The fuzzy weighted median, in contrast, produces sharp edges with minimal artifacts. See again the leg of the chair. Also, no blocking artifacts are seen in the uniform regions.

Fuzzy weighted median filters thus offer improved performance in the interpolation of nonstationary signals, such as images. Visually important cues, such as edges, are well represented in the interpolated signal. Additionally, the blocking artifacts that are present in the output of weighted median filters are noticeably absent in the fuzzy weighted median output.

3.3.5   Image Sharpening

In addition to noise smoothing and artifact removal, another general class of important imaging applications is enhancement. Because edges play a critical role in the human perception of images, one important aspect of image enhancement is edge sharpening. The most widely used method for image sharpening is linear unsharp masking. This method, however, is based on derivative, or highpass, filtering and is therefore highly sensitive to the presence of noise.

One approach to improving the robustness of unsharp masking is to replace the linear derivative operator with a weighted median-based scheme.[46] In this approach, the positive-slope edges are extracted with a highpass weighted median filter. To extract the negative-slope edges, the image is inverted, causing negative-slope edges to become positive-slope images, and the resulting image is processed with the same highpass weighted median filter. Last, the positive and negative edges are scaled and added to the original image, in a procedure similar in concept to that originally developed for unsharp masking.

This sharpening procedure can be made even more robust by utilizing permutation weighted median filters.[46] In this approach, the weights applied to each sample are allowed to vary as a function of the rank of the sample. This allows, for example, a smaller weight to be applied when the sample

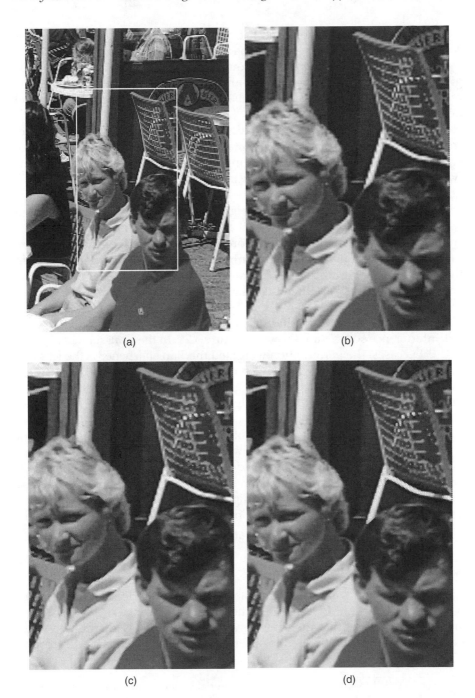

(a)

(b)

(c)

(d)

FIGURE 3.28
(a) Original image with area of interest outlined, and zoomed areas of interest utilizing the
(b) bilinear, (c) weighted median, and (d) fuzzy weighted median methods.

is in the extremes of the ordered set, indicating that it is likely an outlier sample. The permutation weighted median filter, like the standard weighted median filter, can be generalized through the use of fuzzy ordering.[47] This gives the advantage that not only is the sharpening procedure robust to noise, but it is also able to simultaneously sharpen edges and smooth background noise.

To illustrate the performance of each of these methods, results are presented for the image Car, which contains rich edge information. As edge enhancement is particularly challenging for noisy images, we consider four types of common interference: Gaussian noise, Laplacian noise, impulsive noise, and blocking artifacts introduced through JPEG compression. Results are presented for the permutation weighted median and fuzzy permutation weighted median approaches, as well as the well-known noisy image cubic sharpener,[48] which employs a higher-order polynomial edge sensor to ensure that the highpass filter is activated only when strong edges are detected. In the presented results, the scaling factor for the cubic sharpener is chosen to achieve the same enhancement level as the permutation weighted median sharpener.

The sharpened images produced by the compared methods under the various interference sources are shown in Figure 3.29. In the Gaussian, Laplacian, and JPEG cases, the permutation weighted median sharpener is the least robust, producing results that are much noisier than the other methods. In the impulsive noise case, the cubic sharpener yields the worst result as the impulses are amplified as if they are edges.* The cubic sharpener output also contains background noise in the Laplacian noise case. In the Gaussian case, the noise in the cubic sharpener output is located primarily along edges, but is still clearly visible. Moreover, the cubic method distorts the original edges in the JPEG case. In contrast, the fuzzy approach produces sharp, relatively artifact-free images from the JPEG and noisy images. In the impulsive case there are some pulses passed and slightly spread by the fuzzy operation, but these effects could be minimized by simply decreasing the membership spread function.

3.3.6 Multiuser Detection

Rank ordering plays an important role not only in filtering but also in nonparametric detection. In this application, the significance of rank ordering arises from the fact that the sample ranks are uniformly distributed independent of the distribution of the observed samples, implying that detection

---

* Because the scaled edge information is added to the original noisy image in the baseline unsharp scheme, the impulsive noise is still evident in the sharpened image. This can be avoided by replacing the original image by the denoised image. We do not include this modification here in order to compare the effectiveness of the edge extraction.

algorithms based on ranks can be designed that are robust to varying sample statistics. This is very important if little or nothing is known about the noise (or signal) statistics, or if the noise environment is changing rapidly as is often the case in wireless communications. The majority of the detection algorithms developed to date have been based on crisp ranks that contain only very coarse information about the observation samples, and thus lead to detection schemes that exhibit low efficiency when compared to optimal techniques. The use of fuzzy ranks can help to overcome this drawback. We illustrate this through the use of fuzzy rank order (FRO) detectors[49–52] for fast-frequency-hopping multiple access networks.

In a frequency-hopping spread spectrum network, K users transmit sequences of L tones called frames. Each tone is chosen from a set of Q possible tones. At the receiver, the signals of all users are superimposed. The frames of the received signal are noncoherently detected at Q possible frequencies for each of the L time slots. The results are arranged in a $Q \times L$ received matrix, where the rows represent the frequencies and the columns represent the time slots. To detect user K, the rows are de-spread with the Kth user's spreading sequence, yielding a decoded matrix with entries $x_{i,j}$. The decoded matrix will contain one row at the level of the desired user's symbol. This row is referred to as the correct row. The samples in the correct row correspond to the desired signal plus noise and the entries in the remaining rows are due to noise only, where the noise is due to a combination of background noise and contributions from interfering users (multiple access interference). Denoting the cumulative distribution of the samples in the correct row and the spurious rows by F_{S+N} and F_N, and assuming that all elements in the received matrix are mutually independent,[53] the decision of which Q row is the correct row can be put in the framework of a Q-ary hypothesis test:

$$H_k : x_{k,\ell} \sim F_{S+N}, \quad \ell = 1, 2, \ldots, L \quad \text{and}$$
$$x_{i,j} \sim F_N, \quad j = 1, 2, \ldots, L,$$
$$i = 1, \ldots, k-1, k+1, \ldots, Q. \tag{3.24}$$

Thus, the samples in the row under test have to be compared to the remaining pooled data in some fashion.

The conventional detector forms a binary detection matrix by detecting any energy (above the noise floor) in a given time-frequency slot in **x** as a hit, regardless of the actual amount of energy present. To obtain a decision regarding which symbol was transmitted, the number of hits is determined and the row with the maximum number of hits is chosen to be the correct row. Figure 3.30a depicts the energy samples seen in a typical received matrix, for $Q = 8$, $L = 5$, and SNR $= 25$ dB, by gray tones where white is assigned to the maximum energy sample and black is assigned to the element with the smallest value. In this example, the first row is the correct row. This is reflected in the received matrix by two clear hits on the third and fifth hop. Aside from those

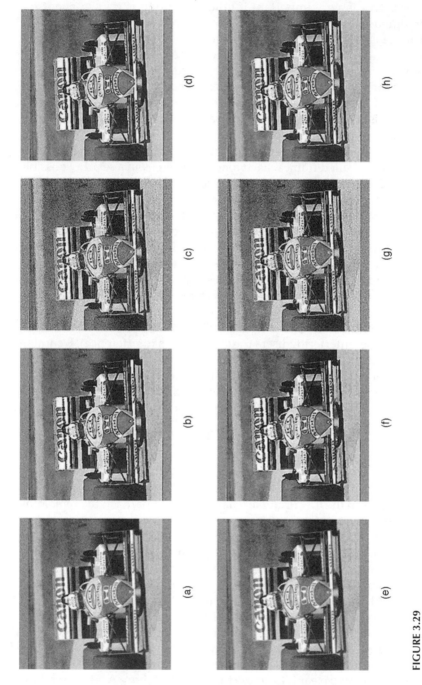

FIGURE 3.29

Image sharpening. From left to right: input image and cubic, permutation WMED, and permutation FWMED outputs. (a to d): Gaussian noise-corrupted input; (e to h): Laplacian noise-corrupted input; (i to l): impulsive noise-corrupted input; (m to p): JPEG blocking artifact-corrupted input. (*Continued*)

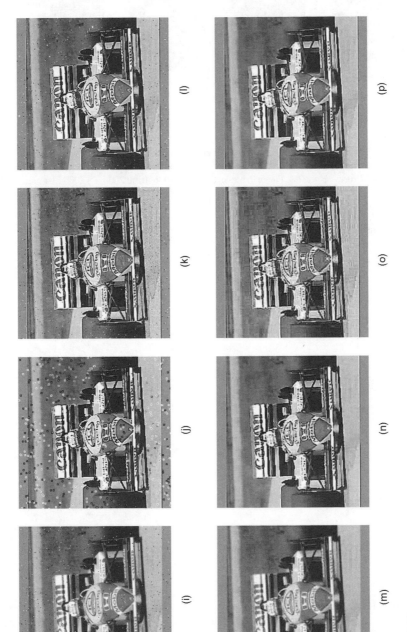

FIGURE 3.29
(Continued)

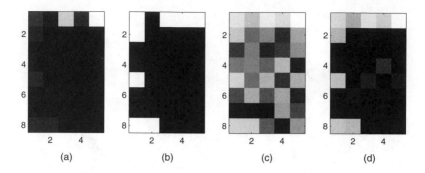

FIGURE 3.30
(a) Decoded matrix **x**, (b) thresholded matrix, (c) crisp rank matrix, and (d) fuzzy rank matrix, for $Q = 8$, $L = 5$, and SNR $= 25$ dB. The first row corresponds to the correct decision. (From A. Flaig et al., *Signal Process. Spec. Iss. Fuzzy Process.*, 80, 1017–1036, June 2000. © 2000 Elsevier. With permission.)

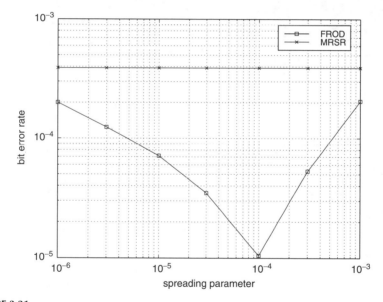

FIGURE 3.31
Bit error rate of FRO detector vs. the fuzzy spreading parameter (λ) for $M = 2$, $Q = 16$, $L = 5$, and SNR $= 25$ dB. The bit error rate of the MRSR is shown for reference. (From A. Flaig et al., *Signal Process. Spec. Iss. Fuzzy Process.*, 80, 1017–1036, June 2000. © 2000 Elsevier. With permission.)

correct hits, there are also somewhat weaker hits scored by interfering users in rows two, five, and eight. Figure 3.30b shows the corresponding threshold matrix, where black (0) and white (1) represent nonhits and hits, respectively. Due to the hard decision the second hit in the first row is irrevocably lost.

The hard-decision majority vote (HDMV) method described above works well in high-SNR environments,[54,55] where samples that originated from signal plus noise and samples that stem from noise only are easily distinguished. A notable loss in efficiency is incurred, however, in low-SNR environments.

Decision rules based on rank ordering are motivated by the observation[53] that the cumulative distributions satisfy

$$F_{S+N}(x) \leq F_N(x) \tag{3.25}$$

for all x. That is, the samples in the correct row are *stochastically larger* than the samples in the remaining rows. The maximum rank sum receiver (MRSR)[56] exploits this property by rank ordering the samples in the decoded matrix x and forming a rank matrix $\mathbf{r} = \{r_{k,\ell}\}$, whose elements are the crisp ranks of the corresponding entries in x, i.e., $r_{k,\ell}$ is the rank of $x_{k,\ell}$. Thus, according to Equation 3.25, hits are typically assigned high ranks, whereas noise samples are assigned low ranks: The decision is made by summing the ranks across each row, and choosing the row with the largest sum as the correct row:

$$H_k : \max_k \sum_{\ell=1}^{L} r_{k,\ell}. \tag{3.26}$$

The main shortcoming of this approach is that the real-valued energy samples in x are replaced by integer ranks that reflect only the relative values of the received energies but not their spread. As an example, consider the integer rank matrix pertaining to x, Figure 3.30c, where bright tones correspond to high ranks and dark tones correspond to low ranks. Comparing Figure 3.30a and Figure 3.30c, it can be observed that, although the two strongest hits are clearly recognizable in both matrices, in the integer rank matrix the weaker hits are difficult to distinguish from the nonhits.

The more general fuzzy ordering-based FRO detectors[50–52] are capable of making a reliable distinction between hits and nonhits due to the clustering property of fuzzy ranks.[38,57] Extending the MRSR detector to the FRO detector is a straightforward fuzzy extension,

$$H_k : \max_k \sum_{\ell=1}^{L} \tilde{r}_{k,\ell}, \tag{3.27}$$

for $k = 1, 2, \ldots, Q$. Figure 3.30d shows the effect of fuzzy ranking when applied to the decoded matrix. The consideration of sample spread results in a clear identification of hits and nonhits indicated by bright and dark gray tones, respectively.

To further illustrate the performance of the various detectors, consider the simplified Rayleigh fading channel.[56] In this model the entries in the decoded matrix obey exponential distributions whose mean value reflects the presence or absence of a signal. More specifically, the samples in a correct row are

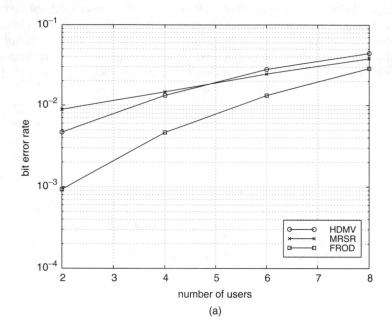

(a)

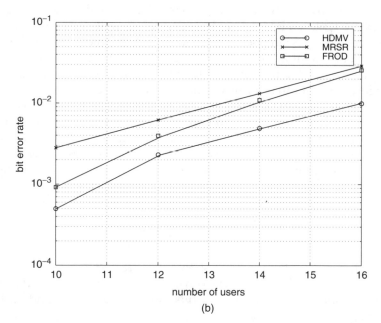

(b)

FIGURE 3.32
Probability of bit error vs. the number of users for the HDMV, MRSR, and the FRO detectors for the $Q = 32$ and $L = 8$ case with (a) SNR = 10 dB and (b) SNR = 25 dB. (From A. Flaig et al., *Signal Process. Spec. Iss. Fuzzy Process.*, 80, 1017–1036, June 2000. © 2003 Elsevier. With permission.)

distributed as $x \sim \lambda_1 e^{-\lambda_1 x}$, while the samples in incorrect rows are distributed as $x \sim p\lambda_1 e^{-\lambda_1 x} + (1 - p)\lambda_0 e^{-\lambda_0 x}$, where $1/\lambda_1 > 1/\lambda_0$. Note that $1/\lambda_1$ and $1/\lambda_0$ are the mean energies of samples that correspond to signal plus noise and noise only. The mixing parameter p is the proportion of samples generated by interfering users and can be found by invoking a randomization argument, which leads to $p = 1 - (1 - 2^{-\log_2(Q)})^{(K-1)}$.

Figure 3.31 depicts the bit error rate of the FRO detector for different values of the fuzzy spreading parameter. As expected, an appropriate amount of fuzzy spreading facilitates the separation of hits from nonhits, and thus reduces the probability of error of the FRO detector in comparison to its (zero spread) limiting case, the MRSR. Figure 3.32 depicts the probability of bit error vs. the number of users for the HDMV, MRSR, and FRO detectors for the $Q = 32$ and $L = 8$ case with SNR = 10 dB and SNR = 25 dB. For the moderate SNR = 10 dB case, the FRO detector clearly outperforms both the HDMV and the MRSR detectors. As the SNR is increased to the SNR = 25 dB case, the HDMV detector achieves a slightly lower bit error rate than the FRO detector. This is not surprising, as it has been shown[55] that, for the simplified fading model, the HDMV detector is asymptotically optimal as the SNR → ∞.

3.4 Conclusions

This two-chapter set develops fuzzy methods in nonlinear signal processing and applies the developed methods to a wide array of signal processing and communications applications. Part I focuses on theoretical development, where the fundamental importance of considering both spatial order and rank order of samples is established. There, the SR concept is formally developed and extended to the fuzzy case in which sample affinity, or spread (diversity), is considered. The concepts of sample affinity and affinity-based ordering are used to establish two broad classes of fuzzy filters: affine filters, which are generalizations of weighted sum filters that incorporate sample spread, and fuzzy weighted median filters, which generalize the standard class of weighted median filters by incorporating fuzzy samples.

In Part II, the affine and fuzzy weighted median filters are applied to a wide range of problems, including robust frequency selective filtering, Synthetic Aperture Radar image filtering, time-frequency plain filtering, multiresolution signal representations, surface smoothing, image smoothing, zooming, and deblocking, and multiuser detection. The results indicate that the inclusion of affinity-based sample spread information in the filtering process yields improved performance. The wide applicability of fuzzy methods and their superior performance show that these concepts constitute important tools for modern signal processing applications.

References

1. L. Yin, R. Yang, M. Gabbouj, and Y. Neuvo, Weighted median filters: a tutorial, *IEEE Trans. Circuits Syst. II: Analog Digital Signal Process.*, 41, May 1996.
2. K.E. Barner, Colored $l-\ell$ filters with tap bias and their application in speech pitch detection, *IEEE Trans. Signal Process.*, 9, 2601–2606, Sept. 2000.
3. P. Gandhi and S.A. Kassam, Design and performance of combination filters, *IEEE Trans. Signal Process.*, 39, 1524–1540, July 1991.
4. F. Palmieri and C.G. Boncelet, Jr., Ll-filters—a new class of order statistic filters, *IEEE Trans. Acousti. Speech Signal Process.*, 37, 691–701, May 1989.
5. M. Shao and C.L. Nikias, Signal processing with fractional lower order moments: stable processes and their applications, *Proc. IEEE*, 81, July 1993.
6. G. Strang and T.Q. Nguyen, *Wavelets and Filter Banks*, Wellesley, MA: Wellesley-Cambridge Press, 1996.
7. I. Daubechies, *Ten Lectures on Wavelets*, Philadelphia: SIAM, 1992.
8. S.G. Mallat, A theory for mulitiresolution signal decomposition: the wavelet representation, *IEEE Trans. Pattern Anal. Mach. Intelligence*, 11, 674–693, July 1989.
9. M. Asghar and K.E. Barner, Nonlinear multiresolution techniques with applications to scientific visualization in a haptic environment, *IEEE Trans. Visualization Comput. Graphics*, 7, 76–93, Mar. 2001.
10. M.W. Asghar, Nonlinear Multiresolution Techniques with Applications to Scientific Visualization in a Haptic Environment, Master's thesis, University of Delaware, Newark, May 1999.
11. A. Zyweck and R.E. Bogner, High-resolution radar imagery of mirage III aircraft, *IEEE Transactions Antennas Propagation*, 42, 1356–1360, Sept. 1994.
12. A.S. Al-Fahoum and A.M. Reza, Combined edge crispiness and statistical differencing for deblocking jpeg compressed images, *IEEE Trans. Image Process.*, 10, 1288–1298, Sept. 2001.
13. S.D. Kim, J. Yi, H.M. Kim, and J.B. Ra, A deblocking filter with two separate modes in block-based video coding, *Trans. Circuits Syst. Video Technol.*, 9, 156–160, 1999.
14. G.R. Arce and S.R. Hasan, Elimination of interference terms of the discrete wigner distribution using nonlinear filtering, *IEEE Trans. Signal Process.*, 48, 2321–2331, Aug. 2000.
15. L. Cohen, Time-frequency distributions—a review, *Proc. IEEE*, 77, 941–981, July 1989.
16. T. Claasen and W. Meclenbrauker, The wigner distribution—a tool for time-frequency signal analysis. II. Discrete time signals, *Philips J. Res.*, 35, 276–300, 1980.
17. A. Janssen, On the locus and spread of pseudo-density functions in the time-frequency plane, *Philips J. Res.*, 37(3), 79–110, 1982.
18. F. Hlawatsch and G. Bourdeaux-Bartels, Linear and quadratic time-frequency signal representations, *IEEE Signal Process. Mag.*, 9, 21–67, April 1992.
19. F. Hlawatsch and P. Flandrin, The interference structure of the Wigner distribution and related time-frequency signal representations, in *The Wigner Distribution—Theory and Applications in Signal Processing*, Amsterdam: Elsevier Science, 1997.

20. L. Atlas, J. Fang, P. Loughlin, and W. Music, Resolution advantages of quadratic signal processing, *Proc. SPIE*, 1566, 134–143, 1991.

21. P. Flandrin, Some features of time-frequency representations of multicomponent signals, in *Proceedings of the International Conference on Acoustics, Speech, and Signal Processing (ICASSP)*, 1984, 41.B.4.1–41.B.4.4.

22. S. Qian and J. Morris, Wigner distribution decomposition and cross–terms deleted representation, *Signal Process.*, 27, 125–144, 1992.

23. D.L. Jones and R.G. Baraniuk, A signal-dependent time-frequency representation: optimal kernel design, *IEEE Trans. Signal Process.*, 41, 1589–1601, Apr. 1993.

24. D.L. Jones and R.G. Baraniuk, An adaptive optimal-kernel time-frequency representation, *IEEE Trans. Signal Process.*, 43, 2361–2371, Oct. 1995.

25. K. Bloch and G.R. Arce, Time-frequency analysis of protein sequence data, in *Proceedings of the IEEE–EURASIP Nonlinear Signal and Image Processing (NSIP) Workshop*, Baltimore, MD, June 2001.

26. K. Tomii and M. Kanehisa, Analysis of amino acids and mutation matrices for sequence comparison and structure prediction of proteins, *Protein Eng.*, 9, Jan. 1996.

27. I. Cosic, Macromolecular bioactivity: is it resonant interaction between macromolecules?—Theory and applications, *IEEE Trans. Biomed. Eng.*, 41, Dec. 1994.

28. V. Veljkovic, I. Cosic, B. Dimitrjevic, and D. Lalovic, Is it possible to analyze DNA and protein sequences by the methods of digital signal processing? *IEEE Trans. Biomed. Eng.*, 32, May 1985.

29. Q. Fang and I. Cosic, Prediction of active sites of fibroblast growth factors using continuous wavelet transforms and the resonant recognition model, in *Proceedings of the Inaugural Conference of the Victorian Chapter of the IEEE EMBS*, 1999.

30. G.R. Arce and N.C. Gallagher, Jr., State description of the root set of median filters, *IEEE Trans. Acoust. Speech Signal Process.*, 30, Dec. 1982.

31. G.R. Arce, N.C. Gallagher, Jr., and T.A. Nodes, Median filters: theory and aplications, in *Advances in Computer Vision and Image Processing*, T.S. Huang, Ed., Vol. 2, Greenwich, CT: JAI Press, 1986.

32. A.C. Bovik, T.S. Huang, and D.C. Munson, Jr., A generalization of median filtering using linear combinations of order statistics, *IEEE Trans. Acoust, Speech Signal Process.*, 31, Dec. 1983.

33. D.R.K. Brownrigg, The weighted median filter, *Commun. Assoc. Comput. Mach.*, 27, Aug. 1984.

34. S.-J. Ko and Y.H. Lee, Center weighted median filters and their applications to image enhancement, *IEEE Trans. Circuits Syst.*, 38, 984–993, Sept. 1991.

35. B. Zeng, Optimal median-type filtering under structural constraints, *IEEE Trans. Image Process.*, 7, July 1995.

36. E.J. Coyle, J.-H. Lin, and M. Gabbouj, Optimal stack filtering and the estimation and structural approaches to image processing, *IEEE Trans. Acoust. Speech Signal Process.*, 37, 2037–2066, Dec. 1989.

37. P. Wendt, E.J. Coyle, and N.C. Gallagher, Jr., Stack filters, *IEEE Trans. Acousti. Speech Signal Process.*, 34, 898–911, Aug. 1986.

38. K.E. Barner and R.C. Hardie, Spatial–rank order selection filters, in *Nonlinear Image Processing*, S.K. Mitra and G. Sicuranza, Eds., San Diego, CA: Academic Press, 2001.

39. J. Astola, J. Haavisto, and Y. Neuvo, Vector median filters, in *Proc. IEEE*, 678–689, Apr. 1990.

40. P. E. Trahanias and A.N. Venetsanopoulos, Vector directional filters—a new class of multichannel image processing filters, *IEEE Trans. Image Process.*, 528–534, Oct. 1993.

41. D.G. Karakos and P.E. Trahanias, Generalized multichannel image-filtering structures, *IEEE Trans. Image Process.*, 6, 1038–1045, July 1997.

42. Y. Shen and K.E. Barner, Marginal fuzzy median and fuzzy vector median filtering of color images, in *Proceedings of the 37th Annual Conference on Information Sciences and Systems (CISS2003)*, Baltimore, MD, Mar. 2003.

43. Y. Shen and K.E. Barner, Fuzzy vector median based surface smoothing in a haptic environment, in *Proceedings of 11th Symposium on Haptic Interfaces for Virtual Environment and Teleoperator Systems*, Los Angeles, CA, Mar. 2003.

44. G. Taubin, IBM Research Report: Linear Anisotropic Mesh Filtering, Tech. Rep. RC22213, IBM Research Division Thomas J. Watson Research Center, Yorktown Heights, NY, Oct. 2001.

45. G.R. Arce and J. Paredes, Image enhancement with weighted medians, in *Nonlinear Image Processing*, S. Mitra and G. Sicuranza, Eds., San Diego, CA: Academic Press, 2000, 27–67.

46. G.R. Arce and J. Paredes, *Image Enhancement and Analysis with Weighted Medians*, San Diego, CA: Academic Press, 2001.

47. Y. Nie and K.E. Barner, Noisy image sharpening using fuzzy weighted median filter, in *Proceedings of the 37th Annual Conference on Information Sciences and Systems (CISS2003)*, Baltimore, MD, Mar. 2003.

48. G. Ramponi, N. Strobel, and S.K. Mitra, Nonlinear unsharp masking methods for image contrast enhancement, *J. Electron. Imaging*, 5, 353–366, July 1996.

49. A. Flaig, A.B. Cooper, G.R. Arce, H. Tayong, and A. Cole-Rhodes, Fuzzy rank-order detectors for frequency hopping networks, in *Proceedings of the 3rd Annual Fedlab Symposium on Advanced Telecommunications/Information Distribution Research Program (ATIRP)*, College Park, MD, Feb. 1999, 121–125.

50. A. Flaig, K.E. Barner, and G.R. Arce, Fuzzy ranking: Theory and applications, *Signal Process. Spec. Iss. Fuzzy Process.*, 80, 1017–1036, June 2000.

51. H. Tayong, A. Cole-Rhodes, B. Cooper, A. Flaig, and G. Arce, A reduced complexity detector for fast frequency hopping, in *Proceedings of the Annual Fedlab Symposium on Advanced Telecommunications/Information Distribution Research Program (ATIRP)*, College Park, MD, Mar. 2000.

52. H. Tayong, A. Beasley, A. Cole-Rhodes, B. Cooper, and G. Arce, Parametric diversity combining in fast frequency hopping, in *Proceedings of the Annual Fedlab Symposium on Advanced Telecommunications/Information Distribution Research Program (ATIRP)*, College Park, MD, Mar. 2001.

53. M.N. Woinsky, Nonparametric detection using spectral data, *IEEE Trans. Inf. Theor.*, 18, 110–118, Jan. 1972.

54. D. Goodman, P. Henry, and V. Prabhu, Frequency-hopped multilevel FSK for mobile radio, *Bell Syst. Tech. J.*, 59, 1257–1275, Sept. 1980.

55. R. Viswanathan and S.C. Gupta, Performance comparison of likelihood, hard-limited, and linear combining receiver, *IEEE Trans. Commun. Theor.*, 31, 670–677, May 1983.

56. R. Viswanathan and S.C. Gupta, Nonparametric receiver for FH-MFSK mobile radio, *IEEE Trans. Commun. Theor.*, 33, 178–184, Feb. 1985.

57. K.E. Barner, A. Flaig, and G.R. Arce, Fuzzy time–rank relations and order statistics, *IEEE Signal Process. Lett.*, 5, 252–255, Oct. 1998.

58. A. Flaig, G.R. Arce, and K.E. Barner, Fuzzy order statistic filters, a data-adaptive filtering framework for nonstationary signals, *IEEE Trans. Signal Process.*, 46, 2102–2112, Aug. 1998.

4

Time-Frequency Wigner Distribution Approach to Differential Equations

Lorenzo Galleani and Leon Cohen

CONTENTS

4.1 Introduction

In the field of signal analysis, time-frequency distributions have historically been used as a means of analyzing signals for their time-varying spectra.[6,7,11] In physics, however, these distributions have been used to understand the solution of the Schrödinger equation, which is a partial differential equation. The idea is to obtain the equation of motion for the Wigner distribution corresponding to the solution of the Schrödinger equation. The basic reasons for doing so is that one gains considerable insight into the nature of the solution, and that it leads to new analysis and approximation methods. Wigner, Moyal, Kirkwood, and many others made significant contributions toward this approach.[4,5,8,12] Equations of motion have been written for both the Wigner distribution and other distributions.

In contrast, in signal analysis, the Wigner distribution has not been used in such a manner. We describe methods that allow one to write a dynamical equation for the Wigner distribution that corresponds to the solution of an ordinary or partial differential equation. We have found many advantages in this approach, both from the point of view of insight and also in devising new methods of solution and approximation to the original equation. As is well known, a differential equation where the dependent variable is time can be converted into an equivalent equation in the frequency domain. Our approach converts the differential equation in the combined time-frequency domain. In this chapter we present these ideas and methods and we do so using concrete examples. While we give the general approach, we emphasize and develop the ideas for the harmonic oscillator, with constant and time varying coefficients, and also for the deterministic and random case. In the appendices we give the general results that can be applied to arbitrary differential equations.

4.2   Harmonic Oscillator with Chirp Driving Force

Consider the harmonic oscillator with a deterministic driving force,

$$\frac{d^2x(t)}{dt^2} + 2\mu\frac{dx(t)}{dt} + \omega_0^2 x = f(t), \tag{4.1}$$

where $x(t)$ is the state variable (e.g., position, current) and $f(t)$ is the driving term. If we want to study the time-frequency properties we could solve this equation and substitute the answer into the Wigner distribution,

$$W_{x,x}(t, \omega) = \frac{1}{2\pi}\int x^*\left(t - \frac{1}{2}\tau\right)x\left(t + \frac{1}{2}\tau\right)e^{-i\tau\omega}\,d\tau. \tag{4.2}$$

We want to write the equation of motion for the Wigner distribution and solve that directly. As mentioned above we have devised a general procedure to do that which is reviewed in Appendix 4.2. For the harmonic oscillator it is

$$\left[a_4\frac{\partial^4}{\partial t^4} + a_3\frac{\partial^3}{\partial t^3} + a_2\frac{\partial^2}{\partial t^2} + a_1\frac{\partial}{\partial t} + a_0\right]W_{x,x}(t, \omega) = W_{f,f}(t, \omega), \tag{4.3}$$

where

$$a_0 = \left(\omega_0^2 - \omega^2\right)^2 + 4\mu^2\omega^2, \tag{4.4}$$

$$a_1 = 2\mu\left(\omega_0^2 + \omega^2\right), \tag{4.5}$$

$$a_2 = \frac{1}{2}\left(\omega_0^2 + \omega^2 + 2\mu^2\right), \tag{4.6}$$

$$a_3 = \frac{1}{2}\mu, \tag{4.7}$$

$$a_4 = 1/16. \tag{4.8}$$

Now consider an important example first considered by Barber and Ursell[1] and Hok[2] and called the *gliding tone problem*. It is the response of a harmonic oscillator to a "gliding tone," that is,

$$f(t) = e^{i\omega_1 t + i\beta t^2/2}. \tag{4.9}$$

The reason this is called the gliding tone problem is because the instantaneous frequency of the driving force increases linearly,

$$\omega_i(t) = \omega_1 + \beta t. \tag{4.10}$$

In the gliding tone problem one wants to ascertain the instantaneous frequency of the response. There have been a number of studies made by examining *approximate* solutions of Equation 4.1, because indeed an exact solution to Equation 4.1 with $f(t)$ given by Equation 4.9 has not been achieved. However, we have been able to solve Equation 4.3 exactly. The answer is given in Appendix 4.3.

We now give some graphical examples to illustrate the results. We first consider the underdamped case, that is, when $\mu < \omega_0$. In Figure 4.1 through Figure 4.3 we plot the Wigner $W_{x,x}(t, \omega)$ for the three cases $\mu = 0.5$, $\mu = 1$,

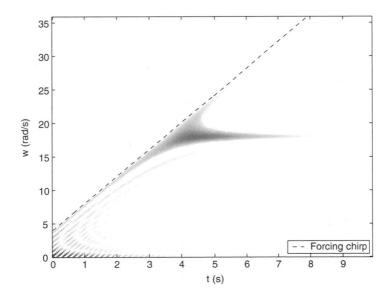

FIGURE 4.1
Wigner distribution of the solution to the gliding tone problem. Underdamped case with $\mu = 0.5$.

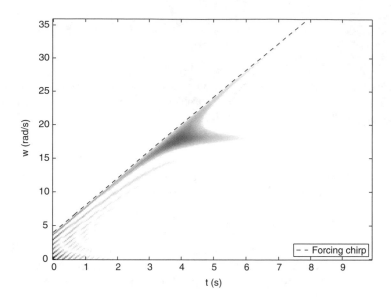

FIGURE 4.2
Wigner distribution of the solution to the gliding tone problem. Underdamped case with $\mu = 1$.

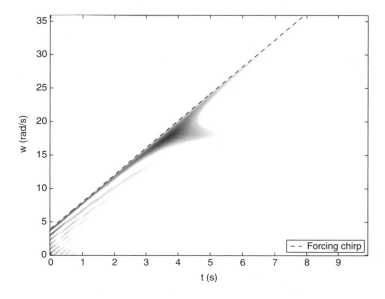

FIGURE 4.3
Wigner distribution of the solution to the gliding tone problem. Underdamped case with $\mu = 1.5$.

and $\mu = 1.5$, respectively. In both of the three cases we choose $\omega_0 = 18$ and $\omega_1 = \beta = 4$. The grayscale image in every picture is the exact Wigner distribution of $x(t)$; the dashed line represents the instantaneous frequency of the forcing chirp, that is, $\omega_i(t) = \omega_1 + \beta t$. The input chirp $f(t)$ is concentrated only along this line, because its representation in the Wigner distribution domain is $\delta(\omega - \omega_1 - \beta t)$. We see that the response of the system is mainly concentrated around the critical frequency ω_c (it is $\omega_c \approx \omega_0$), while it is weaker at all the other frequencies. Also, observing the limit at $\omega = \omega_c$, one can see that the Wigner distribution has an exponential damping factor, where the damping coefficient is 2μ, which is twice the damping of the free oscillation factor μ of the system. Comparing the three pictures we see how changing the damping factor μ influences the system response. Smaller values of μ imply less damping and hence longer tails in the response along ω_c. Increasing μ forces the system to have stronger damping and that is reflected in the shorter tail of the main response located around the resonant frequency ω_c.

In Figure 4.4 we give an example of an overdamped case where $\mu > \omega_0$, and in particular we take $\mu = 30$. Here the system response is anharmonic, and we do not have any special resonant frequency. Notice that the output is greater for small times t, while when $t \to \infty$ the response goes to zero. This is in complete agreement with the result obtained considering the system transfer function.

Finally, in Figure 4.5 we show a critically damped case, with $\mu = \omega_c = 18$. Considerations on this case are similar to those for the overdamped case.

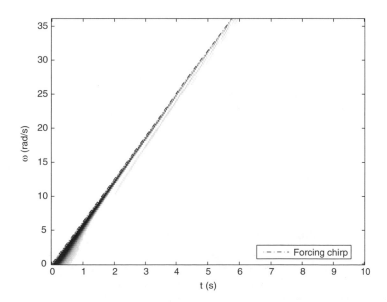

FIGURE 4.4
Wigner distribution of the solution to the gliding tone problem for an overdamped case, $\mu > \omega_0$.

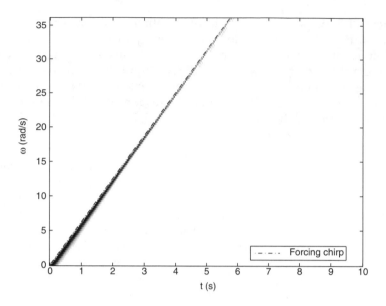

FIGURE 4.5
Wigner distribution of the solution to the gliding tone problem for a critically damped case, $\mu = \omega_0$.

We point out that we have obtained the Wigner distribution from Equation 4.3 by choosing the following initial conditions:

$$W_{x,x}(-\infty, \omega) = \frac{\partial W_{x,x}(-\infty, \omega)}{\partial t} = \frac{\partial^2 W_{x,x}(-\infty, \omega)}{\partial t^2} = \frac{\partial^3 W_{x,x}(-\infty, \omega)}{\partial t^3} = 0.$$

(4.11)

We have proved[15] that this choice corresponds to finding the Wigner distribution of the solution $x(t)$ that has zero initial conditions at $t = -\infty$.

4.3   Approximation Method

Of course, it was fortuitous that the exact solution to Equation 4.3 was achieved, even though one cannot write an exact solution for $x(t)$. We were able to achieve an exact solution because the Wigner distribution of the driving force $f(t) = e^{i\omega_1 t + i\beta t^2/2}$ was a delta function. We now observe that a general class of signals that are very interesting from a physical point of view are the so-called monocomponent signals, which are signals of the form $s(t) = e^{j\varphi(t)}$ (with certain restrictions on the phase). Although the Wigner distribution of a driving force of the form $s(t) = e^{j\varphi(t)}$ is not in general a delta function along the instantaneous frequency $\omega_i(t) = \varphi'(t)$, the delta function is nonetheless a very good approximation to it. That is, the Wigner distribution is concentrated along $\varphi'(t)$. The case of a multicomponent driving force (the sum of several mono-component signals) can be handled by writing it as a sum of delta functions, each centered about the instantaneous frequency of that component.

We have devised a method whose key idea is to replace the Wigner distribution of the forcing function with a linear chirp centered at its instantaneous frequency.[14] Details are described in Reference 16 but basically we first compute the exact Wigner $W(t, \omega)$ of the solution when a linear quadratic phase signal is the input. As can be seen from Equation 4.93 the solution can be written as $W(t, \omega) = W(\tau)$, where $\tau = t - \omega/\beta$. Then, substitute $\tau = t - \Phi(\omega)$ where $\Phi(\omega)$ is evaluated by inversion of $\varphi'(t)$. This is simply obtained by setting $\omega = \varphi'(t)$ and then solving $t = \Phi(\omega)$. The approach will work as long as the instantaneous frequency $\varphi'(t)$ of the input driving force is slowly varying. As an example, consider the forcing function

$$f(t) = Ae^{j\beta t^2/2 + j\gamma t^3/3}, \tag{4.12}$$

where

$$\omega_f(t) = \beta t + \gamma t^2. \tag{4.13}$$

Following the steps described above, we have

$$W_{f,f}(t, \omega) = A\delta(t + \beta/\gamma - \sqrt{(\beta/\gamma)^2 + \omega/\gamma}), \tag{4.14}$$

where A is the normalizing constant and where we first consider positive times. The Wigner is obtained from Equation 4.93 (the underdamped case of the gliding tone problem) by substituting

$$\tau = t + \beta/\gamma - \sqrt{(\beta/\gamma)^2 + \omega/\gamma} \tag{4.15}$$

into it. For negative times we use

$$\tau = t + \beta/\gamma + \sqrt{(\beta/\gamma)^2 + \omega/\gamma}. \tag{4.16}$$

In Figure 4.6 we plot the result using this method and in Figure 4.7 we compare to the exact result. The exact result was obtained by solving the differential equation numerically for $x(t)$ and then numerically calculating the Wigner distribution of the solution. The approximation is very good, and this behavior can be observed experimentally to be true in general for increasing values of β. We notice that the goodness of the approximation increases as $t, \omega \to +\infty$. At low values of time and frequency, we notice that the "anticausal" terms cannot be neglected. By anticausal terms we mean the energy located in the time-frequency region $n = 20 - 40$, $f = 0.05 - 0.15$. We call these oscillatory terms anticausal because they arise well before the main interaction between the input driving force and the system, located on the instantaneous frequency of $f(t)$. They represent the well-known cross terms, artifacts generated by Wigner distribution due to its quadratic formulation.

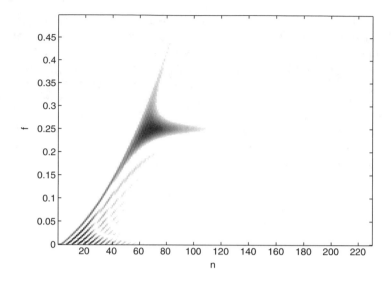

FIGURE 4.6
Approximation of the Wigner distribution $W_{x,x}(t, \omega)$ of the solution $x(t)$ of the harmonic oscillator when the forcing function is $f(t) = Ae^{j\beta t^2/2 + j\gamma t^3/3}$. The parameters are $\beta = 1.2\pi$, $\gamma = 1.5$.

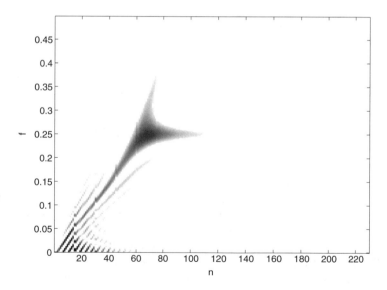

FIGURE 4.7
Numerical evaluation of the Wigner $W_{x,x}(t, \omega)$.

4.4   Stochastic Processes

We now consider the case of random processes. The Wigner distribution of a random process $x(t)$ is defined by

$$W_{x,x}(t, \omega) = \frac{1}{2\pi} \int E\left[x^*\left(t - \frac{1}{2}\tau\right) x\left(t + \frac{1}{2}\tau\right)\right] e^{-i\tau\omega} d\tau, \qquad (4.17)$$

and is called the *Wigner spectrum*.[9,10,13] In Equation 4.17, $E[\ ]$, represents the ensemble averaging operator. We proceed as we did in the previous section by studying the case of the harmonic oscillator

$$\frac{d^2x(t)}{dt^2} + 2\mu \frac{dx(t)}{dt} + \omega_0^2 x(t) = F(t), \qquad (4.18)$$

where now $F(t)$ is a Gaussian random white noise with autocorrelation function

$$R_{F,F}(t_1, t_2) = N_0 \delta(t_1 - t_2). \qquad (4.19)$$

This problem has been widely studied and a fundamental quantity is the power spectrum of $x(t)$.[3] For this case we expect the spectrum to be constant in time if we consider the system to be in "stationary" phase, that is, far away from the initial conditions. Standard methods of stochastic analysis allow one to compute analytically the power spectrum of $x(t)$. However, to show the effectiveness of our method we consider the problem where we make one of the coefficients time varying, a case that the standard methods cannot handle. But first we consider the constant coefficient case, and we show that our method is in perfect agreement with the classical result: it returns an instantaneous spectrum that does not change with time.

The standard result for the power spectrum, $G_x(\omega)$ of $x(t)$, defined as the Fourier transform of the autocorrelation function $R_x(\tau)$, is

$$G_x(\omega) = |H(\omega)|^2 G_F(\omega), \qquad (4.20)$$

where $H(\omega)$ is the transfer function of the system. It is obtained by evaluating the polynomial defined in Equation 4.84 in $i\omega$, that is,

$$H(\omega) = P(i\omega), \qquad (4.21)$$

Notice that with constant coefficients the polynomial is no longer a function of time. Because the power spectrum of the Gaussian noise is constant and equal to $G_F(\omega) = N_0$, we obtain from Equations 4.20 and 4.21

$$G_x(\omega) = \frac{N_0}{\left(\omega_0^2 - \omega^2\right)^2 + 4\mu^2\omega^2}. \qquad (4.22)$$

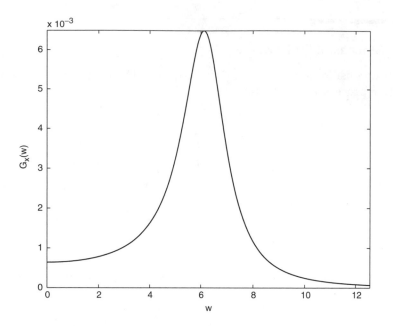

FIGURE 4.8
Power spectrum of the state variable $x(t)$ of the harmonic oscillator of Equation 4.18. Notice the bandpass behavior of the function.

This is the result obtained by Wang and Uhlenbeck.[3] In Figure 4.8 we represent the power spectrum, $G_x(\omega)$, when the following set of parameters is chosen

$$\mu = 1, \quad \omega_0 = 2\pi, \quad N_0 = 1. \tag{4.23}$$

We now obtain the same result using our method. We rewrite Equation 4.18 in polynomial notation

$$\left[D^2 + 2\mu D + \omega_0^2\right] x(t) = F(t). \tag{4.24}$$

The governing equation for the Wigner distribution of $x(t)$ is, from Equation 4.91,

$$\left[A^2 + 2\mu A + \omega_0^2\right]\left[B^2 + 2\mu B + \omega_0^2\right] W_x(t, \omega) = W_F(t, \omega), \tag{4.25}$$

or

$$\left[b_4 \frac{\partial^4}{\partial t^4} + b_3 \frac{\partial^3}{\partial t^3} + b_2 \frac{\partial^2}{\partial t^2} + b_1 \frac{\partial}{\partial t} + b_0\right] W_x(t, \omega) = W_F(t, \omega). \tag{4.26}$$

One can show that the coefficient b_0 is inversely proportional to the square modulus of the transfer function

$$b_0 = (|H(\omega)|^2)^{-1}. \tag{4.27}$$

This means that in general Equation 4.26 can be written as

$$\left[b_4 \frac{\partial^4}{\partial t^4} + b_3 \frac{\partial^3}{\partial t^3} + b_2 \frac{\partial^2}{\partial t^2} + b_1 \frac{\partial}{\partial t} + (|H(\omega)|^2)^{-1} \right] W_x(t, \omega) = W_F(t, \omega). \quad (4.28)$$

Now, it is known that the Wigner spectrum of white noise is

$$W_F(t, \omega) = N_0. \quad (4.29)$$

By substituting Equation 4.29 into the equation for the Wigner, Equation 4.28, we readily find that the solution is constant with respect to time, and in particular

$$W_x(t, \omega) = N_0 |H(\omega)|^2. \quad (4.30)$$

We note that we have solved Equation 4.28 putting the initial conditions at $t = -\infty$. Because we are considering a stable system, at any finite time t we will assume that any initial condition has come to a steady-state solution and hence use the term *stationary solution*. Equation 4.30 is the same as the classical power spectrum but we emphasize that the Wigner spectrum is a two-dimensional quantity and what we have shown is that it does not change in time. A simple explanation is that, considering Equation 4.18 as a filtering problem, the system filters the input noise $F(t)$ with a bandpass filter to produce an output signal $x(t)$. But because both the input random noise and the filter/system are stationary in time, then the output is also stationary.

4.4.1 Harmonic Oscillator with Time-Dependent Coefficients

We now consider the considerably more difficult problem

$$\frac{d^2 x}{dt^2} + 2\mu \frac{dx}{dt} + K(t)x = F(t) \quad (4.31)$$

with

$$K(t) = \omega_0^2 + \epsilon t, \qquad \epsilon \ll 1. \quad (4.32)$$

The equation for the Wigner spectrum is

$$\left[A_t^2 + 2\mu A_t + K(\mathcal{E}_t) \right] \left[B_t^2 + 2\mu B_t + K(\mathcal{F}_t) \right] W_x(t, \omega) = W_F(t, \omega), \quad (4.33)$$

and if we take into account the fact that $K(t)$ is slowly varying, we can approximate this equation by

$$\left[\frac{1}{16} \frac{\partial^4}{\partial t^4} + \frac{\mu}{2} \frac{\partial^3}{\partial t^3} + \left(\mu^2 + \frac{1}{2}(K(t) + \omega^2) \right) \frac{\partial^2}{\partial t^2} \right.$$
$$\left. + 2\mu(K(t) + \omega^2) \frac{\partial}{\partial t} + (K(t) - \omega^2)^2 + 4\mu^2 \omega^2 \right] W_x = W_f \quad (4.34)$$

This equation contains no derivatives with respect to ω and can hence be solved as an ordinary differential equation in t. We choose as initial conditions the following:

$$W_x(0, \omega) = \frac{N_0}{\left(\omega_0^2 - \omega^2\right)^2 + 4\mu^2\omega^2}, \tag{4.35}$$

$$\frac{\partial W_x}{\partial t}(0, \omega) = \frac{\partial^2 W_x}{\partial t^2}(0, \omega) = \frac{\partial^3 W_x}{\partial t^3}(0, \omega) = 0. \tag{4.36}$$

This choice establishes that the system is in the condition $K(t) = \omega_0^2$ for all negative times, and that at $t = 0$, $K(t)$ starts varying with time according to Equation 4.32. In Figure 4.9 we show the solution to the approximate differential equation, Equation 4.34. Notice that the computed Wigner spectrum $W_x(t, \omega)$ looks exactly as we expected, in the sense that it is exactly a time-varying bandpass function. Such solutions have not been obtained before by any means, and therefore we undertook a major simulation to ascertain the accuracy of our answer. We do not describe the numerical simulations here, but the result is shown in Figure 4.10. The agreement is excellent.

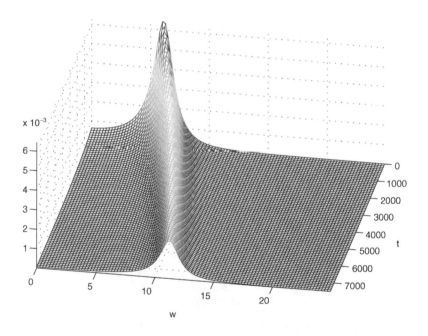

FIGURE 4.9
Behavior of the instantaneous power spectrum obtained by solving the equation for the Wigner spectrum.

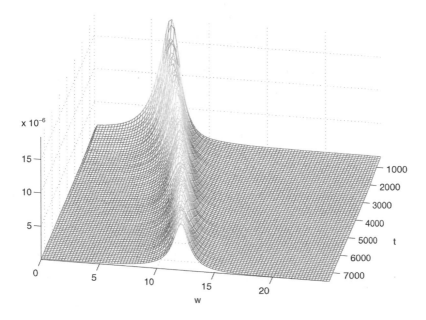

FIGURE 4.10
Behavior of the instantaneous power spectrum obtained by simulation of the harmonic oscillator with stochastic input.

4.5 Partial Differential Equations

As with the case of ordinary differential equations, we will illustrate our method with specific examples. The examples we use are the Schrödinger free particle equation and the diffusion equation:

$$\frac{\partial \psi}{\partial t} = ia \frac{\partial^2 \psi}{\partial x^2} \qquad \text{Schrödinger free particle; } a = \hbar/(2m) \qquad (4.37)$$

$$\frac{\partial u}{\partial t} = D \frac{\partial^2 u}{\partial x^2} \qquad \text{Diffusion equation; } D = \text{diffusion coefficient.} \qquad (4.38)$$

We have chosen these equations because they are fundamental. In addition, they are superficially similar. We want to illustrate how these equations compare in the Wigner representation, that is, in phase space. The Wigner distribution for the Schrödinger equation has been studied for more than 70 years, but we believe it has not been applied to the diffusion equation and equations of that type. The respective momentum functions defined by

$$\phi(p, t) = \frac{1}{\sqrt{2\pi}} \int \psi(x, t) e^{-ixp} dx \qquad (4.39)$$

$$U(p, t) = \frac{1}{\sqrt{2\pi}} \int u(x, t) e^{-ixp} dx \qquad (4.40)$$

satisfy the following equations of motion:

$$\frac{\partial \phi}{\partial t} = -ia p^2 \phi, \qquad (4.41)$$

$$\frac{\partial U}{\partial t} = -Dp^2 U. \qquad (4.42)$$

The Wigner distribution combines both representations; that is, it is a joint representation of position and momentum. However, we point out a fundamental difference between the interpretation of the solution of the Schrödinger and diffusion equations. In the case of the Schrödinger equation, $|\psi(x, t)|^2$ and $|\phi(p, t)|^2$ are the densities of position and momentum. However, in the case of diffusion, $u(x, t)$ and $U(p, t)$ are the densities. Thus, in the case of the Schrödinger equation the Wigner distribution satisfies the so-called marginal conditions, but that is not the case for the diffusion equation. Nonetheless the Wigner distribution gives an indication of how momentum and position are jointly related. More precisely, one should think of the representation as a joint representation in position and spatial frequency.

The Wigner distribution for a field, $u(x, t)$, is

$$W_u(x, p, t) = \frac{1}{2\pi} \int u^*\left(x - \frac{1}{2}\tau, t\right) u\left(x + \frac{1}{2}\tau, t\right) e^{-i\tau p} d\tau \qquad (4.43)$$

$$= \frac{1}{2\pi} \int U^*\left(p + \frac{1}{2}\theta, t\right) U\left(p - \frac{1}{2}\theta, t\right) e^{-i\theta x} d\theta. \qquad (4.44)$$

We will use ψ and u to signify the solution of Schrödinger and diffusion equations, respectively, and use $W_\psi(x, p, t)$ and $W_u(x, p, t)$ for their respective Wigner distributions. In Appendix 4.4 we give the general procedure for obtaining the equations of motion for the Wigner distribution for a partial differential equation. When these methods are applied to the above equations, one obtains

$$\frac{\partial W_\psi}{\partial t} = -2pa \frac{\partial W_\psi}{\partial x}, \qquad (4.45)$$

$$\frac{\partial W_u}{\partial t} = \frac{D}{2} \frac{\partial^2 W_u}{\partial x^2} - 2Dp^2 W_u. \qquad (4.46)$$

Equation 4.45 was first obtained by Wigner and Moyal and its properties have been studied for many years.

It is quite interesting that, while the only difference between the two original equations, Equations 4.37 and 4.38, is an i, the difference in the Wigner distribution equation of motion is quite dramatic. More importantly, we will see that in the Wigner domain both the mathematics and insight become clearer. We point out that the results we present for the Schrödinger equation are classic in the work of Wigner, Moyal, and many others, but the results we present for the diffusion equation we believe to be new. We mention that these equations may be related to the respective Fokker–Planck equations, but that will not be pursued here.

We point out that the equation for diffusion with drift is

$$\frac{\partial u}{\partial t} + c\frac{\partial u}{\partial x} = D\frac{\partial^2 u}{\partial x^2}, \tag{4.47}$$

and the respective Wigner equation of motion is

$$\frac{\partial W_u}{\partial t} + c\frac{\partial W_u}{\partial x} = \frac{D}{2}\frac{\partial^2 W_u}{\partial x^2} - 2Dp^2 W_u. \tag{4.48}$$

However, no generality is lost by taking the drift term equal to zero because if $u(x, t)$ solves the no-drift equation, Equation 4.38, then $u(x - ct, t)$ will solve the equation with drift. Similarly, if $W_u(x, p, t)$ satisfies Equation 4.46, then $W_u(x - ct, p, t)$ satisfies Equation 4.48.

4.5.1 Green's Function

4.5.1.1 Schrödinger Equation

Suppose we want to solve the initial value problem for the Schrödinger equation. That is, given $\psi(x, 0)$ we want $\psi(x, t)$, where $t > 0$. The solution is

$$\psi(x, t) = \int G_\psi(x, x', 0)\psi(x', 0)dx', \tag{4.49}$$

where $G_\psi(x, x', t)$ is the Green's function:

$$G_\psi(x, x', t) = \frac{1}{\sqrt{4\pi i a t}} \exp\left[-\frac{(x - x')^2}{4iat}\right]. \tag{4.50}$$

In momentum space the initial value problem becomes particularly easy. From Equation 4.41 we have

$$\phi(p, t) = e^{-iap^2 t}\phi(p, 0). \tag{4.51}$$

Now consider the same problem for the Wigner distribution, that is, given $W(x, p, 0)$ we want $W(x, p, t)$. From Equation 4.45 it follows that

$$W_\psi(x, p, t) = W_\psi(x - 2apt, p, 0), \tag{4.52}$$

a result first obtained by Wigner and Moyal. Thus, a remarkable simplification is achieved in phase space. But furthermore in phase space we understand what is going on. It shows that as time progresses the phase space point moves with a constant velocity in the x direction but does not move at all in the p direction. The velocity in the x direction is $2ap$.

4.5.1.2   Diffusion Equation

Now consider the diffusion equation. Using the Green's function approach, one has

$$u(x, t) = \int G_u(x, x', t)u(x', 0)dx', \tag{4.53}$$

where

$$G_u(x, x', t) = \frac{1}{\sqrt{4\pi Dt}} \exp\left[-\frac{(x - x')^2}{4Dt}\right], \tag{4.54}$$

and in momentum space

$$U(p, t) = e^{-Dp^2 t}U(p, 0). \tag{4.55}$$

Now consider the Wigner distribution. In Appendix 4.6 we show that

$$W_u(x, p, t) = \frac{1}{\sqrt{2\pi Dt}}e^{-Dp^2 t}\int \exp\left[-\frac{(x - x')^2}{2Dt}\right]W_u(x', p, 0)dx'. \tag{4.56}$$

In Figure 4.11 through Figure 4.14 we show the Wigner distribution computed at times $t = 0.01, 0.1, 1, 10$ and with $D = 100$.

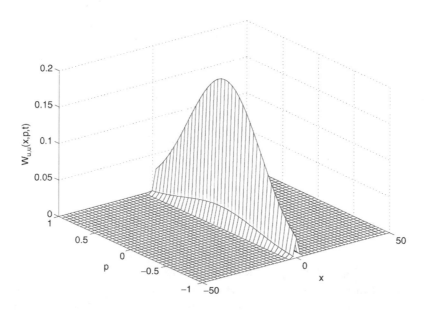

FIGURE 4.11

Wigner distribution of the Green's function for the diffusion equation, for $t = 0.01$.

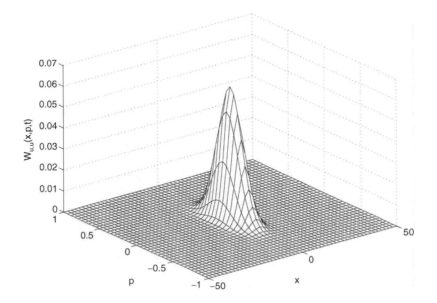

FIGURE 4.12
Wigner distribution of the Green's function for the diffusion equation, for $t = 0.1$.

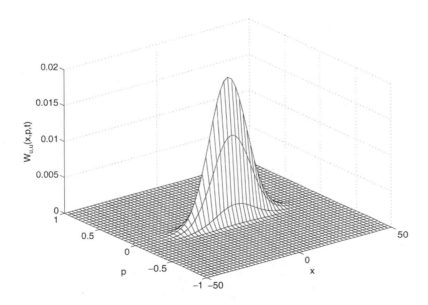

FIGURE 4.13
Wigner distribution of Green's function for the diffusion equation, for $t = 1$.

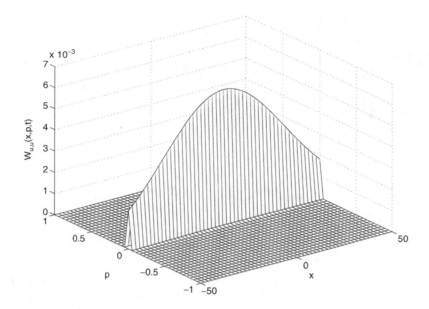

FIGURE 4.14
Wigner distribution of Green's function for the diffusion equation, for $t = 10$.

4.5.2   Wigner Distribution of Green's Function

It is of interest to calculate the Wigner distribution of Green's function for each case. For the Schrödinger case

$$G_\psi(x, x', t, t') = \frac{1}{\sqrt{4\pi i a t}} \exp\left[-\frac{(x - x')^2}{4iat}\right], \tag{4.57}$$

the Wigner distribution is (time is assumed to be positive)

$$W_{G_\psi}(x, p, t) = \frac{1}{2\pi}\delta(x' - x - 2apt). \tag{4.58}$$

For the diffusion equation where

$$G_u(x, x', t) = \frac{1}{\sqrt{4\pi Dt}} \exp\left[-\frac{(x - x')^2}{4Dt}\right], \tag{4.59}$$

the Wigner distribution is

$$W_{G_u}(x, p, t) = \frac{1}{\sqrt{8\pi^3 Dt}} \exp\left[-\frac{(x - x')^2}{2Dt} - 2Dtp^2\right]. \tag{4.60}$$

Thus we see the Wigner distribution shows a significant physical difference between the two Green's functions although superficially the original wave equations and Green's functions are very similar. In the case of $W_{G_\psi}(x, p, t)$, Equation 4.58 shows that each spatial point is transformed by a translation in phase space, the translation being $x' \rightarrow x + 2apt$. But for the case $G_u(x, x', t)$ we see that each point becomes spread/contracted. The position spreads, but the momentum contracts!

EXAMPLE
We take a specific example and work out the two cases side by side. Take

$$\psi(x, 0) = \frac{1}{(2\pi\sigma^2)^{1/4}} \exp\left[-\frac{(x - x_0)^2}{4\sigma^2} + ip_0 x\right], \tag{4.61}$$

$$u(x, 0) = \frac{1}{\sqrt{2\pi\sigma^2}} \exp\left[-\frac{(x - x_0)^2}{2\sigma^2}\right], \tag{4.62}$$

and we note that the densities in both cases are the same. The respective initial Wigner distributions are calculated to be

$$W_\psi(x, p, 0) = \frac{1}{\pi} \exp\left[-\frac{(x - x_0)^2}{2\sigma^2} - 2\sigma^2(p - p_0)^2\right], \tag{4.63}$$

$$W_u(x, p, 0) = \frac{1}{2\pi\sqrt{\pi}\sigma} \exp\left[-\frac{(x - x_0)^2}{\sigma^2} - \sigma^2 p^2\right]. \tag{4.64}$$

Thus, initially, as expected both Wigner distributions are essentially the same. However, as time evolves

$$W_\psi(x, p, t) = \frac{1}{2\pi^2} \exp\left[-\frac{(x - 2apt - x_0)^2}{2\sigma^2} - 2\sigma^2(p - p_0)^2\right], \tag{4.65}$$

$$W_u(x, p, t) = \frac{e^{-(2Dt+\sigma^2)p^2}}{4\pi^3\sigma\sqrt{2Dt}} \int \exp\left[-\frac{(x' - x)^2}{2Dt}\right] \exp\left[-\frac{(x' - x_0)^2}{\sigma^2}\right] dx' \tag{4.66}$$

$$= \frac{1}{4\pi^{5/2}} \frac{1}{\sqrt{2Dt + \sigma^2}} \exp\left[-\frac{(x - x_0)^2}{2Dt + \sigma^2} - (2Dt + \sigma^2)p^2\right], \tag{4.67}$$

where for the last step we have used

$$\int e^{-\alpha(x-x_1)^2 - \beta(x-x_2)^2} dx = \sqrt{\frac{\pi}{\alpha + \beta}} \exp\left[-\frac{\alpha\beta}{\alpha + \beta}(x_1 - x_2)^2\right]. \tag{4.68}$$

We now discuss the physical meaning. In the case of the Schrödinger equations, the Wigner distribution is just rotating. However, for the diffusion case they are spreading in the x direction and contracting in the p direction.

4.6 Conclusion

We have presented a method to write equations of motion for the Wigner distribution that corresponds to the solution of an ordinary or partial differential equation.

Although the differential equation for the Wigner distribution may seem more complicated than the original equation, it may be easier to solve in certain circumstances because it avoids two procedures, namely, to first solve for $x(t)$ and then obtain the Wigner distribution using it. However, more significantly, phase space considerations are essential to nonstationary situations. By considering the equation of motion for the Wigner distribution one can directly understand the physical situation and also devise new methods of approximation.

Acknowledgment

Work supported by the U.S. Air Force Information Institute Research Program (Rome, NY) and the NSA HBCU/MI program.

Appendix 4.1: Notations and Definitions

All the integrals without limits mean integration from $-\infty$ to ∞. We define the Wigner distribution for a signal $x(t)$ by

$$W_{x,x}(t, \omega) = \frac{1}{2\pi} \int x^*\left(t - \frac{1}{2}\tau\right) x\left(t + \frac{1}{2}\tau\right) e^{-i\tau\omega} \, d\tau \qquad (4.69)$$

and the cross Wigner distribution between two signals, $x(t)$ and $y(t)$, by

$$W_{x_1,x_2}(t, \omega) = \frac{1}{2\pi} \int x_1^*\left(t - \frac{1}{2}\tau\right) x_2\left(t + \frac{1}{2}\tau\right) e^{-i\tau\omega} \, d\tau. \qquad (4.70)$$

The Wigner distribution is nonlinear in the signal in that the Wigner of the sum of two signals $x_1(t)$, $x_2(t)$ is different from the sum of their respective Wigner distributions. In fact,

$$W_{x_1+x_2,x_1+x_2}(t, \omega) = W_{x_1,x_1}(t, \omega) + W_{x_2,x_2}(t, \omega) + W_{x_1,x_2}(t, \omega) + W_{x_2,x_1}(t, \omega). \qquad (4.71)$$

It is, however, linear with respect to one of the arguments:

$$W_{x,y_1+y_2}(t, \omega) = W_{x,y_1}(t, \omega) + W_{x,y_2}(t, \omega). \qquad (4.72)$$

When we deal with partial differential equations, we need to consider multi-dimensional signals (fields), $u(x, t)$. For a field the Wigner distribution is

$$W_{u,u}(x, p, t) = \frac{1}{2\pi} \int u^*\left(x - \frac{1}{2}\tau_x, t\right) u\left(x + \frac{1}{2}\tau_x, t\right) e^{-i\tau_x p} \, d\tau_x, \qquad (4.73)$$

and analogously the cross Wigner distribution between two fields, $u_1(x, t)$ and $u_2(x, t)$, is

$$W_{u_1,u_2}(x, p, t) = \frac{1}{2\pi} \int u_1^*\left(x - \frac{1}{2}\tau_x, t\right) u_2\left(x + \frac{1}{2}\tau_x, t\right) e^{-i\tau_x p} \, d\tau_x. \qquad (4.74)$$

For partial differential equations it is generally not possible to write an equation for the Wigner distribution, $W_{u,u}(x, p, t)$, corresponding to an arbitrary equation governing the field. It is nevertheless possible to always derive such an equation for the more general Wigner distribution, which we define by

$$Z_{u,u}(x, p, t, \omega)$$
$$= \frac{1}{(2\pi)^2} \int u^*\left(x - \frac{1}{2}\tau_x, t - \frac{1}{2}\tau\right) u\left(x + \frac{1}{2}\tau_x, t + \frac{1}{2}\tau\right) e^{-i\tau\omega - i\tau_x p} \, d\tau \, d\tau_x.$$
$$(4.75)$$

We note that the ordinary Wigner distribution, $W(x, p, t)$, may be obtained from $Z(x, p, t, \omega)$ by way of

$$W_{u,u}(x, p, t) = \int Z_{u,u}(x, p, t, \omega) \, d\omega. \qquad (4.76)$$

A significant simplification in notation is achieved by defining the following operators:

$$A_t = \frac{1}{2}\frac{\partial}{\partial t} - i\omega, \qquad B_t = \frac{1}{2}\frac{\partial}{\partial t} + i\omega, \qquad (4.77)$$

$$\mathcal{E}_t = t + \frac{1}{2i}\frac{\partial}{\partial \omega}, \qquad \mathcal{F}_t = t - \frac{1}{2i}\frac{\partial}{\partial \omega}. \qquad (4.78)$$

Operators of this kind were defined by Moyal. When dealing with a field $u(x, t)$ one has to introduce the additional operators

$$A_x = \frac{1}{2}\frac{\partial}{\partial x} - ip, \qquad B_x = \frac{1}{2}\frac{\partial}{\partial x} + ip. \qquad (4.79)$$

$$\mathcal{E}_x = x + \frac{1}{2i}\frac{\partial}{\partial p}, \qquad \mathcal{F}_x = x - \frac{1}{2i}\frac{\partial}{\partial p}. \qquad (4.80)$$

Also, we indicate ordinary differentiation in the following alternative ways:

$$\dot{g}(t) = \frac{d}{dt}g(t), \qquad g^{(n)} = \frac{d^n}{dt^n}g(t), \qquad (4.81)$$

and we use the differential operator:

$$D = \frac{d}{dt}. \qquad (4.82)$$

Appendix 4.2: Ordinary Differential Equations

Here, we give the general method for writing the Wigner distribution equation for ordinary differential equations. For

$$a_n \frac{d^n x}{dt^n} + a_{n-1} \frac{d^{n-1} x}{dt^{n-1}} + \cdots + a_1 \frac{dx}{dt} + a_0 x = f(t), \tag{4.83}$$

which we rewrite in polynomial notation

$$P(D)x(t) = f(t), \tag{4.84}$$

where

$$P(D) = a_n D^n + a_{n-1} D^{n-1} \ldots a_1 D + a_0, \tag{4.85}$$

the equation for the Wigner distribution is

$$P^*(A_t)P(B_t)W_{x,x}(t, \omega) = W_{f,f}(t, \omega). \tag{4.86}$$

If the driving force is zero, then one can show that

$$P^*(A_t)W_{x,x} = P(B_t)W_{x,x} = 0. \tag{4.87}$$

Time-Dependent Coefficients

For

$$a_n(t)\frac{d^n x}{dt^n} + a_{n-1}(t)\frac{d^{n-1} x}{dt^{n-1}} + \cdots + a_1(t)\frac{dx}{dt} + a_0(t)x = f(t) \tag{4.88}$$

or

$$P(D, t)x(t) = f(t), \tag{4.89}$$

where

$$P(D, t) = a_n(t)D^n + a_{n-1}(t)D^{n-1} + \cdots + a_1(t)D + a_0(t), \tag{4.90}$$

we have shown that

$$P^*(A_t, \mathcal{E}_t)P(B_t, \mathcal{F}_t)W_{x,x}(t, \omega) = W_{f,f}(t, \omega). \tag{4.91}$$

For the zero driving force case, one obtains

$$P^*(A_t, \mathcal{E}_t)W_{x,x}(t, \omega) = P(B_t, \mathcal{F}_t)W_{x,x}(t, \omega) = 0. \tag{4.92}$$

Appendix 4.3: Exact Solution to the Gliding Tone Problem

The Wigner distribution of $x(t)$, which satisfies Equation 4.1, is

$$W(t, \omega) = \frac{2}{|\beta|} \frac{u(\tau)}{z_2 - z_1} \frac{1}{\bar{z}_1 - z_1} \left(\frac{e^{-2z_1\tau} - e^{-2\bar{z}_2\tau}}{\bar{z}_2 - z_1} - \frac{e^{-2\bar{z}_1\tau} - e^{-2\bar{z}_2\tau}}{\bar{z}_2 - \bar{z}_1} \right)$$

$$- \frac{1}{\bar{z}_1 - z_2} \left(\frac{e^{-2z_2\tau} - e^{-2\bar{z}_2\tau}}{\bar{z}_2 - z_2} - \frac{e^{-2\bar{z}_1\tau} - e^{-2\bar{z}_2\tau}}{\bar{z}_2 - \bar{z}_1} \right) \tag{4.93}$$

with

$$\tau = t - \omega/\beta \tag{4.94}$$

and

$$z_1 = -j\omega + \mu - \sqrt{\mu^2 - \omega_0^2}, \qquad \bar{z}_1 = j\omega + \mu - \sqrt{\mu^2 - \omega_0^2}, \tag{4.95}$$

$$z_2 = -j\omega + \mu + \sqrt{\mu^2 - \omega_0^2}, \qquad \bar{z}_2 = j\omega + \mu + \sqrt{\mu^2 - \omega_0^2}, \tag{4.96}$$

and where $u(t)$ is the step function,

$$u(t) = \begin{cases} 1 & t \geq 0. \\ 0 & \text{otherwise} \end{cases} \tag{4.97}$$

In deriving Equation 4.93, we have used the initial conditions given in Equation 4.11. We now explicitly list the results for the underdamped, overdamped, and critically damped case:

Underdamped:

$$W(t, \omega) = \frac{1}{2|\beta|\omega_c} u(\tau) e^{-2\mu\tau} \left[\frac{\sin(2(\omega - \omega_c)\tau)}{\omega(\omega - \omega_c)} - \frac{\sin(2(\omega + \omega_c)\tau)}{\omega(\omega + \omega_c)} \right]. \tag{4.98}$$

Overdamped:

$$W(t, \omega) = \frac{1}{|\beta|} u(\tau) e^{-2\mu\tau} \left[\frac{\sin(2\omega\tau)\cosh(2\omega_c\tau)}{\omega(\omega^2 + \omega_c^2)} - \frac{\cos(2\omega\tau)\sinh(2\omega_c\tau)}{\omega_c(\omega^2 + \omega_c^2)} \right].$$

Critically Damped:

$$W(t, \omega) = \frac{1}{|\beta|} u(\tau) e^{-2\mu\tau} \frac{\sin(2\omega\tau) - 2\omega\tau\cos(2\omega\tau)}{\omega^3}. \tag{4.99}$$

In the solutions presented there are singularities at some values of ω that can be computed by evaluating the corresponding limits. We show the computed limits for the three cases:

Underdamped:

$$\lim_{\omega \to \pm \omega_c} W(t, \omega) = \frac{1}{2|\beta|\omega_c} u(\tau) e^{-2\mu\tau} \left[\frac{4\omega_c\tau - \sin(4\omega_c\tau)}{2\omega_c^2} \right], \quad (4.100)$$

$$\lim_{\omega \to 0} W(t, \omega) = \frac{1}{|\beta|} u(t) e^{-2\mu t} \left[\frac{\sin(2\omega_c t) - 2\omega_c t \cos(2w_c t)}{\omega_c^3} \right]. \quad (4.101)$$

Overdamped:

$$\lim_{\omega \to 0} W(t, \omega) = \frac{1}{|\beta|} u(t) e^{-2\mu t} \left[\frac{2\omega_c t \cosh(2\omega_c t) - \sinh(2\omega_c t)}{\omega_c^3} \right].$$

$$(4.102)$$

Critically Damped:

$$\lim_{\omega \to 0} W(t, \omega) = \frac{8}{3} \frac{1}{|\beta|} u(t) t^3 e^{-2\mu t}. \quad (4.103)$$

Appendix 4.4: Partial Differential Equations

An equation for the Wigner distribution can be associated with *any* linear partial differential equation with varying coefficients, and in this appendix we show how that is done. We first define the multidimensional independent variable x as

$$x = (x_1, x_2, \ldots, x_m), \quad (4.104)$$

where m represents the number of dimensions. The field will be indicated by

$$u = u(x) = u(x_1, x_2, \ldots, x_m). \quad (4.105)$$

A compact way of writing a general partial differential equation is the *multi-index notation*. It is based on the index α,

$$\alpha = (\alpha_1, \alpha_2, \ldots, \alpha_m), \quad (4.106)$$

built with integer numbers α_r. We introduce the further notation

$$|\alpha| = \alpha_1 + \alpha_2 + \cdots + \alpha_m. \quad (4.107)$$

With this notation we can write in a compact form a general partial derivative

$$D_\alpha = \frac{\partial^{|\alpha|}}{\partial x_1^{\alpha_1} \partial x_2^{\alpha_2} \cdots \partial x_m^{\alpha_m}}. \tag{4.108}$$

The entire class of linear partial differential equations of order N with arbitrary varying coefficients can then be written as

$$\sum_{|\alpha| \le N} a_\alpha(x) D^\alpha u(x) = f(x), \tag{4.109}$$

where $f(x)$ is the forcing term, and $a_\alpha(x)$ is the general coefficient.

To write the equation for the Wigner distribution, we first need to extend its definition to the case of m-dimensional fields,

$$Z_{u,u}(x, p)$$
$$= \left(\frac{1}{2\pi}\right)^m \int u^*(x_1 - 1/2\tau_1, \ldots, x_m - 1/2\tau_m) u(x_1 + 1/2\tau_1, \ldots, x_m + 1/2\tau_m)$$
$$\times e^{-i\tau_1 p_1 - i\tau_2 p_2 + \cdots - i\tau_m p_m} \, d\tau_1 \, d\tau_2 \cdots d\tau_m, \tag{4.110}$$

where the momentum p is given by

$$p = (p_1, p_2, \ldots, p_m). \tag{4.111}$$

We now define the following operators:

$$A = (A_1, \ldots, A_m), \qquad B = (B_1, \ldots, B_m), \tag{4.112}$$
$$\mathcal{E} = (\mathcal{E}_1, \ldots, \mathcal{E}_m), \qquad \mathcal{F} = (\mathcal{F}_1, \ldots, \mathcal{F}_m), \tag{4.113}$$

where

$$A_r = \frac{1}{2}\frac{\partial}{\partial x_r} - ip_r, \qquad B_r = \frac{1}{2}\frac{\partial}{\partial x_r} + ip_r, \tag{4.114}$$

$$\mathcal{E}_r = x_r + \frac{1}{2i}\frac{\partial}{\partial p_r}, \qquad \mathcal{F}_r = x_r - \frac{1}{2i}\frac{\partial}{\partial p_r}, \tag{4.115}$$

for any integer $r = 1, \ldots, m$. Using these definitions one can write the equation for the Wigner distribution Z associated with the general partial differential equation, Equation 4.109,

$$\left(\sum_{|\alpha| \le N} a_\alpha^*(\mathcal{E}) A^\alpha\right) \left(\sum_{|\beta| \le N} a_\beta(\mathcal{F}) B^\beta\right) Z_{u,u}(x, p) = Z_{f,f}(x, p), \tag{4.116}$$

where the meaning of the powers of the operators A, B is the following:

$$A^\alpha = A_1^{\alpha_1} A_2^{\alpha_2} \cdots A_m^{\alpha_m}. \tag{4.117}$$

When the forcing term, $f(x)$, is zero, one can show that the above reduces to an equation that is of the same order, N, as the initial equation. In particular,

$$\sum_{|\alpha| \leq N} \left(a_\alpha^*(\mathcal{E}) A^\alpha \pm a_\alpha(\mathcal{F}) B^\alpha \right) Z_{u,u}(x, p) = 0. \tag{4.118}$$

Appendix 4.5: Derivation of the Wigner Equation of Motion for Diffusion

We show the derivation of the equation of motion for the Wigner distribution for the diffusion equation. We work out the case of diffusion with drift:

$$\frac{\partial u}{\partial t} + c \frac{\partial u}{\partial x} = D \frac{\partial^2 u}{\partial x^2}, \tag{4.119}$$

where $u = u(x, t)$ is the field, c the drift coefficient, and D the diffusion coefficient. To apply our method we first rewrite the equation as

$$\left[\frac{\partial}{\partial t} + c \frac{\partial}{\partial x} - D \frac{\partial^2}{\partial x^2} \right] u(x, t) = 0. \tag{4.120}$$

We now apply the method described in Appendix 4.4 and obtain two equations for $Z(x, p, t, \omega)$

$$\left[A_t + c A_x - D A_x^2 \right] Z(x, p, t, \omega) = 0, \tag{4.121}$$

$$\left[B_t + c B_x - D B_x^2 \right] Z(x, p, t, \omega) = 0. \tag{4.122}$$

Expanding the operators, we have

$$\left[\frac{1}{2} \frac{\partial}{\partial t} - i\omega + \frac{c}{2} \frac{\partial}{\partial x} - icp - \frac{D}{4} \frac{\partial^2}{\partial x^2} + Dp^2 + iDp \frac{\partial}{\partial x} \right] Z(x, p, t, \omega) = 0, \tag{4.123}$$

$$\left[\frac{1}{2} \frac{\partial}{\partial t} + i\omega + \frac{c}{2} \frac{\partial}{\partial x} + icp - \frac{D}{4} \frac{\partial^2}{\partial x^2} + Dp^2 - iDp \frac{\partial}{\partial x} \right] Z(x, p, t, \omega) = 0. \tag{4.124}$$

We add the two equations to have a real equation for the Wigner $Z(x, p, t, \omega)$

$$\left[\frac{\partial}{\partial t} + c \frac{\partial}{\partial x} - \frac{D}{2} \frac{\partial^2}{\partial x^2} + 2Dp^2 \right] Z(x, p, t, \omega) = 0. \tag{4.125}$$

This is the equation of motion for $Z(x, p, t, \omega)$. However, since ω does not appear in the equation, we can integrate it out to obtain an equation for the standard Wigner distribution, $W(x, p, t)$,

$$\frac{\partial W}{\partial t} + c\frac{\partial W}{\partial x} = \frac{D}{2}\frac{\partial^2 W}{\partial x^2} - 2Dp^2 W, \tag{4.126}$$

which is Equation 4.48 of the text.

Appendix 4.6: Green's Function for the Wigner Distribution

First, we obtain Green's function for the Wigner distribution in terms of Green's function for the field. Suppose we propagate the field by

$$u(x, t) = \int G_u(x, x', t)u(x', 0)dx', \tag{4.127}$$

and the Wigner distribution by (done by Cohen[8] and Moyal,[12] who did it for the Schrödinger equation)

$$W(x, p, t) = \int G_W(x, x', p, p', t)W(x', p', 0)dx'dp'. \tag{4.128}$$

We want to express G_W in terms of G_u. Substituting Equation 4.127 into the definition of the classic Wigner distribution, Equation 4.43, we obtain

$$W_u(x, p, t) = \frac{1}{2\pi}\int\int\int G_u^*\left(x - \frac{1}{2}\tau, x', t\right)G_u\left(x + \frac{1}{2}\tau, x'', t\right)$$

$$\times u^*(x', 0)\, u(x'', 0)e^{-i\tau p}d\tau\, dx'dx''. \tag{4.129}$$

But from the definition of the Wigner distribution, we have

$$u^*(x', t)u(x'', t) = \int W((x' + x'')/2, p, t)e^{-i(x'+x'')p}dp, \tag{4.130}$$

and inserting this into Equation 4.129 we obtain

$$W_u(x, p, t) = \frac{1}{2\pi}\int\int\int\int G_u^*\left(x - \frac{1}{2}\tau, x', t\right)G_u\left(x + \frac{1}{2}\tau, \tau' - x', t\right) \tag{4.131}$$

$$\times W(\tau'/2, p', 0)e^{+i\tau'p'}e^{-i\tau p}d\tau\, d\tau'\, dp'\, dx'. \tag{4.132}$$

Therefore,

$$G_W(x, x', p, p', t) = \frac{1}{2\pi} \int \int \int \int G_u^* \left(x - \frac{1}{2}\tau, x', t \right) G_u \tag{4.133}$$

$$\times \left(x + \frac{1}{2}\tau, \tau' - x', t \right) e^{+i\tau' p'} e^{-i\tau p} d\tau \, d\tau' \, dx'. \tag{4.134}$$

This is a general result that always holds for any field equation.

However, for the cases we are considering it is easier to solve the problem in momentum space

$$\phi(p, t) = e^{-iap^2 t} \phi(p, 0), \tag{4.135}$$

$$U(p, t) = e^{-Dp^2 t} U(p, 0), \tag{4.136}$$

and substituting into the definition of the Wigner distribution we have

$$W_u(x, p, t) = \frac{1}{2\pi} e^{-2Dtp^2} \int e^{-Dt\theta^2/2} \left(p + \frac{1}{2}\theta, 0 \right) U \left(p - \frac{1}{2}\theta, 0 \right) e^{-i\theta x} d\theta, \tag{4.137}$$

$$W_\psi(x, p, t) = \frac{1}{2\pi} \int e^{2ia\theta pt} \phi^* \left(p + \frac{1}{2}\theta, 0 \right) \phi \left(p - \frac{1}{2}\theta, 0 \right) e^{-i\theta x} d\theta. \tag{4.138}$$

But from the definition of the Wigner distribution,

$$U^* \left(p + \frac{1}{2}\theta, 0 \right) U \left(x - \frac{1}{2}\theta, 0 \right) = \int W_u(x, p, 0) e^{i\theta x} dx, \tag{4.139}$$

$$\phi^* \left(p + \frac{1}{2}\theta, 0 \right) \phi \left(x - \frac{1}{2}\theta, 0 \right) = \int W_\psi(x, p, 0) e^{i\theta x} dx, \tag{4.140}$$

and therefore

$$W_u(x, p, t) = \frac{1}{2\pi} e^{-2Dp^2 t} \int \int e^{-D\theta^2 t/2} e^{i\theta(x'-x)} W_u(x', p, 0) dx' d\theta, \tag{4.141}$$

$$W_\psi(x, p, t) = \frac{1}{2\pi} \int \int e^{i2a\theta pt} e^{i\theta(x'-x)} W_\psi(x', p, 0) dx' d\theta. \tag{4.142}$$

In both cases the θ integration can be done to yield

$$W_u(x, p, t) = \frac{1}{\sqrt{2\pi Dt}} e^{-2Dp^2 t} \int \exp \left[-\frac{(x - x')^2}{2Dt} \right] W_u(x', p, 0) dx', \tag{4.143}$$

$$W_\psi(x, p, t) = \int \delta(x' - x + 2apt)W_\psi(x', p, 0)dx' \tag{4.144}$$

$$= W_\psi(x - 2apt, p, 0). \tag{4.145}$$

Therefore, the phase space Green's functions defined by

$$W_\psi(x, p, t) = \int G_{W_\psi}(x, x', p, p', t)W_\psi(x', p', 0)dx'dp', \tag{4.146}$$

$$W_u(x, p, t) = \int G_{W_u}(x, x', p, p', t)W_u(x', p', 0)dx'dp' \tag{4.147}$$

are

$$G_{W_\psi}(x, x', p, p', t) = \delta(x' - x + 2apt)\delta(p' - p), \tag{4.148}$$

$$G_{W_u}(x, x', p, p', t) = \frac{e^{-2Dp^2t}}{\sqrt{2\pi Dt}} \exp\left[-\frac{(x - x')^2}{2Dt}\right]\delta(p' - p). \tag{4.149}$$

References

1. N.F. Barber and F. Ursell, The response of a resonant system to a gliding tone, *Philos. Mag.*, 39, 345–361, 1948.
2. G. Hok, Response of linear resonant systems to excitation of a frequency varying linearly with time, *J. Appl. Phys.*, 19, 242–250, 1948.
3. M.C. Wang and G.E. Uhlenbeck, On the theory of the Brownian motion. II, *Rev. Mod. Phys.*, 17 (2 and 3), 1945.
4. E.P. Wigner, On the quantum correction for thermodynamic equilibrium, *Phys. Rev.*, 40, 749–759, 1932.
5. J.G. Kirkwood, Quantum statistics of almost classical ensembles, *Phys. Rev.*, 44, 31–37, 1933.
6. L. Cohen, Time-frequency distributions—a review, *Proc. IEEE*, 77, 941–981, 1989.
7. L. Cohen, *Time-Frequency Analysis*, Englewood Cliffs, NJ: Prentice-Hall, 1995.
8. L. Cohen, Generalized phase–space distribution functions, *J. Math. Phys.*, 7, 781–786, 1966.
9. M.G. Amin, Time–varying spectrum estimation of a general class of nonstationary processes, *Proc. IEEE*, 74, 1800–1802, 1986.
10. J. Pitton, The statistics of time-frequency analysis, *J. Franklin Inst.*, 337, 379–388, 2000.
11. Special Issue on Applications of Time-Frequency Analysis, P. Loughlin (Ed.), *Proc. IEEE*, 84(9), 1996.
12. J.E. Moyal, Quantum mechanics as a statistical theory, *Proc. Cambridge Philos. Soc.*, 45, 99, 1949.
13. W. Martin and P. Flandrin, Wigner-Ville spectral analysis of nonstationary processes, *IEEE Trans. Acoust. Speech Signal Process.*, 33, 1461–1470, Dec. 1985.

14. L. Galleani and L. Cohen, Direct time-frequency approach to dynamical systems, IEEE-EURASIP NSIP, Baltimore, MD, June 3–6, 2001.
15. L. Galleani and L. Cohen, Two approaches to nonstationary linear invariant systems governed by differential equations, IEEE DSP 2000, Hunt, TX, Oct. 15–18, 2000.
16. L. Galleani and L. Cohen, Approximation of the Wigner distribution for dynamical systems governed by differential equations, *EURASIP J. Appl. Signal Process.*, 2002(1), 67–72, 2002.

5

Weighted Myriad Filters

Gonzalo R. Arce, Juan G. Gonzalez, and Yinbo Li

CONTENTS

5.1 Introduction

In recent years, there has been considerable interest in signal processing based on α-stable distributions. The motivations are simple yet profound. First, good empirical fits are often found through the use of stable distributions on data exhibiting skewness and heavy tails. Second, there is solid theoretical justification that non-Gaussian stable processes emerge in practice, e.g., multiple access interference in a Poisson distributed communication network,[1] Internet traffic,[2] and numerous other examples as described in Uchaikin and Zolotarev[3] and in Feller.[4] The third argument for modeling with stable distributions is perhaps the most significant and compelling. Stable distributions satisfy an important generalization of the central limit theorem, which states that the only possible limit of normalized sums of independent and identically distributed terms is stable.[5] A wide variety of impulsive processes found in these applications arise as the superposition of many small independent effects. While Gaussian models are clearly inappropriate, stable distributions

0-8493-1427-5/04/$0.00+$1.50

thus have the theoretical underpinnings to accurately model these types of impulsive processes.[6,7] Stable models are appealing because the generalization of the central limit theorem explains the apparent contradictions of its "ordinary" version, which could not naturally explain the presence of heavy-tailed signals.

Stable models are thus increasingly being used broadly, including in finance and economics[8,9] and in communication systems.[6,10] A number of monographs providing in-depth discussion of stable processes have recently appeared: Zolotarev,[7] Samorodnitsky and Taqqu,[5] Nikias and Shao,[6] Uchaikin and Zolotarev,[3] Adler et al.,[11] and Nolan.[12]

Stable distributions have a parameter α ($0 < \alpha \leq 2$), called the *characteristic exponent*, which controls the heaviness of their tails; a smaller α signifies a heavier-tailed distribution. For $0 < \alpha < 2$, α-stable random variables have infinite variance. The limiting case $\alpha = 2$ leads to the Gaussian distribution, while the case $\alpha = 1$ corresponds to the Cauchy distribution. The Gaussian and Cauchy distributions are the only *symmetric* α-stable distributions having closed-form expressions for their density functions. In addition to the generalized central limit theorem, stable distributions also satisfy the so-called *stability property*: the sum of two independent stable random variables with the same characteristic exponent is also stable with the same characteristic exponent.

Having the rich modeling characteristics of stable distributions at hand, how can signal processing algorithms, suitable for processing and analyzing stable processes, be constructed? An approach that received considerable attention is based on the concept of *fractional lower-order moments*.[6] In this chapter we describe a completely different approach based on the so-called *weighted myriad* filters derived from the theory of *M*-estimation. Much as the Gaussian assumption has motivated the development of linear filtering theory, weighted myriad filters are motivated by the need for a flexible filter class with increased efficiency in non-Gaussian impulsive environments that appear in engineering practice.

The foundation of the proposed algorithms lies in the definition of the *sample myriad* as an *M*-estimator derived from tunable cost functions of the form

$$\rho(X) = \log[K^2 + X^2], \tag{5.1}$$

where K is the tunable parameter. Along the range of tuning values of K, the sample myriad enjoys optimality properties in several practical impulsive models, including the α-stable family. The possibility of tuning the parameter K provides the myriad filter with a rich variety of modes of operation that range from *highly resistant* mode-type estimators to the very efficient class of linear FIR filters.

Weighted myriad filters enjoy the richness of the theory of *M*-estimators, and provide a flexible yet simple framework inherently more powerful than weighted median filters. Because the myriad filter class subsumes that of linear FIR filters, weighted myriad filters can also be tuned to operate efficiently under the Gaussian model. Much as the mean and median have had a profound impact on signal processing, the sample myriad and its related myriad filtering framework lead to a powerful theory upon which *efficient* signal processing algorithms can be developed for applications exhibiting impulsive processes. We begin our discussion with a brief introduction to the practical impulsive models that motivate the definition of the myriad.

5.2 α-Stable Distributions

A common characterization of an α-stable random variable is given by the generalized form of the central limit theorem. Informally,

> *A random variable X is α-stable if it can be the limit of a normalized sum of (possibly shift corrected) i.i.d. random variables*.*

Thanks to the central limit theorem, the Gaussian distribution is an obvious member of the α-stable family. An analytic characterization of the full class of symmetric α-stable distributions was first introduced by Lévy in 1925.[13] According to it, symmetric α-stable random variables follow a characteristic function of the form

$$\phi(\omega) = e^{-\gamma|\omega|^\alpha}. \tag{5.2}$$

The parameter γ, usually called the *dispersion*, is a positive constant related to the scale of the distribution. The parameter α is usually called the *characteristic exponent* or *index*. It can be proved that, in order for Equation 5.2 to define a characteristic function, the values of α must be restricted to the interval (0, 2]. Conceptually speaking, α determines the impulsiveness or tail heaviness of the distribution (smaller values of α indicate increased levels of impulsiveness). The limit case, $\alpha = 2$, corresponds to the zero-mean Gaussian distribution with variance 2γ. All other values of α correspond to heavy-tailed distributions with infinite variance and algebraic tail behavior of the form[†]

$$\Pr(|X| > x) \sim x^{-\alpha}, \quad \text{as } x \to \infty. \tag{5.3}$$

The case $\alpha = 1$ corresponds to the "zero-centered" Cauchy distribution, which has density

$$f(x) = \frac{\gamma}{\pi} \frac{1}{\gamma^2 + x^2}. \tag{5.4}$$

---

[*] The limit here is taken as the number of elements in the sum tends to ∞.

[†] The symbol $\sim$ denotes asymptotic similarity. Formally, X has algebraic tails if there exist positive constants α and c such that $\lim_{x \to \infty} x^\alpha \Pr(|X| > x) = c$.

When $\alpha \neq 1, 2$, no closed expressions exist for the density functions, making it necessary to resort to series expansions or integral transforms to describe them.[7]

Symmetric α-stable densities maintain many of the features of the Gaussian density. They are smooth, unimodal, symmetric with respect to the mode, and bell-shaped. Figure 5.1 illustrates the impulsive behavior of symmetric α-stable processes as the characteristic exponent α is varied. Each of the plots shows an i.i.d. "zero-centered" symmetric α-stable signal with unitary geometric power.* To give a better feeling of the impulsive structure of the data, the signals are plotted twice under two different scales. As can be appreciated, the Gaussian signal ($\alpha = 2$) does not show impulsive behavior. For values of α close to 2 ($\alpha = 1.9$), the structure of the signal is still similar to the Gaussian, although little impulsiveness can now be observed. As the value of α is decreased, the impulsive behavior of the α-stable process increases progressively, exhibiting the highest levels of impulsiveness for very small values of α.

5.3   Running Myriad Smoothers

The sample myriad emerges as the maximum likelihood estimate of location under a set of several distributions within the family of α-stable distributions, including the well-known Cauchy distributions. Since their introduction by Fisher in 1922,[15] myriad-type estimators have been studied and applied under very different contexts as an efficient alternative to cope with the presence of impulsive noise.[16–21] The most general form of the myriad, where the potential of tuning the so-called *linearity parameter* in order to control its behavior is fully exploited, was first introduced by Gonzalez and Arce in 1996.[22] Depending on the value of this free parameter, the sample myriad can present drastically different behaviors, ranging from *highly resistant* mode-type estimators to the familiar (Gaussian-efficient) sample average. This rich variety of operation modes is the key concept explaining important optimality properties of the myriad in the class of symmetric α-stable distributions.

Given an observation vector $\mathbf{X}(n) = [X_1(n), X_2(n), \ldots, X_N(n)]$ and a fixed positive (tunable) value of K, the running myriad smoother output at time n is computed as

$$Y_K(n) = \text{MYRIAD}[K; X_1(n), X_2(n), \ldots, X_N(n)]$$

$$= \arg\min_{\beta} \prod_{i=1}^{N}[K^2 + (X_i(n) - \beta)^2]. \tag{5.5}$$

---

*The geometric power is an indicator of signal strength suited to the class of processes with infinite variance.[14]

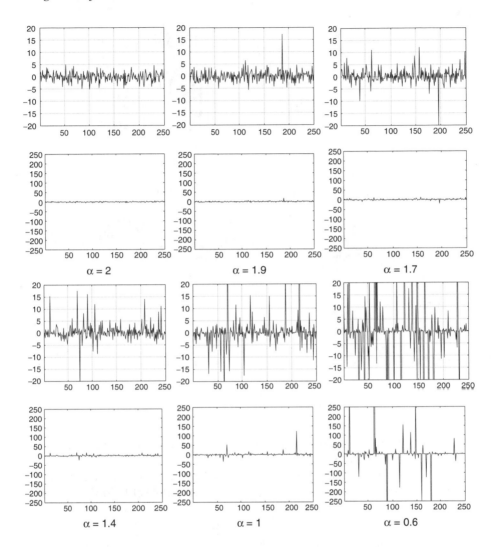

FIGURE 5.1
Impulsive behavior of i.i.d. α-stable signals as the tail constant α is varied. Signals are plotted twice under two different scales. (*Continued*)

The myriad $Y_K(n)$ is thus the value of β that minimizes the cost function in Equation 5.5. Unlike the sample mean or median, the definition of the sample myriad in Equation 5.5 involves the free-tunable parameter K. This parameter will be shown to play a critical role in characterizing the behavior of the myriad. For reasons that will become apparent shortly, the parameter K is referred to as the *linearity parameter*. Because the log function is monotonic,

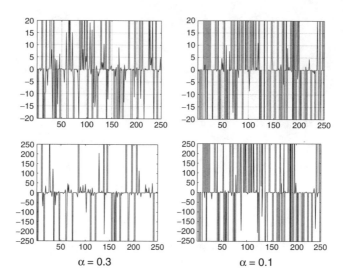

$\alpha = 0.3$ $\qquad$ $\alpha = 0.1$

FIGURE 5.1
(*Continued*)

the myriad is also defined by the equivalent expression

$$Y_K(n) = \arg\min_{\beta} \sum_{i=1}^{N} \log[K^2 + (X_i(n) - \beta)^2]. \tag{5.6}$$

In general, for a fixed value of K, the minima of the cost functions in Equations 5.5 and 5.6 lead to a unique value.

To illustrate the calculation of the sample myriad and the effect of the linearity parameter, consider the sample myriad of the set $\{-3, 10, 1, -1, 6\}$:

$$\hat{\beta}_K = \text{MYRIAD}(K; -3, 10, 1, -1, 6) \tag{5.7}$$

for $K = 20, 2, 0.2$. The myriad cost functions in Equation 5.5, for these three values of K, are plotted in Figure 5.2. The corresponding minima are attained at $\hat{\beta}_{20} = 1.8$, $\hat{\beta}_2 = 0.1$, and $\hat{\beta}_{0.2} = 1$, respectively. The different values taken on by the myriad as the parameter K is varied are best understood by the results provided in the following properties.

PROPERTY 1 (Linear Property)
Given a set of samples, $X_1, X_2, \ldots, X_N$, the sample myriad $\hat{\beta}_K$ converges to the sample average as $K \to \infty$. This is,

$$\lim_{K \to \infty} \hat{\beta}_K = \lim_{K \to \infty} \text{MYRIAD}(K; X_1, \ldots, X_N) = \frac{1}{N} \sum_{i=1}^{N} X_i. \tag{5.8}$$

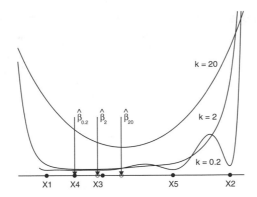

FIGURE 5.2
Myriad cost functions for different values of K.

To prove this property, first note that* $\hat{\beta}_K \leq X_{(N)}$ by checking that for any i, and for $\beta > X_{(N)}$, $K^2 + (X_i - \beta)^2 > K^2 + (X_i - X_{(N)})^2$. In the same way, $\hat{\beta}_K \geq X_{(1)}$. Hence,

$$\hat{\beta}_K = \arg\min_{X_{(1)} \leq \beta \leq X_{(N)}} \prod_{i=1}^{N}[K^2 + (X_i - \beta)^2] \tag{5.9}$$

$$= \arg\min_{X_{(1)} \leq \beta \leq X_{(N)}} \left\{ K^{2N} + K^{2N-2} \sum_{i=1}^{N}(X_i - \beta)^2 + O(K^{2N-4}) \right\}, \tag{5.10}$$

where O denotes the *asymptotic order* as $K \to \infty$. Because adding or multiplying by constants does not affect the arg min operator, Equation 5.10 can be rewritten as

$$\hat{\beta}_K = \arg\min_{X_{(1)} \leq \beta \leq X_{(N)}} \left\{ \sum_{i=1}^{N}(X_i - \beta)^2 + \frac{O(K^{2N-4})}{K^{2N-2}} \right\}. \tag{5.11}$$

Letting $K \to \infty$, the term $O(K^{2N-4})/K^{2N-2}$ becomes negligible, and

$$\hat{\beta}_K \to \arg\min_{X_{(1)} \leq \beta \leq X_{(N)}} \left\{ \sum_{i=1}^{N}(X_i - \beta)^2 \right\} = \frac{1}{N}\sum_{i=1}^{N} X_i.$$

Plainly, an infinite value of K converts the myriad into the sample average. This behavior explains the name *linearity* given to this parameter: the larger the value of K, the closer the behavior of the myriad to a linear estimator. As the myriad moves away from the linear region (large values of K) to

---

* Here, $X_{(i)}$ denotes the ith order statistic of the sample set.

lower linearity values, the estimator becomes more resistant to the presence of impulsive noise. In the limit, when K tends to zero, the analysis of the myriad leads to the discovery of a novel location estimator with particularly good performance in the presence of *very* impulsive noise. In this case, the estimator treats every observation as a possible outlier, assigning more credibility to the most repeated values in the sample. This "mode-type" characteristic is reflected in the name *mode-myriad* given to this estimator.

DEFINITION 1 (Sample Mode-Myriad)
Given a set of samples $X_1, X_2, \ldots, X_N$, the mode-myriad estimator, $\hat{\beta}_0$, is defined as

$$\hat{\beta}_0 = \lim_{K \to 0} \hat{\beta}_K, \tag{5.12}$$

where $\hat{\beta}_K = \text{MYRIAD}(K; X_1, X_2, \ldots, X_N)$.

The following property explains the behavior of the mode-myriad as a kind of generalized sample mode, and provides a simple method for determining the mode-myriad without recurring to the definition in Equation 5.5.

PROPERTY 2 (Mode Property)
The mode-myriad $\hat{\beta}_0$ is always equal to one of the most repeated values in the sample. Furthermore,

$$\hat{\beta}_0 = \arg\min_{X_j \in \mathcal{M}} \prod_{i=1,\, X_i \neq X_j}^{N} |X_i - X_j|, \tag{5.13}$$

where $\mathcal{M}$ is the set of most repeated values.

Proof: Because K is a positive constant, the definition of the sample myriad in Equation 5.5 can be reformulated as $\hat{\beta}_K = \arg\min_\beta P_K(\beta)$, where

$$P_K(\beta) = \prod_{i=1}^{N} \left[1 + \frac{(X_i - \beta)^2}{K^2} \right]. \tag{5.14}$$

When K is very small, it is easy to check that

$$P_K(\beta) = O\left(\frac{1}{K^2}\right)^{N - r(\beta)},$$

where $r(\beta)$ is the number of times the value β is repeated in the sample, and O denotes the asymptotic order as $K \to 0$. In the limit, the exponent $N - r(\beta)$ must be minimized in order for $P_K(\beta)$ to be minimum. Therefore, the mode-myriad $\hat{\beta}_0$ will lie on a maximum of $r(\beta)$, or in other words, $\hat{\beta}_0$ will be one of the most repeated values in the sample.

Now, let $r = \max_j r(X_j)$. Then, for $X_j \in \mathcal{M}$, expanding the product in Equation 5.14 gives

$$P_K(X_j) = \left\{ \prod_{i,\, X_i \neq X_j} \frac{(X_i - X_j)^2}{K^2} \right\} + O\left(\frac{1}{K^2}\right)^{N-r-1}. \tag{5.15}$$

Because the first term in Equation 5.15 is $O(1/K^2)^{N-r}$, the second term is negligible for small values of K, and $\hat{\beta}_0$ can be calculated as

$$\hat{\beta}_0 = \arg\min_{X_j \in \mathcal{M}} P_K(X_j)$$

$$= \arg\min_{X_j \in \mathcal{M}} \prod_{i,\, X_i \neq X_j} \frac{(X_i - X_j)^2}{K^2}$$

$$= \arg\min_{X_j \in \mathcal{M}} \prod_{i,\, X_i \neq X_j} |X_i - X_j|.$$

An immediate consequence of the mode property is the fact that running-window smoothers based on the mode-myriad are *selection-type*, in the sense that their output is always, by definition, one of the samples in the input window. This "selection" property, shared also by the median, makes mode-myriad smoother a suitable framework for image processing, where the application of selection-type smoothers has been shown to be convenient.[23,24]

5.3.1 Geometrical Interpretation

Myriad estimation, defined in Equation 5.5, can be interpreted in a more intuitive manner. As depicted in Figure 5.3a, it can be shown that the sample myriad, $\hat{\beta}_K$, is the value that minimizes the product of distances from point A to the sample points $X_1, X_2, \ldots, X_6$. Any other value, such as $X = \beta'$,

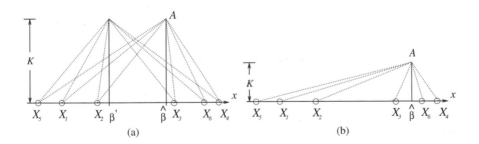

FIGURE 5.3
(a) The sample myriad, $\hat{\beta}$, minimizes the product of distances from point A to all samples. Any other value, such as $x = \beta'$, produces a higher product of distances; (b) the myriad as K is reduced.

produces a higher product of distances. As K is reduced, the myriad searches clusters as shown in Figure 5.3b. If K is made large, all distances become close and it can be shown that the myriad tends to the sample mean.

5.3.2 The Tuning of *K*

The linear and mode properties indicate the behavior of the myriad estimator for large and small values of K. From a practical point of view, it is important to determine whether a given value of K is large (or small) enough for the linear (or mode) property to hold approximately. With this in mind, it is instructive to look at the myriad as the maximum likelihood location estimator generated by a Cauchy distribution with dispersion K (geometrically, K is equivalent to half the interquartile range). Given a fixed set of samples, the maximum likelihood (ML) method locates the generating distribution in a position where the probability of the specific sample set to occur is maximum.

When K is large, the generating distribution is highly dispersed, and its density function looks flat (see the density function corresponding to K_2 in Figure 5.4). If K is large enough, *all* the samples can be accommodated inside the interquartile range of the distribution, and the ML estimator visualizes them as "well behaved" (no outliers.) In this case, a desirable estimator would be the sample average, in complete agreement with the linear property. From this consideration, it should be clear that a fair approximation to the

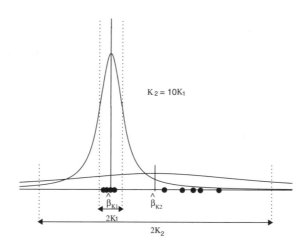

FIGURE 5.4
The role of the linearity parameter when the myriad is viewed as an ML estimator. When K is large, the generating density function is spread and the data are visualized as "well behaved" (the optimal estimator is the sample average). For small values of K, the generating density becomes highly localized, and the data are visualized as very impulsive (the optimal estimator is a cluster locator).

linear property can be obtained if K is large enough so that *all* the samples can be seen as "well-behaved" under the generating Cauchy distribution. It has been observed experimentally that values of K on the order of the data range, $K \sim X_{(N)} - X_{(1)}$, often make the myriad an acceptable approximation to the sample average.

On the other side, when K is small, the generating Cauchy distribution is highly localized, and its density function looks similar to a positive impulse. The effect of such a localized distribution is conceptually equivalent to observing the samples through a magnifying lens. In this case, most of the data look like possible outliers, and the ML estimator has trouble locating a large number of observations inside the interquartile range of the density (see the density function corresponding to K_1 in Figure 5.4.) Placing in doubt most of the data at hand, a desirable estimator would tend to maximize the number of samples inside the interquartile range, inducing to position the density function in the vicinity of a data cluster. In the limit case, when $K \to 0$, the density function becomes infinitely localized, and the only visible clusters will be made of repeated value sets. In this case, one of the most crowded clusters (i.e., one of the most repeated values in the sample) will be located by the estimator, in accordance with the mode property. From this consideration, it should be clear that a fair approximation to the mode property can be obtained if K is made significantly smaller than the distances between sample elements. Empirical observations show that K on the order of

$$K \sim \min_{i,j} |X_i - X_j|, \tag{5.16}$$

is often enough for the myriad to be considered approximately a mode-myriad.

The myriad estimator thus offers a rich class of modes of operation that can be easily controlled by tuning the linearity parameter K. When the noise is Gaussian, for example, large values of the linearity can provide the optimal performance associated with the sample mean, whereas for highly impulsive noise statistics, the resistance of mode-type estimators can be achieved by using myriads with low linearity. The trade-off between efficiency at the Gaussian model and resistance to impulsive noise can be managed by designing appropriate values for K (Figure 5.5).

FIGURE 5.5
Functionality of the myriad as K is varied. Tuning the linearity parameter K adapts the behavior of the myriad from impulse-resistant mode-type estimators (small K) to the Gaussian-efficient sample mean (large K).

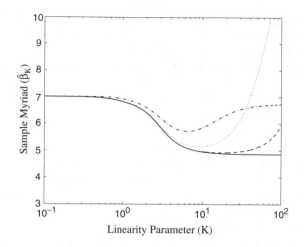

FIGURE 5.6

Values of the myriad as a function of K for the following data sets: (solid) original data set = 0, 1, 3, 6, 7, 8, 9; (dash-dot) original set plus an additional observation at 20; (dotted) additional observation at 100; (dashed) additional observations at 800, −500, and 700.

To illustrate the above, it is instructive to look at the behavior of the sample myriad shown in Figure 5.6. The solid line shows the values of the myriad as a function of K for the data set {0, 1, 3, 6, 7, 8, 9}. It can be observed that, as K increases, the myriad tends asymptotically to the sample average. On the other hand, as K is decreased, the myriad favors the value 7, which indicates the location of the cluster formed by the samples 6, 7, 8, 9. This is a typical behavior of the myriad for small K: it tends to favor values where samples are more likely to occur or cluster. The term *myriad* has been coined as a result of this characteristic.

The dotted line shows how the sample myriad is affected by an additional observation of value 100. For large values of K, the myriad is very sensitive to this new observation. In contrast, for small K, the variability of the data is assumed to be small, and the new observation is considered an outlier, not influencing significantly the value of the myriad.

More interestingly, if the additional observations are the very large data 800, −500, 700 (dashed curve), the myriad is practically unchanged for moderate values of K ($K < 10$). This behavior exhibits a very desirable outlier rejection property, not found, for example, in median-type estimators.

5.3.3   Scale-Invariant Operation

Unlike the sample mean or median, the operation of the sample myriad is not scale invariant; i.e., for fixed values of the linearity parameter, its behavior can vary depending on the units of the data. This is formalized in the following property stated without proof (see Reference 25).

PROPERTY 3 (Scale Invariance)
Let $\hat{\beta}_K(\mathbf{X})$ denote the myriad of order K of the data in the vector $\mathbf{X}$. Then, for $c > 0$,

$$\hat{\beta}_K(c\mathbf{X}) = c\hat{\beta}_{K/c}(\mathbf{X}). \qquad (5.17)$$

According to Equation 5.17, a change of scale in the data is preserved in the myriad only if K experiences the same change of scale. Thus, the scale dependence of the myriad can be easily overcome if K carries the units of the data, or in other words, if K is a *scale parameter* of the data.

5.4   Optimality of the Sample Myriad in the α-Stable Model

In addition to its optimality in the Cauchy distribution ($\alpha = 1$), the sample myriad presents important optimality properties compelling the use of myriad-based methods in the α-stable framework. First, it is well known that the sample mean is the optimal location estimator at the Gaussian model; thus, by assigning large values to the linearity parameter, the linear property guarantees the optimality of the sample myriad in the Gaussian distribution ($\alpha = 2$). The following result states the optimality of the myriad when $\alpha \to 0$, i.e., when the impulsiveness of the distribution is very high. The proof of proposition 1 can be found in Reference 26.

PROPOSITION 1
Let $T_{\alpha,\gamma}(X_1, X_2, \ldots, X_N)$ denote the maximum likelihood location estimator derived from a symmetric α-stable distribution with characteristic exponent α and dispersion γ. Then,

$$\lim_{\alpha \to 0} T_{\alpha,\gamma}(X_1, X_2, \ldots, X_N) = \text{MYRIAD}\{0; X_1, X_2, \ldots, X_N\}. \qquad (5.18)$$

This proposition states that the ML estimator of location derived from an α-stable distribution with small α behaves like the sample mode-myriad. Proposition 1 completes what is called the α-stable triplet of optimality points satisfied by the myriad. On one extreme ($\alpha = 2$), when the distributions are very well behaved, the myriad reaches optimal efficiency by making $K = \infty$. In the middle ($\alpha = 1$), the myriad reaches optimality by making $K = \gamma$, the dispersion parameter of the Cauchy distribution. On the other extreme ($\alpha \to 0$), when the distributions are extremely impulsive, the myriad reaches optimality again, this time by making $K = 0$.

The α-stable triplet demonstrates the central role played by myriad estimation in the α-stable framework. The very simple tuning of the linearity parameter empowers the myriad with good estimation capabilities under markedly different types of impulsiveness, from the very impulsive ($\alpha \to 0$) to the non-impulsive ($\alpha = 2$). Because lower values of K correspond to increased resistance to impulsive noise, it is intuitively pleasant that, for *maximal* impulsiveness ($\alpha \to 0$), the optimal K takes precisely its *minimal* value, $K = 0$. The same condition occurs at the other extreme: *minimal* levels of impulsiveness ($\alpha = 2$) correspond to the *maximal* tuning value, $K = \infty$. Thus, as α is increased from 0 to 2, it is reasonable to expect, somehow, a progressive increase of the optimal K, from $K = 0$ to $K = \infty$. The following proposition provides information about the general behavior of the optimal K. Its proof is a direct consequence of Proposition 3 and the fact that $\gamma^{1/\alpha}$ is a scale parameter of the α-stable distribution.

PROPOSITION 2
Let α and γ denote the characteristic exponent and dispersion parameter of a symmetric α-stable distribution. Let $K_o(\alpha, \gamma)$ denote the optimal tuning value of K in the sense that $\hat{\beta}_{K_o}$ minimizes a given performance criterion (usually the variance) among the class of sample myriads with nonnegative linearity parameters. Then,

$$K_o(\alpha, \gamma) = K_o(\alpha, 1)\gamma^{1/\alpha}. \tag{5.19}$$

Proposition 2 indicates a "separability" of K_o in terms of α and γ, reducing the optimal tuning problem to that of determining the function $K(\alpha) = K_o(\alpha, 1)$. This function is of fundamental importance for the proper operation of the myriad in the α-stable framework, and it will be referred to as the α-K *curve*. Its form is conditioned to the performance criterion chosen, and it may even depend on the sample size. In general, as discussed above, the α-K curve is expected to be monotonically increasing, with $K(0) = 0$ (very impulsive point) and $K(2) = \infty$ (Gaussian point). If the performance criterion is the asymptotic variance, for example, then $K(1) = 1$, corresponding to the Cauchy point of the α-stable triplet. The exact computation of the α-K curve for α-stable distributions* is still under study. A simple empirical form that has consistently provided efficient results in a variety of conditions is

$$K(\alpha) = \sqrt{\frac{\alpha}{2-\alpha}}\gamma^{1/\alpha}, \tag{5.20}$$

which is plotted in Figure 5.7.

The α-K curve is a valuable tool for estimation and filtering problems that must *adapt* to the impulsiveness conditions of the environment. α-K curves in the α-stable framework have been used, for example, to develop myriad-based adaptive detectors for channels with uncertain impulsiveness.[27]

---

* The concept of α-K curve can also be extended to distribution families other than the α-stable.

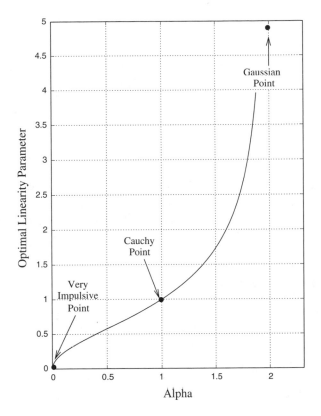

FIGURE 5.7
Empirical α-K curve for α-stable distributions. The curve values at $\alpha = 0$, 1, and 2 constitute the optimality points of the α-stable triplet.

5.5 Weighted Myriad Smoothers

The sample myriad can be generalized to the weighted myriad smoother by assigning positive weights to the input samples (observations); the weights reflect the varying levels of "reliability." To this end, the observations are assumed to be drawn from independent Cauchy random variables, which are, however, *not* identically distributed. Given N observations $\{X_i\}_{i=1}^N$ and nonnegative weights $\{W_i \geq 0\}_{i=1}^N$, let the input and weight vectors be defined as $\mathbf{X} \triangleq [X_1, X_2, \ldots, X_N]^T$ and $\mathbf{W} \triangleq [W_1, W_2, \ldots, W_N]^T$, respectively. For a given *nominal* scale factor K, the underlying random variables are assumed to be independent and Cauchy distributed with a common location parameter β, but varying scale factors $\{S_i\}_{i=1}^N$: $X_i \sim \text{Cauchy}(\beta, S_i)$, where the density function of X_i has the form

$$f_{X_i}(X_i; \beta, S_i) = \frac{1}{\pi} \frac{S_i}{S_i^2 + (X_i - \beta)^2}, \quad -\infty < X_i < \infty, \qquad (5.21)$$

and where

$$S_i \overset{\triangle}{=} \frac{K}{\sqrt{W_i}} > 0, \quad i = 1, 2, \dots, N. \qquad (5.22)$$

A larger value for the weight W_i (smaller scale S_i) makes the distribution of X_i more concentrated around β, thus increasing the reliability of the sample X_i. Note that the special case when all the weights are equal to unity corresponds with the sample myriad at the nominal scale factor K, with all the scale factors reducing to $S_i = K$.

Again, the location estimation problem considered here is closely related to the problem of filtering a time series $\{X(n)\}$ using a sliding window. The output $Y(n)$, at time n, can be interpreted as an estimate of location based on the input samples $\{X_1(n), X_2(n), \dots, X_N(n)\}$. Further, the aforementioned model of independent but not identically distributed samples captures the temporal correlations usually present among the input samples. To see this, note that the output $Y(n)$, as an estimate of location, would rely more on (give more weight to) the sample $X(n)$, when compared with samples that are further away in time. By assigning varying scale factors in modeling the input samples, leading to different weights (reliabilities), their temporal correlations can be effectively accounted for.

The weighted myriad smoother output $\hat{\beta}_K(\mathbf{W}, \mathbf{X})$ is defined as the value β that maximizes the likelihood function $\prod_{i=1}^N f_{X_i}(X_i; \beta, S_i)$. Using Equation 5.21 for $f_{X_i}(X_i; \beta, S_i)$ leads to

$$\hat{\beta}_K(\mathbf{W}, \mathbf{X}) = \arg\max_\beta \prod_{i=1}^N \frac{S_i}{S_i^2 + (X_i - \beta)^2},$$

which is equivalent to

$$\hat{\beta}_K(\mathbf{W}, \mathbf{X}) \overset{\triangle}{=} \arg\min_\beta \prod_{i=1}^N \left[1 + \left(\frac{X_i - \beta}{S_i} \right)^2 \right]$$

$$= \arg\min_\beta \prod_{i=1}^N [K^2 + W_i(X_i - \beta)^2] \qquad (5.23)$$

$$\overset{\triangle}{=} \arg\min_\beta P(\beta). \qquad (5.24)$$

Alternatively, we can write $\hat{\beta}_K(\mathbf{W}, \mathbf{X}) \triangleq \hat{\beta}_K$ as

$$\hat{\beta}_K = \arg\min_\beta Q(\beta) \triangleq \arg\min_\beta \sum_{i=1}^N \log[K^2 + W_i(X_i - \beta)^2]; \qquad (5.25)$$

thus $\hat{\beta}_K$ is the global minimizer of $P(\beta)$ as well as of $Q(\beta) \triangleq \log(P(\beta))$. Depending on the context, we refer to either of the functions $P(\beta)$ and $Q(\beta)$ as the *weighted myriad smoother objective function*. Note that when $W_i = 0$, the corresponding term drops out of $P(\beta)$ and $Q(\beta)$; thus a sample X_i is effectively ignored if its weight is zero.

The definition of the weighted myriad is then formally stated as follows.

DEFINITION 2 (Weighted Myriad)
Let $\mathbf{W} = [W_1, W_2, \ldots, W_N]$ be a vector of nonnegative weights. Given $K > 0$, the weighted myriad of order K for the data $X_1, X_2, \ldots, X_N$ is defined as

$$\hat{\beta}_K = \text{MYRIAD}\{K; W_1 \circ X_1, \ldots, W_N \circ X_N\}$$

$$= \arg\min_\beta \sum_{i=1}^N \log[K^2 + W_i(X_i - \beta)^2], \qquad (5.26)$$

where $W_i \circ X_i$ represents the weighting operation in Equation 5.26. In some situations, the following equivalent expression can be computationally more convenient

$$\hat{\beta}_K = \arg\min_\beta \prod_{i=1}^N [K^2 + W_i(X_i - \beta)^2]. \qquad (5.27)$$

It is important to note that the weighted myriad has only N independent parameters (even though there are N weights and the parameter K). By using Equation 5.27, it can be inferred that if the value of K is changed, the same smoother output can be obtained provided the smoother weights are appropriately scaled. Thus, the following is true:

$$\hat{\beta}_K(\mathbf{W}, \mathbf{X}) = \hat{\beta}_1\left(\frac{\mathbf{W}}{K^2}, \mathbf{X}\right) \qquad (5.28)$$

or

$$\hat{\beta}_{K_1}(\mathbf{W}_1, \mathbf{X}) = \hat{\beta}_{K_2}(\mathbf{W}_2, \mathbf{X}) \quad \text{iff} \quad \frac{\mathbf{W}_1}{K_1^2} = \frac{\mathbf{W}_2}{K_2^2}. \qquad (5.29)$$

Hence, the output depends only on $\mathbf{W}/K^2$.

The objective function $P(\beta)$ in Equation 5.27 is a polynomial in β of degree $2N$, with well-defined derivatives of all orders. Therefore, $P(\beta)$ (and the equivalent objective function $Q(\beta)$) can have at most $(2N - 1)$ local extrema.

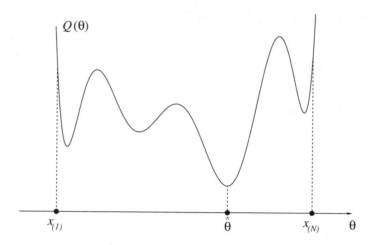

FIGURE 5.8
Sketch of a typical weighted myriad objective function $Q(\beta)$.

The output is thus one of the local minima of $Q(\beta)$:

$$Q'(\hat{\beta}) = 0. \tag{5.30}$$

Figure 5.8 depicts a typical objective function $Q(\beta)$, for various values of K. Note in the figure that the number of local minima in the objective function $Q(\beta)$ depends on the value of the parameter K. In particular, when K is very large only one extremum exists.

As K becomes larger, the number of local minima of $G(\beta)$ decreases. In fact, it can be proved (by examining the second derivative $G''(\beta)$) that a *sufficient* (but not *necessary*) condition for $G(\beta)$, and $\log(G(\beta))$, to be convex and, therefore, have a unique local minimum is that

$$K > \sqrt{\max\{W_j\}_{j=1}^{N}}(X_{(N)} - X_{(1)}).$$

This condition is, however, not necessary; the onset of convexity could be at a much lower K.

As stated in the next property, in the limit as $K \to \infty$, with the weights $\{W_i\}$ held constant, it can be shown that $Q(\beta)$ exhibits a single local extremum. The proof is a generalized form of that used to prove the linear property of the unweighted sample myriad.

PROPERTY 4 (Linear Property)
In the limit as $K \to \infty$, the weighted myriad reduces to the normalized linear estimate

$$\lim_{K \to \infty} \hat{\beta}_K = \frac{\sum_{i=1}^{N} W_i X_i}{\sum_{i=1}^{N} W_i}. \tag{5.31}$$

Again, because of the linear structure of the weighted myriad as $K \to \infty$, the name *linearity parameter* is used for the parameter K. Equation 5.31 provides the link between the weighted myriad and a constrained linear FIR filter: the weighted myriad smoother is analogous to the weighted mean smoother having its weights *constrained* to be nonnegative and normalized (summing to unity).

Figure 5.8 also depicts that the output $\hat{\beta}$ is restricted to the dynamic range of the input samples. This indicates that the output $\hat{\beta}$ is restricted to the dynamic range of the input weighted myriad smoother, thus unable to amplify the dynamic range of an input signal.

PROPERTY 5 (No Undershoot/Overshoot)
The output of a weighted myriad smoother is always bracketed by

$$X_{(1)} \leq \hat{\beta}_K(X_1, X_2, \ldots, X_N) \leq X_{(N)}, \tag{5.32}$$

where $X_{(1)}$ and $X_{(N)}$ denote the minimum and maximum samples in the input window.

Proof: For $\beta < X_{(1)}$,

$$K^2 + W_i\left(X_i - X_{(1)}\right)^2 < K^2 + W_i(X_i - \beta)^2$$

and consequently

$$\prod_{i=1}^{N} \left[K^2 + W_i\left(X_i - X_{(1)}^2\right)\right] < \prod_{i=1}^{N}[K^2 + W_i(X_i - \beta)^2].$$

This implies that any value of β smaller than $X_{(1)}$ leads to a larger value of the myriad objective function. Therefore, the weighted myriad cannot be less than $X_{(1)}$. A similar argument can be constructed for $X_{(N)}$ leading to the conclusion that the weighted myriad cannot be larger than $X_{(N)}$.

At the other extreme of linearity values ($K \to 0$), the weighted myriad becomes what is referred to as the *weighted mode-myriad*. Weighted mode-myriad smoothers maintain the same mode-like behavior of the unweighted mode-myriad, as stated in the following.

PROPERTY 6 (Mode Property)
Given a vector of positive weights, $\mathbf{W} = [W_1, \ldots, W_N]$, the weighted mode-myriad $\hat{\beta}_0$ is always equal to one of the most repeated values in the sample. Furthermore,

$$\hat{\beta}_0 = \arg\min_{X_j \in \mathcal{M}} \left(\frac{1}{W_j}\right)^{\frac{r}{2}} \prod_{i=1, X_i \neq X_j}^{N} |X_i - X_j|, \tag{5.33}$$

where $\mathcal{M}$ is the set of most repeated values, and r is the number of times a member of $\mathcal{M}$ is repeated in the sample.

Proof: Following the steps of the proof for the unweighted version, it is straightforward that

$$\hat{\beta}_0 = \underset{X_j \in \mathcal{M}}{\arg\min} \prod_{i=1,\, X_i \neq X_j}^{N} W_i (X_i - X_j)^2. \tag{5.34}$$

Dividing by $\prod_{i=1}^{N} W_i$, and applying square root to the expression to be minimized, the desired result is obtained.

PROPERTY 7 (Shift and Sign Invariance)
Let $Z_i = X_i + b$. Then, for any K and $\mathbf{W}$,

(i) $\hat{\beta}_K(Z_1, \ldots, Z_N) = \hat{\beta}_K(X_1, \ldots, X_N) + b$;
(ii) $\hat{\beta}_K(-Z_1, \ldots, -Z_N) = -\hat{\beta}_K(Z_1, \ldots, Z_N)$.

Proof: Follows from the definition of the weighted myriad in Equation 5.26.

PROPERTY 8 (Unbiasedness)
Let $X_1, X_2, \ldots, X_N$ be all independent and symmetrically distributed around the point of symmetry c. Then, $\hat{\beta}_K$ is also symmetrically distributed around c. In particular, if $E\hat{\beta}_K$ exists, then $E\hat{\beta}_K = c$.

Proof: If X_i is symmetric about c, then $2c - X_i$ has the same distribution as X_i. It follows that $\hat{\beta}_K(X_1, X_2, \ldots, X_N)$ has the same distribution as $\hat{\beta}_K(2c - X_1, 2c - X_2, \ldots, 2c - X_N)$, which from Property 7, is identical to $2c - \hat{\beta}_K(X_1, X_2, \ldots, X_N)$. It follows that $\hat{\beta}_K(X_1, X_2, \ldots, X_N)$ is symmetrically distributed about c.

5.5.1   Geometrical Interpretation

Weighted myriads as defined in Equation 5.27 can be interpreted in a more intuitive manner. Allow a vertical bar to run horizontally through the real line as depicted in Figure 5.9a. Then the sample myriad, $\hat{\beta}_K$, indicates the position of the bar for which the product of distances from point A to the sample points $X_1, X_2, \ldots, X_N$ is minimum. If weights are introduced, each sample point X_i is assigned a different point A_i in the bar, as illustrated in Figure 5.9b.

The geometrical interpretation of the myriad is intuitively insightful. When K approaches 0, it gives a conceptually simple pictorial demonstration of the mode-myriad formula in Equation 5.13.

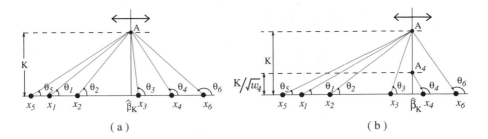

(a) (b)

FIGURE 5.9

(a) The sample myriad, $\hat{\beta}_K$, indicates the position of a moving bar such that the product of distances from point A to the sample points $X_1, X_2, \ldots, X_N$ is minimum. (b) If the weight $W_4 > 1$ is introduced, the product of distances is more sensitive to the variations of the segment $\overline{X_4 A_4}$, very likely resulting in a weighted myriad $\hat{\beta}_K$ closer to X_4.

5.6 Fast Weighted Myriad Computation

Unlike the weighted mean or weighted median, the computation of the *weighted myriad* is not available in explicit form. Its direct computation is therefore a nontrivial task because it involves the minimization of the *weighted myriad objective function*, $Q(\beta)$ in Equation 5.25. The myriad objective function, however, has a number of characteristics that can be exploited to construct fast iterative methods to compute its minimum.

Recall that the weighted myriad is given by

$$\hat{\beta}_K = \arg\min_{\beta} \log(P(\beta)) \overset{\triangle}{=} \arg\min_{\beta} Q(\beta)$$

$$= \arg\min_{\beta} \sum_{i=1}^{N} \log\left[1 + \left(\frac{x_i - \beta}{S_i}\right)^2\right],$$ (5.35)

where $Q(\beta)$ is the *weighted myriad objective function*. Having well-defined derivatives, $\hat{\beta}_K$ is one of the local minima of $Q(\beta)$, i.e., $Q'(\hat{\beta}) = 0$. Because $Q(\beta) = \log(P(\beta))$, the derivative of $Q(\beta)$ can be written as

$$Q'(\beta) = \frac{P'(\beta)}{P(\beta)},$$ (5.36)

where the derivative of $P(\beta) = \prod_{i=1}^{N}(K^2 + W_i(X_i - \beta)^2)$ is given by

$$P'(\beta) = 2\sum_{i=1}^{N} \frac{W_i(\beta - X_i)}{K^2 + W_i(X_i - \beta)^2} P(\beta).$$ (5.37)

From Equation 5.36, it follows that

$$Q'(\beta) = 2 \sum_{i=1}^{N} \frac{W_i(\beta - X_i)}{K^2 + W_i(X_i - \beta)^2}.$$

(5.38)

Using the fact that $S_i = K/\sqrt{W_i}$, the above can be written as

$$Q'(\beta) = 2 \sum_{i=1}^{N} \frac{\left(\dfrac{\beta - X_i}{S_i^2}\right)}{1 + \left(\dfrac{X_i - \beta}{S_i}\right)^2}.$$

(5.39)

Defining

$$\psi(v) \triangleq \frac{2v}{1 + v^2},$$

(5.40)

and referring to Equation 5.39, the following equation is obtained for the local extrema of $Q(\beta)$:

$$Q'(\beta) = -\sum_{i=1}^{N} \frac{1}{S_i} \cdot \psi\left(\frac{X_i - \beta}{S_i}\right) = 0.$$

(5.41)

By introducing the *positive* functions

$$h_i(\beta) \triangleq \frac{1}{S_i^2} \cdot \varphi\left(\frac{X_i - \beta}{S_i}\right) > 0,$$

(5.42)

for $i = 1, 2, \ldots, N$, where

$$\varphi(v) \triangleq \frac{\psi(v)}{v} = \frac{2}{1 + v^2},$$

(5.43)

the local extrema of $Q(\beta)$ in Equation 5.41 can be formulated as

$$Q'(\beta) = -\sum_{i=1}^{N} h_i(\beta) \cdot (X_i - \beta) = 0.$$

(5.44)

This formulation implies that the *sum of weighted deviations* of the samples is zero, with the (positive) weights themselves functions of β. This property, in turn, leads to a simple iterative approach to compute the weighted myriad, as detailed next.

5.6.1   Fixed Point Formulation

Equation 5.44 can be written as

$$\beta = \frac{\sum_{i=1}^{N} h_i(\beta) \cdot X_i}{\sum_{i=1}^{N} h_i(\beta)}.$$

(5.45)

It can be seen that each local extremum of $Q(\beta)$, including the weighted myriad $\hat{\beta}$, can be written as a *weighted mean* of the input samples X_i. Because the weights $h_i(\beta)$ are always positive, the right-hand side of Equation 5.45 is in $(X_{(1)}, X_{(N)})$, confirming that all the local extrema lie within the range of the input samples. By defining the mapping

$$T(\beta) \triangleq \frac{\sum_{i=1}^N h_i(\beta) \cdot X_i}{\sum_{i=1}^N h_i(\beta)} \qquad (5.46)$$

the local extrema of $Q(\beta)$, or the roots of $Q'(\beta)$, are seen to be the *fixed points* of $T(\cdot)$:

$$\beta^* = T(\beta^*). \qquad (5.47)$$

The following *fixed point iteration* results in an efficient algorithm to compute these fixed points:

$$\beta_{m+1} \triangleq T(\beta_m) = \frac{\sum_{i=1}^N h_i(\beta_m) \cdot X_i}{\sum_{i=1}^N h_i(\beta_m)}. \qquad (5.48)$$

In the classical literature, this is also called the *method of successive approximation* for the solution of the equation $\beta = T(\beta)$.[28] It has been proved that the iterative method of Equation 5.48 converges to a fixed point of $T(\cdot)$; thus,

$$\lim_{m \to \infty} \beta_m = \beta^* = T(\beta^*). \qquad (5.49)$$

The speed of convergence of the iterative algorithm (Equation 5.48) depends on the initial value β_0. A simple approach to selecting $\hat{\beta}_0$ is to assign it the value equal to that of the input sample X_i, which leads to the smallest cost $P(X_i)$.

5.6.2 Fixed Point Weighted Myriad Search

Step 1: Select the initial point $\hat{\beta}_0$ among the values of the input samples:

$$\hat{\beta}_0 = \arg\min_{X_i} P(X_i).$$

Step 2: Using $\hat{\beta}_0$ as the initial value, perform L iterations of the fixed point recursion $\beta_{m+1} = T(\beta_m)$ of Equation 5.48. The final value of these iterations is then chosen as the weighted myriad: $\hat{\beta}_{FP} = T^{(L)}(\hat{\beta}_0)$.

This algorithm can be compactly written as

$$\hat{\beta}_{FP} = T^{(L)}\left(\arg\min_{X_i} P(X_i) \right). \qquad (5.50)$$

Note that for the special case $L = 0$ (meaning that no fixed point iterations are performed), the above algorithm computes the selection weighted myriad.

5.7   Weighted Myriad Smoother Design

5.7.1   Center Weighted Myriad Smoothers in Image Denoising

Median smoothers are well known to be very effective at image denoising, especially when the noise is of "salt and pepper" type, which often results from bit errors in the transmission stage or in the acquisition stage. As a subset of traditional weighted median smoothers, center weighted (CW) median smoothers provide similar performance with much less complexity. In CW medians, only the center sample in the processing window is assigned weight, and all other samples are treated equally without emphasis. The larger the center weight, the less smoothing is achieved. Increasing the center weight beyond a certain threshold, CW medians become an identity operation. On the other hand, when the center weight is set to unity (the same as other weights), the CW median becomes a sample median operation.

The same notion of "center weighting" can be applied to the myriad structure as well, thus leading to the following definition of the center weighted myriad smoother (CWMy):

$$Y = \text{MYRIAD}\{K; X_1, \ldots, W_c \circ X_c, \ldots, X_N\}. \tag{5.51}$$

The cost function in Equation 5.25 is now modified to

$$Q(\beta) = \log[K^2 + W_c(X_c - \beta)^2] + \sum_{X_i \neq X_c} \log[K^2 + (X_i - \beta)^2]. \tag{5.52}$$

While similar to a CW median, the above CWMy smoother has significant differences. First, in addition to the center weight W_c, the CWMy has one more free parameter K that controls the impulsiveness rejection. This provides a simple mechanism to attain better smoothing performance. Second, the center weight in the CWMy smoother is inevitably data dependent according to the definition of the objective function in Equation 5.52. For different applications, based on their data ranges, the center weight should be adjusted accordingly.

Particularly for gray-scale image denoising applications where pixel values are normalized between 0 and 1, the two parameters of the CWMy smoother can be chosen as follows:

1. Choose $K = (X_{(U)} + X_{(L)})/2$, where $1 \leq L < U \leq N$, with $X_{(U)}$ the Uth smallest sample in the window and $X_{(L)}$ the Lth smallest sample.
2. Set $W_c = 10,000$.

The linear parameter K is dynamically calculated based on the samples in the processing window. When there is "salt" noise in the window (outliers having large values), the myriad structure assures that they are deemphasized due to the outlier rejection property of K. The center weight W_c is chosen balancing between outlier rejection and detail preservation. It should be large enough to emphasize the center sample and preserve signal details, but not large enough to let through impulsive noise.

It can also be shown that the CWMy smoother with K and W_c defined as above has the capability of completely rejecting "pepper" type noise (having values close to 0).[29] This can be seen as follows. For a single "pepper" outlier sample, the cost function (Equation 5.52) evaluated at $\beta = K$ will always be smaller than that at $\beta = 0$. This means that pepper noise will never go through the smoother as the output given that the parameters K and W_c are correctly chosen. Denote **X** as the corrupted image, **Y** the output smoothed image, and CWMy smoother operation. A special two-pass CWMy smoother can be defined as follows:

$$\mathbf{Y} = 1 - \text{CWMy}(1 - \text{CWMy}(\mathbf{X})). \tag{5.53}$$

Figure 5.10 depicts results of the algorithm defined in Equation 5.53. Figure 5.10a is a noise-free image with 256 gray levels. Pixel values are normalized to be in [0, 1], with 0 representing the brightest and 1 the darkest intensity. Figure 5.10b is Figure 5.10a corrupted by 5% salt and pepper noise. The impulses occur randomly and were generated by MATLAB's imnoise function. Figure 5.10c is the output of a 5×5 CWM smoother with $W_c = 15$, and Figure 5.10d that of a CWMy smoother with $W_c = 10,000$ and $K = (X_{(21)} + X_{(5)})/2$. The superior performance of the CWMy smoother can be readily seen in this figure. The CWMy smoother preserves the original image features significantly better than the CWM smoother (notice the hair details). The mean square error of the CWMy output is consistently less than half of that of the CWM output for this particular image.

5.7.2 Myriadization

The linear property indicates that for very large values of K, the weighted myriad smoother reduces to a constrained linear FIR smoother. The meaning of K suggests that a linear smoother can be provided with resistance to impulsive noise by simply reducing the linearity parameter from $K = \infty$ to a finite value. This would transform the linear smoother into a myriad smoother with the same weights. In the same way as the term *linearization* is commonly used to denote the transformation of an operator into a linear one, the above transformation is referred to as *myriadization*.

Myriadization is a simple but powerful technique that brings impulse resistance to constrained linear filters. It also provides a simple methodology to design suboptimal myriad smoothers in impulsive environments. Basically, a constrained linear smoother can be designed for Gaussian or noiseless

FIGURE 5.10
(a) Noise-free image, (b) with 5% salt and pepper noise, (c) smoothed with 5×5 center weighted median $W_c = 15$, (d) smoothed with 5×5 center weighted myriad $W_c = 10,000$, $K = (X_{(21)} + X_{(5)})/2$.

environments using FIR filter (smoother) design techniques, and then provide the smoother with impulse resistance capabilities by means of myriadization. The value to which K is to be reduced can be designed according to the impulsiveness of the environment, for example, by means of an α-K curve.

It must be taken into account that a linear smoother has to be in "constrained form" before myriadization can be applied. This means that the smoother coefficients W_i must be nonnegative and satisfy the normalization condition $\sum_{i=1}^{N} W_i = 1$. A smoother for which $\sum_{i=1}^{N} W_i \neq 1$ must be first "decomposed"

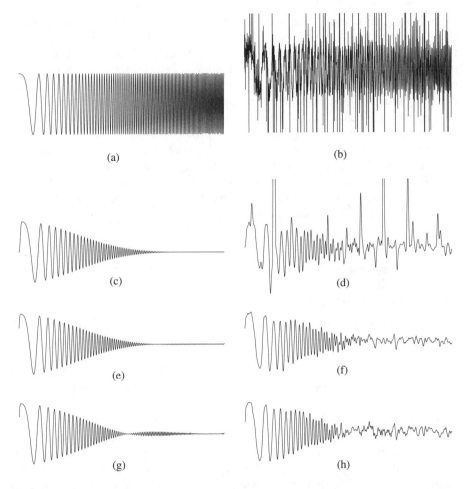

FIGURE 5.11
Myriadizing a linear lowpass smoother in an impulsive environment: (a) chirp signal, (b) chirp in additive impulsive noise, (c) ideal (no-noise) myriad smoother output with $K = \infty$, (e) $K = 0.5$, and (g) $K = 0.2$. Myriad smoother output in the presence of noise with (d) $K = \infty$, (f) $K = 0.5$, and (h) $K = 0.2$.

into the cascade of its normalized version with an amplifier of gain $\sum_{i=1}^{N} W_i$. Design by myriadization is illustrated in the following example.

EXAMPLE: Robust Lowpass Filter Design
Figure 5.11a depicts a unit-amplitude linearly swept-frequency cosine signal spanning instantaneous frequencies ranging from 0 to 400 Hz. The chirp was generated with MATLAB's chirp function having a sampling interval of 0.0005 s. Figure 5.11b shows the chirp immersed in additive Cauchy noise ($\gamma = 1$). The plot is truncated to the same scale as the other signals in

the figure. A lowpass linear FIR smoother with 30 coefficients processes the chirp with the goal of retaining its low-frequency components. The FIR lowpass smoother weights were designed with MATLAB's fir1 function with a normalized frequency cutoff of 0.05. Under ideal, no-noise conditions, the output of the linear smoother would be that of Figure 5.11c. However, the impulsive nature of the noise introduces severe distortions to the actual output, as depicted in Figure 5.11d. Myriadizing the linear smoother by reducing K to a finite value of 0.5 significantly improves the smoother performance (Figure 5.11e and f). Further reduction of K to 0.2 drives the myriad closer to a selection mode where some distortion on the smoother output under ideal conditions can be seen (Figure 5.11g). The output under the noisy conditions is not improved by further reducing K to 0.2, or lower, as the smoother in this case is driven to a "selection" operation mode.

5.8   Weighted Myriad Filters with Real-Valued Weights

Myriad smoothers admitting positive weights only are in essence "lowpass" type filters. Weighted myriad smoothers are thus analogous to normalized linear FIR filters with nonnegative weights summing to unity. There is a clear need to extend these smoothers into a general filter structure, comparable to linear FIR filters, that admit real-valued weights. In the same way that weighted median smoothers are extended to the weighted median filter, a generalized weighted myriad filter structure that admits real-valued weights can be developed. This section describes the structure and properties of such a class of filters admitting positive as well as negative weights. Adaptive optimization algorithms are presented. As would be expected, weighted myriad filters reduce to weighted myriad smoothers whenever the filter coefficients are constrained to be positive.

The approach used to generalize median smoothers to a general class of median filters can be used to develop a generalized class of weighted myriad filters. To this end, the set of real-valued weights are first decoupled in their sign and magnitude. The sign of each weight is then attached to the corresponding input sample and the weight magnitude is used as a positive weight in the weighted myriad smoother structure. The class of weighted myriad filters admitting real-valued weights emerges as follows.

DEFINITION 3 (Weighted Myriad Filters)
Given a set of N real valued weights $(W_1, W_2, \ldots, W_N)$ and the observation vector $\mathbf{X} = [X_1, X_2, \ldots, X_N]^T$, the weighted myriad filter output is defined as

$$\hat{\beta}_K(\mathbf{W}, \mathbf{X}) \stackrel{\triangle}{=} \text{MYRIAD} \left(|W_i| \circ \text{sgn}(W_i) X_i \right)\big|_{i=1}^N = \arg\min_\beta Q(\beta), \qquad (5.54)$$

where

$$Q(\beta) \overset{\triangle}{=} \sum_{i=1}^{N} \log[K^2 + |W_i| \cdot (\text{sgn}(W_i) X_i - \beta)^2] \qquad (5.55)$$

is the objective function of the weighted myriad filter.

The weighted myriad filter is also defined by the equivalent expression

$$\hat{\beta}_K(\mathbf{W}, \mathbf{X}) = \arg\min_{\beta} \prod_{i=1}^{N} [K^2 + W_i(X_i - \beta)^2] \qquad (5.56)$$

$$\overset{\triangle}{=} \arg\min_{\beta} P(\beta); \qquad (5.57)$$

thus $\hat{\beta}_K$ is the global minimizer of $P(\beta)$ as well as of $Q(\beta) \overset{\triangle}{=} \log(P(\beta))$.

Just like in the weighted myriad smoother, the weighted myriad filter also has only N independent parameters. By using Equation 5.55, it can be inferred that if the value of K is changed, the same filter output can be obtained provided the filter weights are appropriately scaled. The following is true:

$$\hat{\beta}_K(\mathbf{W}, \mathbf{X}) = \hat{\beta}_1\left(\frac{\mathbf{W}}{K^2}, \mathbf{X}\right) \qquad (5.58)$$

or

$$\hat{\beta}_{K_1}(\mathbf{W}_1, \mathbf{X}) = \hat{\beta}_{K_2}(\mathbf{W}_2, \mathbf{X}) \quad \text{iff} \quad \frac{\mathbf{W}_1}{K_1^2} = \frac{\mathbf{W}_2}{K_2^2}. \qquad (5.59)$$

Hence, the output depends only on $\mathbf{W}/K^2$.

The objective function $P(\beta)$ in Equation 5.57 is a polynomial in β of degree $2N$, with well-defined derivatives of all orders. Therefore, $P(\beta)$ (and the equivalent objective function $Q(\beta)$) can have at most $(2N-1)$ local extrema. The output is thus one of the local minima of $Q(\beta)$:

$$Q'(\hat{\beta}) = 0. \qquad (5.60)$$

As in the smoother case, the number of local minima in the objective function $Q(\beta)$ depends on the value of the parameter K. When K is very large, only one extremum exists.

In the limit as $K \to \infty$, with the weights $\{W_i\}$ held constant, it can be shown that $Q(\beta)$ exhibits a single local extremum. The proof is a generalized form of that used to prove the linear property of weighted myriad smoothers.

PROPERTY 9 *(Linear Property)*
In the limit as $K \to \infty$, the weighted myriad filter reduces to the normalized linear FIR filter

$$\lim_{K \to \infty} \hat{\beta}_K = \frac{\sum_{i=1}^{N} W_i X_i}{\sum_{i=1}^{N} |W_i|}. \qquad (5.61)$$

Once again, the name "linearity parameter" is used for the parameter K.

At the other extreme of linearity values ($K \to 0$), the weighted myriad filter maintains a mode-like behavior as stated in the following:

PROPERTY 10 (Mode Property)
Given a vector of real-valued weights, $\mathbf{W} = [W_1, \ldots, W_N]$, the weighted mode-myriad $\hat{\beta}_0$ is always equal to one of the most repeated values in the signed sample set $\mathcal{M} = [\mathrm{sgn}(W_1) X_1, \mathrm{sgn}(W_2) X_2, \ldots, \mathrm{sgn}(W_N) X_N]$. Furthermore,

$$\hat{\beta}_0 = \underset{\mathrm{sgn}(W_j) X_j \in \mathcal{M}}{\arg \min} \left(\frac{1}{|W_j|} \right)^{\frac{r}{2}} \prod_{i=1, X_i \neq X_j}^{N} |\mathrm{sgn}(W_i) X_i - \mathrm{sgn}(W_j) X_j|, \qquad (5.62)$$

where $\mathcal{M}$ is the set of most repeated signed values, and r is the number of times a member of $\mathcal{M}$ is repeated in the signed sample set.

5.9 Fast Real-Valued Weighted Myriad Computation

Using the same technique developed in Section 5.6, we can work out the fast computation for the real-valued weighted myriad. Remember that the myriad objective function now is

$$Q(\beta) = \sum_{i=1}^{N} \log[K^2 + |W_i| \cdot (\mathrm{sgn}(W_i) X_i - \beta)^2]. \qquad (5.63)$$

Its derivative with respect to β can be easily found as

$$Q'(\beta) = 2 \sum_{i=1}^{N} \frac{|W_i|(\beta - \mathrm{sgn}(W_i) X_i)}{K^2 + |W_i|(\mathrm{sgn}(W_i) X_i - \beta)^2}. \qquad (5.64)$$

By introducing the *positive* functions

$$h_i(\beta) \triangleq \frac{2|W_i|}{K^2 + |W_i|(\mathrm{sgn}(W_i) X_i - \beta)^2} > 0, \qquad (5.65)$$

for $i = 1, 2, \ldots, N$, the local extrema of $Q(\beta)$ satisfy the following condition:

$$Q'(\beta) = -\sum_{i=1}^{N} h_i(\beta) \cdot (\mathrm{sgn}(W_i) X_i - \beta) = 0. \qquad (5.66)$$

Since $\hat{\beta}_K$ is one of the local minima of $Q(\beta)$, we have $Q'(\hat{\beta}_K) = 0$. Equation 5.66 can be further written as *weighted mean* of the "signed" samples:

$$\beta = \frac{\sum_{i=1}^{N} h_i(\beta) \cdot \text{sgn}(W_i) X_i}{\sum_{i=1}^{N} h_i(\beta)}. \tag{5.67}$$

By defining the mapping

$$T(\beta) \triangleq \frac{\sum_{i=1}^{N} h_i(\beta) \cdot \text{sgn}(W_i) X_i}{\sum_{i=1}^{N} h_i(\beta)}, \tag{5.68}$$

the local extrema of $Q(\beta)$, or the roots of $Q'(\beta)$, are seen to be the *fixed points* of $T(\cdot)$:

$$\beta^* = T(\beta^*). \tag{5.69}$$

We use the following efficient *fixed point iteration* algorithm to compute the fixed points:

$$\beta_{m+1} \triangleq T(\beta_m) = \frac{\sum_{i=1}^{N} h_i(\beta_m) \cdot \text{sgn}(W_i) X_i}{\sum_{i=1}^{N} h_i(\beta_m)}. \tag{5.70}$$

In the limit, the above iteration will converge to one of the fixed points of $T(\cdot)$

$$\lim_{m \to \infty} \beta_m = \beta^* = T(\beta^*). \tag{5.71}$$

It is clear that the "global convergence" feature of this fixed point iteration can be assured from the analysis in Section 5.6 because the only difference is the sample set.

5.9.1   Fixed Point Weighted Myriad Search Algorithm

Step 1: Couple signs of the weights and samples to form the "signed" sample vector $[\text{sgn}(W_1) X_1, \text{sgn}(W_2) X_2, \ldots, \text{sgn}(W_N) X_N]$.

Step 2: Compute the selection weighted myriad:

$$\hat{\beta}_0 = \arg \min_{\beta \in \{\text{sgn}(W_i) X_i\}} P(\text{sgn}(W_i) X_i).$$

Step 3: Using $\hat{\beta}_0$ as the initial value, perform L iterations of the fixed point recursion $\beta_{m+1} = T(\beta_m)$ of Equation 5.70. The final value of these iterations is then chosen as the weighted myriad: $\hat{\beta}_{FP} = T^{(L)}(\hat{\beta}_0)$.

The compact expression of the above algorithm is

$$\hat{\beta}_{\text{FP}} = T^{(L)} \left(\underset{\beta \in \{\text{sgn}(W_i) X_i\}}{\arg\min} \ P(\text{sgn}(W_i) X_i) \right). \tag{5.72}$$

5.10   Weighted Myriad Filter Design

5.10.1   Myriadization

The linear property indicates that for very large values of K, the weighted myriad filter reduces to a constrained linear FIR filter. This characteristic of K suggests that a linear FIR filter can be provided with resistance to impulsive noise by simply reducing the linearity parameter from $K = \infty$ to a finite value. This would transform the linear FIR filter into a myriad filter with the same weights. This transformation is referred to as *myriadization* of linear FIR filters. Myriadization is a simple but powerful technique that brings impulse resistance to linear FIR filters. It also provides a simple methodology to design suboptimal myriad filters in impulsive environments. A linear FIR filter can be first designed for Gaussian or noiseless environments using FIR filter design tools, and then provide the filter with impulse resistance capabilities by means of myriadization. The value to which K is to be reduced can be designed according to the impulsiveness of the environment.

Design by myriadization is illustrated in the following example.

EXAMPLE: Robust Bandpass Filter Design
Figure 5.12a depicts a unit-amplitude linearly swept-frequency cosine signal spanning instantaneous frequencies ranging from 0 to 400 Hz. The chirp was generated with MATLAB's chirp function having a sampling interval of 0.0005 s. Figure 5.12b shows the chirp immersed in additive Cauchy noise ($\gamma = 1$). The plot is truncated to the same scale as the other signals in the figure. A bandpass linear FIR filter with 30 coefficients processes the chirp with the goal of retaining its low-frequency components. The FIR bandpass filter weights were designed with MATLAB's fir1 function with a normalized frequency cutoff of 0.05. Under ideal, no-noise conditions, the output of the FIR filter would be that of Figure 5.12c. However, the impulsive nature of the noise introduces severe distortions to the actual output, as depicted in Figure 5.12d. Myriadizing the linear filter by reducing K to a finite value of 0.5 significantly improves the filter performance (see Figure 5.12e and f). Further reduction of K to 0.2 drives the myriad closer to a selection mode where some distortion on the filter output under ideal conditions can be seen (see Figure 5.12g). The output under the noisy conditions is not improved

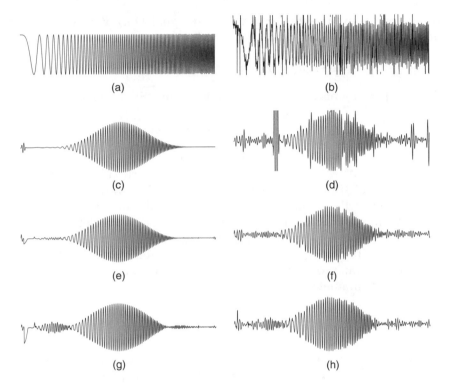

FIGURE 5.12
Myriadizing a linear bandpass filter in an impulsive environment: (a) chirp signal, (b) chirp in additive impulsive noise, (c) ideal (no-noise) myriad smoother output with $K = \infty$, (e) $K = 0.5$, and (g) $K = 0.2$; myriad filter output in the presence of noise with (d) $K = \infty$, (f) $K = 0.5$, and (h) $K = 0.2$.

by further reducing K to 0.2, or lower, as the filter in this case is driven to a "selection" operation mode.

5.10.2 Optimization

The *optimization* of the weighted myriad filter parameters for the case when the linearity parameter K satisfies $K > 0$ was first described in Reference 30. The goal is to design the set of weighted myriad filter weights that optimally estimate a desired signal according to a *statistical error criterion*. Although the mean absolute error (MAE) criterion is used here, the solutions are applicable to the mean square error (MSE) criterion with simple modifications.

Given an input (observation) vector $\mathbf{X} \triangleq [X_1, X_2, \ldots, X_N]^T$, a weight vector $\mathbf{W} \triangleq [W_1, W_2, \ldots, W_N]^T$, and linearity parameter K, denote the weighted myriad filter output as $Y \equiv Y_K(\mathbf{W}, \mathbf{X})$, sometimes abbreviated as $Y(\mathbf{W}, \mathbf{X})$.

The filtering error, in estimating a desired signal $D(n)$, is then defined as $e(n) = Y(n) - D(n)$. Under the MAE criterion, the cost function is defined as

$$J_1(\mathbf{W}, K) \triangleq E\{|e|\} = E\{|Y_K(\mathbf{W}, \mathbf{X}) - D|\}, \qquad (5.73)$$

where $E\{\cdot\}$ represents statistical expectation. The MSE is defined as

$$J_2(\mathbf{W}, K) \triangleq E\{e^2\} = E\{(Y_K(\mathbf{W}, \mathbf{X}) - D)^2\}. \qquad (5.74)$$

When the error criterion adopted is clear from the context, the cost function is written as $J(\mathbf{W}, K)$. Further, the optimal filtering action is independent of K (the filter weights can be scaled to keep the output invariant to changes in K). The cost function is therefore sometimes written simply as $J(\mathbf{W})$, with an assumed arbitrary choice of K. Obtaining conditions for a global minimum that are both necessary and sufficient is quite a formidable task. Necessary conditions, on the other hand, can be attained by setting the gradient of the cost function equal to zero. The *necessary* conditions to be satisfied by the optimal filter parameters are obtained as

$$\frac{\partial J(\mathbf{W})}{\partial W_i} = 2E\left\{e \frac{\partial Y}{\partial W_i}\right\} = 0, \quad i = 1, 2, \ldots, N. \qquad (5.75)$$

The nonlinear nature of Equations 5.75 prevents a closed-form solution for the optimal parameters. The *method of steepest descent*, which continually updates the filter parameters in an attempt to converge to the global minimum of the cost function $J(\mathbf{W})$, is thus applied:

$$W_i(n+1) = W_i(n) - \frac{1}{2}\mu \frac{\partial J}{\partial W_i}(n), \quad i = 1, 2, \ldots, N, \qquad (5.76)$$

where $W_i(n)$ is the ith parameter at iteration n, $\mu > 0$ is the *step size* of the update, and the gradient at the nth iteration is given by

$$\frac{\partial J}{\partial W_i}(n) = 2E\left\{e(n)\frac{\partial Y}{\partial W_i}(n)\right\}, \quad i = 1, 2, \ldots, N. \qquad (5.77)$$

When the underlying signal statistics are unavailable, *instantaneous estimates* for the gradient are used because the expectation in Equation 5.77 cannot be evaluated. Thus, removing the expectation operator in Equation 5.77 and using the result in Equation 5.76, the following weight update is found:

$$W_i(n+1) = W_i(n) - \mu e(n)\frac{\partial Y}{\partial W_i}(n), \quad i = 1, 2, \ldots, N. \qquad (5.78)$$

All that remains is to find an expression for

$$\frac{\partial Y(n)}{\partial W_i} = \frac{\partial}{\partial W_i}\hat{\beta}_K(\mathbf{W}, \mathbf{X}), \quad i = 1, 2, \ldots, N. \qquad (5.79)$$

Recall that the output of the weighted myriad filter is $\hat{\beta}_K(\mathbf{W}, \mathbf{X}) = \arg\min_{\beta} Q(\beta)$, where $Q(\beta)$ is given by

$$Q(\beta) = \sum_{i=1}^{N} \log[K^2 + |W_i| \cdot (\text{sgn}(W_i) X_i - \beta)^2]. \tag{5.80}$$

The derivative of $\hat{\beta}_K(\mathbf{W}, \mathbf{X})$ with respect to the weight W_i, holding all other quantities constant, can be shown to be[30]

$$\frac{\partial}{\partial W_i} \hat{\beta}_K(\mathbf{W}, \mathbf{X}) = \frac{-\left[\dfrac{K^2 \text{sgn}(W_i)(\hat{\beta} - \text{sgn}(W_i) X_i)}{(K^2 + |W_i| \cdot (\hat{\beta} - \text{sgn}(W_i) X_i)^2)^2} \right]}{\left[\sum_{j=1}^{N} |W_j| \dfrac{K^2 - |W_j| \cdot (\hat{\beta} - \text{sgn}(W_j) X_j)^2}{(K^2 + |W_j| \cdot (\hat{\beta} - \text{sgn}(W_j) X_j)^2)^2} \right]}. \tag{5.81}$$

Using Equation 5.78, the following adaptive algorithm is obtained to update the weights $\{W_i\}_{i=1}^{N}$:

$$W_i(n+1) = W_i(n) - \mu e(n) \frac{\partial \hat{\beta}}{\partial W_i}(n), \tag{5.82}$$

with $(\partial \hat{\beta} / \partial W_i)(n)$ given by Equation 5.81.

Considerable simplification of the algorithm can be achieved by just removing the denominator from the update term above; this does not change the direction of the gradient estimate or the values of the final weights. This leads to the following computationally attractive algorithm:

$$W_i(n+1) = W_i(n) - \mu e(n) \left[\frac{K^2 \text{sgn}(W_i)(\hat{\beta} - \text{sgn}(W_i) X_i)}{(K^2 + |W_i| \cdot (\hat{\beta} - \text{sgn}(W_i) X_i)^2)^2} \right]. \tag{5.83}$$

It is important to note that the optimal filtering action is independent of the choice of K; the filter only depends on the value of $\mathbf{w}/K^2$. In this context, one might ask how the algorithm scales as the value of K is changed and how the step-size μ and the initial weight vector $\mathbf{w}(0)$ are changed as K is varied. To answer this, let $\mathbf{g}_o \stackrel{\triangle}{=} \mathbf{w}_{o,1}$ denote the optimal weight vector for $K = 1$. Then, from Equation 5.29, $\mathbf{w}_{o,K}/K^2 = \mathbf{g}_o/(1)^2$ or $\mathbf{g}_o = \mathbf{w}_{o,K}/K^2$. Now consider two situations. In the first, the algorithm in Equation 5.83 is used with $K = 1$, step-size $\mu = \mu_1$, weights denoted as $g_i(n)$, and initial weight vector $\mathbf{g}(0)$. This is expected to converge to the weights $\mathbf{g}_o$. In the second, the algorithm uses a general value of K, step-size $\mu = \mu_K$, and initial weight vector $\mathbf{w}_K(0)$. Rewrite Equation 5.83 by dividing throughout by K^2 and writing the algorithm in terms of an update of w_i/K^2. This is expected to converge to $\mathbf{w}_{o,K}/K^2$ because Equation 5.83 should converge to $\mathbf{w}_{o,K}$. Because $\mathbf{g}_o = \mathbf{w}_{o,K}/K^2$, the above two situations can be compared and the initial weight vector $\mathbf{w}_K(0)$ and the step-size μ_K can be chosen such that the algorithms have the *same behavior* in both cases and converge, as a result, to the *same filter*. This means that $g_i(n) = w_i(n)/K^2$ at each iteration n. It can be shown that this results in

$$\mu_K = K^4 \mu_1 \quad \text{and} \quad \mathbf{w}_K(0) = K^2 \mathbf{w}_1(0). \tag{5.84}$$

This also implies that if K is changed from K_1 to K_2, the new parameters should satisfy

$$\mu_{K_2} = \left(\frac{K_2}{K_1}\right)^4 \mu_{K_1} \quad \text{and} \quad \mathbf{w}_{K_2}(0) = \left(\frac{K_2}{K_1}\right)^2 \mathbf{w}_{K_1}(0). \tag{5.85}$$

EXAMPLE: Robust Highpass Filter Design

Figure 5.13 illustrates some highpass filtering operations with various filter structures over a two-tone signal corrupted by impulsive noise. The signal has two sinusoidal components with normalized frequency 0.02 and 0.4, respectively. The sampling frequency is 1000 Hz. Figure 5.13a shows the two-tone signal in stable noise with exponent parameter $\alpha = 1.4$, and dispersion $\gamma = 0.1$. The result of filtering through a highpass linear FIR filter with 30 taps is depicted in Figure 5.13b. The FIR highpass filter coefficients are designed with MATLAB's fir1 function, with a normalized cutoff frequency of 0.2. It is clear that the impulsive noise has a strong effect on the linear filter output, and the high-frequency component of the original two-tone signal is severely distorted. The myriadization of the above linear filter gives a slightly better performance as shown in Figure 5.13c in the sense that the impulsiveness is greatly reduced, but the frequency characteristics are still not satisfactorily recovered. Figure 5.13c is the result of the optimal weighted myriad filtering with step-size $\mu = 3$. As a comparison, the optimal weighted median filtering result is shown in Figure 5.13d, step-size $\mu = 0.15$. Although these two nonlinear optimal filters perform significantly better than the linear filter, the optimal weighted myriad filter undoubtedly is the best approach, because its output has no impulsiveness presence and no perceptual distortion (except magnitude fluctuations) as in the myriadization and the optimal weighted median cases. Moreover, signal details are better preserved in the optimal weighted myriad realization as well.

Another interesting observation is depicted in Figure 5.14, where ensemble performances are compared. Although both median and myriad filters will converge faster when the step size is large and slower when the step size is small, they reach the same convergence rate with different step sizes. This is expected from their different filter structures. As shown in the plot, when the step-size μ is chosen to be 0.15 for the median, the comparable performance can be found when the step-size μ is in the vicinity of 3 for the myriad. The slight performance improvement can be seen from the plot in the stable region where the myriad has lower excess error floor than the median.

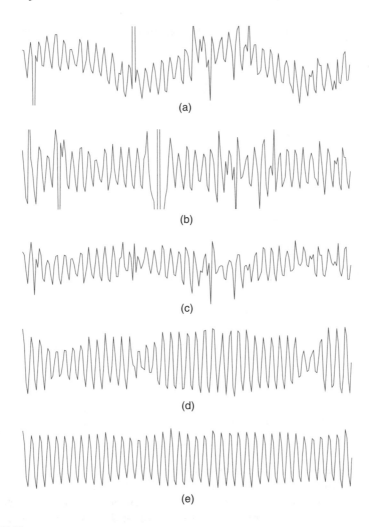

FIGURE 5.13
Robust highpass weighted myriad filter: (a) two-tone signal corrupted by α-stable noise, $\alpha = 1.4$, $\gamma = 0.1$, (b) output of the 30-tap linear FIR filter, (c) myriadization, (d) optimal weighted median filter, (e) optimal weighted myriad filter.

EXAMPLE: Robust Blind Equalization

The *constant modulus algorithm* (CMA) may be the most analyzed and deployed blind equalization algorithm. Many researchers believe that the CMA has become the workhorse for the blind channel equalization just as the least-mean square (LMS) has for the supervised adaptive filtering.[31] Some newly emerging communication technologies, such as digital cable TV, DSL, etc., are found in great favor of blind equalization implementation, mainly because their training is extremely costly if not impossible to carry out. In CMA applications, the linear FIR filter structure is assumed by default. But for

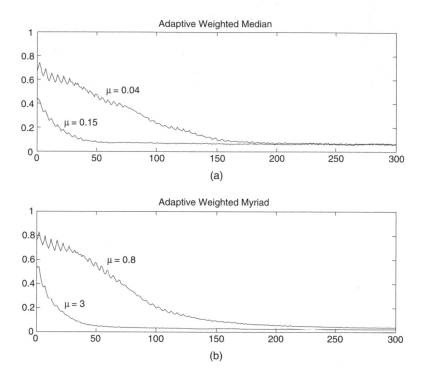

FIGURE 5.14
Comparison of convergence rate of the optimal weighted median and the optimal weighted myriad: (a) optimal weighted median at $\mu = 0.04, 0.15$, (b) optimal weighted myriad at $\mu = 0.8, 3$.

applications like DSL, it has been shown that impulsive noise is prevalent,[32] where inevitably, CMA blind equalization using FIR structure collapses. Here we describe a novel real-valued blind equalization algorithm, which is the combination of the constant modulus criterion and the weighted myriad filter structure. Using myriad filters, we should expect a very close to linear performance when we set the linear parameter K to be far larger than the data samples. When the noise contains impulses, by reducing K to a suitable level, we can manage to remove their influence greatly without losing the capability of keeping the communication eye open.

Consider a pulse amplitude modulation (PAM) communication system; signal and channel are all real. The constant modulus cost function is defined as follows:

$$J(\mathbf{W}, K) \triangleq \frac{1}{4} E\{(|Y(n)|^2 - R_2)^2\}, \tag{5.86}$$

where

$$R_2 = \frac{E|S(n)|^4}{E|S(n)|^2}$$

and $S(n)$ are the signal constellations and $Y(n)$ the filter output. The gradient of the above cost function can be calculated as

$$\nabla_{\mathbf{W}} J (\mathbf{W}, K) = E \left\{ Y(n)(|Y(n)|^2 - R_2) \frac{\partial Y(n)}{\partial \mathbf{W}} \right\}. \tag{5.87}$$

We adopt a scaled version of a real weighted myriad (WMy) filter where the sum of the magnitudes of the filter weights are used as the scaling factor.

$$Y(n) = \left(\sum_{i=1}^{N} |W_i| \right) \cdot \text{MYRIAD}\left(|W_i| \circ \text{sgn}(W_i) X_i \big|_{i=1}^{N}; K\right) \tag{5.88}$$

$$= \left(\sum_{i=1}^{N} |W_i| \right) \cdot \hat{\beta}. \tag{5.89}$$

This is particularly important for equalization applications because the signal energy needs to be considered in these cases. The derivative of the filter output with respect to a single weight is expressed as

$$\frac{\partial Y}{\partial W_i} = \text{sgn}(W_i)\hat{\beta} + \left(\sum_{i=1}^{N} |W_i| \right) \frac{\partial \hat{\beta}}{\partial W_i}, \tag{5.90}$$

and the derivative of $\hat{\beta}$ has already been shown in Equation 5.81. Finally, the weight update can be carried out using the following equation:

$$W_i(n+1) = W_i(n) + \mu Y(n) (|Y(n)|^2 - R_2) \left(\text{sgn}(W_i)\hat{\beta} + \left(\sum_{i=1}^{N} |W_i| \right) \frac{\partial \hat{\beta}}{\partial W_i} \right). \tag{5.91}$$

Unlike the regular WMy filters, which have only N independent parameters as described in Reference 30, all $N + 1$ parameters of the scaled WMy filters (that is, N weights and one linear parameter) are independent. Thus, to best exploit the proposed structure, K needs to be updated adaptively as well. This time, we need to reconsider the objective function of the weighted myriad because K is a free parameter now.

$$\hat{\beta} \triangleq \text{MYRIAD}\left(|W_i| \circ \text{sgn}(W_i) X_i \big|_{i=1}^{N}; K\right)$$

$$= \arg\max_{\beta} \prod \frac{K}{K^2 + |W_i| \cdot (\text{sgn}(W_i) X_i - \beta)^2}$$

$$= \arg\min_{\beta} \sum \left[\log(K^2 + |W_i| \cdot (\text{sgn}(W_i) X_i - \beta)^2) - \log K \right] \tag{5.92}$$

$$= \arg\min_{\beta} Q(\beta, K).$$

Denote

$$G(\hat{\beta}, K) \triangleq Q'(\hat{\beta}, K)$$

$$= 2 \sum_{i=1}^{N} \frac{|W_i|(\hat{\beta} - \text{sgn}(W_i) X_i)}{K^2 + |W_i|(\text{sgn}(W_i) X_i - \hat{\beta})^2} = 0. \qquad (5.93)$$

Following a similar analysis as in the weight update, we can develop a K update algorithm. However, two reasons make it more attractive to update the *squared linearity parameter* $\mathcal{K} \triangleq K^2$ instead of K itself. First, in myriad filters, K always occurs in its squared form. Second, the adaptive algorithm for K might have an ambiguity problem in determining the sign of K. Rewrite Equation 5.93 as

$$G(\hat{\beta}, \mathcal{K}) = 2 \sum_{i=1}^{N} \frac{|W_i|(\hat{\beta} - \text{sgn}(W_i) X_i)}{\mathcal{K} + |W_i|(\text{sgn}(W_i) X_i - \hat{\beta})^2} = 0.$$

Implicitly differentiating both sides with respect to $\mathcal{K}$, we have

$$\left(\frac{\partial G}{\partial \hat{\beta}} \right) \cdot \left(\frac{\partial \hat{\beta}}{\partial \mathcal{K}} \right) + \left(\frac{\partial G}{\partial \mathcal{K}} \right) = 0, \qquad (5.94)$$

thus,

$$\frac{\partial \hat{\beta}}{\partial \mathcal{K}} = -\frac{\dfrac{\partial G}{\partial \mathcal{K}}}{\dfrac{\partial G}{\partial \hat{\beta}}}. \qquad (5.95)$$

Finally, the update for $\mathcal{K}$ can be expressed as

$$\mathcal{K}_i(n+1) = \mathcal{K}_i(n) - \frac{1}{2} \mu \frac{\partial J\,(\mathbf{W}, \mathcal{K})}{\partial \mathcal{K}}(n)$$

$$= \mathcal{K}(n) - \mu E \left\{ e(n) \frac{\partial \hat{\beta}_{\mathcal{K}}}{\partial \mathcal{K}}(n) \right\}. \qquad (5.96)$$

Figure 5.15 depicts a blind equalization experiment where the constellation of the signal is BPSK, and the channel impulse response is simply [1 0.5]. Additive stable noise with $\alpha = 1.5$, $\gamma = 0.002$ corrupts the transmitted data. Figure 5.15a is the traditional linear CMA equalization, while Figure 5.15b is the proposed myriad CMA equalization. It can be seen that, under the influence of impulsive noise, the linear equalizer diverges, but the myriad equalizer is more robust and still gives very good performance. Figure 5.16 shows the adaptation of parameter K in the corresponding realization.

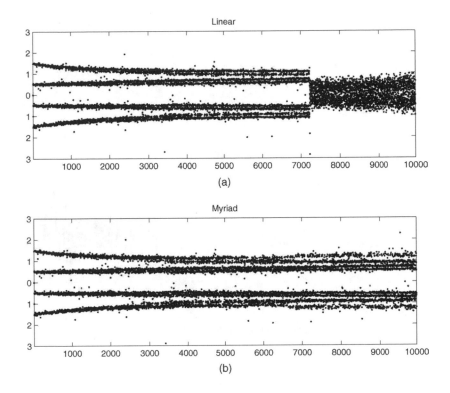

FIGURE 5.15
(a) Linear blind equalization, (b) myriad blind equalization.

5.11 Conclusions

Weighted myriad filtering is a flexible filtering framework that derives important robustness properties from the impulsive characteristics of symmetric α-stable distributions. In the same way that linear and median filters are related to the Gaussian and Laplacian distributions, respectively, myriad filter theory is based on the definition of the sample myriad as the maximum likelihood location estimator of the Cauchy distribution—the only non-Gaussian symmetric α-stable distribution for which a closed-form density is available. When weights (especially real-valued ones) are introduced in the definition, the weighted myriad filters appear as a rich and flexible class of filters that can range, by simply varying a tuning parameter, from highly robust mode-like filter forms to simple and Gaussian-efficient linear FIR filters. This chapter incorporates the latest developments on weighted myriad filters including algorithms for fast calculation, filter design, and optimization. A complete

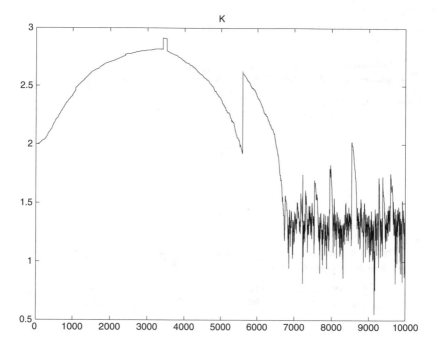

FIGURE 5.16
Adaptation of K.

set of tools for weighted myriad filters, as well as their applications in image denoising, bandpass/highpass filtering, and blind equalization, are provided so that immediate extensions to other applications can be readily performed.

References

1. E. Sousa, Performance of a spread spectrum packet radio network link in a Poisson field of interferers, *IEEE Trans. Inf. Theor.*, 38, 1743–1754, Nov. 1992.
2. W.E. Leland, M.S. Taqqu, W. Willinger, and D.V. Wilson, On the self-similar nature of ethernet traffic (extended version), *IEEE/ACM Trans. Networking*, 2, 1–15, Feb. 1994.
3. V.V. Uchaikin and V.M. Zolotarev, *Chance and Stability: Stable Distributions and Their Applications*, VSP, Zeist, the Netherlands, 1999.
4. W. Feller, *An Introduction to Probability Theory and Its Applications*, New York: John Wiley & Sons, 1971.
5. G. Samorodnitsky and M.S. Taqqu, *Stable Non-Gaussian Random Processes: Stochastic Models with Infinite Variance*, New York: Chapman & Hall, 1994.

6. C.L. Nikias and M. Shao, *Signal Processing with Alpha-Stable Distributions and Applications*, New York: Wiley Interscience, 1995.
7. V. Zolotarev, *One-Dimensional Stable Distributions*, Providence, RI: American Mathematical Society, 1986.
8. B. Mandelbrot, Long-run linearity, locally Gaussian processes, H-spectra, and infinite variances, *Interant. Econ. Rev.*, 10, 82–111, 1969.
9. J.H. McCulloch, Simple consistent estimators of stable distribution parameters, *Commun. Stat. Simulation Comput.*, 15(4), 1109–1136, 1986.
10. B.W. Stuck, Minimum error dispersion linear filtering of scalar symmetric stable processes, *IEEE Trans. Autom. Control*, 23(3), 507–509, 1978.
11. R.J. Adler, R.E. Feldman, and M.S. Taqqu, Analysing stable time series, in *A Practical Guide to Heavy Tails: Statistical Techniques and Applications*, New York: Springer-Verlag, 1998.
12. J.P. Nolan, *Stable Distributions*, Boston: Birkhauser, 2002.
13. P. Lévy, *Calcul des probabilités*, Paris: Gauthier-Villars, 1925.
14. J.G. Gonzalez, D.W. Griffith, and G.R. Arce, Zero-Order Statistics: A Signal Processing Framework for Very Impulsive Environments, Proc. IEEE Signal on Higher Order Statistics, Banff, Alberta, Canada, July 1997.
15. R.A. Fisher, On the mathematical foundation of theoretical statistics, *Philos. Trans. R. Soc. London*, 222, 1922.
16. S. Ambike and D. Hatzinakos, A new filter for highly impulsive α-stable noise, in *Proc. of the 1995 Int. Workshop on Nonlinear Signal and Image Proc.*, Halkidiki, Greece, June 1995.
17. D. Andrews, D. Bickel, P. Hampel, F. Huber, P. Rogers, and J. Tukey, *Robust Estimates of Location: Survey and Advances*, Princeton, NJ: Princeton University Press, 1972.
18. K.L. Boyer, M.J. Mirza, and G. Ganguly, The robust sequential estimator: a general approach and its application to surface organization in range data, *IEEE Trans. PAMI*, 16, Oct. 1994.
19. H.M. Hall, A new model for "impulsive" phenomena: application to atmospheric-noise communication channels, Technical Reports 3412-8 and 7050-7, Stanford Electronics Lab., Stanford University, Palo Alto, CA, Aug. 1966.
20. L.K.S. Rappaport, An optimal nonlinear detector for digital data transmission through non-Gaussian channels, *IEEE Trans. Commun.*, 14, Mar. 1966.
21. F. Steiner, Most frequent value and cohesion of probability distributions, *Acta Geod. Geophys. Mont. Acad. Sci. Hung.*, 8(3–4), 381–395, 1973.
22. J.G. Gonzalez and G.R. Arce, Weighted myriad filters: a robust filtering framework derived from alpha-stable distributions, in *Proceedings of the 1996 IEEE International Conference on Acoustics, Speech, and Signal Processing*, Atlanta, GA, May 1996.
23. G.R. Arce, J.G. Gonzalez, and P. Zurbach, Weighted myriad filters in imaging, in *Proc. Asilomar Conf. on Signals, Systems, and Computers*, Nov. 1996.
24. J.G. Gonzalez, D.L. Lau, and G.R. Arce, Towards a general theory of robust nonlinear filtering: selection filters, in *Proc. IEEE ICASSP'97*, Munich, Germany, Apr. 1997.
25. J.G. Gonzalez and G.R. Arce, Statistically-efficient filtering in impulsive environments: weighted myriad filters, *EURASIP J. Appl. Signal Process.*, 4–20, Jan. 2002.

26. J.G. Gonzalez and G.R. Arce, Optimality of the myriad filter in practical impulsive-noise environments, *IEEE Trans. Signal Process.*, 49, 438–441, Feb. 2001.
27. J.G. Gonzalez, D.W. Griffith, Jr., A.B. Cooper III, and G.R. Arce, Adaptive reception in impulsive noise, in *Proc. IEEE Int. Symp. on Information Theory*, Ulm, Germany, June 1997.
28. D.G. Luenberger, *Optimization by Vector Space Methods*, New York: Wiley, 1969.
29. Y. Li and G.R. Arce, Center Weighted Myriad Smoother in Image Denoising, Technical report, Department of Electrical and Computer Engineering, University of Delaware, Newark, 2003.
30. S. Kalluri and G.R. Arce, Robust frequency-selective filtering using weighted myriad filters admitting real-valued weights, *IEEE Trans. Signal Process.*, 49, 2721–2733, Nov. 2001.
31. S. Haykin, *Unsupervised Adaptive Filtering*, Vol. I and II, New York: Wiley-Interscience, 2000.
32. W. Yu, D. Toumpakaris, J.M. Cioffi, D. Gardan, and F. Gauthier, Performance of asymmetric digitial subscriber lines (adsl) in an impulse noise environment, *IEEE Trans. Commun.*, in press.

6

Data Traffic Modeling—A Signal Processing Perspective

Athina P. Petropulu and Xueshi Yang

CONTENTS

6.1   Introduction

Before the appearance of the Internet, teletraffic engineering had evolved around telephone communications. Telephone calls had been observed to arrive in a Poisson fashion, and the calls' holding times were exponentially distributed. As the bandwidth requirements for a telephone call was fixed, meeting certain quality of service (QoS) requirement for telephone communication was equivalent to guaranteeing a certain call-blocking probability, which in turn could be achieved by appropriate allocation of network resources. The resource allocation could be carried out based on the classical queuing theory that is applied to a Poisson arrival process with exponential holding or service times.

With the invention of the Internet and the explosive advances in computer networks, however, network traffic characteristics have changed

0-8493-1427-5/04/$0.00+$1.50
© 2004 by CRC Press LLC

195

dramatically, causing the classical teletraffic theory to be challenged by several new aspects. The traffic (data traffic) that is defined here as bytes per unit time corresponds to packetized voice, video, images, and computer data files. The data transfer sessions no longer fit the Poisson model well. For example, during an Internet browsing session, the end user and the server may exchange small and large files in a bursty fashion. The number of transfer requests may come in bursts during a sustained time period. That may be followed by a period of "silence," and then again by another "busy" period. Thus, the times between transfer requests are no longer drawn from an exponential distribution. Second, because of the multimedia nature of the data that are being exchanged, there is usually high variability in bandwidth requirements during a single session. These characteristics of data traffic give rise to some very interesting properties, namely, *self-similarity* and *impulsiveness*, both of which can have significant consequences in network design and management.

Developing analytical models that capture the new characteristics of data traffic has considerable significance in several aspects of traffic engineering, such as admission control, flow control, and congestion control. For example, in data networks, switches allocate a certain bandwidth to a group of admitted traffic flows. In general, the allocated bandwidth is not the sum of the peak rate of all flows, but is computed according to the statistical features of the incoming traffic and some predefined QoS. Such practice is referred to as "statistical multiplexing." An analytical model for traffic would be indispensable in allocating bandwidth to meet QoS guarantees. Analytic models would also be used to evaluate the performance of large-scale networks, whose analytic characterization is rather intractable, before they are deployed. This is typically done using traces collected from real networks. However, those traffic traces are shaped by the characteristics of the network in which they were collected, and thus could be inappropriate when used in a different network. Furthermore, real traffic traces may require enormous storage space, given that today's broadband data traffic flows at a speed of giga bits per second. Analytical models can allow testing of large-scale networks via on-the-fly generation of traffic traces. Also, by providing traffic traces with various controllable characteristics, they can offer more insight into the reasons for certain network behavior.

In this chapter, we concentrate on statistical analysis and modeling of data traffic. In particular, we will focus on the two important salient features of data traffic, namely, self-similarity and impulsiveness. The chapter is organized as follows. Mathematical preliminaries are introduced in Section 6.2. Section 6.3 illustrates the self-similar and impulsive nature of data traffic based on real traffic measurements. Traffic models are discussed in Section 6.4, and parameter estimation techniques are reviewed in Section 6.5. Finally, Section 6.6 provides some concluding remarks.

6.2 Preliminaries—Self-Similarity, Long-Range Dependence, and Impulsiveness

In this section, we formulate the concepts of long-range dependence and self-similarity, and point out how they are related. In a communication context, self-similarity is closely related to heavy-tail distributions. We provide the definition of heavy-tailed distributions, and discuss two special classes, namely, the Pareto distributions and the α-stable distributions.

6.2.1 Long-Range Dependence and Self-Similarity

6.2.1.1 Long-Range Dependence

Let $\{X_k\}_{k \in \mathbb{Z}}$ be a discrete-time second-order stationary stochastic process with finite second-order statistics, mean $\mu = \mathrm{E}X_k$, and variance $\sigma^2 = \mathrm{E}\{(X_k - \mu)^2\}$. The autocovariance of $\{X_k\}$ is denoted as

$$r(\tau) = \frac{1}{\sigma^2}\mathrm{E}\{(X_k - \mu)(X_{k+\tau} - \mu)\}. \tag{6.1}$$

DEFINITION 1
$\{X_k\}_{k \in \mathbb{Z}}$ is a long-range dependent process with Hurst parameter H, if for $\tau \in \mathbb{Z}$,

$$\lim_{k \to \infty} \frac{r(\tau)}{k^{2H-2}} = c_r, \quad 1/2 < H < 1 \tag{6.2}$$

where $0 < c_r < \infty$ is a constant.

An equivalent definition[1] of long-range dependence (LRD) is based on the spectrum of the process (provided it exists), $f(\lambda)$. A process is long-range dependent if its spectrum satisfies

$$\lim_{\lambda \to 0} f(\lambda)/|\lambda|^{1-2H} = c_f, \quad 1/2 < H < 1, \tag{6.3}$$

where c_f is a positive constant. Equation 6.2 implies that for long-range dependent processes, it holds that:

$$\sum_{\tau=-\infty}^{\infty} r(\tau) = \infty, \tag{6.4}$$

for $\frac{1}{2} < H < 1$. Thus, although at high lags the autocovariance is small, its cumulative effect is large, giving rise to a behavior that is distinctly different

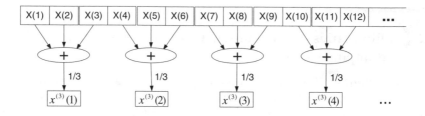

FIGURE 6.1
The aggregate process of $\{X_k\}_{k\in\mathbb{Z}}$ of degree 3.

from that of standard processes. For example, for the well-known ARMA processes, the autocovariance decreases geometrically fast as the lag increases. Such processes are referred to as short-range dependent and have a summable autocovariance.

Further insight into the properties of long-range dependent processes can be gained by looking at their *aggregate processes*. The aggregate process of X_k of degree m, denoted by $X_k^{(m)}$, is a running average of non-overlapping blocks of X_k with length m, i.e.,

$$X_i^{(m)} = \frac{1}{m} \sum_{k=m(i-1)+1}^{mi} X_k. \tag{6.5}$$

Figure 6.1 illustrates the aggregation process of X_k of degree 3.

For a long-range dependent process $\{X_k\}_{k\in\mathbb{Z}}$ it holds that:[2]

$$\lim_{m\to\infty} \frac{\text{Var}\, X_k^{(m)}}{m^{(2H-2)}} = c, \tag{6.6}$$

where c is a constant. Because aggregation is equivalent to time scaling, Equation 6.6 implies that, under time scaling, long-range dependent processes are smoothed out much more slowly in comparison to an independent identically distributed (i.i.d.) process, for which the variance of the aggregate process decays proportional to m^{-1} as m tends to infinity. Thus, in a sense, the time-scaled long-range dependent process maintains similarity to the original process, hinting at a relationship between long-range dependence and self-similarity—a concept to be defined in the following section.

6.2.1.2 Self-Similarity

DEFINITION 2
A real-valued process $\{Y(t)\}_{t\in\mathcal{R}}$ is self-similar with index $H > 0$, if for all $a > 0$, all finite dimensional distributions of $a^{-H}Y(at)$ are identical to the finite dimensional distributions of $Y(t)$, i.e.,

$$\{a^{-H}Y(at), t \in \mathcal{R}\} \overset{d}{=} \{Y(t), t \in \mathcal{R}\}, \tag{6.7}$$

where $\stackrel{d}{=}$ means equality for all finite dimensional distributions. The index H is referred to as the Hurst parameter of the self-similar process $Y(t)$.

Provided that the qth order moments of $Y(t)$ exist, it can be drawn from Equation 6.7 that

$$E|Y(t)|^q = E|Y(1)|^q |t|^{qH}, \tag{6.8}$$

which shows that a self-similar process is clearly nonstationary. It also holds that

$$Y(0) = 0 \quad a.s., \tag{6.9}$$

and

$$E\{Y(s)Y(t)\} = \frac{\sigma^2}{2}(|s|^{2H} + |t|^{2H} - |s - t|^{2H}). \tag{6.10}$$

An example of self-similar process is the Brownian motion $B(t)$, which is defined as follows:

- $B(t)$ is a zero-mean Gaussian process.
- The correlation of $B(s)$ and $B(t)$ is the minimum of s and t.

We can verify that $B(t)$ is self-similar with Hurst parameter $H = \frac{1}{2}$ because

$$E\{a^{-1/2}B(as)a^{-1/2}B(at)\} = a^{-1}\min(as, at) = \min(s, t) = E\{B(s), B(t)\}. \tag{6.11}$$

A more general form of Brownian motion is the fractional Brownian motion (FBM), defined by

$$B_H(t) \stackrel{\Delta}{=} \frac{1}{\Gamma(H + 1/2)} \left[\int_{-\infty}^{0} (|t - \tau|^{H-1/2} - |\tau|^{H-1/2}) dB(\tau) \right.$$

$$\left. + \int_{0}^{t} |t - \tau|^{H-1/2} dB(\tau) \right], \quad t \in \mathcal{R} \tag{6.12}$$

where $B(t)$ is a standard Brownian motion, $\Gamma(.)$ is the Gamma function, and $\frac{1}{2} < H < 1$. Clearly, $B_H(t)$ is a zero-mean Gaussian process and $B_H(0) = 0$. Furthermore, if we extend the definition to include $H = \frac{1}{2}$, we have $B_{1/2}(t) = B(t)$. It is not difficult to show that, for each $a > 0$,

$$\{B_H(at); t \in \mathcal{R}\} \stackrel{d}{=} a^H\{B_H(t); t \in \mathcal{R}\};$$

that is, $B_H(t)$ is self-similar.

Of particular interest are self-similar processes with stationary increments. A process $Y(t)$ is said to have stationary increments if

$$\{X(\delta, t) := Y(t + \delta) - Y(t), t \in \mathcal{R}\} \stackrel{d}{=} \{Y(\delta) - Y(0)\}, \quad \text{for any } \delta. \tag{6.13}$$

In other words, the finite dimensional distributions of $X(\delta, t)$ are independent of t.

Let $Y(t)$ be a self-similar process with stationary increments. Then, $E\{Y(t)\} = 0$, which implies that $X(\delta, t)$ is also zero-mean. The autocovariance of $X(\delta, s)$ and $X(\delta, t)$ equals

$$
\begin{aligned}
r_X(s, t) &= E\{X(\delta, s) X(\delta, t)\} \\
&= E\{[Y(s + \delta) - Y(s)][Y(t + \delta) - Y(t)]\} \\
&= \frac{\sigma^2}{2}\left(|(s - t) + \delta|^{2H} + |(s - t) - \delta|^{2H} - 2|(s - t)|^{2H}\right). \quad (6.14)
\end{aligned}
$$

Although there are many non-Gaussian self-similar processes with stationary increments, there exists a unique Gaussian self-similar process with stationary increments, namely, the FBM.[29] The FBM process has played a significant role in modeling natural phenomena and in modeling data traffic. The increment of FBM is referred to as fractional Gaussian noise.

6.2.1.3   Relationship between Self-Similarity and LRD

Let us consider a self-similar process with stationary increments and let $X(\delta, k)$ be the discrete-time version of the increment. Taking $\delta = 1$ and based on Equation 6.14, we obtain

$$
r_X(\tau) := r_X(n, n + \tau) = \frac{\sigma^2}{2}(|\tau + 1|^{2H} + |\tau - 1|^{2H} - 2|\tau|^{2H}) \quad \text{for } n, \tau \in \mathbb{Z}.
$$

$$(6.15)$$

It is not difficult to show that the above autocovariance satisfies

$$
\lim_{\tau \to \infty} \frac{r_X(\tau)}{\tau^{2H-2}} = H(2H - 1), \tag{6.16}
$$

that is, $r_X(\tau)$ decays hyperbolically. Thus, the stationary increment of a self-similar process is long-range dependent.

DEFINITION 3
A discrete-time process $\{X_k\}_{k \in \mathbb{Z}}$ is exactly second-order self-similar with Hurst parameter $H \in (1/2, 1)$, if its autocovariance, provided it exists, is given by Equation 6.15.

Consider the aggregated process $X_k^{(m)}$ of an exactly second-order self-similar process $\{X_k\}$, at aggregation level m. It can be shown[3] that the autocovariance of $X_k^{(m)}$ and X_k, i.e., $r_X^{(m)}(\tau)$ and $r_X(\tau)$, respectively, are related as

$$
r_X^{(m)}(\tau) = r_X(\tau), \quad m = 2, 3, \ldots \tag{6.17}
$$

The latter equation implies that the second-order statistics of the process do not change under time scaling, hence justifying the use of the term "exactly second-order self-similar."

DEFINITION 4
A process $\{X_k\}_{k \in \mathbb{Z}}$ is said to be asymptotically second-order self-similar if its autocovariance and the autocovariance of its aggregate process are related as

$$\lim_{m \to \infty} r_X^{(m)}(\tau) = r_X(\tau). \tag{6.18}$$

It is shown in Reference 2 that Equation 6.6 implies Equation 6.18; in other words, a long-range dependent process is also asymptotically second-order self-similar.

The preceding discussion indicates that self-similarity and long-range dependence are different concepts. However, in the case of second-order self-similarity, self-similarity implies long-range dependence, and vice versa. For this reason, in the literature and also in this chapter (unless otherwise specified), the terms *self-similarity* and *long-range dependence* are used in an interchangeable fashion.

6.2.2   Heavy-Tailed Distributions and Impulsiveness

DEFINITION 5
A random variable X is heavy-tail distributed with index α if

$$P(X \geq x) \sim c x^{-\alpha} L(x), \qquad x \to \infty, \tag{6.19}$$

for $c > 0$, and $0 < \alpha < 2$, where $L(x)$ is a slowly varying function such that $L(x)$ is positive for large x and $\lim_{x \to \infty} L(bx)/L(x) = 1$ for any positive b.

Intuitively, a heavy-tail distributed random variable can fluctuate far away from its mean value (defined only when $1 < \alpha < 2$), with nonnegligible probability.

The simplest example of a heavy-tail distribution is the Pareto distribution,* which is defined in terms of its complementary distribution function (survival function) as

$$\bar{F}(x) := P(X \geq x) = \begin{cases} \left(\frac{x}{k_0}\right)^{-\alpha}, & x \geq k_0, \\ 1, & x < k_0, \end{cases} \tag{6.20}$$

where k_0 is positive constant and $0 < \alpha < 2$.

For heavy-tail distributions, pth order statistics are finite if and only if $p \leq \alpha$. A direct consequence of this property is that for heavy-tail distributed random variables, the second-order statistics are infinite (the mean is infinite if $\alpha < 1$).

---

* Technically, Pareto distribution includes a large class of distributions, namely, *Pareto I, II, III, and IV*, respectively, according to the increasing complexity in the format of the distribution functions. Here we concentrate on the Pareto I distribution.

Another popular member of the class of heavy-tailed distributions is the α-stable distribution. A random variable X is said to be α-stable distributed if there are parameters $0 < \alpha \leq 2$, $\sigma \geq 0$, $-1 \leq \eta \leq 1$, and μ real such that its characteristic function is of the form:

$$\Phi(\theta) = \exp\left\{i\mu\theta - \sigma^\alpha |\theta|^\alpha \left(1 - i\eta\, \text{sign}(\theta)\varphi(\theta, \alpha)\right)\right\} \tag{6.21}$$

with

$$\varphi(\theta, \alpha) = \begin{cases} \tan\frac{\alpha\pi}{2} & \text{if } \alpha \neq 1 \\ -\frac{2}{\pi}\log|\theta| & \text{if } \alpha = 1, \end{cases} \tag{6.22}$$

$$\text{sign}(\theta) = \begin{cases} 1, & \text{if } \theta > 0 \\ 0, & \text{if } \theta = 0 \\ -1, & \text{if } \theta < 0. \end{cases} \tag{6.23}$$

The parameter α is the characteristic exponent. In short, we denote X as $S_\alpha(\sigma, \eta, \mu)$-distributed.

The parameter μ is a *shift parameter*; for any real constant a, $X + a$ is $S_\alpha(\sigma, \eta, \mu + a)$-distributed. σ is a *scale parameter*; if a is a positive constant and $\alpha \neq 1$, then $a X$ is $S_\alpha(a\sigma, \eta, a\mu)$-distributed. η is the *skewness parameter*; the distribution is symmetric with respect to μ if $\eta = 0$. For $\alpha = 2$ the characteristic function becomes $\Phi(\theta) = \exp(-\sigma^2\theta^2 + i\mu\theta)$, which is the characteristic function of a Gaussian random variable with mean μ and variance $2\sigma^2$. When $\eta = \mu = 0$, X is symmetric α-stable (SαS) and its characteristic function takes the simple form $\Phi(\theta) = e^{-\sigma^\alpha |\theta|^\alpha}$. Hence, the S$\alpha$S distribution is characterized only by the scale parameter σ and the characteristic exponent α.

It can be shown that if $X \sim S_\alpha(\sigma, \eta, \mu)$ with $0 < \alpha < 2$ it holds

$$\begin{cases} \lim_{\lambda\to\infty} \frac{P\{X>\lambda\}}{\lambda^{-\alpha}} = C_\alpha\frac{1+\eta}{2}\sigma^\alpha, \\ \lim_{\lambda\to\infty} \frac{P\{X<-\lambda\}}{\lambda^{-\alpha}} = C_\alpha\frac{1-\eta}{2}\sigma^\alpha, \end{cases} \tag{6.24}$$

where

$$C_\alpha = \left(\int_0^\infty x^{-\alpha}\sin x\right)^{-1} = \begin{cases} \frac{1-\alpha}{\Gamma(2-\alpha)\cos(\pi\alpha/2)} & \text{if } \alpha \neq 1, \\ 2/\pi & \text{if } \alpha = 1. \end{cases} \tag{6.25}$$

In other words, the survival function of X delays in a power-law fashion with respect to its argument.

One can extend the definition of stable distributions to random vectors.[4] For example, the vector $\mathbf{X} = [X_1, X_2, \ldots, X_n]$ is an SαS random vector if and only if the linear combination $a_1 X_1 + \cdots + a_n X_n$ is SαS for all real $a_1, \ldots, a_n$. A random process $\{X(t)\}_{t\in\mathbb{Z}}$ is said to be α-stable if for any $n \geq 1$ and distinct instances $t_1, \ldots, t_n$, the random variables $X(t_1), \ldots, X(t_n)$ are jointly α-stable with the same α.

DEFINITION 6
A stationary stochastic process $\{X(t)\}$ is referred to as impulsive, if its marginal distribution is heavy tailed.

6.2.3 Measuring Dependence in Impulsive Processes

Traditional measures of dependence, such as covariance, are only applicable to processes with finite second-order statistics. For processes that lack second-order statistics, such as impulsive processes, the *generalized codifference* can be used as a measure of dependence.[5] Long-range dependence can then be defined for impulsive processes.

6.2.3.1 Codifference of α-Stable Random Variables

The generalized codifference originated from the codifference defined specifically for jointly α-stable random variables.

There are two measures of dependence for jointly α-stable random variables, namely, the covariation and the codifference.[4] The *covariation* reduces to the covariance when $\alpha \in (1, 2)$, while it is not defined for $\alpha \in (0, 1)$. If we denote by $[X, Y]_\alpha$ the covariation of the jointly α-stable random variables X and Y, it holds

$$[X, Y]_\alpha \neq [Y, X]_\alpha$$
$$[X, Y_1 + Y_2]_\alpha \neq [X, Y_1]_\alpha + [X, Y_2]_\alpha. \tag{6.26}$$

Since the covariation lacks some of the desirable properties of the covariance, it is not as popular as the codifference.

The codifference of two jointly SαS, $0 < \alpha \leq 2$, random variables X_1 and X_2 equals

$$\gamma_{X_1, X_2} = (\sigma_{X_1})^\alpha + (\sigma_{X_2})^\alpha - (\sigma_{X_1 - X_2})^\alpha \tag{6.27}$$

where σ_X is the scale parameter of the SαS variable X. It is symmetric, i.e., $\gamma_{X_1, X_2} = \gamma_{X_2, X_1}$ and for $\alpha = 2$, reduces to the covariance. If X_1 and X_2 are independent, then $\gamma_{X_1, X_2} = 0$.[4] $\gamma_{X_1, X_2} = 0$ *and* $0 < \alpha < 1$ implies that the random variables are independent. An intuitive interpretation of the codifference can be illustrated by the next property. Let $[X_1, X_2]$ and $[X_1', X_2']$ be two SαS random vectors such that $\sigma_{X_1} = \sigma_{X_2} = \sigma_{X_1'} = \sigma_{X_2'}$ and $\gamma_{X_1, X_2} \leq \gamma_{X_1', X_2'}$. Then, for every $c > 0$, it holds that

$$P(|X_1 - X_2| > c) \geq P(|X_1' - X_2'| > c),$$

which shows that X_1' and X_2' are less likely to differ between them than X_1 and X_2, thus X_1' and X_2' are more dependent.

6.2.3.2 Generalized Codifference

As discussed above, the codifference γ_{X_1, X_2} is only defined for jointly SαS random variables. The *generalized codifference* is an extension of codifference, and is defined for processes with general heavy-tailed marginal distributions.[5]

DEFINITION 7

The generalized codifference (GC) of a pair of random variables (X_1, X_2) is given by

$$I(\theta_1, \theta_2; X_1, X_2) \triangleq -\log E[e^{i(\theta_1 X_1 + \theta_2 X_2)}] + \log E[e^{i\theta_1 X_1}]$$
$$+ \log E[e^{i\theta_2 X_2}], \qquad (\theta_1, \theta_2) \in \mathbb{R}^2. \tag{6.28}$$

It is evident that, if X_1 and X_2 are independent, $I(\theta_1, \theta_2; X_1, X_2) = 0$. When (X_1, X_2) are jointly Gaussian variables, the *generalized codifference* can be written as

$$I(\theta_1, \theta_2; X_1, X_2) = -\theta_1 \theta_2 \, \text{cov}(X_1, X_2),$$

which is equal to the covariance when $\theta_1 = -\theta_2 = 1$. More generally, for jointly SαS X_1, X_2, the generalized codifference $I(1, -1; X_1, X_2)$ coincides with the codifference.

For stationary stochastic processes, the generalized codifference can be used to measure dependence structure. In such cases, we use the notation $I(\theta_1, \theta_2; \tau) = I(\theta_1, \theta_2; X(t + \tau), X(t))$. The function $\tau \mapsto I(\theta_1, \theta_2; \tau)$ is the generalized codifference function, which reduces to the autocovariance function when $\{X(t)\}$ is a Gaussian process and $\theta_1 = -\theta_2 = 1$.

6.2.3.3   Long-Range Dependence in the Generalized Sense

Equipped with the generalized codifference, we can define long-range dependence for impulsive processes:

DEFINITION 8

Let $\{X(t)\}_{t \in \mathbb{R}}$ be a stationary process. We say that $X(t)$ is long-range dependent in the generalized sense, if its generalized codifference, $I(\theta_1, \theta_2; \tau)|_{\theta_1 = -\theta_2 = 1}$ satisfies

$$\lim_{\tau \to \infty} I(1, -1; \tau)/\tau^{-\beta} = L(\tau), \tag{6.29}$$

where $L(\tau)$ is slowly varying at infinity and $0 < \beta < 1$.

Note that for Gaussian LRD processes, by setting $\beta = 2(1 - H)$, the above definition reduces to the classical definition of LRD.

6.3   The Self-Similarity and Impulsive Nature of Data Traffic

As supported by extensive studies involving high-definition network measurements, data traffic in high-speed networks exhibits both self-similarity and impulsiveness.[6–8] These characteristics are illustrated here based on real network traffic data. The data (Figure 6.2) corresponds to actual 100-Mbps ethernet traffic, measured on a WWW/Email/FTP/Computing server located at

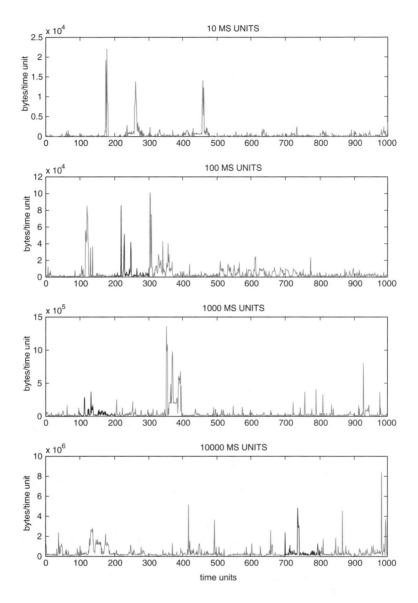

FIGURE 6.2
The Drexel data (10,000 s in total) viewed through four different aggregation intervals: from top to bottom, 10 ms, 100 ms, 1 s, and 10 s.

the ECE Department of Drexel University, using the Snoop program. It consists of all the IP (Internet protocol) packets of private connections with this server, broadcasting, and multicasting, and time-stamped them (at a precision up to 10 ms) during the hours when the traffic was monitored.

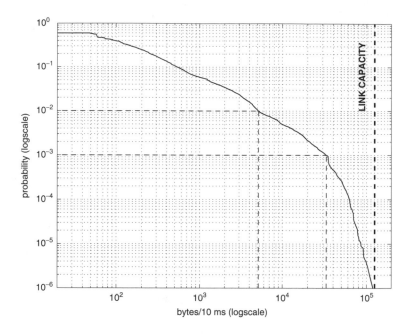

FIGURE 6.3

Empirical complementary distribution function for the traffic data (logarithmic scale on both axes). (From O. Cappe et al., *Signal Process. Mag.*, 19(3), 14–27, May 2002. © 2002 IEEE. With permission.)

Figure 6.2 shows traffic measured every 10 ms. The overall length of the record is about 3 h. The three other plots in Figure 6.2 correspond, from top to bottom, to the "aggregated" data obtained by accumulating the data counts on increasing intervals of 100 ms, 1 s, and finally 10 s (bottom plot in Figure 6.2). After being rescaled in time, each figure corresponds to the segment marked with a black line in the immediately following figure. The striking feature in Figure 6.2 is that the aggregation is not successful in smoothing out the data, which still appears bursty in the bottom plot despite the fact that each point is obtained as the sum of a thousand successive values of the series displayed in the top plot of the same figure. Similar characteristics have been observed in many different experimental setups, including both local area network (LAN) and wide area network (WAN) data (e.g., References 9 through 11, and the references therein). Such behavior hints at self-similarity.

The other important characteristic of the data shown in Figure 6.2 is its extreme impulsiveness. Figure 6.3 illustrates this point by plotting the empirical complementary distribution function (fraction of the data larger than a given value) estimated from the data shown in the top plot of Figure 6.2 (that is, from 1 million 10-ms byte counts). Looking at the right end of Figure 6.3, it is clear that the highest bit-rate one actually observes corresponds to the full capacity of the network link, which is 1 Mbits/10 ms (which corresponds in Figure 6.3

with the vertical dashed bold line at 1.25×10^5 bytes/10 ms, because we use bytes rather than bits to measure the data size). Below the largest value, the empirical complementary distribution function has a very slow decay. Similar observations have been made in other measured data sets (see References 12 through 14). In general, impulsiveness is dominant in data flows generated by a single user or session, and LAN traffic often appears to be impulsive at both single-user and multiple-user levels. In cases where there is intense bandwidth sharing, such as at backbone Internet gateways, individual traffic flows are mainly shaped by the resources allocated to them. In those cases, as the overall traffic approaches the link bandwidth, impulsiveness tends to diminish.

Both self-similarity and impulsiveness have a fundamental impact on network QoS provisioning. In particular, the queue length distribution in infinite buffer systems with self-similar traffic input has a hyperbolically decaying tail, in sharp contrast to Markovian input where the decay is exponential. Consequently, a self-similar traffic source may cause a finite buffer overflow much more frequently than Markovian sources. On the other hand, the marginal distribution of a traffic flow may have a profound impact on network engineering; for example, it can significantly change queueing performance and buffer overflow probability. In Reference 15, the loss rate under different marginal distribution of the traffic flow was investigated, to determine through numerical results that the marginal distribution is a crucial parameter and must be taken into account for loss-rate prediction. Hence, it is very important to take into account the self-similar and impulsive characteristics of the traffic, while performing system performance evaluation for high-speed data networks.

6.4 Modeling Impulsive Self-Similar Data Traffic

To capture the long-range dependence and self-similarity reposed in high-speed network traffic, in recent years, various traffic models have been proposed, e.g., References 11 through 14 and 16 through 30.

In general, traffic models can be classified as either heterogeneous or homogeneous. Heterogeneous traffic models (e.g., References 12 through 14 and 16 through 21) attempt to simulate network traffic that corresponds to aggregation of traffic flows generated by various applications, different users, and different protocols. They are very generic, and can be applied in a wide range of networking scenarios. On the other hand, homogeneous traffic models refer to a specific type of traffic, e.g., MPEG encoded video traffic.[11,22–28] Such traffic is often easier to model more accurately than heterogeneous traffic, as it contains many deterministic features. For example, in MPEG traffic modeling, it is known that the MPEG video sequence often contains the fixed-pattern $IBBPBBPBBPBB$ frames, where I, B, P are intracoded frame, forward motion prediction frame, and forward or backward prediction frames, respectively. I, B, and P frames exhibit distinct statistical properties, such as frame-size distribution, correlations, etc. One can thus model the I, B, and P frames separately and then assemble these components into one complete video sequence.

In the following, we focus on heterogeneous traffic modeling, which is more relevant to the general statistical characterization of today's broadband network traffic. Heterogeneous traffic models can be divided into two classes: behavior models and structural models. Behavior models do not take into account the actual traffic generation mechanism, but rather model statistical characteristics of traffic, e.g., correlation, marginal distribution, or even high-order statistics.[12,14,16,18,19,21] Although, in general, behavior models are amenable to mathematical analysis, their parameters are often not linked to network parameters, and thus cannot provide insight on network behavior. On the other hand, structural models[7,13,17,20] are rooted on the packet/traffic generation mechanism. Their parameters can be easily translated to network parameters, e.g., number of users, user bandwidth, etc., which facilitates understanding of network behavior. However, the analysis of structural models is often more difficult compared to behavior models.

In the following, we present three popular classes of models that have been proposed for broadband heterogeneous network traffic, based on On/Off processes, wavelets, and multifractal processes. Whereas the class of On/Off models are structural models, wavelet models and multifractal models are behavioral models.

6.4.1 On/Off Models

In high-speed networks, the packets are communicated in a *packet train* fashion;[31] once a packet train is triggered, the probability that another packet will follow is very large. Furthermore, the length of the packet train is heavy-tail distributed. This observation led to the celebrated On/Off model,[7] which is sometimes referred to as the Alternating Fractal Renewal Process (AFRP) model.[32] Based on the On/Off model, a single source/destination active pair alternates between two states: the On, during which there is data flow between source and destination, along either way, and the Off, which is the quiet duration. Both the On and Off durations follow a heavy-tail distribution. The self-similar characteristics of the AFRP have been attributed to the heavy-tail properties of the On/Off states' durations. The justification for the heavy-tailed distribution of the On duration lies in the different file size transfer requirements for various applications, empirically observed hyperbolic-tail behavior of the file sizes residing in network file systems, Pareto-like tail behavior of CPU time used by UNIX systems, and also the tendency of multimedia applications to have very large variance (although perhaps not infinite) file residing/transmission systems.[6]

To define alternating On/Off sources, we first need to introduce the two-stage alternating renewal processes. During an On-period, the source generates traffic at a (nominal) constant rate 1. During an Off-period, the source remains silent, and no packets are sent. Let $X_0, X_1, X_2, \ldots$ be i.i.d nonnegative random variables representing the lengths of the On-period and $Y_0, Y_1, Y_2, \ldots,$ be i.i.d nonnegative random variables representing the length of Off-periods. Each of the X and Y sequences is supposed to be i.i.d. The distribution of the X (denoted by $f_1(t)$) and the Y ($f_0(t)$) are heavy-tail distributed with indices α_1 and α_0, respectively, where $1 < \alpha_1 < 2$ and $1 < \alpha_0 < 2$. Hence both distributions have finite mean, which will be denoted by μ_1 and μ_0, respectively.

The On/Off process can be represented by

$$V(t) = \sum_{n=0}^{\infty} 1_{[S_n, S_n + X_{n+1})}(t), \quad t \geq 0,$$

where S_n denotes the time of occurrence of the kth On period, and

$$S_n = S_0 + \sum_{i=1}^{n} T_i, \quad n \geq 1;$$

$1_{[s_1, s_2)}(t)$ is the indicator function, which is defined as nonzero and equal to one only for $t \in [s_1, s_2)$, and $T_j = X_j + Y_j$. The distribution of S_0 is adjusted so that $V(t)$ is strict sense stationary (see Reference 33 for details).

The expected value of $V(t)$ equals $\mu_1/(\mu_0 + \mu_1)$, and its power spectral density is[32]

$$S(\omega) = E\{V(t)\}\delta(\omega/2\pi) + \frac{2\omega^{-2}}{\mu_0 + \mu_1} \text{Re} \left\{ \frac{[1 - Q_0(-j\omega)][1 - Q_1(-j\omega)]}{1 - Q_0(-j\omega)Q_1(-j\omega)} \right\}$$

$$(6.30)$$

where $Q_0(-j\omega), Q_1(-j\omega)$ are the Fourier transforms of $f_0(t)$, and $f_1(t)$, respectively.

For the special case of Pareto-distributed On and Off durations, it was shown in Reference 20 that for $\omega \sim 0$ it holds

$$1 - Q(-j\omega k) = k^{\alpha} e^{j\alpha \frac{\pi}{2}} \Gamma(1 - \alpha) w^{\alpha} - j\frac{k\alpha}{1 - \alpha}\omega. \quad (6.31)$$

Inserting Equation 6.31 into Equation 6.30, ignoring the higher-order terms and considering $\omega \to 0^+$, we obtain

$$S_w(\omega) \sim \frac{2}{\mu_0 + \mu_1}(C_1\omega^{\alpha_1 - 2} + C_2\omega^{\alpha_2 - 2}) \quad (6.32)$$

where C_1, C_2 are constants.

Equivalently, the autocorrelation function of the AFRP becomes

$$R_w(\tau) \sim \tau^{-(\alpha_i - 1)}, \quad \text{as } \tau \to \infty \quad (6.33)$$

where

$$\alpha_i \overset{\triangle}{=} \min\{\alpha_0, \alpha_1\}. \tag{6.34}$$

Equation 6.33 indicates that the AFRP is a long-range dependent process with *Hurst* parameter

$$H = \frac{3 - \min\{\alpha_0, \alpha_1\}}{2}. \tag{6.35}$$

Let us consider a superposition of M i.i.d. On/Off sources. The workload process $\{N_M(t)\}_{t>0}$ is the number of active sources at time t. $\{N_M(t)\}$ is also LRD. The cumulative input of work to the server equals

$$I_M(t) = \int_0^t N_M(s)ds,$$

It was shown in Reference 33 that the cumulative input process (properly normalized) corresponding to an increasing number of i.i.d. On/Off sources converges to FBM, in the sense of convergence in finite dimensional distributions.* This result was important because it established that properly aggregated and rescaled source traffic is not only long-range dependent but also asymptotically self-similar.

While the AFRP model provides insight on the essential self-similar characteristics of modern high-speed network traffic, its Gaussian aggregated result is inconsistent with real traffic data, which depart greatly from Gaussianity as shown in Section 6.3.

The extended alternating fractal renewal process (EAFRP) was proposed in Reference 20 as a simple way to introduce heavy-tailness in the overall traffic model. According to the EAFRP model, the single-user bit rate is treated as a random variable with heavy-tailed characteristics, instead of a constant. Mathematically, the EAFRP can be expressed as

$$W(t) = \sum_{n=0}^{\infty} G_n 1_{[S_n, S_n + X_n)}(t), \tag{6.36}$$

where $\{G_k\}_{k \geq 0}$ is i.i.d. heavy-tail distributed with tail index α, and is independent of the On/Off intervals X_n, Y_n. Figure 6.4 illustrates a sample path of an EAFRP process, which represents the bit-rate process of a single user. It can be shown that $W(t)$ is heavy-tail distributed[20] with tail index α. Determining the dependence structure of EAFRP requires attention, because by letting the reward be heavy-tailed, the second-order statistics of the process become infinite. In Reference 20, the dependence structure of EAFRP was studied through the generalized sense (see Equation 6.28), where it was shown that

---

*The result is formulated as a double limit, and the order of taking the limit matters (first we let $M \to \infty$ and then we let $t \to \infty$).[33]

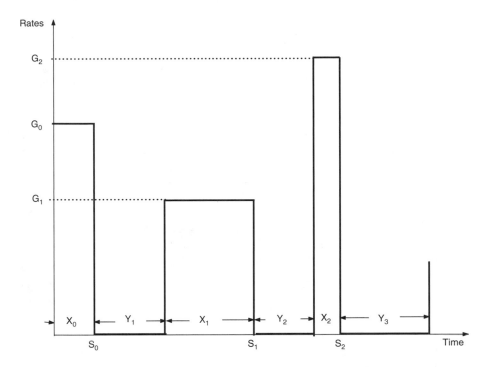

FIGURE 6.4
The sample path of the EAFRP process.

its generalized codifference is asymptotically a power-law function, i.e.,

$$-I(1, -1; \tau) \sim c\tau^{1-\min\{\alpha_1, \alpha_0\}}, \quad c > 0. \tag{6.37}$$

The above equation suggests that the EAFRP model is long-range dependent in the generalized sense.

Figure 6.5a and b demonstrate the effectiveness of the EAFRP model in data traffic modeling, where a segment of real single-user traffic trace (extracted from the same data set as shown in Figure 6.2) as well as a synthesized trace by the EAFRP are illustrated. The parameters of EAFRP are estimated from the real data (for parameter estimation, the reader is referred to Section 6.5). We observe that both traces exhibit impulsiveness, which is confirmed by their linear log-log complementary distribution (LLCD) plots shown in Figure 6.5c and d, respectively. Also shown in Figure 6.5d are the LLCD plots of four independent synthesized traces with the same parameters, where similar observations can be made. A further check of the "goodness-of-modeling" is to estimate the generalized codifferences of both traces. As shown in Figure 6.5e and f (note that five Monte Carlo simulations are overlapped in Figure 6.5f), both the real traffic and the synthesized data traces exhibit similar long-range dependence in the generalized sense.

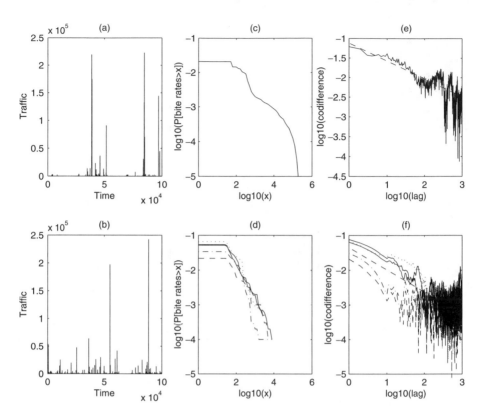

FIGURE 6.5

The actual single-user network traffic (first row), the synthesized traffic traces (second row), and their corresponding LLCD plots and codifference estimates (only one synthesized trace is shown in b). (From X. Yang and A.P. Petropulu, *IEEE Trans. Signal Process.*, 49(7), 1349–1363, July 2001. © 2001 IEEE. With permission.)

Because EAFRP is constructed for modeling single-user traffic, aggregated traffic modeling can then be realized by superimposing single EAFRPs. Indeed, it was shown in Reference 20 that the superposition of EAFRPs is marginally heavy-tail distributed and long-range dependent in the generalized sense. Therefore, the EAFRPs and their superpositions can capture the two most relevant statistical characteristics of high-speed network traffic, i.e., impulsiveness and self-similarity.

In Reference 34, a further modification of the EAFRP model was proposed. This was motivated by the distinctive two-slope appearance of the log-log complementary distribution of real traffic data (see Figure 6.3), and the nature of the true bounds on the user transmission rates in real networks was proposed. Limits on the sender's and the receiver's TCP (Transport Control Protocol) window sizes, TCP congestion avoidance strategies, and bandwidth bottlenecks within the end systems are among the many reasons that lead to an independent limit on each individual user's transmission rate. In reality, therefore, if R is the peak rate of the link onto which traffic from multiple users is multiplexed, the sum of the user transmission rates is bounded by R and each user's transmission rate is bounded by an even smaller quantity, L ($L < R$). In view of the above discussion, a more realistic On/Off model can be obtained by letting the rewarding process G_n in Equation 6.36 be cutoff-Pareto distributed, i.e., the pdf (probability density function) of G_n is given by

$$f_L(x; \alpha, K) = f(x; \alpha, K)(1 - u(x - L)) + \left(\frac{K}{L}\right)^{\alpha} \delta(x - L) \qquad (6.38)$$

where $f(.)$ denotes the Pareto density function (refer to Equation 6.20), $u(.)$ is the unit step function, $\delta(.)$ is the Dirac function, and L represents a limit imposed to the random variable. It can be easily verified that the integral of $f_L(x; \alpha, K)$ taken for x between $-\infty$ to ∞ is one. The existence of these two rate limits, L and R, was shown in References 34 and 35 to result in the distinctive two slope behavior of the LLCD of the overall traffic.[34] The LLCD of synthesized traffic based on this model and that of real traffic are shown in Figure 6.6. The synthesized traffic was constructed as a superposition of 50 On/Off processes with cutoff Pareto distributed rates according to $f_{10^{4.5}}(x; 1.13, 50*10^{1.78})$.

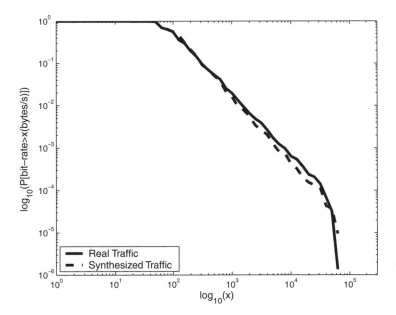

FIGURE 6.6
LLCD of real traffic and synthesized traffic as superposition of 50 On/Off processes with cutoff Pareto distributed rates.

6.4.2   Wavelet Models

As shown in Section 6.2, self-similarity or long-range dependence is closely related to the concept of time/frequency scaling. In this respect, wavelet analysis emerges as a natural framework.

Wavelets are complete orthonormal bases that can be used to represent signals as a function of time. Let $\{X_k\}_{k \in \mathbb{Z}}$ be a discrete-time process. Then X_k can be represented by the inverse wavelet transform

$$X_k = \sum_{j=1}^{K} \sum_{m=0}^{2^{(K-j)}-1} d_j^m \phi_j^m(k) + \phi_0, \tag{6.39}$$

where $0 < k < 2^K$ and ϕ_0 is the mean of X_k over $0 < k < 2^K - 1$. The discrete wavelets $\phi_j^m(t)$ are defined as

$$\phi_j^m(t) = 2^{-j/2}\phi(2^{-j}t - m), \quad \phi < t < 2^k - 1 \tag{6.40}$$

where j and m are positive integers and $\phi(t)$ is the so-called mother wavelet. It is clear that index j characterizes the timescales, while m represents the time translation. The wavelet coefficients d_j^m can be obtained through the wavelet transform

$$d_j^m = \sum_{k=0}^{2^K-1} X_k \phi_j^m(k). \tag{6.41}$$

In Reference 21, a computationally efficient method for modeling heterogeneous network traffic based on wavelet analysis was proposed. To be specific, Reference 21 employs the simplest Haar wavelet as the mother wavelet, i.e.,

$$\phi(t) = \begin{cases} 1 & \text{if } 0 \le t < 1/2, \\ -1 & \text{if } 1/2 \le t < 1, \\ 0 & \text{otherwise.} \end{cases} \tag{6.42}$$

When applying wavelet analysis to data traffic, although the original traffic has a complicated short- and long-range temporal dependence structure, the corresponding wavelet coefficients are only short-range dependent.[21] Thus, a low-order Markov model would suffice for traffic modeling in wavelet domain. To capture the non-Gaussian behavior of broadband heterogeneous traffic, a scheme referred to as timescale-shaping algorithm was proposed.[21] This employs training sequences (real-traffic trace) to shape the wavelet coefficients generated by Gaussian wavelet models, and thus match their empirical distributions.

Wavelet models can be implemented with low complexity. They are behavioral models, however, and as such their parameters are not linked to network parameters.

6.4.3 Multifractal Models

The self-similar or fractal properties of network traffic we have discussed so far refer to monofractality, where the regularity of the time sequence is assumed to be time invariant. More precisely, we will consider the second-order statistics of the incremental process of a self-similar process. Based on Equation 6.14, we have

$$E|X(\delta, t)|^2 = \sigma^2 |\delta|^{2H}. \tag{6.43}$$

If we allow the constant exponent $2H$ to be a function of t, or even a random process rather than a constant or a fixed deterministic function, then the process $X(t)$ is referred to as multifractal. Multifractal processes are more flexible in describing locally irregular phenomena than monofractal processes.

It is argued[36,37] that traffic in wide area networks (WAN) exhibits self-similarity at sufficiently large timescales, and multifractality at typical packet roundtrip timescales. The latter seems to result in network protocols and the end-to-end congestion control mechanism that determines the flow of the packet in the network protocol hierarchy. Furthermore, it was shown[36,37] that the packet arrival pattern inside a connection session (e.g., TCP sessions) matches the multiplicative cascade model, which is constructed as illustrated in Figure 6.7. The model starts with a unit interval that is assigned unit mass. The unit interval is divided into two segments of equal length, which are assigned with weight r and $1 - r$, respectively. The parameter r is in general a random variable in $[0, 1]$ with mean $\frac{1}{2}$. Then, each subinterval and its associated weight are further divided into two parts following the same rule. The weights at stage N can be interpreted as the network load at the time interval $2^{-N}T$, where T is the total time concerned. Via wavelet analysis, it is

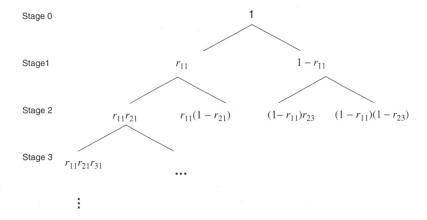

FIGURE 6.7
Schematic diagram of construction rule of multiplicative multifractal.

shown[37] that such a model can capture the relevant characteristics of WAN traffic, i.e., self-similarity and multifractality.

Multiplicative cascade is one of the many ways of constructing multifractal models. For example, in References 19 and 38, wavelets are directly employed to compose multifractal signals for WAN traffic modeling.[30]

6.5   Parameter Estimation

One critical step in data traffic modeling is the estimation of the model parameters. Assuming that data network traffic is indeed self-similar and impulsive, we are interested in estimating the Hurst parameter H and tail index α. In the case of finite variance processes, LRD is described by the Hurst parameter H and its estimation can be done in various ways.[1] Far fewer results exist for the quantification of LRD in the case of infinite variance processes. In the following, we summarize some frequently used techniques for estimating H and α, for LRD and/or impulsive processes.

6.5.1   Estimation of the Hurst Parameter for LRD Processes

As discussed in Section 6.2, long-range dependent processes are characterized by a power-law decay covariance function (see Equation 6.2), or a power-law singularity of the spectral density at zero frequency (see Equation 6.3). Thus, the problem of estimating LRD parameters can be approached by studying either its correlations or power-spectrum density function. Here, we focus on methods in the spectral domain.

The Hurst parameter can be obtained by least-squares regression in the spectral domain, which exploits the form of the spectral density at low frequencies, i.e.,

$$f(\lambda) \sim c_f |\lambda|^{1-2H} \ (|\lambda| \to 0). \tag{6.44}$$

Taking the logarithm on both sides of Equation 6.44 we obtain

$$\log f(\lambda) \sim \log c_f + (1 - 2H) \log |\lambda|. \tag{6.45}$$

In practice, $f(\lambda)$ is estimated by the periodogram $I(\lambda)$, defined as

$$I(\lambda_j) = \frac{1}{2\pi n} \left| \sum_{k=1}^{n} (X_k - \bar{X}_k) e^{ik\lambda_j} \right|^2, \tag{6.46}$$

where $\lambda_j = 2\pi j/n$, $j = 1, \ldots, n/2$ and $\bar{X}_k$ is the sample mean of the data $X_k, k = 1, \ldots, n$. It can be shown[39] that the log-periodogram exhibits the same dependence on H as $f(\lambda)$ in Equation 6.45. The Hurst parameter can thus be estimated by least-squares regression. This method is usually referred to as the GPH estimator,[40] named after the authors of Reference 40.

It should be noted that the relation of Equation 6.44 is valid only in a neighborhood of the zero frequency, and thus the regression line should be computed using only a subset of M log-periodogram samples, where M is some integer less than $n/2$. The choice of M is a difficult task and affects the accuracy of the estimation. Small M results in small bias but high variance.

In practice, to minimize leakage effects, a window function can be used when computing periodograms from real data, i.e.,

$$I_h(\lambda_j) = \frac{1}{2\pi n} \left| \sum_{k=1}^{n} h\left(\frac{k}{n}\right)(X_k - \bar{X}_k)e^{ik\lambda_j} \right|^2, \qquad (6.47)$$

where h is the window function.

For the case of infinite variance LRD linear processes, the above-described methods have been modified in Reference 41 to rely on the use of the normalized periodogram:

$$\tilde{I}_h(\lambda) = \frac{|\sum_{k=1}^{n} h(k/n)(X_k - \bar{X}_k)e^{-ik\lambda}|^2}{\sum_{k=1}^{n} h(k/n)^2(X_k - \bar{X}_k)^2},$$

where h is a window function and $\bar{X}_k$ is the sample mean of the data X_k.

An alternative estimator of the Hurst parameter has been proposed in Reference 20. It results from a linear regression applied to an empirical estimate of the logarithm of the generalized codifference.

6.5.2 Estimation of the Tail Index

Let $\{X_k\}_{k \geq 0}$ be a strict-sense stationary process that takes only positive values and is heavy-tailed with index α. We next consider estimation of the tail index from a finite observed sequence $[X_1, X_2, \ldots, X_n]$.

6.5.2.1 *Log-Log Complementary Distribution Plot*

To determine the presence or absence of the heavy-tail effects, the most commonly used method is the LLCD graph. For heavy-tailed distribution, as $x \to \infty$, it holds

$$\log \bar{F}(x) \sim \log L(x) - \alpha \log x, \qquad (6.48)$$

where $0 < \alpha < 2$ and $\bar{F}(x)$ is the survival function of X. Plotting the complementary empirical distribution function of data that are drawn from a heavy-tailed distribution gives an approximately straight line for large x values. An estimate of α can be obtained by fitting a line through least-squares regression.

6.5.2.2 Hill Estimator

Let $X_{(1)} \leq X_{(2)} \leq \cdots \leq X_{(n)}$ be the order statistics of the sample $X_1, \ldots, X_n$. The Hill estimator[42] is defined for $k < n$ as:

$$\hat{\alpha}_{k,n}^{-1} = k^{-1} \sum_{j=1}^{k} \log \left[\frac{X_{(n-k+j)}}{X_{(n-k)}} \right]. \tag{6.49}$$

where k is the number of upper-order statistics used in the estimation.

It was shown in Reference 43, for i.i.d sequences $\{X_j\}$, $\hat{\alpha}_{k,n}^{-1}$ is a consistent estimator of α^{-1}. In practice, the choice of k is a difficult issue. A useful graphical tool is the so-called *dynamic Hill plot*, which consists of plotting $\hat{\alpha}_{k,n}$ as a function of k. Because $\hat{\alpha}_{k,n}^{-1}$ is a consistent estimator of α^{-1}, the Hill estimator should stabilize around α for certain large k.

6.5.2.3 Quantile-Quantile Regression Plot (qq-plot)

Assuming $[X_1, \ldots, X_n]$ is drawn from Pareto distribution with support $[1, \infty)$, then the jth sample $X_{(j)}$ from its order statistics $[X_{(1)} \leq x_{(2)} \leq \cdots \leq X_{(n)}]$ constitutes the empirical quantile corresponding to $\{(1 - j/n)^{-1/\alpha}, 1 \leq j \leq n\}$, which is the quantile of their probability distribution. Hence, an estimate α can be obtained by plotting the empirical quantiles vs. the theoretical quantiles (qq-plot) in log scales for both axes. The graph $\{(-\log(1-j/(n)), \log(X_{(j)})), 1 \leq j \leq n\}$ appears approximately as a line with slope α^{-1}.

By choosing k such that $k^{-1} + k/n \to 0$, the slope of regression $1 - \log(1 - j/(k+1))$ through the points $\log(X_{(n-k+j)}/X_{(n-k)})$ can be used as another estimator of the tail index. This estimator is referred to as the *static qq-plot* estimator.[44]

The choice of k is again a difficult issue. To avoid it, one can use the so-called *dynamic qq-plot*. As in the dynamic Hill plot, the dynamic qq-plot is obtained by plotting the regression slopes as a function of k for $l < k < n$, and consequently identifying the stable regime of k.

6.6   Concluding Remarks

In this chapter, we have focused on analysis and modeling of broadband heterogeneous data network traffic. Data traffic is fundamentally different from traditional voice-only traffic, which can be essentially characterized by the Poisson inter-arrivals and exponential holding/session times and can thus be accurately modeled by memoryless or Markov models. In contrast, data traffic exhibits persistent correlations, i.e., long-range dependence, when the traffic load process is considered. Furthermore, the marginal distribution of the traffic exhibits impulsiveness, which combined with the long-range dependence poses great challenges for network traffic engineering. For example, it has been shown[20] that both marginal impulsiveness and long-range dependence can change the packet loss behavior radically. In particular, for a single-server queue with a constant service rate and a long-range dependent and impulsive input source, the buffer overflow probability becomes a power-law function of buffer size; that is, the probability that a packet is lost during transmission, due to overflow, decays polynomially fast as buffer size increases, and the speed of decay is dominated either by the Hurst parameter or the tail index of the marginal distribution of the traffic. Note that for traffic of Markov type, e.g., voice traffic, the buffer overflow probability is an exponential function of buffer size.

We have introduced the basic concepts of self-similarity, long-range dependence, heavy-tail distributions, and impulsiveness, along with the mathematical tools to characterize them. Motivated by the intuitive observation of the self-similar and impulsive nature of real-data traffic, we discussed various traffic models that can capture the relevant statistical characteristics, i.e., self-similarity and impulsiveness, of data traffic. In particular, we studied the On/Off models in detail, because they not only can account for the two salient features of impulsive self-similar data traffic, but also provide insights into the physical understanding of these statistical properties. We also reviewed wavelet models and multifractal models, as they represent another perspective on describing the time-scaling variant/invariant specialties of data traffic. We also provided a review of various techniques for estimating the Hurst parameter and tail indexes for long-range dependent and/or impulsive processes, as they play an indispensable role in statistical analysis and modeling.

Acknowledgment

The authors thank Dr. Harish Sethu for his helpful discussions with them.

References

1. J. Beran, *Statistics for Long-Memory Processes*, New York: Chapman & Hall, 1994.
2. B. Tsybakov and N.D. Georganas, On self-similar traffic in ATM queues: definitions, overflow probability bound, and cell delay distribution, *IEEE Trans. Networking*, 5, 397–409, June 1997.
3. D.R. Cox, Long-range dependence: a review, in *Statistics: An Appraisal*, H. David and H. David, Eds., Ames: Iowa State University Press, 1984, 55–74.

4. G. Samorodnitsky and M.S. Taqqu, *Stable Non-Gaussian Random Processes: Stochastic Models with Infinite Variance*, New York: Chapman & Hall, 1994.

5. A.P. Petropulu, J.-C. Pesquet, X. Yang, and J. Yin, Power-law shot noise and relationship to long-memory processes, *IEEE Trans. Signal Process.*, 48(7), July 2000.

6. M.E. Crovella and A. Bestavros, Self-similarity in World Wide Web traffic: evidence and possible causes, *IEEE/ACM Trans. Networking*, 5(6), Dec. 1997.

7. W. Willinger, M.S. Taqqu, R. Sherman, and D.V. Wilson, Self-similarity through high-variability: statistical analysis of ethernet LAN traffic at the source level, *IEEE/ACM Trans. Networking*, 5(1), Feb. 1997.

8. W. Willinger, V. Paxson, and M.S. Taqqu, Self-similarity and heavy tails: structural modeling of network traffic, in *A Practical Guide to Heavy Tails: Statistical Techniques and Applications*, R.J. Adler, R. Feldman, and M.S. Taqqu, Eds., Boston: Birkhauser, 1998.

9. W.E. Leland, M.S. Taqqu, W. Willinger, and D.V. Wilson, On the self-similar nature of ethernet traffic (extended version), *IEEE/ACM Trans. Networking*, 2, 103–115, 1994.

10. V. Paxson, Wide-area traffic: the failure of Poisson modeling, *IEEE/ACM Trans. Networking*, 3, 226–244, 1995.

11. J. Beran, R. Sherman, M.S. Taqqu, and W. Willinger, Long-range dependence in variable-bit-rate video traffic, *IEEE Trans. Commun.*, 43(2), 1995.

12. J. Ilow, Forecasting network traffic using Farima models with heavy tailed innovations, in *ICASSP 2000*, Istanbul, Turkey, June 2000.

13. T. Mikosch, S.I. Resnick, H. Rootzén, and A. Stegeman, Is network traffic approximated by stable Lévy motion or fractional Brownian motion? *Ann. Appl. Prob.*, 12(2002), 23–68, 2001.

14. A. Karasaridis and D. Hatzinakos, Network heavy traffic modeling using α-stable self-similar processes, *IEEE Trans. Commun.*, 49, 1203–1214, July 2001.

15. M. Grossglauser and J. Bolot, On the relevance of long-range dependence in network traffic, *IEEE/ACM Trans. Networking*, 7(5), Oct. 1999.

16. B.K. Ryu and S.B. Lowen, Point process approaches to the modeling and analysis of self-similar traffic. I. Model construction, in *Proc. IEEE INFOCOM'96*, 1468–1475, 1996.

17. A.T. Andersen and B.F. Nielsen, An application of superpositions of two state Markovian sources to the modeling of self-similar behavior, *Proc. IEEE INFOCOM'97*, pp. 196–204, 1997.

18. X. Yang, A.P. Petropulu, and V. Adams, Ethernet traffic modeling based on the power-law Poisson model, in *Proc. 33rd Annu. Conf. Inform. Sci. Syst.*, Baltimore, MD, Mar. 1999.

19. R.H. Riedi, M.S. Crouse, V.J. Ribeiro, and R.G. Baraniuk, A multifractal wavelet model with application to network traffic, *IEEE Trans. Inf. Theor.*, 45(3), Apr. 1999.

20. X. Yang and A.P. Petropulu, The extended alternating fractal renewal process for modeling traffic in high-speed communication networks, *IEEE Trans. Signal Process.*, 49(7), 1349–1363, July 2001.

21. S. Ma and C. Ji, Modeling heterogeneous network traffic in wavelet domain, *IEEE/ACM Trans. Networking*, 9(5), 634–649, 2001.

22. P.-R. Chang and J.-T. Hu, Optimal nonlinear adaptive prediction and modeling of MPEG video in ATM networks using pipelined recurrent neural networks, *IEEE J. Select. Areas Commun.*, 15(8), 1087–1100, 1997.

23. A.M. Dawood and M. Ghanbari, Content-based MPEG video traffic modeling, *IEEE Trans. Multimedia*, 1(3), 77–87, 1999.
24. M.R. Frater, J.F. Arnold, and P. Tan, A new statistical model for traffic generated by VBR coders for television on broadband ISDN, *IEEE Trans. Circuits Syst. Video Technol.*, 4(12), 521–526, 1994.
25. D.P. Heyman and T.V. Lakshman, Source models for VBR broadcast video traffic, *IEEE Trans. Networking*, 4(2), 40–48, 1996.
26. C. Huang, M. Devetsikiotis, I. Lambadaris, and A.R. Kaye, Modeling and simulation of self-similar variable bit rate compressed video: a unified approach, *Comput. Commun. Rev.*, 25, 114–125, 1995.
27. B. Jabbari, F. Yegenoglu, Y. Kuo, S. Zafar, and Y.-Q. Zhang, Statistical characterization and block-based modeling of motion-adaptive coded video, *IEEE Trans. Circuits Syst. Video Technol.*, 3(6), 199–207, 1993.
28. M.M. Krunz and A.M. Makowski, Modeling video traffic using m/g/1 input processes: a compromise between Markovian and LRD models, *IEEE J. Select. Areas Commun.*, 16(6), 733–748, 1998.
29. O. Cappe, E. Moulines, J.C. Pesquet, A.P. Petropulu, and X. Yang, Long-range dependence and heavy-tail modeling for teletraffic data, *Signal Process. Mag.*, 19(3), 14–27, May 2002.
30. P. Abry, R. Baraniuk, P. Flandrin, R. Riedi, and D. Veitch, Multiscale nature of netwrok traffic, *IEEE Signal Process. Mag.*, 28–45, May 2002.
31. R. Jain and S.A. Routhier, Packet trains: measurements and a new model for computer network traffic, *IEEE J. Select. Areas Commun.*, 4, 986–995, 1986.
32. S.B. Lowen and M.C. Teich, Fractal renewal processes generate $1/f$ noise, *Phys. Rev. E*, 47(2), Feb. 1993.
33. M.S. Taqqu, W. Willinger, and R. Sherman, Proof of a fundamental result in self-similar traffic modeling, *Comput. Commun. Rev.* 27, 5–23, 1997.
34. J. Yu, A. Petropulu, and H. Sethu, Rate-limited EAFRP: a new improved model for high-speed network traffic, *Int. Conf. on Acoustics Speech and Signal Processing*, 2003.
35. J. Yu, A. Petropulu, and H. Sethu, Rate-limited EAFRP: a new improved model for high-speed network traffic, *IEEE Trans. Signal Process.*, submitted.
36. A. Feldmann, A.C. Gilbert, and W. Willinger, Data networks as cascade: investigating the multifractal nature of Internet WAN traffic, in *Proc. ACM/ SIGCOMM'98*, Vancouver, BC, Canada, pp. 25–38, 1998.
37. A.C. Gilbert, W. Willinger, and A. Feldmann, Scaling analysis of conservative cascades, with applications to network traffic, *IEEE Trans. Inf. Theor.*, 45(3), 971–991, Apr. 1999.
38. A. Arneodo, E. Bacry, and J. F. Muzy, Random cascades on wavelet dyadic trees, *J. Math. Phys.*, 39(8), 4124–4164, 1998.
39. Y. Yajima, A central limit theorem of Fourier transforms of strongly dependent stationary processes, *J. Time Ser. Anal.*, 10, 375–383, 1989.
40. J. Geweke and S. Porter-Hudak, The estimation and application of long memory time series models, *J. Time Ser. Anal.*, 4, 221–238, 1983.
41. M.S. Taqqu and V. Teverovsky, On estimating the intensity of long-range dependence in finite and infinite variance time series, in *A Practical Guide to Heavy Tails: Statistial Techniques and Applications*, R. Adler, R. Feldman, and M.S. Taqqu, Eds., Boston: Birkhauser, 1998.

42. B.M. Hill, A simple general approach to inference about the tail of a distribution, *Ann. Stat.*, 3, 1163–1174, 1975.
43. S. Resnick and C. Starica, Consistency of Hill's estimator for dependent data, *J. Appl. Prob.*, 32, 139–167, 1995.
44. M. Kratz and S. Resnick, The qq-estimator and heavy tails, *Commun. Stat. Stochastic Models*, 12(4), 699–724, 1996.

7

Nonlinear Adaptive Filters for Acoustic Echo Cancellation in Mobile Terminals

Giovanni L. Sicuranza, Alberto Carini, and Andrea Fermo

CONTENTS

7.1   Introduction

The modern trend in the field of communications toward mobile terminal equipment has attracted, in recent years, the increasing interest of a number of researchers in industry and academia. One of the current topics in this research and application area is adaptive filtering for acoustic echo cancellation.[1,2] In fact, high-quality mobile services require compensation for acoustic echoes using effective, low-cost adaptation algorithms.[3,10,11,19] Application-oriented research topics range from the development of specific adaptation algorithms,[17] including, in particular, control of the adaptation step size and realization of efficient implementations, to other relevant aspects such as double-talk control, background noise reduction, and de-reverberation.[6]

Acoustic echoes in mobile communications originate from sound propagation between the loudspeaker and the microphone of one terminal, here called the *near-end* terminal. The acoustic echoes propagate to the other terminal, i.e., the *far-end* terminal, where they are perceived by the far speaker as annoying disturbances. This effect is particularly relevant in modern ultra-compact GSM and third-generation (3-G) handset receivers, because of their small size, the relative closeness between loudspeaker and microphone, and the long propagation delay between the two communication sides. In current practice, echoes are reduced by using acoustic echo cancelers, which are digital adaptive filters that estimate the echo signal to cancel it by subtraction. For this purpose, linear adaptive filters are generally used because their low complexity allows real-time implementations. However, they often have to identify an acoustic echo path that may contain significant nonlinearities. In fact, while microphones operate in general in a linear condition because of the small amplitude of the signals, amplifiers and loudspeakers originate the most relevant nonlinear distortions.[13,15] Low-cost terminals employ amplifiers that may often be operated with large signals and, consequently, may be driven to saturation. Even if this extreme working condition is not reached, there are other nonlinear effects that require accurate compensation in high-quality terminals. In loudspeakers some significant distortions of this kind come from the nonlinearities due to inhomogeneity of the magnetic flux and nonlinear stiffness of the suspension.[15] In addition, nonlinear effects are possible due to the vibrations of the enclosure. In all these cases, a nonlinear filter clearly guarantees better system identification, and thus greater echo cancellation, than does a linear filter.

To combat the saturation effects of the amplifier, the adaptive nonlinear filter may employ a nonlinearity that depends only on the present signal sample, i.e., a memoryless nonlinearity, followed by a linear finite impulse response (FIR) filter.[27,28,34,35,37–40] Polynomial functions, memoryless neural networks, and saturation functions have been proposed for implementing memoryless nonlinearity. However, the adaptation of the parameters of this nonlinearity may often require very long convergence times. Moreover, the resulting filters are not able to cope with the nonlinearities of the loudspeaker, which have a non-negligible memory.[5,13] A more general solution resorts to nonlinear filters with memory in which the present output sample depends on input and previous output samples.[5,8,12,13,21,31–33,36,42] To this class of nonlinear filters belong also the *polynomial filters*, which have been widely studied in the scientific literature. A recent complete account can be found in Reference 24. In general, unless these filters model mild nonlinearities, polynomial filters require too many computational resources for real-time implementation. Nevertheless, in consideration of the significant advantages obtained by such nonlinear models, one of the research lines in this area is the study of effective, low-cost realizations and low-complexity adaptation algorithms. Taking into account both these aspects simultaneously allows significant improvements in many applications, including acoustic echo cancellation in mobile terminals.

In this chapter we discuss the problem of the nonlinear acoustic echo cancellation using polynomial filters. We show how it is possible to obtain significant improvements over linear adaptation techniques even with simple nonlinear structures. In Section 7.2 we first briefly review the state of the art in the field of nonlinear cancellation of acoustic echoes for handset receivers. Then we introduce the class of polynomial and Volterra filters and describe some low-complexity structures that are able to achieve good performance with a reduced increase in computational complexity with respect to linear filters. In particular, the recursive bilinear filter [22] and different realizations of the nonrecursive quadratic filters are considered. The latter filters, also known as second-order Volterra filters, admit, within some simplifying conditions, implementation structures with reduced complexity. In particular, the realizations considered here are the cascade structure with a memoryless preprocessor and a linear filter,[38] the multi-memory-decomposition[14] structure, the parallel-cascade[30] structure, and the simplified Volterra filters.[12] To make the whole nonlinear echo canceler efficient, it is necessary to equip it with an adaptation algorithm granting fast adaptation and tracking rates. In Section 7.3, extension of the theory of the affine projection (AP) algorithms from linear to quadratic filters is presented. We compare the performances of such adaptation algorithms with those of the classical least-mean square (LMS) and normalized least-mean square (NLMS) algorithms and show that the AP technique offers better convergence and tracking capabilities. We complete the chapter by comparing in Section 7.4 the performances of the above-mentioned realizations for nonrecursive quadratic filters and those of the bilinear filter. Our final remarks, reported in Section 7.5, note that the reduced complexity of specific realizations and the fast adaptation and tracking rates of the AP algorithms make simple polynomial echo cancelers attractive for real-time implementations in high-quality handset devices.

7.2 Low-Complexity Nonlinear Adaptive Acoustic Echo Cancelers

In recent years, cellular vendors have been greatly concerned about the quality of the audio of handset receivers. One of the main issues concerning audio quality is the need for suppression of the sound signal that propagates between the loudspeaker and the microphone of the handset terminal and then reaches the far end. Therefore, the far-end speaker hears a delayed replica of his or her voice, which results in a quite annoying disturbance. This effect may be particularly relevant in hands-free systems for vehicular or teleconference applications. The problem also exists in modern ultracompact GSM and 3-G handset receivers because of their small size and, consequently, the close proximity of the loudspeaker and the microphone, and because of the long propagation delay between the two communication sides. The technical solution to this problem consists of using an acoustic echo canceler (AEC), which is a digital adaptive filter whose output $y(n)$ estimates the reference signal $d(n)$, i.e., the echo signal, in order to cancel it by subtraction, as shown in Figure 7.1. The adaptivity of the AEC is a key point because compensations obtained with filters having fixed coefficients are not able to model

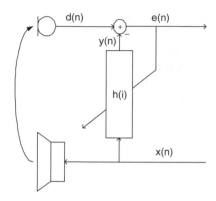

FIGURE 7.1
Acoustic echo canceler.

correctly the highly time-variant behaviors of the echo paths in different working conditions.

Usually, in commercial applications, the AECs used are linear because they can be implemented with simple adaptive algorithms and a reduced computational complexity, thus permitting low-cost, low-power realizations. However, it is well known that the acoustic echo path in handset receivers can be highly nonlinear.[4] The nonlinearities arise mainly from nonlinear distortions introduced by the amplifier and the loudspeaker when operated with large signals. From a qualitative point of view, this situation is encountered when the speaker sets the loudspeaker volume to a value much higher than normal. Indeed, it is not uncommon that the user in a noisy environment sets the loudspeaker volume 6 to 12 dB higher than in a silent environment. Other nonlinear effects, in general quite difficult to characterize, arise because of the vibrations of the enclosure. In all the above-mentioned cases, a nonlinear filter is able to achieve better system identification, and consequently greater echo suppression, than a linear filter. However, in most cases the implementation of these filters is computationally expensive because of the large number of parameters required to characterize the system. In fact, the memory of the AEC filter should be as long as the acoustic echo path response, which typically requires, with a sampling rate of 8 kHz, ~128 samples for handset receivers, ~256 samples for car-hands-free receivers, and ~1024 to 2048 samples for teleconference systems.

In this section we review some low-complexity nonlinear adaptive AECs proposed in the literature. In particular, we describe a few structures belonging to the class of polynomial filters,[24] which are able to model the nonlinear acoustic echo path with a reduced number of coefficients. We consider in some detail the cascade structures of a memoryless preprocessor and a linear filter,[38] the multi-memory-decomposition filter structure,[14] the parallel-cascade structure,[30] the simplified Volterra filter,[12] and the bilinear filter.[22]

7.2.1   Introduction to Polynomial and Volterra Filters

Discrete-time causal polynomial filters[24] are described by the input–output relationship:

$$y(n) = \sum_{i=0}^{P} f_i\left[x(n), x(n-1), \ldots, x(n-N+1), y(n-1), \ldots, y(n-M+1)\right],$$

$$(7.1)$$

where the function $f_i[\cdots]$ is a polynomial of order i in the variables within the parentheses. The linear filter is a particular case of polynomial filters because the relationship in Equation 7.1 becomes linear if $f_i[\cdots] = 0$ for all $i \neq 1$. With reference to Equation 7.1, polynomial filters can be classified into recursive and nonrecursive filters. Recursive filters are characterized by (possibly nonlinear) feedback terms in their input–output relationships and, as infinite impulse response linear filters, possess an infinite memory. A simple example of a recursive filter is the bilinear filter represented by the following equation:

$$y(n) = \sum_{i=0}^{N_1} a_i x(n-i) + \sum_{j=1}^{N_2} b_j y(n-j) + \sum_{i=0}^{N_3} \sum_{j=1}^{N_4} c_{ij} x(n-i) y(n-j). \quad (7.2)$$

In consideration of its infinite memory, a recursive polynomial filter admits, within some stability constraints, a convergent *Volterra series expansion* of the form[24]

$$y(n) = h_0 + \sum_{m_1=0}^{\omega} h_1(m_1)x(n-m_1) + \sum_{m_1=0}^{\infty} \sum_{m_2=0}^{\infty} h_2(m_1, m_2)x(n-m_1)x(n-m_2)$$

$$+ \cdots + \sum_{m_1=0}^{\infty} \sum_{m_2=0}^{\infty} \cdots \sum_{m_p=0}^{\infty} h_p(m_1, m_2, \ldots, m_p)x(n-m_1)$$

$$\times x(n-m_2) \cdots x(n-m_p) + \cdots, \quad (7.3)$$

where $h_p(m_1, m_2, \ldots, m_p)$ denotes the pth order *Volterra kernel* of the nonlinear filter.

Nonrecursive polynomial filters, commonly called Volterra filters, are characterized by input–output relationships that result from a double truncation of the Volterra series, i.e., a memory truncation, by limiting the memory of the filters to a finite number of terms in the summations of Equation 7.3, and an order truncation by limiting the number of Volterra kernels

$$y(n) = h_0 + \sum_{m_1=0}^{N_1-1} h_1(m_1)x(n-m_1) + \sum_{m_1=0}^{N_2-1} \sum_{m_2=0}^{N_2-1} h_2(m_1, m_2)x(n-m_1)x(n-m_2)$$

$$\cdots + \sum_{m_1=0}^{N_p-1} \sum_{m_2=0}^{N_p-1} \cdots \sum_{m_p=0}^{N_p-1} h_p(m_1, m_2, \ldots, m_p)x(n-m_1)x(n-m_2)$$

$$\cdots x(n-m_p). \quad (7.4)$$

The main advantages of polynomial filters are related to the following facts:[24]

- The theory of polynomial filters can be viewed as an extension of linear filter theory. Consequently, many results related to the analysis and design of linear filters can be derived in a quite straightforward manner.
- It can be shown that, under relatively mild conditions, polynomial models are capable of approximating a large class of nonlinear systems with a finite number of coefficients.
- There are many processes in the real world that can be described using polynomial models.

The main disadvantages of polynomial filters are related to the following facts:[24]

- Polynomial models may require a large number of coefficients to adequately represent many real-world nonlinear systems. As the memory span and the order of nonlinearity increase to even moderately large values, the implementation complexity becomes overwhelming.
- The stability properties of recursive polynomial systems are not completely understood at this time.

In spite of these disadvantages, the performance improvements possible with polynomial models are significant in a large number of applications. To avoid a huge complexity increase, most current applications employ truncated Volterra models. The simplest polynomial filter of such a class is the quadratic filter obtained by limiting the terms in Equation 7.4 to the linear and the second-order kernels. By using a common alternative notation, the filter output can be expressed using a vector-matrix representation in the form

$$y(n) = \mathbf{h}^T \mathbf{x}(n) + \mathbf{x}^T(n)\mathbf{H}\mathbf{x}(n), \qquad (7.5)$$

where $\mathbf{x}(n)$ is the vector of the past N input samples, $\mathbf{h}$ is the vector formed with the N coefficients of the linear filter, and $\mathbf{H}$ is the $N \times N$ matrix representing the second-order Volterra kernel. Without loss in generality, the second-order Volterra kernel can be assumed as symmetric[24] and thus even the matrix $\mathbf{H}$ is symmetric. For a more efficient realization, $\mathbf{H}$ can be modified to assume the upper triangular form. In this case the number of multiplications needed to obtain one output sample is equal to $N(N+1)/2$.

It is worth noting that the use of a quadratic term in addition to the linear term is often sufficient to offer performance improvements in several applications. This is also the situation encountered when modeling the acoustic echo path in handset receivers. As already mentioned, the analysis of typical acoustic echo paths clearly shows that the nonlinearities are mainly due to the amplifier, the loudspeaker, and the nonlinear vibration of the enclosure. The amplifier can be modeled by a nonlinear filter without memory and its nonlinear effects can be compensated for by a fixed nonlinear filter.[27,28] The loudspeaker, in contrast, is better modeled by a nonlinear system with memory. Neglecting the nonlinearities due to the enclosure vibrations, the whole echo path can be represented by the cascade of a nonlinear and a linear system. In the simplest case, this structure is actually represented by a second-order Volterra model with finite memory. This model has in principle a large number of coefficients according to the long memory of the whole system, resulting from the long reverberation time of the echo path. However, in a typical second-order kernel modeling of the acoustic echo path, the coefficient values quickly decay moving far from the main diagonal. The measured second-order kernel of the acoustic echo path of a commercial GSM receiver is shown in Figure 7.2. Because the kernel is assumed to be symmetric, only half of the coefficients are depicted in this figure. Similar shapes also have been reported by other researchers.[36] It is possible to see that at 8-kHz sampling frequency the coefficients with significant values actually cover a memory span of less than 30 samples. Moreover, they lie on the kernel diagonals adjacent to the main one so that approximate realizations can be derived using even a

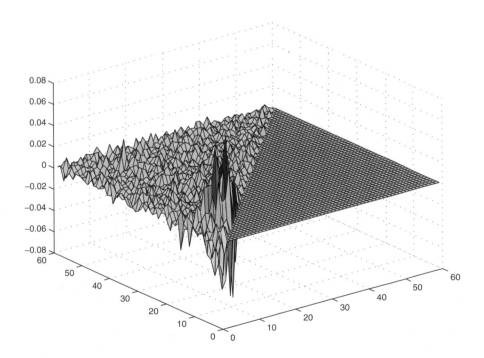

FIGURE 7.2
Second-order kernel of the acoustic echo path. (From A. Fermo et al., *J. Comput. Inf. Technol.*, 8(4), 333, Dec. 2000. With permission.)

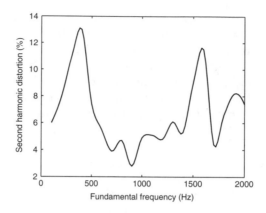

FIGURE 7.3
Second-order harmonic distortion of the echo path plotted vs. the fundamental frequency. (From A. Fermo et al., *J. Comput. Inf. Technol.*, 8(4), 333, Dec. 2000. With permission.)

small number of diagonals. This fact has been exploited in one of the models used to represent the acoustic echo path, as shown below. It is also worth noting that the other Volterra-based models described in the following have been derived on the basis of suitable simplifications so that low-complexity realizations can be devised.

Finally, to demonstrate the relevance of a second-order model, the measured second-order harmonic distortion of the echo path in a commercial GSM receiver operated with a high loudspeaker volume is plotted in Figure 7.3 vs. the fundamental frequency.

7.2.2   Cascade of a Memoryless Preprocessor and a Linear Filter

One of the most computationally efficient structures that have been proposed in the literature for compensating for the nonlinearities in the acoustic echo path is the cascade of a memoryless preprocessor and a linear filter, as shown in Figure 7.4. This filter can efficiently compensate for the nonlinearities introduced by the amplifier. In contrast, only a rough compensation is provided for the nonlinear components generated by the loudspeaker.

As we have already mentioned, the main sources of the nonlinearities in the acoustic echo path of the mobile phone are the amplifier and the loudspeaker. The loudspeaker should be modeled as a nonlinear system with memory,[15] while the amplifier can be assumed to be a memoryless nonlinear system. Therefore, in all cases where the loudspeaker nonlinearities are neglectable, it is possible to model the acoustic echo path with the cascade of a memoryless nonlinear preprocessor and a linear filter, as shown in Figure 7.4.

Different structures have been proposed for the memoryless preprocessor realization; they range from neural networks to saturation curves to

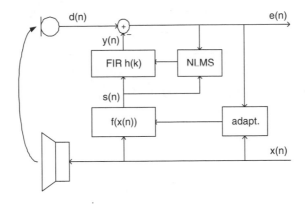

FIGURE 7.4
Cascade of a memoryless preprocessor and a linear filter.

polynomial filters. In the last case the memoryless nonlinearity of Figure 7.4 is given by

$$f(x(n)) = 1 + \sum_{i=2}^{P} a_i x^i(n), \tag{7.6}$$

where an order $P = 7$ is often sufficient to compensate for the amplifier nonlinearities.

Both linear filters and the memoryless preprocessor could be adapted with the NLMS adaptive algorithm. Nevertheless, because of the cascade structure and the eigenvalue spread of the autocorrelation matrix of the memoryless preprocessor input signal, a very low convergence rate would be experienced in this case (see Section 7.3). To overcome this problem, the memoryless preprocessor can be directly adapted with the recursive least squares (RLS) algorithm.[38] Indeed, because of the low order of the memoryless preprocessor, the RLS algorithm does not severely affect the overall computational budget of the acoustic echo canceler.

7.2.3 Multi-Memory-Decomposition Structure

One of the first low-complexity realizations proposed in the literature is the *multi-memory-decomposition* (MMD) structure. This structure, studied in Reference 14, is obtained by means of the interconnection of three linear FIR filters, as shown in Figure 7.5.

The output of the filter is given by

$$y(n) = \sum_{k=0}^{N_p-1} h_p(k) \sum_{i=0}^{N_a-1} h_1(i)x(n-i-k) \sum_{j=0}^{N_a-1} h_2(j)x(n-j-k), \tag{7.7}$$

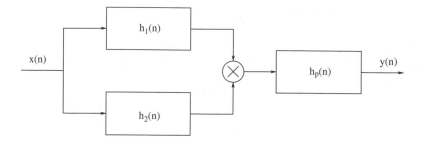

FIGURE 7.5
MMD filter.

where N_a is the memory length of the filters h_1 and h_2, and N_p is the memory length of the filter h_p. This MMD structure is a quadratic filter whose second-order kernel has nonzero coefficients only in N_a diagonals adjacent to the main diagonal. The complexity associated with the implementation of this kind of filter is very low if compared to that of a complete Volterra filter. In particular, the number of multiplications needed is equal to $2N_a + N_p + 1$, and thus is substantially proportional to the memory $N = N_a + N_p$ of the filter. However, the MMD structure can just roughly approximate the generic second-order nonlinear system because of the reduced number of parameters involved. Furthermore, in the adaptive version of this filter the adaptation strategy becomes quite critical because it is necessary to decide in what order the coefficients of the single filters are updated. As a consequence, the complexity of the adaptation algorithm significantly increases and in fact becomes proportional to $N_a \times N_p$. Moreover, the adaptation procedure suffers from the existence of local minima because the filter output is not linear with respect to the coefficients of the linear filters. For this reason the approximation achieved with the MMD filter is often not adequate.

7.2.4 Parallel-Cascade Structures

Parallel-cascade structures[24] are based on the assumption that any $N \times N$ symmetric matrix H of rank $r \leq N$ can be decomposed into a finite sum of r rank one matrices:[7]

$$\mathbf{H} = \sum_{i=1}^{r} q_i \mathbf{r}_i \mathbf{r}_i^T , \tag{7.8}$$

where $q_i, i = 1, 2, \ldots, r$, are scalar numbers and $\mathbf{r}_i, i = 1, 2, \ldots, r$, are vectors formed with N elements. Using this decomposition for the coefficient matrix $\mathbf{H}$ in Equation 7.5 results in the following expression for the output of the

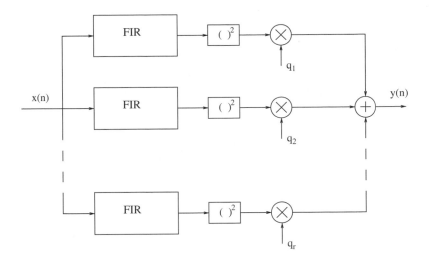

FIGURE 7.6
Parallel-cascade filter.

quadratic term:

$$y(n) = \sum_{i=1}^{r} q_i [x^T(n)r_i] [r_i^T x(n)] = \sum_{i=1}^{r} q_i y_i^2(n), \qquad (7.9)$$

where

$$y_i(n) = x^T(n)r_i = r_i^T x(n) \qquad (7.10)$$

is the output of a linear FIR filter with impulse response described by the coefficients of the vector r_i. This implies that if the kernel matrix describing a second-order Volterra filter has rank equal to r, then it can be implemented by a filter bank with r channels. Each channel is formed with a linear filter followed by a squaring function and a multiplier, as shown in Figure 7.6.

A good approximation of the second-order Volterra filter can be frequently obtained by using only m channels, with $m \ll r$, and thus allowing significant computational savings.[30] Two problems arise with the adaptation procedures applied to this type of filter. The first is that the minimum-error solution is not unique and, consequently, the filter coefficients can oscillate between the different solutions. The second problem is the presence of local minima, again because the output of the filter is not linear with respect to the parameters to be updated. One remedy to the first problem consists in constraining the first $i-1$ coefficients of the ith branch to zero and setting the ith coefficient to 1. These coefficients do not need updating. The decomposition obtained in this way is known as the LDL^T decomposition. It is obtained by expressing the

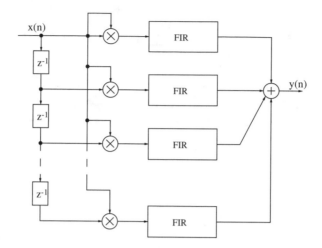

FIGURE 7.7
Structure of simplified Volterra filters.

symmetric matrix $\mathbf{H}$ of the second-order kernel as

$$\mathbf{H} = \sum_{i=1}^{r} d_i \mathbf{l}_i \mathbf{l}_i^T , \qquad (7.11)$$

where the d_i are scalar numbers and the vectors $\mathbf{l}_i$ have $i - 1$ leading zeros.[24]

However, the problem associated with the presence of local minima is not completely solved.[25] For this reason the quality of the approximation is often limited, even when a large number of channels are used.

7.2.5   Simplified Volterra Filters

By analyzing the second-order kernel of several acoustic echo paths, it has been noted that the coefficients with the most significant amplitude lie on the diagonals adjacent to the main diagonal. Therefore, an approximation of the second-order system can be achieved by setting to zero the coefficients on some diagonals far from the main one. This is the idea underlying the *simplified Volterra filter* (SVF) structure[12] reported in Figure 7.7. Each channel of this structure implements a quadratic filter whose kernel matrix has nonzero coefficients only in one diagonal. By arranging N of these filters in parallel, the complete second-order Volterra filter is implemented. By deleting some of the channels, i.e., those corresponding to diagonals far from the main one, a suitably simplified structure of the second-order Volterra filter is obtained. As a result, significant computational savings can be achieved.

From a more formal point of view, the output of the SVF structure can be expressed as

$$y(n) = \sum_{i=1}^{M} y_i(n), \tag{7.12}$$

where M is the number of channels actually used and $y_i(n)$ is the output of the generic channel in Figure 7.7, given by the following equation:

$$y_i(n) = \sum_{k=0}^{N-i} h(k, k+i-1)x(n-k)x(n-k-i+1), \tag{7.13}$$

where N is the memory length of the filter.

7.2.6 Bilinear Filters

Just as linear infinite impulse response (IIR) filters can represent many linear systems with far fewer coefficients than their FIR counterparts, recursive polynomial models can accurately represent many nonlinear systems with greater efficiency than the truncated Volterra series representation.[24] For sake of simplicity, the bilinear model of Equation 7.2 can be further simplified in the following form:

$$y(n) = \sum_{i=0}^{N_1} a_i x(n-i) + \sum_{i=1}^{N_2} b_i y(n-i) + \sum_{j=0}^{N_1}\sum_{i=1}^{N_2} c_{i,j} y(n-i)x(n-j). \tag{7.14}$$

It is worth noting that this type of system not only has infinite memory, but also may represent very large orders of nonlinearity.[22] On the other hand, the implementation of the adaptive version of this type of filter suffers from two main difficulties. First of all, the filter stability is not guaranteed according to the recursive nature of the structure. Second, the adaptation method should be carefully chosen. In fact, there are two fundamentally different classes of methods for the adaptation of recursive filters, known as *output error* methods and *equation error* methods, respectively.[18] The output error methods estimate the reference signal $d(n)$ using a truly recursive system by actually feeding back the estimate $y(n)$ to generate the adaptive estimates of the reference signal. In a large number of situations, the output error algorithms are able to obtain unbiased estimates in the steady state. However, they are not simple to implement. Moreover, the error surfaces are nonlinear functions of the coefficient values, and thus they may contain local minima and the adaptive filters may not necessarily converge to the global minimum. On the contrary, the equation error methods form the estimates using samples of the input $x(n)$ and the reference signal $d(n)$. Consequently, they are not truly recursive estimators in the sense that the output of the estimator is not actually used to compute the

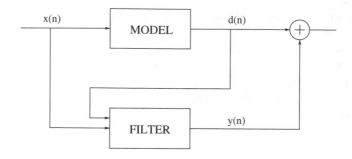

FIGURE 7.8
Equation error adaptive bilinear filter.

estimates. Their greatest advantage is their inherent simplicity. In fact, the recursive estimation problem can be converted into a two-channel nonrecursive estimation problem, as shown in Figure 7.8. In practice, the delayed samples of the reference signal $d(n)$ are used in Equation 7.14 in place of the delayed samples of the actual output $y(n)$. Equation 7.14 can then be written as

$$y(n) = \mathbf{a}^T \mathbf{x}(n) + \mathbf{b}^T \mathbf{d}(n) + \mathbf{d}^T(n)\mathbf{C}\mathbf{x}(n), \tag{7.15}$$

where $\mathbf{x}(n)$ is the vector of the N_1 past input samples, $\mathbf{d}(n)$ is the vector of the N_2 past samples of the reference signal, $\mathbf{a}$ is the vector formed with the N_1 coefficients of the nonrecursive part of the linear filter, $\mathbf{b}$ is the vector formed with the N_2 coefficients of the recursive part of the linear filter, and $\mathbf{C}$ is the $N_2 \times N_1$ matrix formed with the coefficients $c_{i,j}$ of Equation 7.14. According to the nonrecursive nature of the estimation, the stability problems of the recursive structures are avoided. Moreover, the error surface is quadratic in the coefficients, and thus it has a unique minimum, unless the autocorrelation matrix of the input vector is singular. The drawback is that the estimation procedure leads to a biased estimate of the optimal solution, thus limiting the performances in the identification procedure.

7.3　Adaptive Algorithms for Nonlinear Acoustic Echo Cancelers

The most popular algorithms for the adaptation of linear filters are the well-known LMS and NLMS algorithms.[18,43] However, LMS and NLMS algorithms, when applied to acoustic echo cancelers, usually exhibit low convergence and tracking rates because of the high degree of correlation of the speech signal and the high number of coefficients needed for accurate modeling of the echo path. To increase the convergence rate in the presence of highly correlated signals, various families of adaptive algorithms have been proposed in the literature.[18] RLS algorithms, for example, can offer faster convergence rates, but in general require computational resources, often too large for a real-time implementation. On the other hand, their fast versions, i.e., the fast recursive least squares (FRLS) algorithms, still suffer from numerical problems when implemented with finite-word-length arithmetics. Another class of algorithms that are able to offer better convergence and tracking rates than LMS algorithms, i.e., the so-called AP algorithms, were proposed by Ozeki and Umeda about two decades ago. It has been reconsidered quite recently and fast versions of AP algorithms, i.e., fast affine projection (FAP) algorithms, have been developed.[16,26,41] These algorithms retain the good performance of AP algorithms but with a computational load close to that of LMS algorithms. The significant aspect of this class of algorithms is to offer, in the presence of correlated signals, convergence rates higher than LMS algorithms and tracking capabilities better than RLS algorithms.[17] For this reason, they are particularly suitable for AEC implementation; fast convergence and tracking capability are important features for AECs because of the highly time-varying nature of the echo path.

An advantage of polynomial models is that all these adaptation techniques can, in principle, be extended to Volterra filters as their output is still linear with respect to the filter coefficients.[24] The drawback is that the computational requirements are often too high even for a truncated second-order Volterra filter. Even when the most economic LMS and NLMS algorithms are adopted, the implementation complexity of a nonlinear AEC is not suitable for acoustic echo cancellation in real time. In addition, when these algorithms are applied to Volterra filters, the convergence rate becomes too low because of the ill-conditioned nature of the input autocorrelation matrix.[22] The NLMS algorithm may improve such a rate, but it is not the ultimate solution.

In this section, after a brief introduction to the classical LMS and NLMS algorithms, we show how it is possible to extend the basic AP algorithm to quadratic Volterra filters. To obtain an efficient implementation, the diagonal realization structure employed by SVFs is exploited.

7.3.1 LMS and NLMS Algorithms

The objective of an adaptive filter is to process the input signal $x(n)$ so that its output signal $y(n)$ is close to a reference signal $d(n)$. In theory, this objective is reached by iteratively minimizing the statistical expectation of some convex cost function of the error signal

$$e(n) = d(n) - y(n) \tag{7.16}$$

using a steepest-descent technique. This technique consists of updating the filter coefficients at the time $n+1$ using a small fraction of the partial derivatives of the cost function taken with respect to the filter coefficients computed at time n. In practice, to obtain a realizable algorithm, so-called stochastic gradient approximation is used; this consists of simply replacing the expectation of the error cost function with its actual value at time n. Even though this is a rough approximation, the resulting algorithm works reasonably well for small steps in updating the filter coefficients.

The LMS adaptive filter uses a cost function defined by the mean-square estimation error, i.e.,

$$J(n) = E\{e^2(n)\}. \tag{7.17}$$

Within the stochastic gradient approximation, the partial derivatives required to update the coefficients can be easily computed,[18,23,43] and eventually the LMS linear FIR adaptive filter is described by the following equations:

$$e(n) = d(n) - \mathbf{x}^T(n)\mathbf{h}(n), \tag{7.18}$$

$$\mathbf{h}(n+1) = \mathbf{h}(n) + \mu\mathbf{x}(n)e(n), \tag{7.19}$$

where $\mathbf{x}(n)$ is the vector of the past N input values, $\mathbf{h}(n)$ is the vector of the corresponding N coefficients, and μ is a small positive constant that controls the speed of convergence. Because the output of a truncated Volterra filter is linear with respect to its coefficients, the same relationships hold. The only difference is the way in which the input and the coefficient vectors are defined. For a pure quadratic filter, the following definitions apply:[24]

$$\mathbf{x}(n) = [x^2(n) \ x(n)x(n-1) \cdots x(n)x(n-N+1) \ x^2(n-1) \cdots x^2(n-N+1)]^T, \tag{7.20}$$

$$\mathbf{h}(n) = [h_2(0,0;n)h_2(0,1;n) \cdots h_2(0,N-1;n)h_2(1,1;n) \cdots h_2(N-1,N-1;n)]^T, \tag{7.21}$$

where $\mathbf{x}(n)$ can be considered an augmented input vector and $\mathbf{h}(n)$ is the corresponding vector formed with the coefficients $h_2(m_1, m_2)$, $m_1 = 0, 1, \ldots, N-1$, $m_2 = m_1, \ldots N - 1$ at time n. Equations 7.20 and 7.21 take into account the symmetry property of the quadratic kernel. The difficulties with LMS adaptive linear filters arise from the correlation in the input signal because the convergence speed of the coefficients is controlled by the largest eigenvalue of the autocorrelation matrix of the input vector. The drawbacks in the nonlinear case come from the same problems related to the eigenvalues spread. However, in such a case, even in the presence of a white input signal, the eigenvalues may exhibit quite a large distribution. This fact is due to the presence in the autocorrelation matrix of the augmented input signal vector of entries that depend on products of $2p$ input samples, where p is the order of the truncated Volterra filter. Therefore, in the nonlinear case, the convergence rate of LMS algorithms is, in general, not as fast as desired.

From the analysis of the updating Equation 7.19 it is possible to see that the performance of the algorithm varies with the input signal amplitude. Consequently, the modes of convergence that depend on the eigenvalues of the autocorrelation matrix change when the input signal amplitude changes. A common remedy for these difficulties is to use an NLMS adaptive algorithm of the form

$$\mathbf{h}(n+1) = \mathbf{h}(n) + \frac{\mu}{||\mathbf{x}(n)||^2}\mathbf{x}(n)e(n), \tag{7.22}$$

where $||\mathbf{x}(n)||^2 = \mathbf{x}^T(n)\mathbf{x}(n)$ is the squared Euclidean norm of the input vector. Using for the input vector the definition in Equation 7.20, the extension of this method to the Volterra filters can be easily derived. However, because the powers associated with the entries of the augmented input vector may differ significantly, the NMLS algorithms are not as effective as in the linear case.

7.3.2 The Affine Projection Algorithm

To overcome the problem of low convergence rates in the presence of highly correlated signals, various families of adaptive algorithms have been proposed in recent years;[18] for example, RLS and FRLS. In contrast to the LMS adaptive filters, which find an approximate solution to the minimization of the mean-squared estimation error, RLS adaptive filters yield the exact solution to the deterministic problem of finding the minimum of a weighted sum of the errors. RLS algorithms grant a faster convergence rate than LMS algorithms, but this improvement is obtained at the expense of an increase in computational complexity. On the other hand, the fast versions of these algorithms may suffer from numerical instabilities and require suitable rescue procedures.

To overcome the limitations in the convergence rate of LMS algorithms, RLS and FRLS algorithms have been adapted to truncated Volterra filters, and in particular to quadratic filters.[20] However, even in the best cases, the computational complexity remains proportional to $O(N^3)$ operations per iteration where N is the memory length of the filter, i.e., an order of magnitude greater than the computational complexity $O(N^2)$ of LMS filters. A more detailed discussion of all these adaptive filters can be found in Reference 24.

Recently, increasing attention has been paid to the AP algorithms initially proposed by Ozeki and Umeda.[29] The significant feature of this class of algorithms is the ability to offer, in the presence of correlated signals, convergence rates higher than those of LMS algorithms and tracking capabilities better than those of RLS algorithms,[17] as shown in the following. This capability is extremely relevant for truncated Volterra filters because of the ill-conditioned nature of the autocorrelation matrix of the augmented input signal. The extension to such filters of the basic AP algorithms is feasible in principle and has actually been proposed for quadratic filters.[12] The main idea in Reference 12 is to extend the AP algorithm to a quadratic filter implemented by means of its diagonal representation of Equations 7.12 and 7.13. This choice allows us to greatly simplify the derivation of the AP algorithm. On the other hand, the adaptation algorithm can be easily applied to the whole filter and to its simplified version, as described in Section 7.2.

Let us rewrite Equation 7.13 in the vector form

$$y_i(n) = \mathbf{h}_i^T(n)\mathbf{x}_i(n),\tag{7.23}$$

where $\mathbf{h}_i(n)$ is the vector formed with the $N - i + 1$ coefficients of the ith channel

$$\mathbf{h}_i(n) = [h(0, i-1)\ h(1, i)\ \cdots\ h(N-i, N-1)]^T.\tag{7.24}$$

The input vector $\mathbf{x}_i(n)$ formed with $N - i + 1$ entries, with $1 \le i \le M$, is defined as

$$\mathbf{x}_i(n) = \begin{bmatrix} x(n)x(n-i+1) \\ x(n-1)x(n-i) \\ \vdots \\ x(n-N+i)x(n-N+1) \end{bmatrix}.\tag{7.25}$$

Let us define two vectors of $K = \sum_{k=1}^{M}(N-k+1)$ elements

$$\mathbf{h}(n) = \left[\mathbf{h}_1^T(n) \cdots \mathbf{h}_M^T(n)\right]^T\tag{7.26}$$

$$\mathbf{x}(n) = \left[\mathbf{x}_1^T(n) \cdots \mathbf{x}_M^T(n)\right]^T\tag{7.27}$$

formed with the partial vectors $\mathbf{h}_i(n)$ and $\mathbf{x}_i(n)$. Then, the output of the pure quadratic filter can be written as

$$y(n) = \mathbf{h}^T(n)\mathbf{x}(n).\tag{7.28}$$

The aim of the AP algorithm of order L is to find the minimum norm of the coefficient increments that set to zero the last L *a posteriori* errors at time $n - j + 1$

$$\epsilon_{n+1}(n - j + 1) = d(n - j + 1) - \mathbf{h}^T(n+1)\mathbf{x}(n - j + 1),\tag{7.29}$$

where $j = 1, \ldots, L$. In a more explicit form, the following L constraints should be verified for a Lth order AP algorithm

$$\mathbf{h}^T(n+1)\mathbf{x}(n) = d(n),$$

$$\vdots$$

$$\mathbf{h}^T(n+1)\mathbf{x}(n - L + 1) = d(n - L + 1).\tag{7.30}$$

The function $J(n)$ to be minimized is

$$J(n) = \delta\mathbf{h}^T(n+1)\delta\mathbf{h}(n+1) + \sum_{j=1}^{L}\lambda_j\left(d(n-j+1) - \mathbf{h}^T(n+1)\mathbf{x}(n-j+1)\right),$$

$$\tag{7.31}$$

where

$$\delta \mathbf{h}(n+1) = \mathbf{h}(n+1) - \mathbf{h}(n), \qquad (7.32)$$

and λ_j are Lagrange multipliers. By differentiating $J(n)$ with respect to $\delta \mathbf{h}(n+1)$, and setting to zero the partial derivatives, the following set of K equations is obtained

$$2\delta \mathbf{h}(n+1) = \sum_{j=1}^{L} \lambda_j \mathbf{x}(n-j+1) = \mathbf{G}(n)\boldsymbol{\Lambda}, \qquad (7.33)$$

where the $K \times L$ matrix $\mathbf{G}(n)$ is defined as

$$\mathbf{G}(n) = [\mathbf{x}(n)\ \mathbf{x}(n-1)\cdots\mathbf{x}(n-L+1)] \qquad (7.34)$$

and

$$\boldsymbol{\Lambda} = [\lambda_1 \cdots \lambda_L]^T . \qquad (7.35)$$

An alternative expression for the vector $\boldsymbol{\Lambda}$ is derived premultiplying Equation 7.33 by $\mathbf{G}^T(n)$

$$\boldsymbol{\Lambda} = \left(\mathbf{G}^T(n)\mathbf{G}(n)\right)^{-1}\mathbf{G}^T(n)2\delta \mathbf{h}(n+1). \qquad (7.36)$$

The $L \times 1$ vector $\mathbf{G}^T(n)2\delta \mathbf{h}(n+1)$ in the last equation can be written as

$$\mathbf{G}^T(n)2\delta \mathbf{h}(n+1) = 2\begin{bmatrix} \mathbf{x}^T(n)\delta \mathbf{h}(n+1) \\ \mathbf{x}^T(n-1)\delta \mathbf{h}(n+1) \\ \vdots \\ \mathbf{x}^T(n-L+1)\delta \mathbf{h}(n+1) \end{bmatrix} = 2\mathbf{e}(n), \qquad (7.37)$$

where $\mathbf{e}(n)$ is the vector of the *a priori* estimation errors whose generic jth element is given by

$$\mathbf{x}^T(n-j+1)\delta \mathbf{h}(n+1) = \mathbf{x}^T(n-j+1)\mathbf{h}(n+1) - \mathbf{x}^T(n-j+1)\mathbf{h}(n)$$
$$= d(n-j+1) - \mathbf{x}^T(n-j+1)\mathbf{h}(n), \qquad (7.38)$$

for $1 \leq j \leq L$. Finally, by using Equations 7.33, 7.36, and 7.37, the following relation is derived:

$$\delta \mathbf{h}(n+1) = \mathbf{G}(n)\left(\mathbf{G}^T(n)\mathbf{G}(n)\right)^{-1}\mathbf{e}(n). \qquad (7.39)$$

The vector $\delta \mathbf{h}(n+1)$ can be expressed by means of its M components, i.e.,

$$\delta \mathbf{h}(n+1) = \left[\delta \mathbf{h}_1^T(n+1)\cdots\delta \mathbf{h}_M^T(n+1)\right]^T . \qquad (7.40)$$

By partitioning the matrix $\mathbf{G}(n)$ in submatrices $\mathbf{G}_i(n)$ of dimensions $(N - i + 1) \times L$, the following set of equations for $1 \leq i \leq M$ is then obtained:

$$\delta \mathbf{h}_i(n + 1) = \mathbf{G}_i(n)\left(\mathbf{G}^T(n)\mathbf{G}(n)\right)^{-1}\mathbf{e}(n), \tag{7.41}$$

which refer to the generic channel i in the diagonal realization. Then, the updating relations for the coefficients of the channel i are given by

$$\mathbf{h}_i(n + 1) = \mathbf{h}_i(n) + \mu_i \mathbf{G}_i(n)\left(\mathbf{G}^T(n)\mathbf{G}(n)\right)^{-1}\mathbf{e}(n), \tag{7.42}$$

for $1 \leq i \leq M$, where μ_i is a parameter that controls both the convergence rate and the stability of the AP algorithm. The most relevant problem here is the estimate of the $L \times L$ matrix $\mathbf{G}^T(n)\mathbf{G}(n)$. This matrix needs to be computed as an estimate of the autocorrelation matrix of the augmented input signal corresponding to the last L input vectors. This estimate is often ill-conditioned and suffers from noise amplification effects.[17] As a consequence, the computation of its inverse, required in Equation 7.42, is a critical step. To avoid these difficulties, it is convenient to replace the matrix $\mathbf{G}^T(n)\mathbf{G}(n)$ in Equation 7.42 with a simpler and more stable estimate obtained using the $L \times L$ matrix $\tilde{\mathbf{R}}_i(n)$, exploiting only input samples involved in the ith channel. In other words, we assume

$$\tilde{\mathbf{R}}_i(n) = \mathbf{l}_i(n)\mathbf{l}_i^T(n), \tag{7.43}$$

where $\mathbf{l}_i(n)$ is the vector

$$\mathbf{l}_i(n) = \begin{bmatrix} x(n)x(n - i + 1) \\ x(n - 1)x(n - i) \\ \vdots \\ x(n - L + 1)x(n - L - i + 2) \end{bmatrix}, \tag{7.44}$$

and we recursively estimate the autocorrelation matrix for the ith channel with the exponential mean of Equation 7.45:

$$\mathbf{R}_i(n) = \lambda \mathbf{R}_i(n - 1) + (1 - \lambda)\tilde{\mathbf{R}}_i(n), \tag{7.45}$$

where λ is the forgetting factor $(0 < \lambda < 1)$, which determines the temporal memory length in the estimation of the autocorrelation matrix. To avoid the inversion, it is convenient to directly update the inverse matrix $\mathbf{P}_i = \mathbf{R}_i^{-1}(n)$, as is done for RLS algorithms. By using the matrix inversion lemma,[9] it results in

$$\mathbf{P}_i(n) = \frac{1}{\lambda}\mathbf{P}_i(n - 1) - \frac{1}{\lambda}\frac{\mathbf{P}_i(n - 1)\mathbf{l}_i(n)\mathbf{l}_i^T(n)\mathbf{P}_i(n - 1)}{\frac{\lambda}{1 - \lambda} + \mathbf{l}_i^T(n)\mathbf{P}_i(n - 1)\mathbf{l}_i(n)}. \tag{7.46}$$

TABLE 7.1

AP Adaptive Algorithm of Order L for a Volterra Filter Implemented with M Diagonals

---

$y_i(n) = \mathbf{h}_i^T(n)\mathbf{x}_i(n)$

$y(n) = \sum_{i=1}^{M} y_i(n)$

$e_j(n) = d(n - j + 1) - \sum_{i=1}^{M} \mathbf{h}_i^T(n)\mathbf{x}_i(n - j + 1)$

$\mathbf{e}(n) = [e_1(n)e_2(n)\cdots e_L(n)]^T$

$\mathbf{G}_i(n) = [\mathbf{x}_i(n)\mathbf{x}_i(n - 1)\cdots\mathbf{x}_i(n - L + 1)]$

$\mathbf{l}_i(n) = [x(n)x(n - i + 1)\ x(n - 1)x(n - i)\cdots x(n - L + 1)x(n - L - i + 2)]^T$

$\mathbf{k}_i(n) = \dfrac{\mathbf{P}_i(n - 1)\mathbf{l}_i(n)}{\frac{\lambda}{1 - \lambda} + \mathbf{l}_i^T(n)\mathbf{P}_i(n - 1)\mathbf{l}_i(n)}$

$\mathbf{P}_i(n) = \dfrac{1}{\lambda}(\mathbf{P}_i(n - 1) - \mathbf{k}_i(n)\mathbf{l}_i^T(n)\mathbf{P}_i(n - 1))$

$\mathbf{h}_i(n + 1) = \mathbf{h}_i(n) + \mu_i\mathbf{G}_i(n)\mathbf{P}_i(n)\mathbf{e}(n)$

Initialization: $\mathbf{P}_i = \delta\mathbf{I}, \mathbf{h}_i = \mathbf{0} \quad \forall i$

By defining the gain vector $\mathbf{k}_i(n)$ as

$$\mathbf{k}_i(n) = \frac{\mathbf{P}_i(n - 1)\mathbf{l}_i(n)}{\frac{\lambda}{1 - \lambda} + \mathbf{l}_i^T(n)\mathbf{P}_i(n - 1)\mathbf{l}_i(n)}, \tag{7.47}$$

the following recursive estimate for $\mathbf{P}_i(n)$ is derived:

$$\mathbf{P}_i(n) = \frac{1}{\lambda}\left(\mathbf{P}_i(n - 1) - \mathbf{k}_i(n)\mathbf{l}_i^T(n)\mathbf{P}_i(n - 1)\right). \tag{7.48}$$

By replacing $(\mathbf{G}^T(n)\mathbf{G}(n))^{-1}$ in Equation 7.42 with $\mathbf{P}_i(n)$ given in Equation 7.48, the final updating expression for the ith channel of the diagonal realization is

$$\mathbf{h}_i(n + 1) = \mathbf{h}_i(n) + \mu_i\mathbf{G}_i(n)\mathbf{P}_i(n)\mathbf{e}(n) \tag{7.49}$$

for $1 \leq i \leq M$. The equations employed for updating the coefficients and for filtering the input signal using the diagonal realization are summarized in Table 7.1. The algorithm can be applied to the whole quadratic filter simply by setting $M = N$. Because this adaptive algorithm treats all channels of the nonlinear filter separately, it is easy to extend this algorithm to a generic Volterra filter with different order terms. It has also been shown[12] that the complexity of this AP algorithm is about L times that of the LMS algorithm. It is worth noting that for $L = 1$ the AP algorithm becomes in practice an LMS algorithm, while for increasing value of the number L of the errors minimized it tends to behave as an RLS algorithm. This is the reason for the improved performance that is experienced.

7.4   Experimental Results for Acoustic Echo Cancellation

In this section we first present some results concerning the characteristics of the AP algorithms derived in the previous section. Then we compare the performance of various realizations of nonlinear echo cancelers to those of the linear AEC. The experimental results of this section employ both artificial acoustic echo paths, obtained from the identification of real systems, and echo signals recorded from two commercial handset devices, one of medium quality and one of high quality. The acoustic echo signals have been generated and recorded in different operating conditions, using as input signals both colored noises and real voices with different loudspeaker volumes. Although the simulations have been performed on a large set of echoes and signals, we report here, for the sake of simplicity, only the results of single significant experiments.

7.4.1   Experimental Results for the AP Algorithm

The first set of experiments concerns the convergence characteristics of the AP adaptation algorithms described in Section 7.3. We consider here the identification of a pure second-order Volterra system. In particular, we consider the identification of the second-order kernel represented in Figure 7.2, which is the quadratic term of a real acoustic echo path. The Volterra kernel was truncated to a 30-taps memory length. In all the experiments of this subsection we have employed as input signals some colored noise sequences, which had a spectral content similar to the voice. White Gaussian noise was added to the output of the Volterra filter to achieve a 30-dB signal-to-noise ratio. Equations 7.42 and 7.49 have been tested on the pure second-order Volterra filter with a memory of 30 samples, first using an AP algorithm of order $L = 2$. Different estimates of the autocorrelation matrix have been considered, i.e., the recursive estimate by diagonals of the inverse of the autocorrelation matrix as in Equation 7.49, with $M = N$, the recursive estimate of the inverse matrix $(\mathbf{G}^T(n)\mathbf{G}(n))^{-1}$ and its direct computation at each step. The corresponding adaptation curves are marked in Figure 7.9 with the letters A, B, C, respectively. They have been obtained with colored noise input sequences and are the means of over 100 independent tests. It clearly appears that the algorithm of Equation 7.49 that uses the autocorrelation matrices related to the single channels of the diagonal realization gives the fastest learning curve A. On the other hand, the direct computation of the inverse of the autocorrelation matrix as in Equation 7.42 produces a more erratic behavior of the adaptation curve and a reduced convergence rate, as is clearly shown by the curve C. Improved behavior can be observed when in Equation 7.42 we recursively estimate the inverse autocorrelation matrix in a way similar to Equation 7.48, as shown by the learning curve B. Indeed, the exponential smoothing introduced by the recursive estimate leads to a more regular convergence of the algorithm.

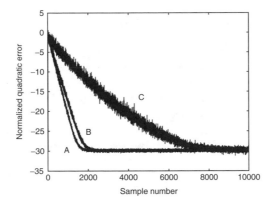

FIGURE 7.9
Adaptation curves for different estimates of the autocorrelation matrix.

It should be noted that a major advantage offered by the algorithm of Equation 7.49 is the possibility of directly applying it to the SVFs described in Section 7.2. According to acoustic echo path behavior, a reduction of the number of channels can be achieved without substantially affecting the general performance, as shown in the last set of experiments in this section. It is also worth noting that similar learning curves have been obtained for AP algorithms of order $L > 2$.

In the second set of experiments, the performance of the AP algorithms of order $L > 2$ has been evaluated. In these experiments, the diagonal realization structure of the pure second-order Volterra filter ($M = N$) with a memory of 30 samples is considered. The adaptation algorithms used are those of Equation 7.49 for $L = 2, 3, 5,$ and 10. As a reference, the standard NLMS algorithm has been chosen. The test conditions are the same as those used for the first set of experiments. The adaptation curves for the AP algorithms are shown in Figure 7.10 together with the curve for the NLMS algorithm. In these experiments the adaptation constants μ_i are chosen to reach the same mean-squared error at convergence for all the algorithms. The improvement obtained in the convergence rate by the AP algorithms is clear. It can also be noted that even the simple AP algorithm of order 2 is sufficient to guarantee a convergence rate more than three times higher than that of the NLMS algorithm. Increasing the order of the AP algorithms produces further increments in the convergence rate. It is worth noting that the better performances of the AP algorithms with respect to the NLMS algorithm are strictly related to the correlations existing in the actual and augmented input signals.

Finally, the good tracking characteristics of the AP algorithms can be appreciated from Figure 7.11, where the learning curve of the AP algorithm of order 2 is shown, together with that for the NLMS algorithm, when a mismatch affects the echo path. The mismatch was obtained by modifying all the

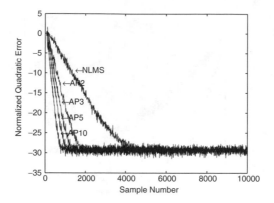

FIGURE 7.10
Adaptation curves for NLMS and AP algorithms of order $L = 2, 3, 5, 10$.

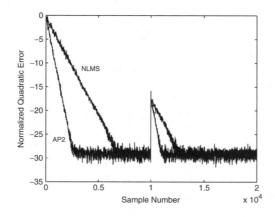

FIGURE 7.11
Adaptation curves for the NLMS and AP-2 algorithms in the presence of a mismatch in the echo path.

coefficients of the quadratic model of an aleatory quantity lower than 0.1% of the coefficient value. We can notice that the AP algorithm is still faster than the NLMS algorithm in recovering the steady-state condition.

7.4.2   Experimental Results for Low-Complexity Nonlinear Filters

We report here the results obtained for some of the structures described in Section 7.2 for the cancellation of nonlinear acoustic echoes in mobile termi-nals. The test conditions are the same as in the set of experiments described above; the only difference is that here we identify an artificial echo path modeled with a second-order Volterra filter, with linear and quadratic ker-nels having memories equal to 100 and 30 samples, respectively. Again, this

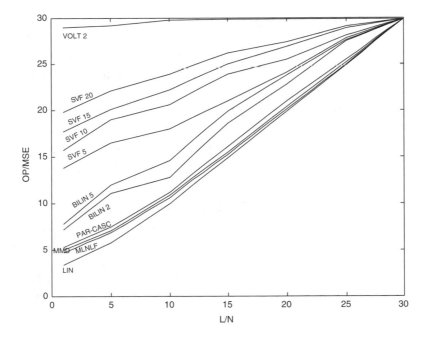

FIGURE 7.12
Echo cancellation level (dB) for a set of echo cancelers vs. second-order nonlinear distortion level (dB).

model was obtained by the identification of the true acoustic echo path of a commercial handset terminal.

To evaluate the relevance and the benefits of a nonlinear approach to acoustic echo cancellation, we consider the results of simulations done in the presence of different levels of quadratic nonlinearities. To this purpose, we define as L/N the ratio between the average power of the signal at the output of the linear part of the system modeling the echo path, and the average power of the signal at the output of the second-order part of the model. As a figure of merit of the echo cancellation achieved, we consider the ratio between the average echo signal power and the average residual echo signal power OP/MSE. This figure of merit is plotted in Figure 7.12 vs. the nonlinear distortion level L/N for different truncated Volterra filters and bilinear filters. The first set of filters always includes a linear part having a memory of 100 samples, whereas the bilinear filter, according to its recursive structure, permits a shorter memory with respect to the input signal, here fixed to $N_1 = 15$ samples. The following filters have been considered:

LIN: Linear FIR filter with 100 coefficients

MLNLF: Cascade of a memoryless polynomial preprocessor with order 7 and a linear FIR filter with 100 coefficients

MMD: Multi-memory-decomposition structure, with $N_a = 21$ and $N_p = 9$

PAR-CAS: Parallel-cascade structure implemented using 10 channels

BILIN 2: Bilinear filter with a memory of $N_2 = 2$ samples with respect to the reference signal

BILIN 5: Bilinear filter with a memory of $N_2 = 5$ samples with respect to the reference signal

SVF 5: Simplified Volterra filter implemented using 5 channels

SVF 10: Simplified Volterra filter implemented using 10 channels

SVF 15: Simplified Volterra filter implemented using 15 channels

SVF 20: Simplified Volterra filter implemented using 20 channels

VOLT 2: Second-order Volterra filter with a 30 × 30 quadratic kernel

All the filters were adapted with the NLMS algorithm, and the step size was chosen to optimize the performance of the echo canceler in terms of residual error. The performance of the pure linear filter clearly decays almost linearly with the increment of the nonlinear distortion, while the second-order Volterra filter behaves well for all levels of distortions. The MLNLF, MMD, and PAR-CAS filters offer little advantage with respect to the linear filter. The bilinear filters show good performance for low levels of distortion but significantly degrade in the presence of higher nonlinearities. The SVFs offer good performance even for high levels of distortion. This is particularly true for SVF 20, but the simpler SVF 5 also has a sufficiently good behavior.

To validate these kinds of results, a further set of experiments has been performed using many real acoustic echo signals recorded in a quiet environment with two commercial handset devices, one of medium quality and one of high quality. The acoustic echo signals were generated and recorded in different operating conditions using as input signals both colored noises and real voices with different loudspeaker volumes. The colored noise sequences have a spectral content similar to that of speech signals. The measured impulse response of the linear system generating the echo has been found to be 100 samples long with a delay of about 25 samples. The measured characteristics of the second-order kernel of a typical echo path like that shown in Figure 7.2 allow us to limit the memory of the quadratic part of the acoustic echo path to 30 samples. In the following experiments, the relevant information is, for all the above-reported structures, the echo return loss enhancement (ERLE) and the number of multiplications needed per output sample. The ERLE is defined as

$$\text{ERLE}(n) = 10 \log_{10} \frac{E[d^2(n)]}{E[e^2(n)]}, \tag{7.50}$$

where $d(n)$ is the echo signal picked up by the microphone and $e(n)$ is the residual echo after cancellation. It is worth noting that in these experiments the

TABLE 7.2

Echo Cancellation Obtained at the Maximum
Loudspeaker Volume and Corresponding
Number of Multiplications per Output Sample

Filter	ERLE (dB)	Mult.
Linear	17.0	202
MLNLF	17.5	289
MMD	17.5	456
PAR-CASC	17.9 ·	528
BILIN 2	19.1	267
BILIN 5	20.0	363
SVF 5	19.7	487
SVF 10	20.8	722
SVF 15	21.5	907
SVF 20	21.8	1042
VOLT 2	22.1	1599

level of the input and output signals is important. In fact, at low loudspeaker volume settings the echo signal picked up by the microphone is relatively small in comparison to the background noise. Because the nonlinear distortions are partly masked by the noise, there is quite a small gain in using a nonlinear echo canceler in place of a linear one. In contrast, when the loudspeaker volume becomes higher, the nonlinear distortions become relevant and are no longer masked by the background noise. In this case, nonlinear filters clearly outperform linear filters. The results reported in Table 7.2 refer to the filters listed above, adapted with the NLMS algorithm, when the maximum loudspeaker volume is set. In these conditions, the second-order Volterra filter provides an improvement of 5 dB over the linear filter but with a corresponding increase in complexity not suitable for real-time implementations. The MLNLF structure provides only poor results because in our experiments the nonlinear effects were mainly generated by the loudspeaker. The performances reported for the MMD structure are the best found for many possible choices of N_a and N_p, for a total memory length of $N = 30$. The poor results obtained are due to the presence of local minima. Also, the PAR-CAS structure does not give particularly good results. For the BILIN structures, we experimentally found that the best choice is to implement a filter with low memory with respect to the reference signal $d(n)$ and longer memory with respect to the input signal $x(n)$. With this choice, intermediate results have been obtained in terms of ERLE at a low computational complexity, so bilinear filters are possible candidates for implementation of low-cost nonlinear echo cancelers. However, the limitations of the estimation procedure mentioned above should be taken into account. Finally, the SVFs give significant gains over the linear filter even with a low number of branches. In general, a trade-off between the ERLE and the number of multiplications can be observed. This feature, together with the good convergence and tracking

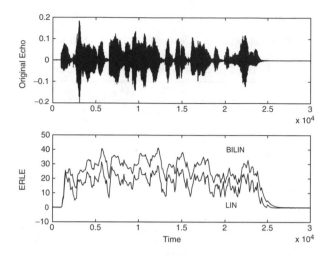

FIGURE 7.13
Original echo and ERLE (dB) vs. iterations for BILIN 5 and LIN.

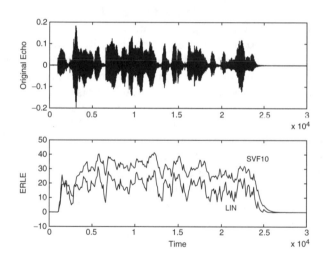

FIGURE 7.14
Original echo and ERLE (dB) vs. iterations for SVF 10 and LIN.

rates offered by the AP algorithms, make these filters an attractive choice for acoustic nonlinear echo cancellation in mobile terminals.

To further validate the quality of the results obtained using a nonlinear approach, the graphics of the ERLE measured using as AEC the linear filter LIN and the nonlinear filters BIL 5, SVF 10, and SVF 20 are compared in Figure 7.13 through Figure 7.15, respectively. The simulations have been done

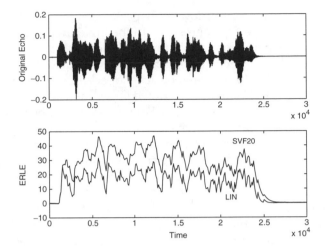

FIGURE 7.15
Original echo and ERLE (dB) vs. iterations for SVF 20 and LIN.

using the measured linear and quadratic kernels of the nonlinear echo path specified above, with memories equal to 100 and 30 samples, respectively. Moreover, a random noise with an average power of −30 dB has been added to the echo signal. In the diagrams, at each time iteration the mean of the previous 500 samples is plotted. The advantage of using a nonlinear AEC in place of the linear AEC is remarkable, as is clearly shown in these figures. Again, a trade-off exists between the implementation complexity and the quality measured by the ERLE of the nonlinear filters.

7.5 Conclusions

Both general and specific signal processing methods and techniques are widely used in modern applications of wireless communication systems. To obtain high-quality services it is often necessary to renounce the most common linear framework and resort to a nonlinear technique. This is the case with the application discussed in this chapter on adaptive filters for acoustic echo cancellation in mobile terminals. In fact, the performance of commercial AECs is often limited by the nonlinearities of the echo path originating from the sound propagation between the loudspeaker and the microphone of the receiver and then reaching the far-end terminal. Here, they are perceived by the far speaker as an annoying disturbance. In general, in ultracompact GSM and 3-G handset receivers, such nonlinearities depend on the loudspeaker volume. In fact, in the case of low loudspeaker volume, linear filters may guarantee acceptable performances due to the high attenuation introduced by the acoustic echo path. Conversely, for high loudspeaker volume, nonlinear filters are required. In particular, it has been demonstrated that second-order Volterra filters are able to offer a good level of echo cancellation. However, the second aspect that should be considered in commercial applications is the need for low-cost, low-power solutions. For this reason, the behaviors of some low-complexity nonlinear filters have been analyzed in this chapter to determine a structure able to reach performance in acoustic echo cancellation near to that of a complete second-order Volterra filter with reduced computational resources. The realizations considered here have been the simple structure formed with a memoryless nonlinearity followed by a linear FIR filter, the multi-memory-decomposition and parallel-cascade structures, the simplified Volterra filters, and the bilinear filters. As noted in the chapter, some of the structures we considered performed poorly because of the presence of local minima or because these structures were inadequate to cope with the nonlinearities originated by the loudspeaker. Eventually, only two structures satisfy our requests, i.e., the simplified Volterra filter and the equation-error bilinear filter. As a consequence of the specific characteristics of the acoustic echo path, the simplified Volterra filters can achieve almost the same results obtained by the complete quadratic filters but with a lower computational load. Bilinear filters need even less computational resources but obtain slightly worse results. This drawback is essentially due to the fact that the estimation procedure leads to a biased estimate of the optimum solution, thus limiting the performance of the identification algorithm.

The overall efficiency of a nonlinear echo canceler is strictly related to the characteristics of the adaptation algorithm used. In fact, fast adaptation and tracking rates are needed. To this purpose, the theory of the AP algorithms for quadratic filters has been reviewed in the chapter. The aspects that make their derivation from the linear case other than trivial have been noted. Moreover, the strategies proposed for the estimation of the augmented input autocorrelation matrix and for the direct updating of its inverse have been investigated by experiments. Finally, the performance of the AP adaptation algorithms have been compared with those of the classical NLMS algorithm. It has been shown that the AP technique offers better convergence and tracking capabilities with use of the simplified Volterra models, without excessive increase of computational complexity.

The experimental results reported for low-complexity nonlinear filters used for acoustic echo cancellation confirm that simple polynomial echo cancelers based on the simplified Volterra models are attractive for real-time and relatively low-cost implementations in high-quality handset devices. In particular, they are suitable for applications where the echo level is quite high and where nonlinear distortions are not masked by the background noise, as happens, for example, with hands-free systems implemented on handset receivers.

References

1. W. Kellermann, Ed., Special Section on Current Topics in Adaptive Filtering for Hands-Free Acoustic Communication and Beyond, *Signal Process.*, 80(8), Sept. 2000.
2. P. Dreiseitel and E. Hänsler, Eds., Special Issue on Acoustic Echo and Noise Control, *Eur. Trans. Telecommun.*, 2, Mar.–Apr. 2002.
3. 3GPP TS 26.115 V4.0.0, 3rd Generation Partnership Project; Technical Specification Group Services and System Aspects; Echo Control for Speech and Multi-Media Services, Mar. 2001.
4. A.N. Birkett and R.A. Goubran, Limitations of handsfree acoustic echo cancellers due to nonlinear loudspeaker distortion and enclosure vibration effects, in *Proc. of IEEE ASSP Workshop on Applications of Signal Processing to Audio and Acoustics*, New Paltz, NY, Oct. 1995, 103–106.
5. A.N. Birkett and R.A. Goubran, Acoustic echo cancellation using NLMS-neural network structures, *Proc. of ICASSP-95, International Conference on Acoustics Speech and Signal Processing*, Detroit, MI, May 8–12, 1995.
6. C. Breining et al., Acoustic echo control, *Signal Process. Mag.*, 16(4), 42–69, July 1999.
7. H.H. Chang, C.L. Nikias, and A.N. Venetsanopoulos, Efficient implementation of quadratic filters, *IEEE Trans. Acoust. Speech Signal Process.*, ASSP-34, 1511–1528, Dec. 1986.
8. J.P. Costa, T. Pitarque, and E. Thierry, Using orthogonal least squares identification for adaptive nonlinear filtering of GSM signals, *Proc. of ICASSP-97, International Conference on Acoustics Speech and Signal Processing*, Munich, Germany, Mar. 1997, 2397–2400.
9. C.F.N. Cowan and P.M. Grant. *Adaptive Filters*, Prentice-Hall, Englewood Cliffs, NJ, 1985.
10. ETSI TS 143 050 V4.0.0, Digital Cellular Telecommunication System (Phase 2+); Transmission Planning Aspects of the Speech Service in the GSM Public Land Mobile Network (PLMN) System, Mar. 2001.
11. ETSI TS 126 132 V4.0.0: Universal Mobile Telecommunication System (UMTS); Terminal Acoustic Characteristics for Telephony; Requirements, Mar. 2001.
12. A. Fermo, A. Carini, and G.L. Sicuranza, Simplified Volterra filters for acoustic echo cancellation in GSM receivers, in *Proc. of EUSIPCO-2000*, Tampere, Finland, Sept. 2000.
13. A. Fermo, A. Carini, and G.L. Sicuranza, Analysis of different low complexity nonlinear filters for acoustic echo cancellation, *J. Comput. Inf. Technol.*, (8)4, 333–339, Dec. 2000.
14. W.A. Frank, An efficient approximation to the 2nd order Volterra filter, *Signal Process.*, 45, 97–113, 1995.
15. X. Y. Gao and W. M. Snelgrove, Adaptive linearization of a loudspeaker, in *Proc. of ICASSP-91, International Conference on Acoustics Speech and Signal Processing*, Toronto, Canada, 3589–3592, May 14–17, 1991.
16. S. Gay, *Fast Projection Algorithms with Application to Voice Excited Echo Cancellers*, Ph.D. dissertation, Rutgers University, Piscataway, NJ, 1994.

17. A. Gilloire, E. Moulines, D. Slock, and P. Duhamel, State of the art in acoustic echo cancellation, in *Digital Signal Processing in Telecommunication*, A.R. Figueiras-Vidal, Ed., Springer-Verlag, New York, 1993.
18. S. Haykin, *Adaptive Filter Theory*, 3rd ed., Prentice-Hall, Englewood Cliffs, NJ, 1996.
19. ITU-T Recommendation G.167: *Acoustic Echo Controllers*, Mar. 1993.
20. J. Lee and V. J. Mathews, A fast recursive least squares second-order Volterra filter and its performance analysis, *IEEE Trans. Signal Process.*, (41)3, 1087–1101, Mar. 1993.
21. S. Lian, P. Su, and F. Guangzeng, Echo cancellation using nonlinear layered bilinear IIR filter, *J. China Univ. Posts Telecommun.*, (5)1, Jun. 1998
22. V.J. Mathews, Adaptive polynomial filters, *IEEE Signal Process. Mag.*, 8(3), 10–26, July 1991.
23. V. J. Mathews and S.C. Douglas, *Adaptive Filters*, Prentice-Hall, Englewood Cliffs, NJ, 2000.
24. V.J. Mathews and G.L. Sicuranza, *Polynomial Signal Processing*, John Wiley & Sons, New York, 2000.
25. S. Marsi and G.L. Sicuranza, On reduced-complexity approximation of quadratic filters, in *Proc. 27th Asilomar Conf. Signals, Syst., Comput.*, Pacific Grove, CA, 1026–1030, Nov. 1993.
26. M. Montazeri and P. Duhamel, A set of algorithms linking NLMS and block RLS algorithms, *IEEE Trans. Signal Process.*, 43(2), 444–453, Feb. 1995.
27. L.S.H. Ngia and J. Sjöberg, Nonlinear Acoustic Echo Cancellation Using a Hammerstein Model, Report CTH-TE-65, Nov. 1997.
28. B.S. Nollett and D.L. Jones, Nonlinear echo cancellation for hands-free speakerphones, in *Proc. of NSIP'97, Workshop on Nonlinear Signal and Image Processing*, Mackinac Island, MI, Sept. 8–10, 1997.
29. K. Ozeki and T. Umeda, An adaptive filtering algorithm using an orthogonal projection to an affine subspace and its properties, *Electron. Commun. Jpn.*, J67-A,5, 126–132, Feb. 1984.
30. T.M. Panicker and V.J. Mathews, Parallel-cascade realizations and approximations of truncated Volterra systems, in *Proc. ICASSP-96*, Atlanta, GA, 1996.
31. G.L. Sicuranza, A. Bucconi, and P. Mitri, Adaptive echo cancellation with nonlinear digital filters, in *Proc. of ICASSP-84, International Conference on Acoustics Speech and Signal Processing*, San Diego, CA, Mar. 1984.
32. G.L. Sicuranza and G. Ramponi, Distributed arithmetic implementation of nonlinear echo cancellers, in *Proc. of ICASSP-85, International Conference on Acoustics Speech and Signal Processing*, Tampa, FL, Mar. 1985, 1617–1620.
33. G.L. Sicuranza, Quadratic filters for signal processing, *Proc. IEEE*, 80(8), 1262–1285, Aug. 1992.
34. A. Stenger and R. Rabenstein, An acoustic echo canceller with compensation of nonlinearities, in *Proc. of EUSIPCO-98, European Signal Processing Conference*, Island of Rhodes, Greece, Sept. 8–11, 1998.
35. A. Stenger, W. Kellermann, and R. Rabenstein, Adaptation of acoustic echo cancellers incorporating a memoryless nonlinearity, in *Proc. of IWAENC-99, Workshop on Acoustic Echo and Noise Control*, Pocono Manor, PA, Sept. 1999, 168–171.
36. A. Stenger, L. Trautmann, and R. Rabenstein, Nonlinear acoustic echo cancellation with 2nd order adaptive Volterra filters, in *Proc. of ICASSP-99, International*

Conference on Acoustics Speech and Signal Processing, Phoenix, AZ, May 15–19, 1999.

37. A. Stenger and R. Rabenstein, Adaptive Volterra filters for nonlinear acoustic echo cancellation, in *Proc. of NSIP'99, Workshop on Nonlinear Signal and Image Processing*, Antalya, Turkey, June 20–23, 1999.

38. A. Stenger and W. Kellermann, Adaptation of a memoryless preprocessor for nonlinear acoustic echo cancelling, *Signal Process.*, 80(8), 1747–1760, Sept. 2000.

39. A. Stenger and W. Kellermann, Nonlinear acoustic echo cancellation with fast converging memoryless preprocessor, in *Proc. of ICASSP-2000, International Conference on Acoustics Speech and Signal Processing*, Istanbul, Turkey, June 2000.

40. A. Stenger and W. Kellermann, RLS-adapted polynomial for nonlinear acoustic echo cancelling, in *Proc. of EUSIPCO-2000, European Signal Processing Conference*, Tampere, Finland, Sept. 2000.

41. M. Tanaka, Y. Kaneda, S. Makino, and J. Kojima, Fast projection algorithm for adaptive filtering, *IEICE Trans. Fundam.*, E-78-A(10), Oct. 1995.

42. E.J. Thomas, Some considerations on the application of Volterra representation of nonlinear networks to adaptive echo cancelers, *Bell Syst. Tech. J.*, 50(8), 2797–2805, Oct. 1971.

43. B. Widrow and S.D. Stearns, *Adaptive Signal Processing*, Prentice-Hall, Englewood Cliffs, NJ, 1985.

8

Blind and Semiblind Channel Estimation

Visa Koivunen, Mihai Enescu, and Marius Sirbu

CONTENTS

8.1 Introduction

In digital wireless communications, the impairment caused by time-varying multipath fading must be compensated for. The transmitted signal propagating via multiple paths experiences various delays due to the differing lengths of the paths. The resulting intersymbol interference (ISI) may distort the received signal so severely that the transmitted symbol sequence cannot be recovered. The effects of the ISI can be mitigated in the receiver with either direct equalization or channel identification together with, for example, maximum likelihood sequence estimation (MLSE). Typically, no prior knowledge of the channel impulse response is available and the channel is time-varying. Hence, it needs to be estimated and the estimates updated on a regular basis. The estimation is commonly performed by transmitting a known sequence of training symbols in a periodic fashion or by using a separate pilot signal. The receiver generates a local copy of the training data and the channel coefficients or equalizer are estimated. In the presence of Gaussian noise, the estimators derived are typically linear. This type of periodic transmission of training data or pilot signals significantly reduces the effective data rates. For example, in the global system for mobile communications (GSM), approximately 20% of the data in a time frame is allocated for the training sequence.

To improve effective data rates and, consequently, effective spectral efficiency, blind receiver structures have been proposed. The term *blindness* means that the receiver has no knowledge of either the transmitted sequence or the channel impulse response. Channel identification, equalization, or demodulation is then performed using only some statistical or structural properties of the transmitted and received signal. Training data can then be either completely excluded or significantly reduced, and information symbols transmitted instead. In the face of deep fades, blind methods may allow for tracking fast variations in the channel and reacquiring operational conditions using information symbols only. The processing in blind receivers is typically nonlinear. Common design goals for blind receiver algorithms are the following: *capability to identify any type of channel, fast convergence to the desired solution, capability of tracking channel time variations, and low computational complexity.*

In blind channel estimation, some channel types may not be identified and some ambiguities may remain, for example, rotation of the constellation pattern. Also, the estimator may converge slowly. Using a limited number of training symbols solves these problems. Limited training data in conjunction with blind algorithms lead to semiblind methods. Semiblind methods provide a more feasible solution for practical communication systems.

Blind receiver structures may be derived for different types of communication systems. Some design goals and the statistical and structural signal properties that may be exploited are different in time-division multiple access (TDMA), code-division multiple access (CDMA), and orthogonal frequency-division multiplexing (OFDM) systems, for example. In single-user equalization, mitigating ISI is the main task and the systems are modeled as single-input single-output (SISO) or single-input multiple-output (SIMO) systems. In wireless multiuser communications, the received signal is further degraded by interference, for example, co-channel interference (CCI) resulting from reusing the same frequency in another cell (TDMA systems, like GSM) or multiuser (MUI) interference caused by differences in received signal power or loss of orthogonality in the spreading codes (DS-CDMA). Receivers employing multiple antennas may cancel interferences by also using spatial processing.

Systems employing multiple transmitters and receivers may be modeled as multiple-input multiple-output (MIMO) systems. Such systems provide major improvement in spectral efficiency (bits/s/Hz) and link quality by exploiting the diversity (multiple independent channels between the transmitter and receiver) and array (signal-to-noise ratio, or SNR) gains. Because the radio spectrum is a scarce and expensive resource, the potentially high capacity of MIMO systems may be achieved only if the channels are reliably estimated. Furthermore, effective spectral efficiency in these systems is reduced by large amount of training data. Hence, there is a strong motivation to develop blind and semiblind receivers for MIMO systems as well.

In this chapter we present the principles of blind channel estimation. Various statistical and structural signal properties used in blind channel equalization and identification are described. Impairment caused by the channel and channel models are also considered. Algorithms for single-user receivers are briefly reviewed. The problem of blind equalization in MIMO systems is considered in more detail. Widely used blind equalization and channel estimation methods for wireless systems such as GSM, DS-CDMA, and OFDM are reviewed. Finally, semiblind methods are considered briefly.

Because it is impossible to cover all the research by the many different researchers in the area of blind receiver structures, we refer readers to the several recent books and papers for more detailed information.[1-6]

8.2   Time-Varying Channel Model

Fundamentally, mobile radio communication channels are time-varying multipath channels. As the performance of digital radio communication systems is strongly affected by impairment caused by scattering, reflection, and diffraction, channel models are of great interest.[7-15]

A time-varying radio channel (TVC) may be represented by a two-dimensional channel impulse response $h(t, \tau)$; Figure 8.1 presents an example. Multipath propagation results in time dispersion of the transmitted signal, which is visible on the τ axis of $h(t, \tau)$ plot. Time variations of the channel are given on the t axis, where T is the symbol duration.

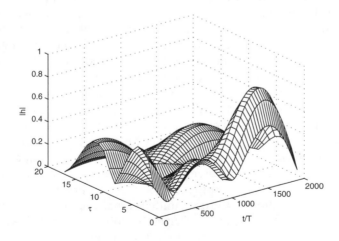

FIGURE 8.1
COST-207 "Hilly Terrain" channel, receiver speed of 90 km/h.

Two types of fading characterize mobile communications: large-scale fading, which is due to the motion over large areas, and small-scale fading, which is due to small changes in position.[14] Small-scale fading is often called Rayleigh fading. If the multiple reflective paths are large in number and if there is no line-of-sight (LOS) signal component, the envelope of the received signal may be statistically described by a Rayleigh pdf. When there is a dominant nonfading component, such as an LOS propagation path, the small-scale fading envelope may be modeled by a Rician pdf. As the amplitude of the nonfading component approaches zero, the Rician pdf approaches a Rayleigh pdf. The Rician distribution is often described using a parameter K known as the Rician factor. It completely specifies the Rician distribution, as it is defined as the ratio between the deterministic signal power and the variance of the multipath.[13]

In wireless communications many physical factors in the radio propagation channel cause fading. Typically there is no LOS path between the mobile units and the base station. Consequently, the received signal consists of multiple copies of the transmitted signal that arrive at the receiver through different indirect paths. When LOS is present, the channel can be modeled as containing an LOS component and also as containing multipath components, as discussed later in this section. The randomly distributed amplitudes, phases, and arrival angles of these multipath copies of the transmitted signal cause fluctuations in the received signal power, thereby introducing fading. In addition to (multipath) fading, multipath propagation also lengthens the time required for the main portion of the transmitted signal to reach the receiver. This phenomenon is quantified by maximum excess delay τ_{max}. In the case of a single transmitted signal waveform τ_{max} represents the time between the first and the last received component. Depending on the relative durations of the maximum excess delay and the symbol period, multipath fading is conventionally classified into either frequency-flat fading or frequency-selective fading.[14] Multipath fading is frequency-selective when the symbol period is smaller than the maximum excess delay. Thus, the channel induces ISI. The fading is flat when all the received multipath components arrive within the symbol period. The coherence bandwidth f_0 is a measure of the range of frequencies over which the channel response can be considered "flat." In other words, coherence bandwidth is the range of frequencies over which two frequency components are likely to have high amplitude correlation.[13]

Other causes of fading are the frequency offsets between two sources. The signal from one source undergoes Doppler shift due to relative motion between a transmitter and a receiver. Also, there may be a carrier frequency mismatch between transmit and receive oscillators. Frequency offsets result in frequency modulation on the transmitted signal, and thereafter cause channel time-variations that are characterized by coherence time. Depending on the relative values of the coherence time T_c and the symbol period, a fading channel can be categorized as time-flat when the symbol period is much less than the channel coherence time. Otherwise, it is time-selective.

Angle spread is an important quantity in characterizing the radio channel. It causes space-selective fading. This effect is due to the spread of directions of the incoming paths and is caused by scattering in the vicinity of the transmitter or receiver. A large angle spread is caused by rich scattering and will allow spatial diversity, meaning that the signals propagate via independent channels to different antenna elements, and the correlation between antenna elements is low. This property is exploited in MIMO systems to improve spectral efficiency via spatial multiplexing and link reliability.

In a macrocell environment, the motion of the mobile will give rise to Doppler spread, which causes time selectivity of the channels. Taking into account both time-selectivity and frequency-selectivity of the channel, fading channels may be categorized into one of the following four types:

1. Flat fading channels (channels are both time and frequency flat)
2. Frequency-selective fading channels (channels are frequency selective but time flat)
3. Time-selective fading channels (channels are time selective but frequency flat)
4. Doubly selective fading channels (channels are both frequency and time selective)

The wide-sense stationary uncorrelated scattering (WSSUS) linear time-variant channel model is widely used to model signal propagation in the mobile communications environment. The WSSUS model was introduced by Bello[7] and was further investigated, for example, in Reference 16. The channel may be modeled as follows:[16]

$$
h_{ij}(t, \tau) = \frac{1}{\sqrt{N_l}} \sum_{l=1}^{N_l} e^{j(2\pi f_{d,l}t + \theta_l)} h_{RF}(\tau - \tau_l), \tag{8.1}
$$

where N_l is the number of echo paths, $f_{d,l}$ is the Doppler spread, θ_l is the angular spread, and $h_{RF}(t)$ is the impulse response of the receive filter. For each delay τ, the channel is given by selecting:

1. N_l Doppler frequencies $f_{d,l}$ from a random variable with classical Jakes pdf in $(-f_{d,\max}, f_{d,\max})$. The maximum Doppler spread can be expressed as $f_{d,\max} = v/\lambda_w$, where v is the mobile station speed and λ_w is the signal wavelength.
2. N_l initial phases θ_l from a uniform distributed random variable in $[0, 2\pi]$.
3. N_l echo delay times τ_l. Each delay spread is a random variable with probability density function proportional to the mean power delay profile of the propagation environment.

The uncorrelated scattering assumption leads to a finite impulse response (FIR) channel model in which all the taps vary independently. That is, the time variations of the tap coefficients are mutually uncorrelated while exhibiting the same time-correlation behavior. The physical situation underlying this model is the existence of a few large scatterers far from the mobile receiver and the existence of a small number of scatterers in the vicinity of the mobile receiver. Classical analysis of digital transmission through a fading medium models $h_{ij}(t, \tau)$ as zero-mean random variables. In certain applications, such as cellular communications, a direct nonfading path may also exist, superimposed on the fading path. In this case, the coefficients $h_{ij}(t, \tau)$ have nonzero mean (Rician fading). The overall nonzero mean channel is then:[17,18]

$$\breve{h}_{ij}(t, \tau) = \tilde{h}_{ij}(\tau) + h_{ij}(t, \tau), \tag{8.2}$$

where $\tilde{h}_{ij}(\tau)$ is a constant mean and $E[h_{ij}(t, \tau)] = 0$.

As pointed out in Reference 19, accurate mathematical channel models are based on collected measurements of actual channels. The channel models that are employed in GSM system are defined in Reference 11. Measurements have been made over typical bandwidths of 10 to 20 MHz at or near 900 MHz. Four propagation environments are described in the project: Typical Urban (TU), Bad Urban (BU), Hilly Terrain (HT), and Rural Area (RA), each of which has specific parameter sets including delay and power profiles. These propagation environments are determined by individual delay distributions that are piecewise exponential functions.

More advanced channel models taking into account the spatial component are being developed. Models for WCDMA and multicarrier systems have been derived and validated by measurements. Classical channel models provide information on signal power level distribution and Doppler shifts of received signals. Modern spatial channel models incorporate concepts as time delay spread, angles of arrival and departure, and different antenna geometries.[20,21] An overview of spatial channel models is presented in Reference 22. Much research and many measurements of channel models have been done at AT&T Laboratories.[23] Several other models are presented in special publications on channel modeling, such as Reference 15.

8.2.1   MIMO Channel Modeling

In the MIMO scenario we have a channel matrix **H**, which models the medium between the m transmit and n receive antennas. Typically, the channels are considered to be independent. This assumption holds if the spacing between the antenna elements is larger than the coherence distance. Real environment measurement campaigns are ongoing in order to derive better models for MIMO channels. Key concepts such as antenna spacing, antenna height, scattering radius at both mobile and base station, and the placement of these scatterers influence the rank and correlation properties of the channel matrix, and they have to be taken into account when building a channel model. An appealing property of MIMO models is that they promise major improvement in spectral efficiency. However, this relies on the correlation properties between antennas as well as on the rank of the MIMO channel matrix. At best, the diversity order is $m \times n$ and there are $\min(m, n)$ independent channels available.

One way of modeling the MIMO channel is based on defining the transmitter and receiver correlation matrices and building a MIMO correlation matrix that can be used along with the i.i.d. channel correlation matrix.[24,25] In Reference 25 spatial correlation matrices are computed for both the transmitter and receiver and the spatial correlation matrix of the MIMO channel is defined as the Kronecker product of the two matrices. This matrix can then be used with the (independently identically distributed) i.i.d. channel matrix. This type of setup also has been validated by measurement campaigns. A similar technique is presented in Reference 24, where correlation matrices are generated based on the angle spread and antenna spacing at both mobile transmitter and base station. These matrices and the i.i.d. channel matrix are then combined to obtain a MIMO channel with correlated receive antennas.

A very useful model classification is presented in Reference 26, where the following models are introduced:

- Uncorrelated high-rank (UHR) model, which is also known as the i.i.d. model where all elements of the channel matrix are i.i.d. complex Gaussian.
- Correlated low-rank (CLR) model, given by $\mathbf{H} = g_{rx}g_{tx}^{*}\mathbf{u}_{rx}\mathbf{u}_{tx}$, where g_{rx} and g_{tx} are the receive and transmit fading coefficients modeled as independent Gaussian, and $\mathbf{u}_{rx}$ and $\mathbf{u}_{tx}$ are fixed deterministic vectors of size $m \times 1$ and $n \times 1$ with unit modulus entries.
- Uncorrelated low-rank (ULR) model, given by $\mathbf{H} = \mathbf{g}_{rx}\mathbf{g}_{tx}^{*}$, where $\mathbf{g}_{rx}$ and $\mathbf{g}_{tx}$ are independent receive and transmit fading vectors containing i.i.d. complex values components.

These three types of channels cover several MIMO scenarios. The rank of the channel matrix is an important property. For example, in the CLR case there are few independent channels and we obtain very little diversity or multiplexing gain. In the ULR case we have diversity but the capacity is lower than in the UHR case because every realization of the channel has rank 1. The above models can describe a variety of realistic scenarios. For example, during measurement campaigns it has been observed that high angular spread can be experienced at the mobile terminal. Consequently, diversity gain may be achieved even with short distances between the antenna elements. The angular spread at the base station is typically smaller and the correlation between the antennas is higher.

8.3   Signal Properties Used in Blind Equalization

In this section we consider various signal properties exploited in blind channel equalization and identification. Most properties are presented in the context of single-user receiver for the sake of simplicity. The extensions to MIMO systems are considered as well.

If no exact knowledge of the transmitted sequence is available, blind algorithms rely on some statistical and structural properties of the transmitted signal. Communication channels typically alter these properties and blind receiver algorithms then try to recover or restore these properties. Temporal signal properties are typically considered but if several receivers are available, spatial dimension may also be exploited. Consequently, spatial or space-time processing also may take place at the receiver.

The following signal properties are commonly used in blind channel estimation:

- Cyclostationarity
- Higher-order statistics (HOS)
- Bussgang statistics
- Finite alphabet property
- Constant modulus
- Shaping statistics at the transmitter
- Uncorrelatedness and independence
- Special matrix structures following from the system model

Each of these properties is discussed in more detail in the following subsections. Typically, blind methods exploit one or a combination of these properties in estimating the channel or the equalizer. In a multiuser communications context, the interference caused by other users is often highly structured. One may dramatically improve the performance by taking into account this property of interference, as in the multiuser detection introduced by Verdu.[27]

There are quite a few underlying assumptions needed in deriving blind receivers. The signals may be assumed to be either random or deterministic. Random sequences are typically assumed to be white and wide-sense stationary (WSS) when sampled at symbol rate. For efficiently source-coded signals where the redundancy is removed, this is a reasonable assumption. It also holds for many widely used channel coding schemes.[28]

In the case of deterministic data models, the sequences are typically required to have linear complexity and sufficient excitation to ensure the identifiability of the channel. In practice, it follows that the sample covariance matrix of the sequence is not rank deficient. The pulse shape employed at the transmitter and the upper bound for channel order are typically assumed to be known as well.

Channel identification or equalization using blind methods requires that information about both the channel amplitude and phase responses can be acquired from the received signal. Second-order statistics of a WSS process contain no phase information. Hence, one cannot distinguish between minimum phase and nonminimum phase channels. Therefore, other statistics of the signal have to be used to extract the phase response. If the multiple-output model resulting from oversampling or employing multiple receivers is used, the received signal typically possesses the cyclostationarity property; i.e., signal statistics such as the autocorrelation function are periodic and the phase information is retained. Moreover, communication signals are typically non-Gaussian. Hence, the higher-than-second-order statistics of the signal are nonzero, which may also be exploited in equalization. HOS retain the phase information as well.

8.3.1   Higher-Order Statistics

Early blind algorithms were either implicitly or explicitly based on HOS. In time domain, HOS are represented by higher-than-second-order cumulants and moments. Their frequency domain counterparts obtained by multidimensional Fourier transforms are called polyspectra and moment spectra. In most HOS-based equalization algorithms proposed so far, polyspectrum is employed. Higher-order statistics and spectra may not be a feasible approach to constructing practical equalizers. HOS have a large variance, and consequently large sample sets are needed to obtain reliable channel estimates. This is a severe drawback, in particular, in applications where the channel is time varying, data rates are high, or low computational complexity is needed. See Reference 4 for more details.

8.3.2   Bussgang Property

The first algorithms developed for *self-recovering* (blind) equalization were Bussgang methods, including Sato's algorithm.[29] Such equalizers employ higher-order statistics implicitly through zero-memory nonlinearities. The Bussgang property of a stochastic process means its autocorrelation function equals the cross-correlation function between the process and the output of a zero-memory nonlinearity $f(\cdot)$, i.e.,

$$E[x(k)x(k-i)] = E[x(k)f(x(k-i))].$$

Bussgang algorithms incorporate a structure including a cascade of an equalizer $d(k)$ and the nonlinearity $f(\cdot)$. The output of Bussgang nonlinearity is considered to be the desired response of the equalizer. An error signal $e(k)$

is formed and a quadratic error criterion is minimized. Consequently, the input–output relationship may be given as

$$y(k) = \sum_i h(i; k)s(k - i) + w(k),$$ (8.3)

where $y(k)$ is the sampled channel output, $h(i; k)$ is the channel impulse response at time k, $s(k)$ is the transmitted data sequence, and $w(k)$ is additive noise. The output of the equalizer is obtained by

$$x(k) = \sum_{i=0}^{L} d(i; k)y(k - i).$$ (8.4)

The equalizer coefficients may then be updated using the well-known LMS (least-mean square) algorithm as follows:

$$d(i; k + 1) = d(i; k) + \mu e(k)y(k),$$ (8.5)

where μ is the step size, $e(k) = f(x(k)) - x(k)$, and $f(\cdot)$ is a Bussgang nonlinearity. In the Sato algorithm the nonlinearity is of form $f(x(k)) = E[s(k)]^2/E[s(k)]\mathrm{sgn}(x(k))$. Also, maximizing the absolute value of the kurtosis of the signal subject to power constraint has been employed.[30]

8.3.3   Constant Modulus

The widely used constant modulus algorithm (CMA) was proposed independently in References 31 and 32. It aims at restoring the constant modulus property of a communication signal by penalizing for the dispersion of squared output magnitude from constant. Many commonly used modulation schemes such as GMSK (Gaussian minimum shift keying), M-PSK, and FM have a constant envelope property. The cost function is defined as follows:

$$J(k) = E[(|x(k)|^2 - \gamma)^2],$$ (8.6)

where γ is dispersion constant $\gamma = E[|s(k)|^4]/E[|s(k)|^2]$. The error term $e(k)$ in the LMS update described above is $e(k) = x(k)(\gamma - |x(k)|^2)$. The dispersion constant is selected depending on the modulation scheme, for example, $\gamma = 1.0$ for uniform M-PSK. The constant modulus cost function may be used to equalize source constellations that do not have a constant envelope, such as 16-QAM. Different dispersion constants then need to be employed. The constant modulus method is robust in the face of carrier-phase offset because the CMA cost function is insensitive to the phase of $x(k)$.

For finite-length symbol rate spaced equalization, the CM cost function is shown to have local minima, and hence convergence is not guaranteed. Fractionally spaced CMA equalizers avoid this problem; see Reference 4.

8.3.4   Second-Order Cyclostationary Statistics

Cyclostationary signals have the property that statistics, such as mean or auto-correlation function, are periodic. Many anthropogenic signals encountered in communications contain such periodicities. Conventional models for random processes such as WSS ignore this valuable information. Gardner[33] discovered the fact that non-minimum phase channel equalization/identification may be obtained from the second-order cyclostationary (CS) statistics of the received signal because cyclic autocorrelation function preserves the phase informa-tion. Cyclostationary statistics may be obtained by taking several samples per symbol (sampling rate P/T, where P is the oversampling factor) or by em-ploying multiple antennas at the receiver. This leads to vector valued signals and multiple-output signal models.

Blind equalizers typically use second-order cyclostationary statistics. Hence, smaller sample sizes are required than in HOS-based receivers, and the algo-rithms converge faster. The main drawback is that some channel types may not be identified; see Reference 34. In particular, the channel cannot be identified if the subchannels resulting from oversampling share common zeros. If the multiple-output model is obtained by using an antenna array at the receiver, this limitation is less severe. There is also a phase ambiguity involved in blind channel estimation that may be resolved, for example, by using differential coding. However, some loss in SNR results from using such a coding scheme.

The continuous-time received signal is

$$y(t) = \sum_{k=-\infty}^{\infty} s(k)h(t - kT) + w(t), \tag{8.7}$$

where $h(t)$ is the channel impulse response, $s(k)$ the sequence of information symbols and $w(t)$ additive white Gaussian noise (AWGN). The response $h(t)$ is assumed to be of finite length. By taking several samples per symbol interval from $y(t)$, the received sequence becomes cyclostationary. Fractionally spaced channel output resulting from oversampling by factor P may be written as

$$y\left(k\frac{T}{P}\right) = \sum_{l} s(l) h\left(k\frac{T}{P} - lT\right) + w\left(k\frac{T}{P}\right). \tag{8.8}$$

The output of the ith subchannel $h_i(k)$ can be written as

$$y_i(k) = \sum_{l=-\infty}^{\infty} s(l)h_i(k - l) + w_i(k), \quad \text{for } i = 0, \ldots, P - 1. \tag{8.9}$$

An equalizer $d_i(k)$ is employed for each subchannel $h_i(k)$.

An alternate vector representation is obtained by stacking P samples taken in each symbol interval to a P-dimensional vector $\mathbf{y}(k) = [y_0(k), \ldots, y_{P-1}(k)]^T$. Similarly, the noise vector may be written as $\mathbf{w}(k) = [w_0(k), \ldots, w_{P-1}(k)]^T$ and the channel impulse response in vector form as $\mathbf{h}(k) = [h_0(k), \ldots, h_{P-1}(k)]^T$ where each vector component is kth tap of each subchannel. By stacking N received vector samples into an $(NP \times 1)$-vector $\mathbf{y}_N(k) = [\mathbf{y}^T(k), \ldots, \mathbf{y}^T(k - N+1)]^T$ and doing the same for the noise vectors $\mathbf{w}_N(k) = [\mathbf{w}^T(k), \ldots, \mathbf{w}^T(k - N + 1)]^T$, we can write a matrix equation (see Reference 3):

$$\mathbf{y}_N(k) = \mathcal{H}_N \mathbf{s}_N(k) + \mathbf{w}_N(k). \tag{8.10}$$

The $(L + N - 1) \times 1$ input signal vector (where L is the length of the channel) is defined as $\mathbf{s}_N(k) = [s(k), s(k - 1), \ldots, s(k - L - N + 2)]^T$ and the channel coefficients are collected into a Sylvester resultant matrix with dimension $NP \times (L + N - 1)$

$$\mathcal{H}_N = \begin{bmatrix} \mathbf{h}(0) & \mathbf{h}(1) & \ldots & \mathbf{h}(L-1) & \mathbf{0} & \ldots & \mathbf{0} \\ \mathbf{0} & \mathbf{h}(0) & \ldots & \mathbf{h}(L-2) & \mathbf{h}(L-1) & \ldots & \mathbf{0} \\ \vdots & \vdots & \ddots & \vdots & \vdots & \ddots & \vdots \\ \mathbf{0} & \mathbf{0} & \ldots & \mathbf{h}(0) & \mathbf{h}(1) & \ldots & \mathbf{h}(L-1) \end{bmatrix}. \tag{8.11}$$

There are quite a few blind equalizers employing the above matrix model; see, for example, References 3, and 34 to 36. Next, two commonly used SOCS-based blind equalization methods are briefly outlined.

8.3.4.1   FS-CMA Algorithm

The FS-CMA algorithm is probably the simplest and very reliable fractionally spaced equalization method.[3,4,32] In case we have oversampling factor P, the equalizer taps d are updated using the stochastic gradient method as follows:

$$\mathbf{d}(k + 1) = \mathbf{d}(k) + \mu \mathbf{y}^*(k)x(k)(|x(k)|^2 - \gamma)^2, \tag{8.12}$$

where γ is the dispersion factor for the modulation scheme employed, $x(k)$ is the equalizer output, and

$$\mathbf{y}(k) = [y_k^1, \ldots, y_{k-(N-1)}^1, y_k^2, \cdots, y_{k-(N-1)}^2, \cdots, y_k^P, \cdots, y_{k-(N-1)}^P]^T, \tag{8.13}$$

where superscript denotes the subchannel, i.e., fractionally sampled data are organized on a subchannel basis.

8.3.4.2   Prediction Error Filtering

Slock proposed using linear prediction for blind equalization.[36] The value of the current received symbol is predicted on the basis of a set of previously received symbols. The desired response is thus the eventually received symbol. The filter coefficients are found by minimizing the prediction error. The transmitted sequence is assumed to be white. A prediction-error filter acts as a whitening filter for the process and restores uncorrelatedness in the sequence. For vector stationary processes in a multiple-output system the whitening achieves the equalization.[37] By solving the multichannel prediction error problem, zero forcing and MMSE equalizers can be obtained. A noiseless FIR SIMO channel can be perfectly equalized by a bank of FIR filters.[36]

8.3.5 Maximum Likelihood Methods

Maximum Likelihood estimators have desirable large sample properties, i.e., they are asymptotically optimal. Blind maximum likelihood channel estimators and equalizers have been derived both for deterministic signal and stochastic signal models. The resulting techniques are called deterministic maximum likelihood (DML) and stochastic maximum likelihood (SML) methods, respectively. In the deterministic case, both the input sequence and channel are deterministic but unknown parameters. Consequently, the number of unknowns grows at the arrival of each new observation. If the stochastic signal model is used, the input is assumed to be a random i.i.d sequence and its distribution is known. Then, the channel or equalizer tap coefficients are the only unknown parameters. The input sequence may be non-Gaussian and the phase information may be obtained from the higher-than-second-order statistics. If the random sequence is assumed to be Gaussian, the multiple-output model has to be used and co-primeness of subchannels is required to ensure the identifiability of the channel; see Reference 3. Noises are commonly assumed to be Gaussian.

Finding the maximum likelihood estimate in practice is difficult because the likelihood function may have multiple local minima. Typically, iterative techniques such as EM-algorithm are used. High-quality initial estimates or several starting values are needed to find the optimal solution instead of the local minimum. In DML, the problem of estimation of channel and input sequence may be simplified by estimating them separately.

8.3.6 Shaping Signal Statistics at Transmitter

Signal statistics may be altered at the transmitter to make channel estimation easier. For example, the cyclostationarity property may be transmitter induced; i.e., the transmitted signal is modified in periodic fashion so that the received discrete time signal is cyclostationary even when using symbol rate sampling.[38] This may be considered a precoding operation, which leads to easier channel estimation. The approach is commonly used in multicarrier systems, such as OFDM, in which case all the channels may be identified and the common-zeros problem of conventional SOCS methods is avoided. Also, special transmit filters may be designed to make interference cancellation (signal separation) easier, or to match the channel.

8.3.7   Finite Alphabet

The finite alphabet (FA) property of digital modulation schemes may be exploited in blind equalization. In the FA structure the source signal is chosen from a finite set such as ±1 for BPSK (binary phase shift keying) or a set of phase shifts for DQPSK signal. Methods using FA property fit the received data to the unknown channel taps and project the estimated symbols onto an FA. Typically, these methods estimate the channel and symbols in an alternating manner using the well-known least-squares method, or advanced versions of EM-algorithm such as SAGE. The FA property has also been used in conjunction with other signal properties such as cyclostationarity.

8.3.8   Special Matrix Structures

Multiple-output SIMO and MIMO models can lead to special matrix structures, for example, Block–Hankel or Block–Toeplitz matrices for channel coefficients or symbols. Blind equalization algorithms can be developed to exploit or enforce such structures. In addition, the full column rank property of channel matrix is a useful property. The column or row space of the data matrix may be used to find the column space of the channel matrix or the row space of the matrix where the symbols are stacked, respectively.

Subspace methods exploit the fact that the low-rank model is applicable, and one can perform the decomposition to signal and noise subspaces based on the pattern of the eigenvalues of the data correlation matrix, or singular values of the data matrix. Signal subspace is spanned by eigenvectors corresponding to dominant eigenvalues of the correlation matrix, or singular vectors corresponding to dominant singular values of the data matrix. The same subspace is spanned by the columns of the channel matrix $\mathcal{H}_N$.[35,39] Well-known signal or noise subspace estimation methods such as MUSIC may be applied to determine the channel coefficients. This approach is explained in more detail in the context of blind MIMO channel identification algorithms.

8.4   Receivers for MIMO Systems

8.4.1   I-MIMO and FIR-MIMO

Over the years, several models[40] of the transmission process have been used. We start from the basic linear model that relates the received signals and the transmitted ones by

$$\mathbf{y}_k = \mathbf{H}\mathbf{s}_k + \mathbf{w}_k, \tag{8.14}$$

where $\mathbf{H}$ is the channel matrix associated with m transmitter, n receiver MIMO system, $\mathbf{s}$ is a column vector of m transmitted signals, $\mathbf{y}$ is a column vector

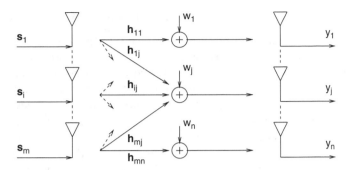

FIGURE 8.2
FIR-MIMO model.

of n received signals, and k is the time index. If the elements of matrix $\mathbf{H}$ are complex scalars, we have an instantaneous MIMO (I-MIMO) model, which is typical in systems with frequency-flat fading. The channel may, however, be time varying.

If the elements of the channel matrix $\mathbf{H}$ are FIR filters, the n-dimensional vector of received signals $\mathbf{y}_k$ is assumed to be produced from the m-dimensional vector of transmitted signals using the following z-domain representation:

$$\mathbf{y}(z) = \mathbf{H}(z)\mathbf{s}(z). \tag{8.15}$$

Assuming that the channels from transmitter i to receiver j have equal order L (Figure 8.2), we have

$$\mathbf{y}_k = \sum_{l=0}^{L} \mathbf{H}^l \mathbf{s}_{k-l} + \mathbf{w}_k. \tag{8.16}$$

This model, shown in Figure 8.2, may be called the finite impulse response multiple-input multiple-output (FIR-MIMO) model. Finally, if the channel matrix is allowed to be time varying, we will use the notation $\mathbf{H}^l = \mathbf{H}_k^l$. The channel matrix is typically assumed to have a full column rank. For individual transmit vector component sequences, similar assumptions as in the SISO case earlier in this chapter are usually employed. Depending on the transmission and channel coding scheme employed, assumptions on dependencies or independence among the transmitted signal component may be employed in deriving the receiver algorithms. Obviously, space-time coding introduces redundancy, whereas in spatial multiplexing schemes the signal components are assumed to be independent.

Many of the blind methods derived for SISO and SIMO models have been extended to MIMO models. For example, there are blind subspace methods, prediction error filtering methods, as well as CMA-related methods for MIMO systems; see References 1 and 2 for more detail.

8.4.2   Subspace Method for Blind Identification of MIMO Systems

In blind MIMO system identification, the input–output relationship may be rewritten in the form of a low-rank model if a sufficient number of received data samples are available. The fact that the column space of the received data matrix and column space of the channel matrix span the same subspace may then be exploited in determining the parameters of the FIR-MIMO model. A well-known noise subspace method is used to determine the channel matrix up to an ambiguity matrix. In blind identification some ambiguities always remain unsolved. In blind subspace methods where the low-rank model is used, these ambiguities may be described in a form of constant full-rank $m \times m$ ambiguity matrix, where m is the number of transmitted sources.

The remaining ambiguity matrix can be considered to perform instantaneous linear mixing of the equalized signals. Consequently, we have an I-MIMO model with non-Gaussian communication signals that may be demixed via blind source separation (BSS).[5] To accomplish the separation, the sources are assumed to be statistically independent, which is reasonable if signals originate from different users.

8.4.2.1   Signal Model

We assume the standard FIR-MIMO baseband signal model[39] with m transmitters and n receivers, in which the received signal $\mathbf{y}(k)$ having n components arranged as a column vector is of the form

$$\mathbf{y}(k) = \sum_{l=0}^{L} \mathbf{H}_l \mathbf{s}(k - l) + \mathbf{w}(k). \tag{8.17}$$

Here $\mathbf{s}(k) = [s_1(k), s_2(k), \ldots, s_m(k)]^T$ is an m-dimensional signal vector ($n > m$), L is the channel order, $\{\mathbf{H}_l\}_{l=0,\ldots L}$ are the unknown $n \times m$ matrix-valued impulse response coefficients associated with the transfer function $\mathbf{H}(z) = \sum_{l=0}^{L} \mathbf{H}_k z^{-l}$, and $\mathbf{w}(k)$ is noise. Let us make the following notation: $\bar{\mathbf{H}} = [\mathbf{H}_0^T, \mathbf{H}_1^T, \ldots, \mathbf{H}_L^T]^T$.

We assume that

$$\text{rank}(\bar{\mathbf{H}}(z)) = m \quad \text{for each } z \tag{8.18}$$

$$\bar{\mathbf{H}}(L) \quad \text{is of full column rank.} \tag{8.19}$$

These assumptions are needed to ensure the channel identifiability by using only second-order statistics of the received signal vector $\mathbf{y}(k)$; see Reference 39. By stacking $N+1$ observations of Equation 8.17 into an $(N+1)n \times 1$ vector $\mathbf{Y}(k) = [\mathbf{y}^T(k), \mathbf{y}^T(k-1), \ldots, \mathbf{y}^T(k-N)]^T$ we may write

$$\mathbf{Y}(k) = \mathcal{H}_N(\bar{\mathbf{H}})\mathbf{S}(k) + \mathbf{W}(k). \tag{8.20}$$

Here, $\mathbf{S}(k) = [\mathbf{s}(k)^T, \mathbf{s}(k-1)^T, \ldots, \mathbf{s}(k-N-L)^T]^T$, $\mathbf{V}(k) = [\mathbf{v}(k)^T, \mathbf{v}(k-1)^T, \ldots, \mathbf{v}(k-N)^T]^T$, and $\mathcal{H}_N(\bar{\mathbf{H}})$ is the $(N+1)n \times m(L+N+1)$ channel matrix (Sylvester matrix) given by

$$\mathcal{H}_N(\bar{\mathbf{H}}) = \begin{bmatrix} \mathbf{H}_0 & \mathbf{H}_1 & \cdots & \mathbf{H}_L & \mathbf{0} & \cdots & \mathbf{0} \\ \mathbf{0} & \mathbf{H}_0 & \mathbf{H}_1 & \cdots & \mathbf{H}_L & \ddots & \mathbf{0} \\ \vdots & \ddots & \ddots & \ddots & \ddots & \ddots & \mathbf{0} \\ \mathbf{0} & \cdots & \mathbf{0} & \mathbf{H}_0 & \mathbf{H}_1 & \cdots & \mathbf{H}_L \end{bmatrix}.$$

Assumptions 8.18 and 8.19 imply that the channel matrix is of full column rank when the stacking parameter N is chosen large enough.

8.4.2.2 Channel Identification

We now review the basic steps of a subspace-based identification method[35] later extended to MIMO systems.[39] Assume that $\mathbf{Y}(k)$ given in Equation 8.20 is a WSS process and the signal $\mathbf{S}(k)$ and noise $\mathbf{W}(k)$ are mutually independent and of zero mean. The covariance matrix of $\mathbf{Y}(k)$ is

$$E[\mathbf{Y}(k)\mathbf{Y}(k)^H] = \mathbf{R_y} = \mathcal{H}_N(\bar{\mathbf{H}})\mathbf{R_s}\mathcal{H}_N(\bar{\mathbf{H}})^H + \mathbf{R_w},$$

where $\mathbf{R_s} = E[\mathbf{S}(k)\mathbf{S}(k)^H]$ is the signal covariance matrix, which is assumed to be of full rank, and $\mathbf{R_w} = E[\mathbf{W}(k)\mathbf{W}(k)^H]$ is the noise covariance matrix of the form $\mathbf{R_w} = \sigma^2\mathbf{I}$, where σ^2 is the noise power. The maximum channel order L is assumed to be known as well.

Based on the pattern of the eigenvalues of the received signal covariance matrix $\mathbf{R_y}$, one can perform the decomposition to signal and noise subspaces. The signal subspace spanned by eigenvectors corresponding to the $m(L + N + 1)$ largest eigenvalues spans the same space as the columns of the channel matrix $\mathcal{H}_N(\bar{\mathbf{H}})$. The remaining $r = n(N + 1) - m(L + N + 1)$ eigenvectors span the noise subspace. The corresponding eigenvalues are all equal to the noise variance σ^2.

Denote the noise subspace eigenvectors by $\mathbf{g}_i$, $i = 1, \ldots, r$. It is a standard result that

$$\mathcal{H}_N^H(\bar{\mathbf{H}})\mathbf{g}_i = \mathbf{0}, \quad i = 1, \ldots, r.$$

This orthogonality of the signal and noise subspaces allows for identification of the channel matrix $\bar{\mathbf{H}}$ up to a right multiplication of an invertible $m \times m$ ambiguity matrix.

To illustrate how the identification is done, partition the noise subspace eigenvectors as

$$\mathbf{g}_i = \left[\mathbf{g}_0^{(i)^T}, \mathbf{g}_1^{(i)^T}, \ldots, \mathbf{g}_N^{(i)^T} \right]^T, \tag{8.21}$$

where $\mathbf{g}_l^{(i)}, l = 0, 1, \ldots, N$ are of size $n \times 1$. Define

$$
\mathcal{G}_i = \begin{bmatrix} \mathbf{g}_0^{(i)} & \mathbf{g}_1^{(i)} & \cdots & \mathbf{g}_N^{(i)} & 0 & \cdots & 0 \\ 0 & \mathbf{g}_0^{(i)} & \mathbf{g}_1^{(i)} & \cdots & \mathbf{g}_N^{(i)} & \ddots & \vdots \\ \vdots & \ddots & \ddots & \ddots & \ddots & \ddots & 0 \\ 0 & \cdots & 0 & \mathbf{g}_0^{(i)} & \mathbf{g}_1^{(i)} & \cdots & \mathbf{g}_N^{(i)} \end{bmatrix}. \tag{8.22}
$$

It can be shown[35,39] that for each column of the matrix $\bar{\mathbf{H}}$, $\mathbf{h}_i$, $i = 1, \ldots, m$,

$$
\mathbf{h}_i^H \left(\sum_{i=1}^{r} \mathcal{G}_i \mathcal{G}_i^H \right) \mathbf{h}_i = 0.
$$

Moreover, the dimension of the null space of the matrix

$$
\mathcal{C} = \sum_{i=1}^{r} \mathcal{G}_i \mathcal{G}_i^H
$$

is m. This implies that

$$
\mathbf{H} = \mathbf{B}\mathbf{R}^{-1},
$$

where $\mathbf{B}$ is a $n(L + 1) \times m$ matrix of the eigenvectors of $\mathcal{C}$ corresponding to the eigenvalues that are equal to zero, and $\mathbf{R}$ is an invertible $m \times m$ matrix. In other words, by using the noise subspace eigenvectors, we may determine the channel matrix up to a right multiplication of an invertible $m \times m$ ambiguity matrix, i.e., determine

$$
\mathbf{B} = \mathbf{H}\mathbf{R}. \tag{8.23}
$$

8.4.2.3 *Equalization Using Subspace Method and Blind Source Separation*

Assume that we know the channel matrix up to a right multiplication of an invertible $m \times m$ matrix $\mathbf{R}$; i.e., we know $\mathbf{B} = \mathbf{H}\mathbf{R}$. Then

$$
\mathcal{H}_N(\mathbf{B}) = \mathcal{H}_N(\mathbf{H})\bar{\mathbf{R}},
$$

where $\bar{\mathbf{R}}$ is an $m(L + N + 1) \times m(L + N + 1)$ block diagonal matrix

$$
\bar{\mathbf{R}} = \begin{bmatrix} \mathbf{R} & 0 & \cdots & 0 \\ 0 & \mathbf{R} & \cdots & 0 \\ \vdots & & \ddots & 0 \\ 0 & 0 & \cdots & \mathbf{R} \end{bmatrix}.
$$

Consider the noise-free case of the signal model (Equation 8.20)

$$\mathbf{Y}(k) = \mathcal{H}_k(\bar{\mathbf{H}})\mathbf{S}(k).$$

Let $\mathcal{H}_N(\mathbf{B})^\dagger$ be the pseudo-inverse of the matrix $\mathcal{H}_N(\mathbf{B})$, i.e.,

$$\mathcal{H}_N(\mathbf{B})^\dagger = \left(\mathcal{H}_N(\mathbf{B})^H \mathcal{H}_N(\mathbf{B})\right)^{-1} \mathcal{H}_N(\mathbf{B})^H.$$

Now it is easy to see that

$$\mathcal{H}_N(\mathbf{B})^\dagger \mathbf{Y}(k) = \mathbf{X}(k) = \bar{\mathbf{R}}^{-1}\mathbf{S}(k)$$

or

$$\mathbf{S}(k) = \bar{\mathbf{R}}\mathcal{H}_N(\mathbf{B})^\dagger \mathbf{Y}(k).$$

Note that

$$\bar{\mathbf{R}}^{-1} = \begin{bmatrix} \mathbf{R}^{-1} & \mathbf{0} & \cdots & \mathbf{0} \\ \mathbf{0} & \mathbf{R}^{-1} & \cdots & \mathbf{0} \\ \vdots & & \ddots & \mathbf{0} \\ \mathbf{0} & \mathbf{0} & \cdots & \mathbf{R}^{-1} \end{bmatrix}$$

and consider, for example, m first components of $\mathbf{Y}(k)$. Denote these components by $\mathbf{y}(k)$. Then

$$\mathbf{y}(k) = \mathbf{R}^{-1}\mathbf{s}(k).$$

This equation shows that by assuming that the m source signals are statistically independent and non-Gaussian, we may estimate the signals $\mathbf{s}(k)$ up to a permutation and complex scaling. In other words, by combining subspace-based channel estimation and BSS, we may find an estimate

$$\hat{\mathbf{s}}(k) = \mathcal{S}\mathcal{P}\mathbf{s}(k),$$

where $\mathcal{S}$ is a complex-valued diagonal matrix and $\mathcal{P}$ is a real-valued permutation matrix. The separation task may be performed by any BSS algorithm, for example, EASI.[41] The remaining rotation and scaling of the constellation pattern may be resolved by using differential coding and the FA property of the communication signals.[42] In QAM modulation, the independence of I and Q components may also be exploited in resolving phase ambiguity.

8.4.3 Multichannel Blind Deconvolution

Much research has been done in the area of multichannel blind deconvolution and BSS.[43,44] Various scenarios have been considered, ranging from the SIMO model, obtained by oversampling at the receiver or by using several receivers,[45] to the MIMO case.[46] Different techniques based on HOS, subspace decomposition, or multichannel frequency-domain deconvolution have been reported in the literature.[43] Typically, assumptions include linear time-invariant (LTI) systems, infinite SNR, and infinite equalizer length.[46] In contrast, in real communication systems channels are time varying (TV), SNR values are rather low in typical operation conditions, and equalizer lengths are finite.

Using the model described by Equation 8.16, the goal of blind deconvolution is to estimate transmitted signals using a multichannel linear filter of the form:

$$\mathbf{x}_k = \sum_{l=0}^{L} \mathbf{W}_k^l \mathbf{y}_{k-l}, \tag{8.24}$$

where $\mathbf{W}_k^l$ are the $(m \times n)$ matrix coefficients of the separating system and L is the filter order.

In the BSS research community, several methods for solving this problem have been proposed; see Reference 44. Transmitted signals are assumed to be statistically independent and non-Gaussian. Typically, the number of transmitters m must be fewer than or equal to the number of receivers n. The convolution in time domain may be changed into multiplication by doing the processing in the frequency domain. Then, in the frequency domain we have an I-MIMO model that may be solved more easily. The I-MIMO mixing matrix needs to be of full rank. A solution based on BSS and natural gradient adaptation is proposed in Reference 48. For more detail, see Reference 44.

8.5   Blind Receivers for GSM/EDGE

GSM is one of the most widely used wireless communication systems. GSM is a TDMA system where each frequency band is shared by eight users, separated in time by their non-overlapping timeframes. Each frame consists of two 58-bit data streams and a mid-amble of 26 bits length used for synchronization and channel estimation. At first glance it can be observed that more than 20% of the data burst is not used for transmitting data bits. Therefore, schemes that can improve the effective data rate are desired.

Blind equalization is typically designed only for linear modulations, like QAM. In such a system the received signal from a channel with the impulse response $h(t)$ is

$$y(t) = \sum_{k=-\infty}^{\infty} a_k h(t - kT) + w(t), \qquad a_k \in \mathcal{A}, \tag{8.25}$$

where T is the symbol period and $\mathcal{A}$ is the input constellation set. The input sequence a_k is i.i.d. sequence while the noise $w(t)$ is WSS, white, and independent of the input sequence but not necessarily Gaussian. The composite channel impulse response $h(t)$ contains the transmitter and receiver filters as well as the physical channel response. In GSM, nonlinear GMSK modulation is used. Therefore, the direct application of the linear blind equalization schemes may not be feasible. In general, when blind equalization is considered for GSM, a linear approximation of the GMSK signal is performed first.[49–52]

8.5.1 Linear Approximation of GMSK Signals

The baseband GMSK signal is given by

$$s(t) = \exp\left[j\frac{\pi}{2}\sum_{n=-\infty}^{\infty}\alpha_n\psi(t-nT)\right], \qquad (8.26)$$

where α_n is the binary data for transmission. The continuous phase modulation pulse $\psi(t)$ is

$$\psi(t) = \int_{-\infty}^{t} c(\tau - 2T)$$

$$c(t) = g(t) * \text{rect}(t/T)$$

$$\text{rect}(t/T) = \begin{cases} 1/T, & |t| \le T/2 \\ 0, & |t| > T/2 \end{cases}$$

$$g(t) = B\sqrt{\frac{2\pi}{\ln 2}}\exp\left[-\frac{2\pi^2 B^2 t^2}{\ln 2}\right]. \qquad (8.27)$$

$$(8.28)$$

The excess bandwidth parameter B is set to 0.3 in GSM so that

$$\psi(t) \approx \begin{cases} 0, & t \le 0; \\ 1, & t \ge 4T. \end{cases} \qquad (8.29)$$

The baseband GMSK signal (Equation 8.26) can be approximated with a sum of 16 terms, each constituting a linear QAM signal (see Reference 50 for details). Among the 16 different pulses, only two pulses are significant while the others contain very little energy.

Retaining these two most significant pulses, the linear approximate model for GMSK with $B = 0.3$ can be written as

$$s(t) \approx \sum_{n=-\infty}^{\infty} a_{0,n} p_0(t - nT) + \sum_{n=-\infty}^{\infty} a_{1,n} p_1(t - nT), \qquad (8.30)$$

where

$$a_{0,n} = \exp\left(j\frac{\pi}{2}\sum_{k=-\infty}^{n}\alpha_k\right) = j\alpha_n a_{0,n-1}$$

$$a_{1,n} = j\alpha_n \exp\left(j\frac{\pi}{2}\sum_{k=-\infty}^{n-2}\alpha_k\right) = j\alpha_n a_{0,n-2},\qquad(8.31)$$

and $p_0(t)$ and $p_1(t)$ are the two most significant pulses. More than 99% of the approximated GMSK signal energy is contained in the $p_0(t)$ pulse; then, the signal can be further simplified taking into account only the first term in the approximation (Equation 8.30)

$$s(t) \approx \sum_{n=-\infty}^{\infty} a_n p_0(t - nT).\qquad(8.32)$$

With this approximation the maximum achievable SNR is 23 dB even in the noiseless case.

8.5.2   SOS-Based Blind Equalization for GSM

For SOS-based blind equalization, the necessary channel diversity can be obtained by oversampling if only one receive antenna is available. The pulse shape p_0, and therefore the linearized GMSK signal, has little excess bandwidth beyond the $1/2T$ limit. Oversampling with a rate higher than $1/T$ will not generate enough diversity. In Reference 50, and later in Reference 52, the required diversity is obtained using a simple derotation scheme on the received signal and then considering the I-Q branches as separate subchannels.

The sequence $a_n = j\alpha_n a_{n-1}$ is a pseudo-quaternary shift keying (QPSK) sequence because, at any given time, a_n can only take two values rather than four. It can be written also as $a_n = j^n\tilde{a}_n$ where $\tilde{a}_n = \pm1$ is a binary-phase shift-keying (BPSK) sequence. With this definition, the received baud rate-sampled signal is given by

$$x_n = \sum_{k=-\infty}^{\infty} h_k j^{n-k}\tilde{a}_{n-k} + w_n = j^n \sum_{k=-\infty}^{\infty} [h_k j^{-k}]\tilde{a}_{n-k} + w_n.\qquad(8.33)$$

At the receiver part we first derotate the received baud-sampled signal:

$$\tilde{x}_n = j^{-n}x_n = \sum_{k=-\infty}^{\infty} [h_k j^{-k}]\tilde{a}_{n-k} + j^{-n}w_n.\qquad(8.34)$$

Because $\tilde{a}_n$ is a real sequence we obtain two subchannel outputs by taking the real and the imaginary part of the received derotated sequence. The derotation scheme results in a single-input two-output (SITO) model, which can be used in blind equalization.

In Reference 50 several blind equalization as well as blind channel identification and MLSE detection schemes have been studied based on the derotation scheme. Three of them are based on second-order statistics (SOS), and the fourth is a SIMO HOS-based blind equalizer. The performance is studied over channels considered to be time invariant during each frame. The COST 207 channel model[16] has been used to generate the channels. The necessary condition for the channel impulse response, such that the two derotated subchannels can be identified using the SOS, is given as well. Among the linear equalizers, CMA gives the best performance. The other algorithms are sensitive to the channel order estimation. The nonlinear blind equalizers based on blind channel identification and MLSE detection perform much worse than the traditional trained channel estimator.

When the subchannels resulting from oversampling or multiantenna reception or derotation share common zeros, the SOS-based blind equalization algorithms fail. In GSM the channels are time varying, and therefore such situations may be encountered. Li and Ding[51] proposed in a semiblind approach to equalization for GSM. The main idea is that we can use some of the training symbols as well as the SOS of the output signal to eliminate the identifiability and ambiguity problems. The channel diversity is obtained using the same derotation scheme. There are two important achievements of semiblind equalizers: the training sequence can be 10 bits shorter than in the standard and they achieve better performance in the presence of channel order mismatch and subchannel common zeros.

Based on the same GMSK signal linear approximation and derotation as in Reference 50, Trigui and Slock[52] derived performance bounds for co-channel interference cancellation in GSM. They showed that whenever the number of channels is equal to or larger than the number of users, then the suboptimal single-user detector, which takes the spatiotemporal correlation structure of the interferences correctly into account, suffers a bounded matched filter loss.

8.5.3 HOS-Based Blind Equalization for GSM

In mobile communications systems, channel estimation and equalization —both blind and nonblind techniques—should satisfy the following requirements:

- Reliable estimation of the channel impulse response has to be obtained from a few samples of the received signal.
- It has to apply to arbitrary channels, whether or not they are singular, critical, etc.

Among the different blind equalization techniques, those based on HOS satisfy the two above conditions.

A comprehensive study of the reliability of HOS-based blind equalizers to GSM is given in Reference 49. They compare the performance of two blind channel estimators (EVI[53] and WS[54]) with that of two nonblind techniques (least-squares and cross-correlation based). For the blind techniques a data burst contains 142 data bits only, without training sequence, whereas for the nonblind techniques a standard GSM burst with a mid-amble of 26 training bits is transmitted. The channels are considered to be time invariant over a data burst duration, and they were generated according to Reference 16. The experimental results show that the blind channel estimator EVI entails a mean SNR loss of 1.2 to 1.3 dB in comparison to the least-squares solution. Furthermore, its performance remains unaffected by the co-channel interferences at signal-to-interference ratios (SIR) larger than 10 dB. The HOS-based blind equalizers converge relatively slowly. The best need an entire GSM data frame to converge.

An improved blind channel estimation method for GSM using EVI has been reported.[55] A technique similar to turbo coding is used to iteratively improve the channel estimate obtained by the blind method. It also solves the (complex) scalar ambiguity inherent in all blind channel estimation approaches.

The GSM standard has been enhanced to allow higher data rates and to provide quality services by allocating a larger bandwidth (e.g., PCS-1900) or by using higher-order modulations (8-PSK in EDGE). Then the equalization task becomes more difficult because the Doppler spread increases, making the channels time variant even within the data burst. Higher-output SNR and high-quality channel estimates have to be ensured in order to deal with higher-order constellations such as 8-PSK. At this moment it is difficult to believe that a completely blind solution can satisfy all these conditions. The semiblind approach may be a more feasible solution for practical systems. It saves a significant part of the transmission spectrum and provides better overall performance. However, a good blind method is the core of a semiblind method. Hence, further development of blind methods is well justified.

8.6   Blind Receivers for CDMA

Radio access in the third generation of mobile communications is based on CDMA. Therefore, there has been a great deal of research work on CDMA in recent years. The basic properties of CDMA will not be considered here because they are presented in many text books, e.g., Reference 56. In the following we describe the characteristics of the downlink and uplink channels in the context of channel estimation. Then blind receiver structures for CDMA systems are briefly reviewed. In particular, blind channel estimation and blind multiuser detection (MUD) are addressed. Blind MUD is a method to cancel the effect of other users' signals in a CDMA system without any knowledge of the transmitted information symbols. Some blind MUD methods also perform blind channel estimation. The problems in developing blind receivers are different in downlink and uplink cases. Furthermore, different methods for periodic (short) code and aperiodic (long) code systems are needed.

8.6.1 Downlink

In downlink, the base station is transmitting signals to all the received users in a synchronous manner. The signal received by a certain user, sampled at the chip rate, can be written as

$$y(k) = \sum_{p=1}^{P} \sum_{l=1}^{L_h} h(k; l) s_p(k - l) + w(k), \tag{8.35}$$

where P is the number of active users in the cell, $h(k; l)$ is the time-varying channel impulse response, L_h is the channel length in chip periods, and $w(k)$ is the additive noise. The transmitted chip sequence of the pth user $s_p(k)$ is given by

$$s_p(k) = \sum_{m} b_p(m) c_p(k - mL_c), \tag{8.36}$$

where $b_p(m)$ is the mth information symbol transmitted by the pth user, $c_p(k)$ is the spreading code of the pth user, and L_c is the processing gain. A particular characteristic of the downlink is that the channels are the same for all users. Mitigation of ISI is an important problem at high data rates, i.e., when the spreading factor is low.

8.6.2 Uplink

In uplink, the base station is receiving a superposition of the active users' signals. The received signal sampled at the chip rate can be written as

$$y(k) = \sum_{p=1}^{P} \sum_{l=1}^{L_h} h_p(k; l) s_p(k - l - \tau_p) + w(k). \tag{8.37}$$

In this case the channels from all the users are different. The users are also characterized by their propagation delays τ_p. The base station has to estimate the channels and the propagation delays simultaneously for all the users. We can easily see that the downlink channel of a CDMA network is a special case where the channel impulse responses are equal for all the users and the relative propagation delays are zero.

8.6.3   Signal Properties

When deriving blind channel estimation, equalization, or MUD techniques, one should exploit as many signal properties and diversity modes as possible. In downlink, only the code of the desired user is typically assumed to be known. In uplink, all the spreading codes are known. This information may be used, for example, to solve the ambiguities inherent in blind channel estimation. The spreading codes are designed to have high autocorrelation and low cross-correlation properties. The system may use short codes where the code is the same for each symbol interval. This assumption simplifies the system model and reduces the algorithm complexity. In practice, however, an aperiodic code (long code) unique for each base station may be used in downlink to identify the users belonging to a certain cell. In uplink, each mobile terminal is given a unique long code. Therefore, the overall spreading codes are in fact aperiodic. Long spreading codes are typically modeled as stochastic processes. Using such codes reduces the intercell interference and randomizes the chip sequences.

Both spatial and time-domain signal properties may be exploited. The principles of the space–time processing for wireless communications as well as the space–time characterization of the transmission channel are reviewed in Reference 21. The spatial diversity obtained by using multiple receive antennas is often considered in uplink. The system model may then be presented as a MIMO system, depicted in Figure 8.3. Because of the size limitations of the mobile terminals, multiple antenna receivers are rarer in downlink. However, mobile terminals are typically in rich scattering environments and would gain significantly from the diversity.

Many blind equalization/identification techniques have been proposed for CDMA systems, especially for short code systems. We focus on blind methods that exploit CDMA-specific characteristics together with statistical properties of the signals.

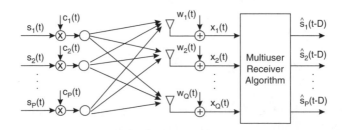

FIGURE 8.3
MIMO system model characterization of uplink CDMA.

8.6.4 Blind Channel Estimation for CDMA

The blind equalization/identification methods for CDMA proposed so far in the literature employ second-order statistics of the received sampled signals. The complexity of the HOS-based methods would significantly increase the already high computational complexity. In the context of long codes, correlation matching, subspace estimation, and maximum likelihood methods have been used.

In Reference 57, both blind and pilot-aided channel estimation is considered in the context of long-code systems. A two-dimensional RAKE receiver is proposed in a scenario where the base station is equipped with multiple antennas. At the output of each antenna array element, there is a linear FIR filter for each user. The filter coefficients for the user of interest are adjusted so that their combined output produces the symbol-modulated spreading sequence. The MMSE equalizer coefficients are found using the Wiener–Hopf equation where the cross-correlation vector is obtained from the correlation matrices of the pre- and post-despread signal. A principal component algorithm is proposed for blind estimation of the cross-correlation vector. The blind identification relies on the property that the pre- and post-despreading covariance matrices of the antenna outputs are different only by a rank-one matrix defined by the desired cross-correlation vector. A deterministic least-squares approach to blind equalization is developed, as well, but only for the case of an underloaded cell.

Related subspace techniques for channel estimation in CDMA with aperiodic spreading codes have been proposed.[58,59] A preprocessing stage using pseudo-inversion of the code matrix is employed in Reference 58. Channel is identified using a noise subspace method. Both blind and data-aided methods are presented and their performance is compared with the theoretical performance bounds. Multicode transmission and multiantenna receivers using subspace channel estimation are considered in Reference 59. A partial despreading preprocessing stage leads to significant savings in computational complexity and improved robustness to code selection in comparison to that described in Reference 58.

8.6.5 Blind MUD

With the traditional RAKE receiver,[60] the user of interest is coherently solving the multipaths of the received signal. It uses the maximum ratio combining (MRC) to deliver optimal performance. Unfortunately, the RAKE receiver is very sensitive to multiuser interference (MUI). Therefore, more powerful techniques are needed to cope with the MUI. The MUD techniques[27] exploit both the code knowledge and the channel characteristics of each user to improve the overall performance.

Madhow[61] has presented an excellent review of the blind adaptive techniques for channel identification and interference suppression in CDMA.

He classifies the receivers based on the amount of prior knowledge available. Adaptive interference suppression techniques are also considered by Honig[63] and Poor and Wornell.[62] An overview of the multiuser receivers for short code CDMA may be found in Reference 63, where different approaches to blind and data-aided methods are compared.

Honig proposed the first blind MUD technique for short-code CDMA; see Reference 63. The constrained minimum output energy approach related to MVDR beamforming was developed. The desired user's channel response was assumed known. Madhow[61] combined blind MUD with blind channel estimation.

A subspace method for blind MUD and channel identification has been proposed by Wang and Poor.[64] When certain conditions are fulfilled, such as the channel matrix has a full column rank, the channel matrix is shown to be blindly identifiable up to a nonsingular ambiguity matrix. A multiuser extension of the well-known algorithm of Reference 35 is developed, and using the knowledge of the user spreading codes, the ambiguity inherent to blind identification may be solved in both uplink and downlink cases. In uplink, the users' propagation delays are assumed known. In the special case of flat fading, a method for estimating the propagation delays is derived as well. Projecting the users' estimated channels into the signal subspace, MMSE and zero-forcing detectors are also derived and their robustness is analyzed. Torlak and Xu[65] have proposed a blind subspace method for blind channel identification in asynchronous CDMA. They show that the subspaces of the data matrix contain sufficient information for unique determination of channel. There are typically three steps in the subspace MUD computations: channel delay estimation, the delayed components' gain estimation, and the actual MUD projection. All steps exploit the signal subspace of the received sequence.

A blind MUD in dispersive channels has been proposed.[66,67] Only the desired user's code needs to be known. The receiver is found as a solution to a special MMSE cost function minimization in two stages. The signal subspace as well as the code subspace properties are used to define the cost function, thus allowing for improved performance at low SNR. Several receive antennas may be used.

Papadias and Huang[68] have proposed two linear adaptive space–time approaches to MUD in CDMA. The optimum space–time MMSE MUD is first derived and its bit error rate (BER) performance is analyzed. Two adaptive implementations of the MMSE detector are developed: a blind multiuser CMA (MU-CMA) and a trained MU-LMS. The global convergence of the blind method is proved as well.

If there are multiple antennas available at the transmitter, space–time codes can be used to improve the system performance. In Reference 69 the properties of a space–time block coding scheme are used with the spreading code properties for canceling various interferences in a CDMA downlink with multiple antennas at the base station. The increased interuser interference due to the increased number of antennas and the self-interference have to be canceled. The ZF and MMSE multiuser receiver are derived assuming knowledge of the channels. The proposed blind Capon receiver for space–time block coded systems requires only knowledge of the spreading codes and the timing of the desired user. It is able to suppress the overall interference (MUI, ICI, and narrowband interference). The basic idea is to design a filter bank such that it passes the signal of interest without distortion while suppressing the interference in a nonparametric manner. In the noise-free case, the Capon channel estimator is unique and exact (up to a scalar). The scalar ambiguity may be solved using a semiblind method.

8.7 OFDM

8.7.1 Transmission Model for OFDM Systems

Multicarrier techniques, and especially OFDM, are considered feasible alternatives for the future, beyond third-generation mobile communications. A great deal of research is devoted to developing these techniques for various communication systems. Currently, OFDM has been used in high-speed modems (used for digital subscriber lines[70]), digital audio and video broadcasting, as well as in 5-GHz broadband wireless local area network standards.[71]

In single-carrier systems, an equalization stage is very important in mitigating ISI. Depending on the techniques applied, this stage may have high computational complexity. In addition, the equalizer coefficients have to be updated frequently if the channel is time varying. The benefit of OFDM is that it turns the frequency-selective channel into a set of parallel narrowband channels, which leads to very simple equalization schemes. A simple OFDM transmission diagram for a SISO case is presented in Figure 8.4.

The principle of OFDM is based on the following setup. The original data stream is split into a number of lower rate data streams $\mathbf{a}$. An N-point IDFT (inverse discrete Fourier transform) is applied to N complex data symbols at the transmitter, creating the kth modulated OFDM block $\tilde{\mathbf{x}}(k) = \mathbf{F}_N \mathbf{a}(k)$.

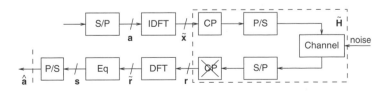

FIGURE 8.4
OFDM transmission diagram, SISO case.

A cyclic prefix (CP) of length L (representing last L samples of the IDFT output) is added to overcome interblock interference and intersymbol interference. As a result, the total length of the OFDM symbol is $M = N+L$. The CP acts as a guard space between two consecutive OFDM symbols and maintains the orthogonality among the subcarriers. If the channel length is less than the CP length, by removing the CP at the receiver we also remove the interblock interference. Hence, in the receiver the CP is removed and an N-point DFT is applied. The received $N \times 1$ signal block after CP insertion, followed by transmission on the wireless channel and CP removal, is expressed as

$$\mathbf{r}(k) = \widetilde{\mathbf{H}}\, \tilde{\mathbf{x}}(k) + \mathbf{w}(k). \tag{8.38}$$

As a result of CP insertion and removal operations, $\widetilde{\mathbf{H}}$ is an $N \times N$ circulant matrix, with the (q, l)th entry given by $h_{(q-l)\mathrm{mod}N}$. The channel taps $\{h_l\}_{l=0,\ldots,L-1}$ are assumed to be constant within the time duration of one ODFM block, and are also supposed to vary independently in time. The channel impulse response is no longer than L, the length of the CP, in order to avoid interblock interference. The noise term $\mathbf{w}$ is typically chosen to be circular white Gaussian.

Because circulant matrices implement circular convolutions, they are diagonalized by DFT and IDFT operations, and thus after having performed the Fourier transform of Equation 8.38, we obtain

$$\tilde{\mathbf{r}}(k) = \mathbf{F}_N^H \widetilde{\mathbf{H}} \mathbf{F}_N \mathbf{a}(k) + \mathbf{F}_N^H \mathbf{w}(k), \tag{8.39}$$

$$\tilde{\mathbf{r}}(k) = \mathbf{D}\, \mathbf{a}(k) + \tilde{\mathbf{n}}(k), \tag{8.40}$$

where $\tilde{\mathbf{n}}(k) = \mathbf{F}_N^H \mathbf{w}(k)$ and the diagonal matrix

$$\mathbf{D} = \mathbf{F}_N^H \widetilde{\mathbf{H}} \mathbf{F}_N = \mathrm{diag}\left\{ \sum_{q=0}^{N-1} h_q(k) \exp\left(-j\frac{2\pi nq}{N}\right) \right\}_{n=0,\ldots,N-1}$$

contains the frequency response of the channel, evaluated at the subcarrier frequencies. Finally, single-tap equalization (denoted by **Eq** in Figure 8.4) can be performed in the frequency domain. The MMSE equalization may be applied to the received data as follows:

$$\mathbf{s}(k) = \left(\mathbf{D} + \sigma^2\mathbf{I}\right)^{-1} \tilde{\mathbf{r}}(k), \tag{8.41}$$

where σ^2 is the variance of the noise. Then, decisions are carried out on $\mathbf{s}(k)$ to obtain the symbol estimate $\hat{\mathbf{a}}(k)$.

8.7.2   Blind Channel Estimation in OFDM

In OFDM systems, the channel estimation can be performed either in the frequency domain or the time domain. Most of the OFDM channel estimators introduced in the literature rely on pilot sequences. Blind methods have also received attention in OFDM because they save bandwidth, reduce overhead signaling, and allow tracking of slow-channel variations. A number of blind-channel estimation techniques have been developed for OFDM, some of them based on cyclostationarity[72] or on subspace decomposition that exploits the CP and zero padding (ZP) structure of OFDM.[73] One problem that may arise is related to HOS-based blind methods. In OFDM, due to the high number of subcarriers, the transmit signals are close to Gaussian. Hence, application of blind separation types of algorithms, which rely on HOS, is difficult. Blind techniques based on SOS have been proposed instead.[73,74]

The algorithm presented in Reference 73 is based on subspace decomposition and takes advantage of the inherent redundancy introduced by the CP in estimating the channel in a blind manner. This technique is insensitive to the constellation size, unlike the decision-directed techniques, which suffer from performance degradation when a higher-order constellation is used. It also guarantees the channel identifiability regardless of the channel zero's location when the entire noise subspace is considered. A semiblind version was introduced,[73] which improves convergence speed of the algorithm. The quality of the initial channel estimates at the beginning of the frame is also improved.

The method proposed in Reference 74 solves the problem of channel estimation and equalization in a blind manner for a MIMO system. SOS are used to identify the channel on a subchannel-by-subchannel basis. This technique does not have restrictions to channel zeros and exhibits low sensitivity to stationary noise. An upper bound on the channel order is assumed known.

In OFDM transmission, CP allows for a simple one-tap equalization scheme. The data rate can be increased by reducing the CP length. This may cause severe ISI, and more powerful equalization schemes are needed to mitigate this effect. Another way of reducing ISI is obtained by using ZP instead of CP. This can be considered to be a form of periodic precoding. A blind channel estimator exploiting the CP has been proposed.[75] The method uses the algebraic structure of the signal model, which is a function of the channel.

Based on the observation that the multicarrier system turns a single wideband frequency-selective channel into a set of correlated narrowband flat-fading channels, Luise et al.[76] propose a blind channel estimation-detection scheme that exploits the frequency correlation between neighboring subcarriers. Channel equalization is accomplished by means of a trellis decoder on a block-by-block basis and is shown to perform significantly better than a differential encoding–decoding scheme.

Compared to the single-transmit antenna case, the channel estimation for multiple-transmit antenna systems is made more difficult by the fact that the received signal is a superposition of the transmitted signals from different antennas. The problem of channel estimation was investigated in SIMO[77] and MIMO[74] scenarios. The SIMO type of scenario can be obtained by oversampling or by using multiple antennas. Typically, the multiple channels (SIMO or MIMO) are considered to be independent. In Reference 78 a generalized space–time coded multicarrier transceiver is proposed. Multicarrier precoding maps the frequency-selective channel into a set of flat-fading channels whereas the space-time block coding facilitates the equalization and exploits the diversity provided by multiple transmit antennas. A blind channel estimation scheme is proposed based on a deterministic constant modulus algorithm. The inherent ambiguities are solved by transmitting few training symbols.

The impact of channel estimation errors in mobile OFDM systems with multiple receive antennas is of great interest. This type of analysis was performed in Reference 79, where a sensitivity analysis using perfect and erroneous channel estimates was performed. The effect of the erroneous channel estimates can be modeled as an additional Gaussian noise source. The goal of the study is to provide a quick performance assessment without the need to run large-scale system simulations featuring practical channel estimation techniques.

Blind OFDM receivers that do not need channel estimation have been proposed.[80] A multiuser equalizer using a vector constant modulus algorithm (VCMA) has been developed. The VCMA relies on the constant mean block energy property of the transmitted signal. To reduce the interuser interference (IUI) that is due to the interference caused by other users, a decorrelation criterion is involved as a second criterion in deriving the equalizer. IUI may be canceled by minimizing the correlation among users' outputs.

A semiblind time-domain channel estimation and tracking method has been introduced,[81] in the case of OFDM transmission. The tracking algorithm is based on a Kalman filter and runs in the time domain, whereas the equalization is performed in the frequency domain. Frequency offsets may also be compensated for using a similar approach. The model is nonlinear so that the extended Kalman filter has to be used instead.

8.8   Semiblind Techniques

Equalization of time-varying FIR-MIMO channels is a very important research topic in communications. In general, the design of an optimal equalizer requires precise knowledge of channel parameter values.[4] Channel parameters are usually estimated using a limited number of data samples. From this perspective we can identify[4] three types of channel estimation approaches: training-based, blind, and semiblind. Assuming perfect knowledge of the MIMO channel, an MLSE is the optimum receiver, but has exponential complexity even when implemented using the Viterbi algorithm.[12]

Another method is to bypass the channel estimation step and to directly compute parameter values for a desired equalizer structure.[30] Still another option is to estimate the channel and then to design an equalizer based on the estimated channel.[18,82] Such equalizers perform *joint tracking* and *equalization*.[83]

Techniques for blind equalization of SIMO and MIMO have been presented.[1] Despite the advantages of blind equalization, such as improved effective data rate, there are also some significant drawbacks. For example, some ambiguities always remain, such as the rotation of the constellation pattern. Some blind methods converge very slowly, and not all channel types can be identified (the problem of common zeros in SOCS). When the channel coherence time is short, the block blind equalization algorithms cannot be used, because the basic assumption on a quasi-stationary channel over the block period does not hold. Therefore, adaptive schemes capable of tracking channels' time variations are preferred. Semiblind methods using a limited amount of known training data solve the above problems. A common definition of the term *semiblind* is given by referring to the use of the known information. Training sequence (TS) methods base the parameter estimation only on the received signal containing known symbols. All the other observations, which may contain unknown data symbols, are ignored.[1] Blind methods are based on all the received data and on knowledge of the structural and statistical properties of the transmitted data, but not on explicit knowledge of input symbols. Semiblind techniques combine the information contained in few known symbols with the statistical and structural properties of the communication signals. The presence of a small number of training signals allows resolution of the ambiguities. Moreover, all the channel types become identifiable[1] and convergence speed improves. Semiblind methods also extend the sample support. Consequently, lower variance estimates of the unknown channel may be obtained or a larger number of parameters may be estimated with reasonable variance.

Semiblind techniques are very appealing from the performance point of view, as their performance can be superior to that of either TS or blind techniques separately. Semiblind techniques are applicable in cases where both TS and blind methods fail in estimating and tracking the channel.[1] This problem occurs in fast, time-varying channels where deep fades may take place. The performance of semiblind methods can be further improved by optimal placing of the training symbols.

In time-frequency selective channels the channel or equalizer parameters should be tracked continuously. Semiblind techniques can be developed by using adaptive channel estimators and equalizers in a decision-directed mode. To ensure the convergence, a small TS has to be transmitted first. This allows the channel estimator to acquire the channel rapidly. After that, the channel or equalizer is tracked using the past decisions from the equalizer. Several schemes including decision feedback equalizers (DFE) have been proposed for MIMO systems.[82,84,85]

8.9   Conclusion

In this chapter, blind and semiblind channel estimation was studied. Blind methods exploit statistical and structural properties of communication signals instead of known training sequences or pilot signals in estimating the channel or equalizer. As a result, significant improvement in effective data rates may be achieved. Models for time- and frequency-selective communication channels were presented. Signal properties, e.g., cyclostationarity and higher-order statistics, used in blind methods, were briefly described. Commonly used single-user blind equalization and channel identification methods were reviewed. Their extensions to MIMO systems were considered, as well. To illustrate the use of blind methods in practical wireless communication systems, blind receiver structures proposed for GSM, CDMA, and OFDM systems were briefly reviewed. Finally, semiblind methods combining the use of a limited number of training symbols with a blind method were considered. Semiblind methods appear to provide a feasible solution to practical communication receivers. They avoid the ambiguities inherent in fully blind methods and the algorithms converge faster. They also exploit the information present both in training symbols and information symbols, which improves the quality of the estimates. Blind methods remain, however, an important research topic because the core of each semiblind method is a powerful blind method.

References

1. G. Giannakis, Y. Hua, P. Stoica, and L. Tong, Eds., *Signal Processing Advances in Wireless and Mobile Communications*, Vol. 1: *Trends in Channel Estimation and Equalization*, Englewood Cliffs, NJ: Prentice-Hall, 2000.
2. G.B. Giannakis, Y. Hua, P. Stoica, and L. Tong, Eds., *Signal Processing Advances in Wireless and Mobile Communications*, Vol. 2: *Trends in Single-User and Multi-User Systems*, Englewood Cliffs, NJ: Prentice-Hall, 2000.
3. L. Tong and S. Perreau, Blind channel estimation: from subspace to maximum likelihood methods, *Proc. IEEE*, 86(10), 1951–1968, 1998.
4. J.K. Tugnait, L. Tong, and Z. Ding, Single-user channel estimation and equalization, *IEEE Signal Process. Mag.*, 17, 17–28, 2000.
5. J.-F. Cardoso, Blind signal separation: statistical principles, *Proc. IEEE*, 86(10), 2009–2025, 1998.
6. A. Paulraj, C.B. Papadias, V.U. Reddy, and A.J. van der Veen, Space-time blind signal processing, in *Wireless Communications: Signal Processing Perspectives*, Englewood Cliffs, NJ: Prentice-Hall, 1998.
7. P.A. Bello, Characterization of randomly time-variant linear channels, *IEEE Trans. Commun. Syst.*, 11(4), 360–393, 1963.

8. W.C. Jakes, *Microwave Mobile Communications*, New York: Wiley, 1974.
9. V. Koivunen, J. Laurila, and E. Bonek, Blind methods for wireless communication receivers, in *Review of Radio Science 1999–2002*, Englewood Cliffs, NJ: John Wiley & Sons, 2002, pp. 247–273.
10. European Communities, Evolution of land mobile radio (including personal communication), Office for Official Publications of the European Communities, Luxembourg, Apr. 1989 to Apr. 1996, http://www.lx.it.pt/cost231/.
11. European Communities, Digital land mobile radio communications—COST 207, Office for Official Publications of the European Communities, Luxembourg, Mar. 14, 1984 to Sept. 13, 1988.
12. J.G. Proakis, *Digital Communications*, 3rd ed., New York: McGraw-Hill, 1999.
13. T.S. Rappaport, *Wireless Communications: Principles Practice*, Englewood Cliffs, NJ: Prentice-Hall, 1999.
14. B. Sklar, Rayleigh fading channels in mobile digital communication systems. I. Characterization, *IEEE Commun. Mag.*, 90–100, 1997.
15. M. Toeltsch, J. Laurila, K. Kalliola, A.F. Molisch, P. Vainikainen, and E. Bonek, Statistical characterization of urban spatial radio channels, *IEEE J. Selected Areas Commun.*, 20(3), 539–549, 2002.
16. P. Hoeher, A statistical discrete-time model for the WSSUS multipath channel, *IEEE Trans. Vehic. Technol.*, 41(4), 461–468, 1992.
17. L.M. Davis, I.B. Collings, and R.J. Evans, Coupled estimators for equalization of fast-fading mobile channels, *IEEE Trans. Commun.*, 46(10), 1262–1265, 1998.
18. M.K. Tsatsanis, G.B. Giannakis, and G. Zhou, Estimation and equalization of fading channels with random coefficients, *Signal Process.*, 53, 211–229, 1996.
19. A.F. Molisch, Ed., *Wideband Wireless Digital Communications*, Englewood Cliffs, NJ: Prentice-Hall, 2000.
20. J.C. Liberti and T.S. Rappaport, *Smart Antennas for Wireless Communications*, Englewood Cliffs, NJ: Prentice-Hall, 1999.
21. A. Paulraj and C.B. Papadias, Space-time processing for wireless communications, *IEEE Signal Process. Mag.*, 14, 49–83, 1997.
22. R.B. Erdel, P. Cardieri, K.W. Sowerby, and T.S. Rappaport, Overview of spatial channel models for antenna array communications, *IEEE Personal Commun. Mag.*, 5(1), 10–22, 1998.
23. J. Winters, Smart antennas for wireless systems, *IEEE Personal Commun.*, 5(1), 23–27, 1998.
24. D. Gesbert and J. Akhtar, Breaking the barriers of Shannon's capacity: an overview of MIMO wireless systems, *Telektronikk Telenor J.*, 2002.
25. J.P. Kermoal, L. Schumacher, K.I. Pedersen, P.E. Mogensen, and F. Frederiksen, A stochastic MIMO radio channel model with experimental validation, *IEEE J. Select. Areas Commun.*, 20(6), 1211–1225, 2002.
26. D. Gesbert, H. Bolcskei, D.A. Gore, and A.J. Paulraj, Outdoor MIMO wireless channels: models and performance prediction, *IEEE Trans. Commun.*, 50(12), 1926–1934, 2002.
27. S. Verdu, Ed., *Multiuser Detection*, New York: Cambridge University Press, 1998.
28. J. Mannerkoski and V. Koivunen, Autocorrelation properties of channel encoded sequences—applicability to blind equalization, *IEEE Trans. Signal Process.*, 12(48), 3501–3507, 2000.
29. Y. Sato, A method of self recovering equalization for multilevel amplitude modulation, *IEEE Trans. Commun.*, 23, 679–682, 1975.

30. O. Shalvi and E. Weinstein, New criteria for blind deconvolution of nonminimum phase systems (channels), *IEEE Trans. Inf. Theor.*, 312–321, 1990.
31. D.N. Godard, Self-recovering equalization and carrier tracking in two-dimensional data communication systems, *IEEE Trans. Commun.*, 28(11), 1867–1875, 1980.
32. J.R. Treichler and B.G. Agee, A new approach to multipath correction of constant modulus signals, *IEEE Trans. Acoust. Speech Signal Process.*, 2(31), 459–471, 1983.
33. W. Gardner, W. Brown, and C.-K. Chen, Spectral correlation of modulated signals. II. Digital modulation, *IEEE Trans. Commun.*, 35(6), 595–601, 1987.
34. H. Liu, G. Xu, L. Tong, and T. Kailath, Recent developments in blind channel equalization: from cyclostationarity to subspaces, *Signal Process.*, 50, 82–99, 1996.
35. E. Moulines, P. Duhamel, J. Cardoso, and S. Mayrargue, Subspace methods for blind identification of multichannel FIR filters, *IEEE Trans. Signal Process.*, 43(2), 526–535, 1995.
36. D.T. Slock, Blind fractionally-spaced equalization, perfect reconstruction filter banks and multichannel linear prediction, *IEEE Int. Conf. Acoust. Speech Signal Process.*, 4, 585–588, 1994.
37. D. Gesbert, C.B. Papadias, and A. Paulraj, Blind equalization of polyphase FIR channels: a whitening approach, in *31st Asilomar Conference on Signals, Systems & Computers*, Vol. 2, 1997, 1604–1608.
38. E. Serpedin and G.B. Giannakis, Blind channel identification and equalization with modulation-induced cyclostationarity, *IEEE Trans. Signal Process.*, 7(46), 1930–1944, 1998.
39. P. Loubaton, E. Moulines, and P. Regalia, Subspace method for blind identification and deconvolution, in *Signal Processing Advances in Wireless & Mobile Communications*, G. Gannakis et al., Eds., Englewood Cliffs, NJ: Prentice-Hall, 2000.
40. S. Haykin, Ed., *Unsupervised Adaptive Filtering*, Vol. 1, *Blind Source Separation*, New York: Wiley, 2000.
41. J.-F. Cardoso, Performance and implementation of invariant source separation algorithms, *IEEE Int. Symp. Circuits Syst.*, 2, 85–88, 1996.
42. S. Visuri and V. Koivunen, Resolving ambiguities in subspace-based blind receiver for MIMO channels, in *36th Asilomar Conference on Signals, Systems and Computers*, 2002, 589–593.
43. A. Cichochi and S. Amari, *Adaptive Blind Signal and Image Processing: Learning Algorithms and Applications*, John Wiley & Sons, New York, 2002.
44. S. Haykin, Ed., *Unsupervised Adaptive Filtering*, Vol. II, *Blind Deconvolution*, New York: Wiley, 2000.
45. Y. Hua, Fast maximum likelohood for blind identification of multiple FIR channels, *IEEE Trans. Signal Process.*, 44(3), 661–672, 1996.
46. A. Gorokhov, P. Loubaton, and E. Moulines, Second order blind equalization in multiple input multiple output FIR systems: a weighted least squares approach, *IEEE Int. Conf. Acoust. Speech Signal Process.*, 5, 2415–2418, 1996.
47. J.K. Tugnait, Blind equalization and channel estimation for multiple-input multiple-output communications systems, *IEEE Int. Conf. Acoust. Speech Signal Process.*, 5, 2443–2446, 1996.

48. S.-I. Amari, S.C. Douglas, A. Cichocki, and H.H. Yang, Multichannel blind deconvolution and equalization using the natural gradient, in *IEEE Workshop on Signal Processing Advances in Wireless Communications*, 1997, 101–104.

49. D. Boss, K.-D. Kammeyer, and T. Petermann, Is blind channel estimation feasible in mobile communication systems? A study based on GSM, *IEEE J. Selected Areas Commun.*, 16(8), 1479–1492, 1998.

50. Z. Ding and G. Li, Singel-channel blind equalization for GSM cellular systems, *IEEE J. Selected Areas Commun.*, 16(8), 1493–1505, 1998.

51. G. Li and Z. Ding, Semi-blind channel identification for individual data bursts in GSM wireless systems, *Signal Process.*, 80(10), 2017–2031, 2000.

52. H. Trigui and D.T. Slock, Performance bounds for cochannel interference cancellation within the current GSM standard, *Signal Process.*, 80(7), 1335–1346, 2000.

53. D. Boss, B. Jelonnek, and K.-D. Kammeyer, Eigenvector algorithm for blind MA system identification, *Signal Process.*, 66(1), 1–26, 1998.

54. J.A.R. Fonollosa and J. Vidal, System identification using a linear combination of cumulant slices, *IEEE Trans. Signal Process.*, 41(6), 2405–2412, 1993.

55. K.-D. Kammeyer, V. Kuhn, and T. Peterman, Blind and nonblind turbo estimation for fast fading GSM channels, *IEEE J. Select. Areas Commun.*, 19(9), 1718–1729, 2001.

56. H. Holma and A. Toskala, *WCDMA for UMTS—Radio Access for Third Generation Mobile Communications*, New York: Wiley, 2000.

57. H. Liu and M.D. Zoltowski, Blind equalization in antenna array CDMA systems, *IEEE Trans. Signal Process.*, 45(1), 161–172, 1997.

58. A. Weiss and B. Friedlander, Channel estimation for DS-CDMA downlink with aperiodic spreading codes, *IEEE Trans. Commun.*, 47(10), 1561–1569, 1999.

59. M. Melvasalo and V. Koivunen, Blind channel estimation in multicode CDMA system with antenna array, in *IEEE Sensor, Array and Multichannel Signal Processing Workshop*, Washington, D.C., Aug. 2002.

60. R. Price and P.E. Green, A communication technique for multipath channels, *Proc. IRE*, 46, 555–570, 1958.

61. U. Madhow, Blind adaptive interference suppression for direct-sequence CDMA, *Proc. IEEE*, 86(10), 2049–2069, 1998.

62. V.H. Poor and G.W. Wornell, Eds., *Wireless Communications—Signal Processing Perspectives*, Englewood Cliffs, NJ: Prentice-Hall, 1998, chap. 2.

63. M. Honig and M. Tsatsanis, Adaptive techniques for multiuser CDMA receivers, *IEEE Signal Process. Mag.*, 17(3), 49–61, 2000.

64. X. Wang and H.V. Poor, Blind equalization and multiuser detection in dispersive CDMA channels, *IEEE Trans. Commun.*, 46(1), 91–103, 1998.

65. M. Torlak and G. Xu, Blind multiuser channel estimation in asynchronous CDMA systems, *IEEE Trans. Signal Process.*, 45(1), 137–147, 1997.

66. D. Gesbert, J. Sorelius, and A. Paulraj, Blind multi-user MMSE detection of CDMA signals, *IEEE Int. Conf. Acoust. Speech Signal Process.*, 3191–3164, 1998.

67. D. Gesbert, J. Sorelius, P. Stoica, and A. Paulraj, Blind multiuser MMSE detector of CDMA signals in ISI channels, *IEEE Commun. Lett.*, 3(8), 233–235, 1999.

68. C.B. Papadias and H. Huang, Linear space-time multiuser detection for multipath CDMA signals, *IEEE J. Selected Areas Commun.*, 19(2), 254–264, 2001.

69. H. Li, X. Lu, and G.B. Giannakis, Capon multiuser receiver for CDMA systems with space-time coding, *IEEE Trans. Signal Process.*, 50(5), 1193–1204, 2002.

70. The DWMT: a multicarrier transceiver for ADSL using M-band wavelets, ANSI Standard T1E1.4 Comm. Contrib., 1991.
71. Broadband radio access networks (BRAN): High Performance Radio Local Area Networks (HIPER-LAN) Type 2: System Overview, ETR101683114, 1999.
72. R.W. Heath and G.B. Giannakis, Exploiting input cyclostationarity for blind identification of OFDM systems, *IEEE Trans. Signal Process.*, 47, 848–856, 1999.
73. B. Muquet, M. de Courville, and P. Duhamel, Subspace-based blind and semi-blind channel estimation for OFDM systems, *IEEE Trans. Signal Process.*, 50(7), 1699–1712, 2002.
74. H. Bölcskei, R.W. Heath, and A.J. Paulraj, Blind channel identification and equalization in OFDM-based multiantenna systems, *IEEE Trans. Signal Process.*, 50(1), 96–109, 2002.
75. U. Tureli and H. Liu, Blind carrier synchronization and channel estimation for OFDM communications, *IEEE Int. Conf. Acoust. Speech Signal Process.*, 6, 3509–3512, 1998.
76. M. Luise, R. Reggiannini, and G.M. Vitetta, Blind equalization/detection for OFDM signals over frequency-selective channels, *IEEE J. Selected Areas Commun.*, 8(16), 1568–1578, 1998.
77. H. Ali, J.H. Manton, and Y. Hua, A SOS subspace method for blind channel identification and equalization in bandwidth efficient OFDM systems based on receive antenna diversity, *IEEE Signal Processing Workshop on Statistical Signal Processing*, 2001, 401–404.
78. Z. Liu, G.B. Giannakis, S. Barbarossa, and A. Scaglione, Transmitt-antennae space-time block coding for generalized OFDM in the presence of unknown multipath, *IEEE J. Select. Areas Commun.*, 19(7), 1352–1364, 2001.
79. A.A. Hutter, E. de Carvalho, and J.M. Cioffi, On the impact of channel estimation for multiple antenna diversity reception in mobile OFDM systems, in *Asilomar Conference on Signals, Systems and Computers*, Vol. 2, 2000, 1820–1824.
80. T. Abrudan, M. Sirbu, and V. Koivunen, A multi-user vector constant modulus algorithm for blind equalization in OFDM, in *IEEE Workshop on Signal Processing Advances in Wireless Communications*, 2003.
81. T. Roman, M. Enescu, and V. Koivunen, Time-domain method for tracking dispersive channels in mimo OFDM systems, *IEEE Int. Conf. Acoust. Speech Signal Processing*, 2003.
82. C. Komninakis, C. Fragouli, A.H. Sayed, and R.D. Wesel, Multi-input multi-output fading channel tracking and equalization using Kalman estimation, *IEEE Trans. Signal Process.*, 50(5), 1065–1076, 2002.
83. M. Enescu, Adaptive Methods for Blind Equalization and Signal Separation in MIMO Systems, Ph.D. thesis, Helsinki University of Technology, Helsinki, Finland, 2002.
84. M. Enescu, M. Sirbu, and V. Koivunen, Adaptive equalization of time-varying MIMO channels, Tech. Rep. 34, Signal Processing Laboratory, Helsinki University of Technology, 2001, submitted to *Signal Processing*, May 2001.
85. M. Sirbu and V. Koivunen, Multichannel stochastic gradient method for adaptive equalization in uplink asynchronous DS/CDMA, *Proc. of IEEE Sensor Array and Multichannel Signal Proc. Workshop, SAM*, 2002, 303–306.

9

Bayesian Image and Video Enhancement Using a Non-Gaussian Prior

Richard R. Schultz and Robert L. Stevenson

CONTENTS

9.1 Introduction

Image and video enhancement involves the extraction of high-frequency details from an observed visual data set. In this chapter, several open research problems are investigated, including region-of-interest image magnification from a digital still image, the postprocessing of still image compression artifacts, and the integration of multiple digital video frames/fields to generate a superresolved video still. Primary real-world applications include the enhancement of remote sensing, reconnaissance, and surveillance images and video. The overall goal of this research is to perform enhancement on the image data as accurately as possible, so computational efficiency is not an issue. Specifically, iterative nonlinear enhancement techniques will be considered for their edge-preserving properties, although linear algorithms related to least squares are generally much faster and easier to implement. However, linear estimation results in smooth (low-frequency) solutions, which do not contain sharp, discontinuous structures.

Magnification algorithms refer to an upsampling factor r, the factor by which a particular spatial region-of-interest is to be enlarged in both dimensions. Low-resolution and high-resolution pixel grids related by this upsampling/downsampling factor are shown superimposed in Figure 9.1.

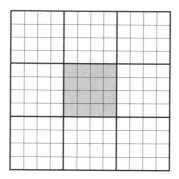

FIGURE 9.1

Superimposed low-resolution and high-resolution pixel grids, related by the downsampling/ upsampling factor r. The shaded low-resolution pixel contains 16 square high-resolution pixels, for an upsampling factor of $r = 4$ in both the horizontal and vertical directions. For the high- to low-resolution mapping $\mathbf{y} = \mathbf{Hx}$ defined in this chapter, each low-resolution pixel value is set equal to the average of the $r \times r$ square high-resolution pixels contained within its boundary. This serves to model the integration of light over each sensor's physical area.

The remotely sensed image of a farm near St. Thomas, North Dakota, is shown in Figure 9.2, captured by the *IKONOS* satellite with a 1-m ground sampling distance (GSD). As the acquired imagery is successively downsampled, an appearance of "blockiness" results because high-frequency structures become more and more aliased. Interpolation or reconstruction techniques are used to recover the original (generally unknown) data from its undersampled, lower-resolution observations. Several popular techniques exist for the interpolation of digital still images. These include one-shot methods such as bilinear and cubic B-spline interpolation, as well as incrementally more accurate iterative techniques such as deterministic Tikhonov regularization and stochastic Bayesian maximum *a posteriori* (MAP) estimation.[1]

Block-based coding is pervasive in commercial image and video compression algorithms. The Joint Photographic Experts Group (JPEG) standard,[2] which applies the discrete cosine transform (DCT) to 8×8 pixel blocks throughout an image, achieves compression ratios on the order of 20:1 through the quantization of high-frequency DCT coefficients with very little loss of visual quality. Because the human visual system does not perceive high-frequency information as readily as low-frequency content, lowpass filtering of the 8×8 frequency squares through the quantization of DCT coefficients is quite effective at reducing the data volume without severely degrading the visual information. For the most part, an arbitrary single-band image coded with a bit rate as low as 0.5 bits per pixel (bpp) generally does not appear to have highly noticeable distortion. Figure 9.3 presents the 1-m *IKONOS* satellite image compressed at successively lower bit rates, in which the image content tends to be degraded. As the bit rate decreases, the resulting coded image loses more and more high-frequency content. Specifically, for

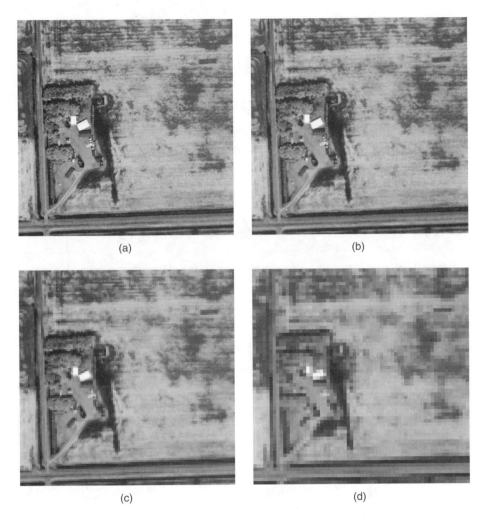

(a)   (b)

(c)   (d)

FIGURE 9.2
Macroview of the *IKONOS* satellite picture, which shows the reduction in the number of constraints available for enhancing a digital image as its resolution is decreased. (a) Original satellite image (1-m ground sampling distance). (b) Original image downsampled by $r = 2$ (effective GSD of 2 m). (c) Original image downsampled by $r = 4$ (effective GSD of 4 m). (d) Original image downsampled by $r = 8$ (effective GSD of 8 m).

extremely low bit rates, block-based compression artifacts appear within the data. Several researchers have studied the problem of postprocessing block-DCT compression artifacts,[3–5] utilizing least squares, projection onto convex sets (POCS), and other iterative methods to restore the high-frequency information lost due to compression. It is important to have complete knowledge of the quantization parameters in this case, so that the many-to-one compression mapping is satisfied while postprocessing the data.

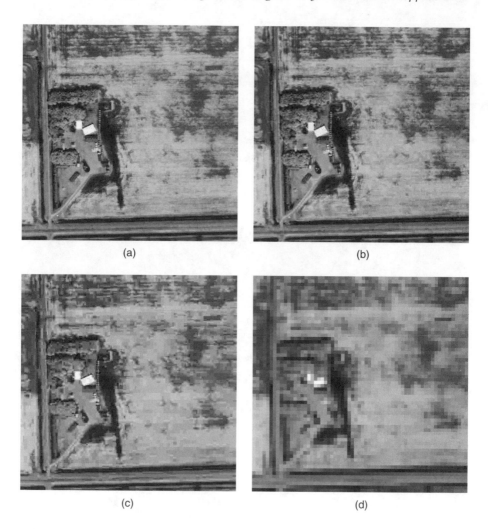

FIGURE 9.3
Macroview of the block DCT-encoded *IKONOS* satellite picture with 8 × 8 pixel blocks, to show the reduction in the number of constraints available for enhancing a digital image as its bit rate decreases. (a) Original satellite picture coded at 0.43 bpp. (b) Block-DCT coding at 0.264 bpp. (c) Block-DCT coding at 0.15 bpp. (d) Block-DCT coding, in which only the DC (lowest-frequency) discrete cosine transform coefficient is retained from each 8 × 8 pixel block. Note that this last case is identical to the downsampled satellite picture shown in Figure 9.2, with an effective ground sampling distance of 8 m.

Whenever a single digital still image is available for enhancement, the relevant interpolation algorithms all generate similar results. To extract additional visual details from image data, a sequence of video frames/fields containing object or scene movement must be utilized. In essence, superresolution enhancement exploits the temporal correlations present between undersampled

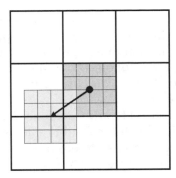

FIGURE 9.4

Illustration of why superresolution video enhancement works. The large (low-resolution) pixel moves from an initial position in one video frame to a new position in the next frame. If the pixel undergoes a *subpixel* displacement with respect to the low-resolution grid, the original element contributes to the intensity of four pixels in the next frame. This subpixel overlap between the two frames allows for additional information to be extracted from the sequence, above and beyond that contained within any of the individual frames. If r denotes the upsampling factor, each low-resolution pixel can be subdivided into $r \times r$ smaller (high-resolution) pixels, and the goal of superresolution video enhancement is to estimate the values of these high-resolution pixels given the low-resolution video frames. A subpixel motion vector is shown, representing a displacement of $(-\frac{3}{4}, -\frac{1}{2})$ elements on the low-resolution grid, or $(-3, -2)$ elements on the corresponding high-resolution grid, with an upsampling factor of $r = 4$.

frames/fields to improve spatial resolution.[6–8] Potentially, substantial additional information can be extracted from the sequence, provided that there is a great deal of subpixel overlap from frame to frame. This concept is explained intuitively in Figure 9.4. The basic idea for superresolution was originally introduced in 1984 by Tsai and Huang,[9] in which an observation model was defined for a sequence consisting of subpixel shifts of the same scene. They relied on a pseudo-inverse (least-squares) technique to regularize the problem when all required subpixel shifts of the scene were not available, and thus a direct inversion could not be calculated. Patti et al.[7] developed a POCS-based superresolution enhancement algorithm to integrate interlaced video frames. Bayesian methods of video frame integration were introduced separately by Cheeseman et al.[10] and Schultz and Stevenson.[8] The multiframe resolution enhancement algorithm originally proposed by Schultz and Stevenson includes a video observation model that accommodates either progressive or interlaced frames, as well as a discontinuity-preserving stochastic image prior. Furthermore, their video observation model uses block matching to incorporate both global scene transformations and independent object displacements, necessary for the representation of movement within a sequence. More accurate optical flow estimation techniques have been compared to block matching for their efficacy with respect to superresolution enhancement quality.[11–13] Subpixel block matching was found to compare favorably with these more sophisticated motion recovery techniques, because block matching generates accurate displacements along spatial edges. Discontinuous image regions also happen to be most affected by video enhancement. Most recently, these superresolution enhancement techniques have been applied to interframe coded video,[14–17] to suppress the DCT block artifacts present in low bit rate sequences.

This chapter is organized as follows. Section 9.2 describes the image and video observation models for the related problems of still image magnification, block-DCT image compression postprocessing, and superresolution video enhancement. Because of the lack of available constraints in real-world images and video, these applications all become ill-posed inverse problems, which require regularization. Popular deterministic and stochastic regularization techniques for a number of inverse imaging problems are presented in Section 9.3, including constrained least squares, projection onto convex sets, and Bayesian maximum *a posteriori* estimation. Several nonquadratic edge penalty functions are identified in Section 9.3, which are utilized in non-Gaussian stochastic image priors that model the data more accurately than the Gaussian. Specifically, the Huber edge penalty function[1,8] is used in the Huber–Markov random field model, a non-Gaussian prior with heavy-tailed behavior that accurately models piecewise smooth image data. The corresponding Bayesian estimation problem involves the optimization of a convex, nonquadratic function with a unique global minimum that can be found through gradient descent methods. An extensive set of simulation results are presented and interpreted in Section 9.4, which highlights the utility of nonlinear image and video enhancement over linear techniques. Finally, Section 9.5 provides a brief summary of this chapter, as well as several future avenues of exploration.

9.2   Image and Video Observation Models

The imaging problems of interest, namely, region-of-interest image magnification, the postprocessing of block-DCT image compression artifacts, and superresolution video enhancement, each attempt to reconstruct an original, unknown image from a set of observed data constraints. The running theme of this chapter is that the number of available (and useful) constraints will limit the ability of any enhancement algorithm to accurately recover the original data. Intuitively, the lower the spatial resolution of the observed data, the more difficult it becomes to recover high-resolution details from a still image or video sequence. Likewise, the lower the bit rate of a coded image or sequence, the more difficult it becomes to properly enhance the data. The image/video observation models for the applications of interest are presented as follows.

9.2.1 Region-of-Interest Image Magnification

The goal of the image magnification problem is to estimate an unknown high-resolution image from a digital still image, denoted as

$$y = Hx, \tag{9.1}$$

where **y** is an $N_1 N_2$-lexicographically ordered vector of the observed image pixels, and the unknown high-resolution image is denoted as **x**.[1] The matrix **H** serves to downsample the unknown high-resolution image through $r \times r$ pixel block averaging, as depicted in Figures 9.1 and 9.2. This quite accurately models the integration of light over a low-resolution charge-coupled device (CCD) sensor, which is assumed to contain a number of smaller high-resolution sensors within its physical area. In essence, as the downsampling factor increases, fewer constraints are available for the recovery of the original high-resolution image through interpolation.

9.2.2 Block Discrete Cosine Transform Image Compression Postprocessing

The goal of the image compression postprocessing problem is to estimate the original, uncompressed image from the given observations, expressed as

$$y = H^{-1} Q[Hx] \tag{9.2}$$

in the spatial (pixel) domain.[3] Block-DCT coding is assumed, in which the DCT and quantization are applied to 8×8 pixel squares. Postprocessing of compressed data serves to suppress the appearance of block artifacts, as well as to recover high-frequency information within each 8×8 element discrete cosine transform frequency square that is lost to the coding process. In this observation model, **H** is a unitary matrix used to transform the pixel-domain blocks into the DCT-domain, such that **Hx** represents the lexicographically ordered block-DCT coefficients. Additionally, $Q[\cdot]$ is a nonlinear operator that performs scalar quantization on the block-DCT coefficients, while H^{-1} represents the inverse block discrete cosine transform matrix. For low bit rate still image coding (less than approximately 0.5 bpp for a single-band image), visual artifacts appear within the uncompressed image along block boundaries. Several low bit rate, JPEG-like images are shown in Figure 9.3. For extremely low bit rates (less than approximately 0.2 bpp), accurate recovery of the original data becomes rather difficult. As the bit rate is increased, postprocessing algorithms can effectively remove the block coding artifacts from the data and extract high-frequency details, because additional constraints are infused into the observation model.

9.2.3 Superresolution Video Enhancement

The goal of the general superresolution video enhancement problem is to estimate an unknown high-resolution video still given the following short, low-resolution image sequence:[8]

$$\mathcal{Y} = \{\mathbf{y}^{(i)}\}, \qquad \text{for } i = 0, \dots, N-1 \tag{9.3}$$

Assume that a spatial region-of-interest within the kth frame/field, mathematically modeled as

$$\mathbf{y}^{(k)} = \mathbf{H}^{(k)}\mathbf{x}, \tag{9.4}$$

has been selected for enhancement. In Equation 9.4, $\mathbf{H}^{(k)}$ serves to downsample the original high-resolution image and incorporate frame interlacing,[7,18] if necessary. In fact, Equations 9.1 and 9.4 are identical when only a single, progressively scanned video frame is available. Superresolution enhancement is achieved through the temporal integration of pictures containing subpixel object or scene transformations, such that an estimate of a region-of-interest within the unknown high-resolution frame $\mathbf{x}$ is computed from the sequence $\mathcal{Y}$. It is assumed that neighboring frames/fields contain similar content, with small movements from frame to frame. The video observation model that takes into account pixel displacements between $\mathbf{y}^{(i)} \in \mathcal{Y}$ for $i \neq k$ and the reference frame selected for enhancement, $\mathbf{y}^{(k)}$, is given as

$$\mathbf{y}^{(i)} = \hat{\mathbf{H}}^{(i)}\mathbf{x} + \mathbf{n}^{(i)}, \qquad \text{for } i \neq k. \tag{9.5}$$

This expression incorporates additional constraints into the observation model, through neighboring motion-compensated frames. In Equation 9.5, $\hat{\mathbf{H}}^{(i)}$ downsamples the high-resolution image, performs frame interlacing (if necessary), and compensates the data using estimated subpixel displacements. Because this linear operator is constructed using estimated displacements, it is also denoted as an estimate. The independent and identically distributed (i.i.d.) Gaussian random vector $\mathbf{n}^{(i)}$ represents the error in estimating subpixel displacements between $\mathbf{y}^{(k)}$ and $\mathbf{y}^{(i)}$ for use in constructing $\hat{\mathbf{H}}^{(i)}$. If the subpixel motion between video frames/fields can be estimated accurately, it becomes possible to extract details from a spatial region-of-interest that are not visible within any single frame or field.

9.3   Regularization of Ill-Posed Inverse Problems

Assume that a blurred and noisy observed image can be mathematically modeled as

$$\mathbf{y} = \mathbf{H}\mathbf{x} + \mathbf{n}. \tag{9.6}$$

The unknown original image $\mathbf{x}$ and additive noise $\mathbf{n}$ are assumed to be uncorrelated, and the matrix $\mathbf{H}$ implements the convolution of the imaging system point spread function (PSF) with the original data. At first glance, the exact solution to this inverse problem appears to be rather simple,

$$x = (H^T H)^{-1} H^T (y - n).$$ (9.7)

However, $H^T H$ is almost always ill-conditioned, and thus calculating the exact solution is not possible. In other words, computing an estimate $\hat{x}$ of the original image given only the observed data y is generally an ill-posed inverse problem, since the solution derived from the observation model in Equation 9.6 is not unique. Various regularization techniques serve to make this general inverse problem well posed through the incorporation of constraints, or *a priori* information, on the solution. Popular regularization techniques from the image deconvolution, reconstruction, and enhancement literature include (1) constrained least squares; (2) projection onto convex sets; and (3) Bayesian maximum *a posteriori* estimation. Although this list is far from comprehensive, most common regularization methods can be placed into these three general categories (e.g., the pseudo-inverse and the Wiener filter are both essentially least-squares techniques). Because image degradation through image blur and additive white Gaussian noise is the quintessential ill-posed inverse problem in digital imaging, all three regularization methods will be presented in terms of the digital image observation model in Equation 9.6.

9.3.1 Constrained Least Squares

Least-squares restoration is a deterministic regularization method in which the feasible solution satisfies a bound E^2 on a stabilizing functional, such that

$$\|Qx\|^2 \leq E^2.$$ (9.8)

The linear regularization operator Q imposes an assumption of global smoothness on the data, and it is implemented by the convolution of a highpass finite impulse response (FIR) filter with the image. Typically, the discrete Laplacian operator is used as the highpass filter mask. The solution is further constrained by the following bound on noise power:

$$\|y - Hx\|^2 \leq \|n\|^2 \leq \epsilon^2$$ (9.9)

Combining the constraint sets in Equations 9.8 and 9.9 results in an objective function,

$$J(x) = \frac{\epsilon^2}{E^2} \|Qx\|^2 + \|y - Hx\|^2 \leq 2\epsilon^2,$$ (9.10)

which is minimized by the Tikhonov–Miller regularized solution. This linear estimate is computed as

$$\hat{x} = \left(H^T H + \frac{\epsilon^2}{E^2} Q^T Q \right)^{-1} H^T y.$$ (9.11)

Note the similarity of this linear solution to Equation 9.7, in which the term $(\epsilon^2 / E^2) Q^T Q$ serves to regularize the matrix inverse. The Wiener filter,[19] which outputs the minimum mean square error estimate, provides a similar linear solution to this particular problem.

9.3.2   Projection onto Convex Sets

In the POCS regularization method, the unknown signal x is assumed to be an element of an appropriate Hilbert space. Each item of *a priori* information, or constraint, restricts the solution to a closed convex set within this Hilbert space. For m pieces of information, there are m corresponding closed convex sets C_i, for $i = 1, 2, \ldots, m$, such that $\hat{x} \in C_0 = \bigcap_{i=1}^{m} C_i$, provided that the intersection C_0 is nonempty. Given the convex sets C_i and their respective projection operators P_i, the sequence generated by the projection

$$\hat{x}_{k+1} = P_m P_{m-1} \cdots P_1 \hat{x}_k, \qquad \text{for } k = 0, 1, \ldots \tag{9.12}$$

converges weakly to a feasible solution in the intersection C_0 of the constraint sets.

Specifically for the image restoration problem, there exists a closed, convex constraint set for each observed blurred image pixel, such that

$$C_{n_1, n_2, k} = \{\hat{x}(i_1, i_2) : |r_k(n_1, n_2)| \le \delta_0\},$$

$$\text{for } 0 \le n_1 \le N_1 - 1 \quad \text{and} \quad 0 \le n_2 \le N_2 - 1. \tag{9.13}$$

In this expression, the residual between the original blurred image and the blurred estimate at iteration k of the POCS algorithm is given as

$$r_k = y - H\hat{x}_k. \tag{9.14}$$

The quantity δ_0 is an *a priori* bound, which represents the statistical confidence of the actual image being a member of the set $C_{n_1, n_2, k}$. Since this value bounds the residual at every pixel, it is determined from the statistics of the noise process. For image deconvolution, the projection operator corresponding to each convex constraint set is given as follows:[19]

$$P_{n_1, n_2, k} [\hat{x}_k (i_1, i_2)]$$

$$= \begin{cases} \hat{x}_k (i_1, i_2) + \dfrac{r_k(n_1, n_2) - \delta_0}{\sum_{(j_1, j_2)} h^2(n_1, n_2; j_1, j_2)} h(n_1, n_2; i_1, i_2), & \text{for } r_k(n_1, n_2) > \delta_0 \\ \hat{x}_k (i_1, i_2), & \text{for } -\delta_0 \le r_k(n_1, n_2) \le \delta_0 \\ \hat{x}_k (i_1, i_2) + \dfrac{r_k(n_1, n_2) + \delta_0}{\sum_{(j_1, j_2)} h^2(n_1, n_2; j_1, j_2)} h(n_1, n_2; i_1, i_2), & \text{for } r_k(n_1, n_2) < -\delta_0 \end{cases}$$

$$\tag{9.15}$$

The initial condition $\hat{\mathbf{x}}_0$ is usually selected as the blurred and noisy observed image $\mathbf{y}$. Additional projections are often applied, including a pixel underflow/overflow constraint. One of the primary advantages of selecting the POCS regularization technique is that it can incorporate spatially varying blurs quite easily into the video observation model. It should be noted that the POCS algorithm is in general nonlinear, because the projection operations are generally nonlinear. However, experimental restoration results obtained through POCS regularization are quite comparable to least-squares (linear) estimates.

9.3.3 Bayesian Maximum *a Posteriori* Estimation

Stochastic regularization uses the Bayesian framework, which requires probability densities for the image and the noise. Prior information is incorporated into the density $p(\mathbf{x})$, representing the likelihood of a particular image $\mathbf{x}$. The value of $\mathbf{x}$ which maximizes the posterior density $p(\mathbf{x}|\mathbf{y})$ is known as the Bayesian maximum *a posteriori* (MAP) estimate. The posterior density can be expanded using Bayes' rule, yielding

$$p(\mathbf{x}|\mathbf{y}) = \frac{p(\mathbf{y}|\mathbf{x})\,p(\mathbf{x})}{p(\mathbf{y})}. \tag{9.16}$$

Assuming that $\mathbf{x}$ and $\mathbf{y}$ are statistically uncorrelated, the conditional density $p(\mathbf{y}|\mathbf{x})$ is a probability density model for the noise:

$$p(\mathbf{y}|\mathbf{x}) = p(\mathbf{y} - \mathbf{H}\mathbf{x}|\mathbf{x}) = p(\mathbf{n}|\mathbf{x}) = p(\mathbf{n}) \tag{9.17}$$

Equivalently, the MAP estimate is found at the maximum of the log-likelihood function,

$$\hat{\mathbf{x}} = \arg\max_{\mathbf{x}} \log p(\mathbf{x}|\mathbf{y}), \tag{9.18}$$

so that the estimate can be expressed as

$$\hat{\mathbf{x}} = \arg\max_{\mathbf{x}} \left\{ \log p(\mathbf{y}|\mathbf{x}) + \log p(\mathbf{x}) \right\}. \tag{9.19}$$

In this general formulation of the Bayesian estimation problem, both the noise and image prior densities must be selected.

An i.i.d. Gaussian density is assumed for the noise, to model the thermal activity present in CCD image acquisition sensors. This zero-mean noise model is given by the conditional density

$$p(\mathbf{y}|\mathbf{x}) = \frac{1}{(2\pi)^2 \sigma^{N_1 N_2}} \exp\left\{ -\frac{1}{2\sigma^2} \|\mathbf{y} - \mathbf{H}\mathbf{x}\|^2 \right\}, \tag{9.20}$$

where σ^2 is defined as the noise variance at each pixel.

Selecting an appropriate prior model for the image data is critically important for the accuracy of Bayesian estimation. Typically, a Gaussian distribution is also assumed for the image, such that

$$p(\mathbf{x}) = \frac{1}{Z} \exp\left\{-\frac{1}{2}(\mathbf{x} - \mu_\mathbf{x})^T \Gamma^{-1}(\mathbf{x} - \mu_\mathbf{x})\right\}. \tag{9.21}$$

Here, $\mu_\mathbf{x}$ represents the image mean (possibly spatially varying), Γ denotes the image covariance matrix, and Z is a normalizing constant. A sparse covariance matrix is often assumed, with a narrow band of nonzero elements along the diagonal to model an image with strong correlations among neighboring pixels but zero correlation among distant pixels. A Gaussian image prior is an accurate model for globally smooth image data, but that is usually not a good assumption for real-world digital imagery.

For Gaussian noise and image models, the Bayesian MAP estimate becomes

$$\hat{\mathbf{x}} = \arg\min_\mathbf{x}\left\{\frac{1}{2\sigma^2}\|\mathbf{y} - \mathbf{Hx}\|^2 + \frac{1}{2}(\mathbf{x} - \mu_\mathbf{x})^T \Gamma^{-1}(\mathbf{x} - \mu_\mathbf{x})\right\}, \tag{9.22}$$

resulting in the linear estimate

$$\hat{\mathbf{x}} = (\mathbf{H}^T\mathbf{H} + \sigma^2\Gamma^{-1})^{-1}(\sigma^2\Gamma^{-1}\mu_\mathbf{x} + \mathbf{H}^T\mathbf{y}). \tag{9.23}$$

Note the similarity between this linear Bayesian estimate and the least-squares estimate in Equation 9.11. Both of these solutions generally possess smooth object edges and discontinuities that appear somewhat blurred.

The advantage of utilizing the Bayesian framework for inverse problems in image processing is that it is possible to stochastically model the image data more accurately. A Gaussian image density is often assumed for mathematical and computational convenience. Non-Gaussian densities can more closely approximate the true underlying image distribution, and the non-linear estimates that result are often more accurate than linear solutions. To incorporate a non-Gaussian image prior into the Bayesian MAP estimation problem formulation, a two-dimensional Markov random field (MRF) must be used to represent the image pixels. The Gibbs distribution is the probability distribution over an MRF, with the corresponding density

$$p(\mathbf{x}) = \frac{1}{Z} \exp\left\{-\frac{1}{2\beta}\sum_{c \in \mathcal{C}} V_c(\mathbf{x})\right\}. \tag{9.24}$$

Equation 9.24 is known as the Gibbs prior model.[20] In this expression, β is the "temperature" parameter of the distribution, $V_c(\mathbf{x})$ is a potential function of a local group of points c known as cliques, and $\mathcal{C}$ denotes the set of all image cliques.

Digitized images of interest are assumed to be piecewise smooth. In other words, only small variations exist between neighboring sample values, with discontinuities separating smooth regions. Intuitively, this is a relatively good assumption, in the context of people or objects moving across a stationary camera's field of view (e.g., surveillance imagery), or a moving camera viewing a stationary scene (e.g., remote sensing imagery). A particular Gibbs prior that incorporates this *a priori* knowledge has the density

$$p(\mathbf{x}) = \frac{1}{Z} \exp\left\{-\frac{1}{2\beta} \sum_{c \in \mathcal{C}} \rho(\mathbf{d}_c^T \mathbf{x})\right\}. \tag{9.25}$$

Four finite differences $\mathbf{d}_{\mathbf{n},m}^T \mathbf{x}$, for $m = 1, \ldots, 4$, are used to approximate second-order spatial derivatives at each pixel location $\mathbf{n} = (n_1, n_2)$, with horizontal, vertical, and diagonal orientations:

$$\mathbf{d}_{n_1,n_2,1}^T \mathbf{x} = x_{n_1,n_2-1} - 2x_{n_1,n_2} + x_{n_1,n_2+1} \tag{9.26}$$

$$\mathbf{d}_{n_1,n_2,2}^T \mathbf{x} = 0.5x_{n_1+1,n_2-1} - x_{n_1,n_2} + 0.5x_{n_1-1,n_2+1} \tag{9.27}$$

$$\mathbf{d}_{n_1,n_2,3}^T \mathbf{x} = x_{n_1-1,n_2} - 2x_{n_1,n_2} + x_{n_1+1,n_2} \tag{9.28}$$

$$\mathbf{d}_{n_1,n_2,4}^T \mathbf{x} = 0.5x_{n_1-1,n_2-1} - x_{n_1,n_2} + 0.5x_{n_1+1,n_2+1}. \tag{9.29}$$

Each finite difference is an argument to an edge penalty function, $\rho(\xi)$, which ensures that smooth signals are mapped to high probability values. Specifically, each directional finite difference is a "spatial activity" measure, with a large value near an object edge and a small value in a smooth image region. Thus, for small values of each finite difference (in smooth regions), the probability should be high with a correspondingly small edge penalty. Conversely, large finite difference values (in discontinuous regions) should be assigned a lower probability, with a correspondingly higher edge penalty.

Stochastic image priors are characterized by different edge penalty functions. For example, the Gauss–Markov random field (GMRF) model uses the quadratic edge penalty $\rho(\xi) = \xi^2$, which severely penalizes edges. Essentially, the GMRF prior models the image data in the same manner as the Gaussian density in Equation 9.21. Non-Gaussian models are characterized by discontinuity-preserving functions, which reduce this edge penalty, dependent on a threshold, shape, or scaling parameter. Several desirable properties of a nonquadratic edge penalty function include the following:

1. Convexity:

$$\rho[\alpha\xi + (1 - \alpha)\psi] \le \alpha\rho(\xi) + (1 - \alpha)\rho(\psi) \tag{9.30}$$

2. Symmetry:

$$\rho(\xi) = \rho(-\xi) \tag{9.31}$$

3. Provides a small penalty for discontinuities:

$$\rho(\xi) < \xi^2, \qquad \text{for } |\xi| \text{ large} \tag{9.32}$$

If the selected prior is characterized by a convex edge penalty function, then the log-likelihood function to be optimized is also convex. In this case, gradient descent techniques may be employed to compute the unique minimum. Symmetry of the penalty function is desirable because symmetric functions treat rising and falling edges in the same manner. Finally, discontinuities are maintained when edges are penalized less than the quadratic function, particularly for large discontinuities that represent distinct object boundaries. Examples of signal models characterized by both nonconvex and convex edge penalty functions which all satisfy the second and third properties listed, are presented next.

Nonconvex edge penalty functions have been used in the past by a number of researchers. Although the edge-preserving properties of nonconvex functions are desirable, they unfortunately result in ill-posed computational problems for computing the Bayesian estimate because the corresponding log-likelihood functions possess multiple minima. Blake and Zisserman[21] introduced the neighbor interaction function, which is quadratic up to a threshold T:

$$\rho_T(\xi) = \begin{cases} \xi^2, & |\xi| \le T \\ T^2, & |\xi| > T \end{cases}. \tag{9.33}$$

This function is shown in Figure 9.5, and it provides a constant edge penalty. Many other nonconvex discontinuity-preserving functions are similar in

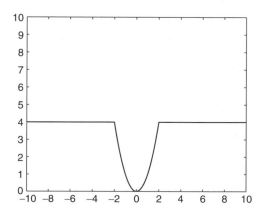

FIGURE 9.5
Blake and Zisserman's neighbor interaction function, with threshold value $T = 2$.

shape to the neighbor interaction function, such as the edge penalty function proposed by Geman and Reynolds,[22]

$$\rho_s(\xi) = -\frac{1}{1 + \left|\dfrac{\xi}{s}\right|},$$ (9.34)

in which s serves as a scaling parameter for the data.

Convex edge penalty functions have also been investigated, primarily because of the difficulty surrounding nonconvex optimization. The generalized Gauss–Markov random field (GGMRF) model was introduced by Bouman and Sauer[23] for edge-preserving image restoration and tomographic image reconstruction. This model is characterized by the edge penalty function,

$$\rho_p(\xi) = |\xi|^p, \qquad \text{for } 1 \le p \le 2,$$ (9.35)

so that the GGMRF model is a Laplacian density for $p = 1$ and a Gaussian density for $p = 2$. The parameter, p, controls the overall shape of this edge penalty function, as shown in Figure 9.6. Schultz and Stevenson[1] proposed the Huber–Markov random field (HMRF) model, which utilizes the piecewise convex function first introduced by Huber to preserve discontinuities,

$$\rho_{T_H}(\xi) = \begin{cases} \xi^2, & |\xi| \le T_H, \\ 2T_H |\xi| - T_H^2, & |\xi| > T_H. \end{cases}$$ (9.36)

The Huber threshold parameter, T_H, separates the quadratic and linear regions of this edge penalty function, serving to control the size of discontinuities within the data. The Huber edge penalty function is displayed in Figure 9.7, superimposed on the quadratic to show the reduced penalty assigned to large discontinuities.

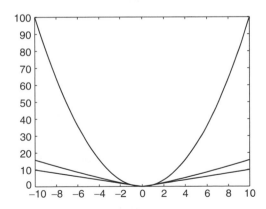

FIGURE 9.6
Family of generalized Gaussian edge penalty functions, with $p = 1$ (absolute value), $p = 1.2$, and $p = 2$ (quadratic).

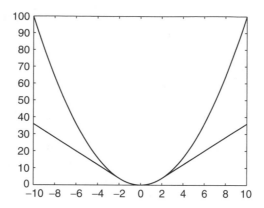

FIGURE 9.7

Huber edge penalty function with threshold value $T_H = 2$, superimposed on the quadratic edge penalty function used in the Gauss–Markov random field model.

For this research, we will assume that the image data is modeled by the Huber–Markov prior. As $T_H \to \infty$, the Huber edge penalty function becomes quadratic. In this case, a Gauss–Markov random field is formulated for the image, in which edges are severely penalized. For smaller values of the threshold parameter, the penalty assigned to edges is reduced, and the Huber–Markov model exhibits a more heavy-tailed behavior than the Gaussian. Throughout this chapter, comparisons will be made between nonlinear estimates computed using the Huber–Markov prior and linear estimates computed with the Gauss–Markov image model.

By selecting the Huber–Markov image prior model and assuming additive white Gaussian noise, the Bayesian estimate in Equation 9.19 becomes

$$\hat{\mathbf{x}} = \arg\min_{\mathbf{x}} \left\{ \sum_{\mathbf{n}} \sum_{m=1}^{4} \rho_{T_H} \left(\mathbf{d}_{\mathbf{n},m}^T \mathbf{x} \right) + \lambda \|\mathbf{y} - \mathbf{H}\mathbf{x}\|^2 \right\}. \tag{9.37}$$

In this expression, the objective function is dependent on two model parameters, the Huber threshold, T_H, which controls the size of reconstructed discontinuities, and a smoothing parameter, $\lambda = \beta/\sigma^2$, which provides a trade-off between piecewise smoothness (in the prior term) and matching the observed data exactly (in the noise term). Because the objective function in Equation 9.37 is convex, a gradient descent algorithm may be utilized to compute the unique nonlinear estimate.[1,8] This is the primary advantage of selecting a convex edge penalty function—the unique minimum of a convex objective function can be found relatively quickly when numerically calculating the estimate, since simulated annealing and related techniques are not required to search for the solution in a nonconvex objective function with local minima.

9.4 Simulations

Bayesian MAP estimation is the enhancement technique of choice in the simulations, because of its versatility in showing the difference between linear and nonlinear solutions. For each image and video observation model of interest in Section 9.2, the corresponding Bayesian MAP estimation problem is first derived. Image magnification, low bit rate block-DCT postprocessing, and superresolution video enhancement all have constraints that must be satisfied, resulting in constrained optimization problems that can be solved using the gradient projection algorithm.[1,8] Empirical simulation studies involving both linear (Gauss–Markov) and nonlinear (Huber–Markov) Bayesian estimates are presented. In the simulations, the Gauss–Markov random field model is implemented by a quadratic edge penalty function with $T_H \rightarrow \infty$, whereas the Huber–Markov random field model uses the convex but nonquadratic Huber edge penalty function with $T_H = 1.5$. The 1-m *IKONOS* satellite picture shown in Figure 9.2 is used throughout all of the simulations, so that each enhanced image can be compared visually to the same original information. Furthermore, only gray scale (8-bit) data are used in the experiments, so that color/multispectral processing does not obfuscate the enhancement results obtained by applying the algorithms to a single visible band. Finally, a quantitative image quality measure such as peak signal-to-noise ratio (PSNR) will not be used to compare the estimates, because the only truly effective evaluation of an image enhancement scheme is a qualitative visual comparison of the results. Unfortunately, qualitative interpretations also happen to be quite subjective, but the true efficacy of any image processing technique lies in the eye of the beholder, and this should not be swayed by potentially flawed quantitative assessment measures.

9.4.1 Region-of-Interest Image Magnification

Assuming a Huber–Markov random field model for piecewise smooth image data, the high-resolution image estimate can be computed from a digital image still as the following Bayesian MAP estimate:

$$\hat{\mathbf{x}} = \arg \min_{\mathbf{x} \in \mathcal{X}} \left\{ \sum_{\mathbf{n}} \sum_{m=1}^{4} \rho_{T_H} \left(\mathbf{d}_{\mathbf{n},m}^{T} \mathbf{x} \right) \right\}. \tag{9.38}$$

This solution is constrained to the set of images

$$\mathcal{X} = \{ \mathbf{x} : \mathbf{y} = \mathbf{H} \mathbf{x} \}, \tag{9.39}$$

such that all candidate high-resolution solutions must retain $r \times r$ block-wise averages that are identical to the original, low-resolution image $\mathbf{y}$.

This constraint set follows directly from the image magnification observation model in Equation 9.1. Because the objective function in Equation 9.38 is convex for both the Gauss–Markov ($T_H \to \infty$) and the Huber–Markov ($T_H < 10$ for most real-world digital images) random field models, the gradient projection constrained optimization algorithm[1] can be used to efficiently compute the unique estimate $\hat{\mathbf{x}}$. This iterative algorithm is initialized with a nearest neighbor upsampling of the low-resolution data, such that

$$\hat{\mathbf{x}}_0 = \mathbf{H}^T \mathbf{y}. \tag{9.40}$$

The linear projection required during each iteration is derived from the image magnification observation model.[1] As a rule of thumb, the fewer available constraints (in other words, the higher the upsampling factor r), the greater the number of iterations required for convergence.

The 1-m *IKONOS* satellite picture has been downsampled by a factor of $r = 4$ in Figure 9.8, so that the effective ground sampling distance becomes 4 m. The intent is to upsample the 4-m data back to its original resolution, to visually compare the interpolated estimate $\hat{\mathbf{x}}$ to the (known, in this synthetic example) high-resolution image $\mathbf{x}$. It is apparent that the linear and nonlinear Bayesian estimates are only slightly more accurate than the bilinear and cubic B-spline interpolations in this case. The nonlinear (Huber–Markov) Bayesian estimate is somewhat more sharply defined than the other enhanced images shown in Figure 9.8, but it is difficult to justify the additional computation required to calculate the solution. In Figure 9.9, data with an effective GSD of 8 m was created synthetically and then upsampled back to its original resolution by a factor of $r = 8$. So much of the original data is lost in this experiment that none of the interpolation methods performs very well. As a rule of thumb, the higher the upsampling factor (in other words, the fewer available constraints), the greater the number of iterations required for convergence. Still image magnification methods that do not have extensive *a priori* information available regarding scene content are capable of limited visual quality for upsampling factors much greater than $r = 4$. Either additional model-based constraints[24] or multiple video frames/fields[8] must be incorporated into the observation model for higher interpolation factors.

9.4.2   Postprocessing of Compression Artifacts in Block-DCT Coding

In this case, the optimization problem required to compute the Bayesian MAP estimate is identical to that used in image magnification,

$$\hat{\mathbf{x}} = \arg\min_{\mathbf{x} \in \mathcal{X}} \left\{ \sum_{\mathbf{n}} \sum_{m=1}^{4} \rho_{T_H} \left(\mathbf{d}_{\mathbf{n},m}^T \mathbf{x} \right) \right\}, \tag{9.41}$$

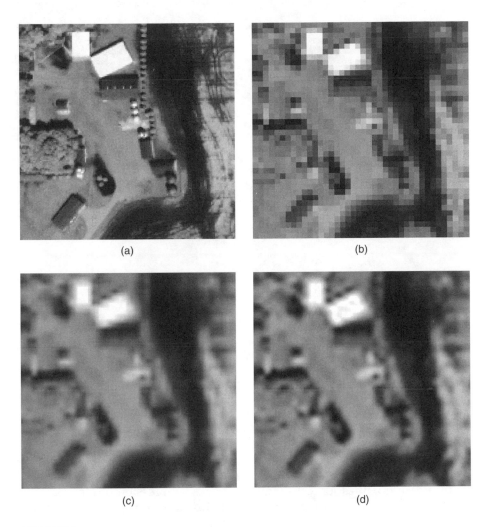

(a)   (b)

(c)   (d)

FIGURE 9.8

Region-of-interest image magnification of the *IKONOS* satellite picture with an effective GSD of 4 m. (a) Microview of the original 1-m image data. (b) 4-m image data. (c) Bilinear interpolation of the 4-m pixels. (d) Cubic B-spline interpolation of the 4-m pixels. (e) Bayesian enhancement of the 4-m data using a Gauss-Markov random field model. (f) Bayesian enhancement of the 4-m data using a Huber-Markov random field model with $T_H = 1.5$. Note that the nonlinear estimate generated using the HMRF model is slightly sharper than the linear interpolation generated by the GMRF model. (*Continued*)

with a different set of constraints required to satisfy the block-DCT coding observation model:

$$X = \{ \mathbf{x} : \mathbf{y} = \mathbf{H}^{-1} Q [\mathbf{Hx}] \} . \tag{9.42}$$

The original, observed image data $\hat{\mathbf{x}}_0 = \mathbf{y}$ is used as the initial condition for the gradient projection algorithm. Because the compressed data is in the block

(e)   (f)

FIGURE 9.8
(*Continued*)

discrete cosine transform frequency domain, the nonlinear projection operation actually takes place in the block-DCT domain. During every gradient projection iteration, the current solution must be transformed into the frequency domain ($\mathbf{H}\hat{\mathbf{x}}_j$), with each frequency value compared to the bounds of its predefined quantization cell. If a particular DCT coefficient is inside its assigned cell, then no action is taken. If a DCT coefficient is outside its assigned cell, however, then the coefficient is set equal to its nearest upper or lower bound.[3] In this manner, the solution is always guaranteed to compress back to the original observed image. Generally, enough variance is available within each quantization cell to allow for the suppression of block coding artifacts and the reconstruction of higher frequency components within each 8×8 DCT frequency square. As a rule of thumb, the lower the bit rate (in other words, the fewer available constraints), the greater the number of iterations required for convergence.

The 1-m *IKONOS* satellite images shown in Figure 9.10 were coded using a JPEG-like block-DCT algorithm. As expected, decreasing the bit rate from 0.43 to 0.264 to 0.15 bpp results in reduced image fidelity when compared to the original data, and thus a reduced number of constraints available for image enhancement. Bayesian postprocessing results using the Gauss–Markov prior are shown in Figure 9.11, and Bayesian solutions computed with the Huber–Markov prior are displayed in Figure 9.12. Block-DCT artifacts are definitely suppressed in all of the estimates, but it becomes rather difficult to recover high-frequency information for bit rates much below 0.2 bpp. It is apparent that as the number of available constraints decreases, the visual quality of the postprocessed estimates also decreases. However, fine structures such

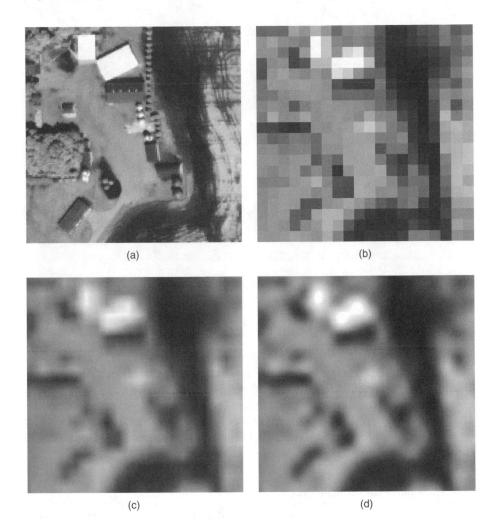

(a)

(b)

(c)

(d)

FIGURE 9.9

Region-of-interest image magnification of the *IKONOS* satellite picture with an effective GSD of 8 m. (a) Microview of the original 1-m image data. (b) 8-m image data. (c) Bilinear interpolation of the 8-m pixels. (d) Cubic B-spline interpolation of the 8-m pixels. (e) Bayesian enhancement of the 8-m data using a Gauss-Markov random field model. (f) Bayesian enhancement of the 8-m data using a Huber-Markov random field model with $T_H = 1.5$. Because there are such a limited number of constraints available in the observed data, the linear and nonlinear Bayesian solutions appear to be very similar, without significantly more detail than the cubic B-spline interpolated image. (*Continued*)

as object edges and other discontinuities are sharper in Figure 9.12, which provides evidence for the superiority of nonlinear over linear processing.

Finally, to compare the postprocessing of block-DCT compressed images to the highly related problem of still image magnification, only the DC

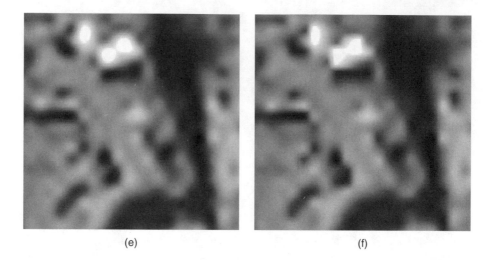

(e)   (f)

FIGURE 9.9
(*Continued*)

coefficient was retained in each 8×8 frequency square. This image has also been included in Figure 9.10, and it is identical to the downsampled *IKONOS* image shown in Figure 9.9 with an effective GSD of 8 m. In comparing the Bayesian estimates in Figures 9.9, 9.11, and 9.12, it is readily apparent that image magnification with $r = 8$ and the postprocessing of extremely low bit rate images coded using the block-DCT generate virtually identical solutions.

9.4.3   Superresolution Video Enhancement

Using a Huber–Markov random field model for piecewise smooth image data, the high-resolution video still can be computed as the following Bayesian MAP estimate:

$$\hat{\mathbf{x}} = \arg\min_{\mathbf{x} \in \mathcal{X}} \left\{ \sum_{\mathbf{n}} \sum_{m=1}^{4} \rho_{T_H}\left(\mathbf{d}_{\mathbf{n},m}^T \mathbf{x}\right) + \sum_{\substack{\mathbf{y}^{(i)} \in \mathcal{Y} \\ i \neq k}} \lambda^{(i)} \left\| \mathbf{y}^{(i)} - \hat{\mathbf{H}}^{(i)} \mathbf{x} \right\|^2 \right\} \qquad (9.43)$$

In this expression, the constraint set is defined as

$$\mathcal{X} = \left\{ \mathbf{x} : \mathbf{y}^{(k)} = \mathbf{H}^{(k)} \mathbf{x} \right\}, \qquad (9.44)$$

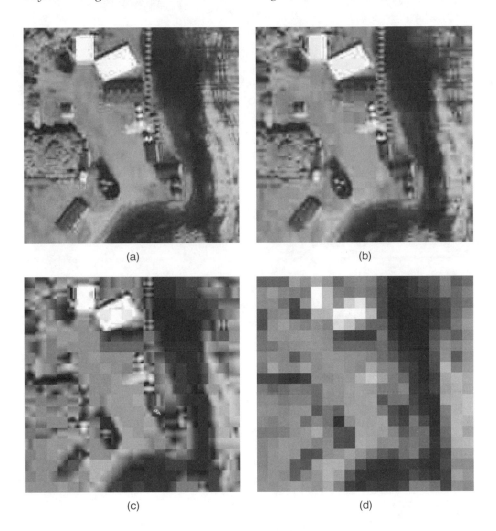

FIGURE 9.10
Microview of the block-DCT-encoded *IKONOS* 1-m satellite picture with 8 × 8 pixel blocks, to show the reduction in the number of constraints available for enhancing a digital image as its bit rate is decreased. (a) Original satellite picture coded at 0.43 bpp. (b) Block-DCT coding at 0.264 bpp. (c) Block-DCT coding at 0.15 bpp. (d) Block-DCT coding, in which only the DC (lowest frequency) discrete cosine transform coefficient is retained from each 8 × 8 frequency square. Note that this last case is identical to the downsampled *IKONOS* image with an effective GSD of 8 m.

in order to constrain the solution to the observed data present in the reference frame or field. $\mathbf{H}^{(k)}$ is not an estimate in this case, because displacement vectors are not required to downsample and/or subsample $\mathbf{x}$. The parameter $\lambda^{(i)}$ represents the confidence in estimating $\hat{\mathbf{H}}^{(i)}$, constructed using the subpixel displacements estimated from $\mathbf{y}^{(i)}$ with respect to $\mathbf{y}^{(k)}$ through the application of block matching. If only a single, progressively scanned video frame

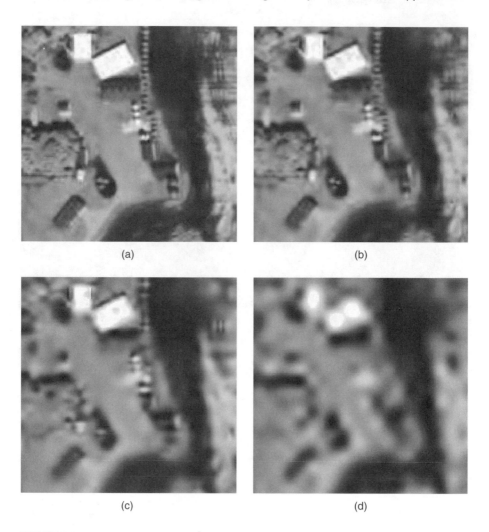

FIGURE 9.11

Postprocessing of block-DCT coding artifacts using Bayesian estimation with the Gauss–Markov random field prior model. (a) 0.43 bpp. (b) 0.264 bpp. (c) 0.15 bpp. (d) DC discrete cosine transform coefficient only.

is available for enhancement, this problem formulation is identical to that derived for still image magnification. In fact, the initial condition $\hat{x}_0 = \mathbf{H}^{(k)^T}\mathbf{y}^{(k)}$ for the gradient projection algorithm is the same as that used in magnification, although both upsampling and interlacing must be taken into account for the video frame. As a rule of thumb, the fewer available constraints (in other words, the fewer available frames/fields with similar scene content), the greater the number of iterations required for convergence.

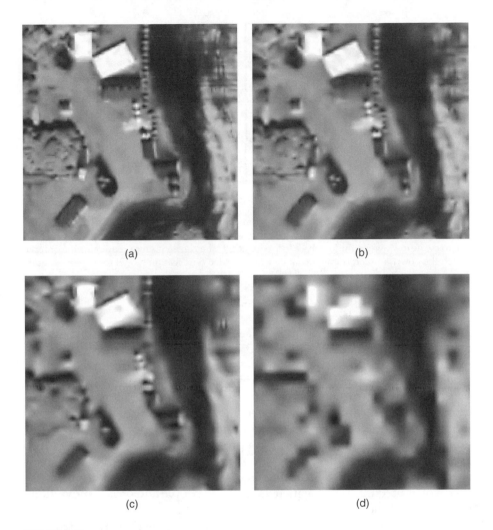

FIGURE 9.12
Postprocessing of block-DCT coding artifacts using Bayesian estimation with the Huber–Markov random field prior model, $T_H = 1.5$. (a) 0.43 bpp. (b) 0.264 bpp. (c) 0.15 bpp. (d) DC discrete cosine transform coefficient only.

The numerous additional constraints available from a video sequence consisting of people, objects, or camera motion between frames make super-resolution one of the most powerful techniques available for spatial region-of-interest detail extraction. As a caveat, scene integration is only possible where lighting conditions remain constant throughout the sequence of interest and when subpixel motion can be estimated very accurately. For the most part, subpixel motion estimation has not been adequately addressed in the super-resolution enhancement literature because it is a very challenging

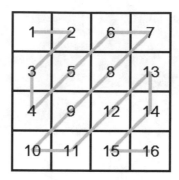

FIGURE 9.13

Pixel shifts used to create the high-resolution frames that comprise the synthetic *IKONOS* satellite image sequence. Each of these 16 progressively scanned frames was downsampled by a factor of r to create the low-resolution observed sequence $\mathcal{Y}$. The zigzag pattern represents the order in which the downsampled frames are added to the video observation model for superresolution enhancement.

problem in and of itself.[11–13,25,26] Displacements were estimated automatically for the experimental results presented in this chapter; however, the selected examples all have global scene shifts between frames, and this makes motion recovery relatively easy. To state this in another way, the sequences that are best suited for superresolution enhancement are those of stationary scenes acquired by moving cameras. Remote sensing, reconnaissance, and surveillance video footage often contain this type of representative motion. The superresolution video enhancement algorithm presented in this chapter does indeed accommodate independent object motion; unfortunately, the difficulty in accurately estimating the displacements of individual pixels makes superresolution of limited utility for sequences in which people and objects move along independent trajectories in front of a camera.

A synthetic sequence of 16 progressively scanned images was created from the 1-m *IKONOS* satellite image, in order to compare the efficacy of superresolution video enhancement with still image magnification. A region-of-interest was first extracted from the original still image, and this was designated as the progressively scanned reference frame. Next, 15 additional images were extracted from the same still image, through all possible combinations of 0, 1, 2, and 3 pixel shifts both horizontally and vertically with respect to the reference frame. Figure 9.13 shows the displacements that were generated, with frame 1 (no horizontal and no vertical shifts) designated as the reference. These 16 frames were subsequently downsampled by a factor of $r = 4$, to create a synthetic video sequence with an effective GSD of 4 m and subpixel shifts of 0.0, 0.25, 0.5, and 0.75 pixels with respect to the low-resolution grid. Low-resolution frames were added to the video observation model in a zigzag pattern, following the path traversed in Figure 9.13. For example, if three frames are integrated to reconstruct the original image data, the downsampled reference picture (frame 1), the downsampled frame with a 0.25 pixel horizontal shift (frame 2), and the downsampled frame with a 0.25 pixel vertical shift (frame 3) are used as the observed sequence y.

Subpixel motion estimation was performed using the block matching technique.[8,25] In this case, all frames were first upsampled independently using cubic B-spline interpolation,[1] and then block matching was applied to each upsampled frame with respect to the upsampled reference picture (frame 1) selected for enhancement. Motion estimation accuracy is critical for superresolution video enhancement, but block matching is only accurate within image areas that contain a large quantity of edges. This is further exacerbated by large values of the upsampling factor r, because the lack of constraints makes the interpolated frames smooth. Since it is known *a priori* that global displacements were used to model a camera panning the scene of interest, it became possible to filter each motion vector field to estimate a more accurate displacement for each frame. The nonlinear α-trimmed mean filter[26] was applied to each set of motion vectors, to remove outliers and to generate a very accurate global displacement estimate for each frame with respect to the reference picture.

Figure 9.14 shows the Bayesian superresolution estimates computed using the Gauss–Markov prior, for the synthetic *IKONOS* sequence with an effective GSD of 4 m. Note that more details become visible in the superresolved estimate as the number of frames used in the video observation model increases. Additionally, it should be mentioned that superresolution video enhancement follows a "law of diminishing returns," in which there is a nonlinear relationship with respect to the amount of extra information provided by each frame. The integration of three frames provides a dramatic increase in detail over single-frame image magnification. Integrating six frames provides only marginally more information to the superresolved estimate, while all 16 frames generate a very accurate representation of the scene of interest, but only slightly more so than the estimate computed from six frames. Theoretically, if all 16 frames are included in the observation model, the superresolution enhancement problem should become a well-posed inverse problem, and the exact data can be reconstructed from the observations. However, due to video model inaccuracies, the Bayesian estimate computed from the 16 observations is not identical to the original data.

Nonlinear Bayesian estimates are shown in Figure 9.15, in which the Huber–Markov model was used to preserve edges. Indeed, edges and discontinuities are significantly sharper than the Gauss–Markov simulation results found in Figure 9.14. Perfect subpixel motion recovery was possible in both the Gauss–Markov and the Huber–Markov simulations, and thus the motion confidence parameter was set to a high value, $\lambda^{(i)} = 1000$. It must be reiterated that the motion was estimated from a synthetic video sequence with frames extracted from the same still image, so that image intensities were constant along scene trajectories. In this case, it is quite simple to recover the motion accurately.

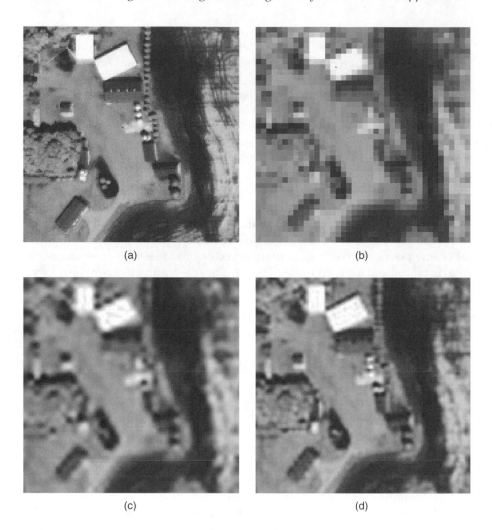

(a)

(b)

(c)

(d)

FIGURE 9.14

Superresolution video enhancement of the *IKONOS* satellite picture with an effective GSD of 4 m, using Bayesian estimation with the Gauss–Markov random field prior model, $\lambda^{(i)} = 1000$. (a) Microview of the original 1-m image data. (b) 4-m image data. (c) Bayesian enhancement of the 4-m pixels using 1 frame. (d) Bayesian enhancement of the 4-m pixels using 3 frames. (e) Bayesian enhancement of the 4-m pixels using 6 frames. (f) Bayesian enhancement of the 4-m pixels using all 16 available frames. Note that a "law of diminishing returns" for adding frames into the video model is observed in super-resolution enhancement. (*Continued*)

Subpixel motion estimation is certainly more challenging for arbitrary, real-world video sequences.

High-resolution frames 1, 5, 12, and 16 along the diagonal of Figure 9.13 were downsampled by a factor of $r = 8$ to generate a four-frame synthetic *IKONOS* sequence with an effective GSD of 8 m. Figure 9.16 compares the

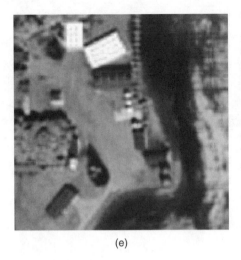

(e)   (f)

FIGURE 9.14
(*Continued*)

corresponding Gauss–Markov and Huber–Markov enhanced images with the original data. Although far from perfect, many more details can be extracted from the sequence because of the additional constraints provided by four progressively scanned frames, even for the relatively large upsampling factor. Moreover, the nonlinear solutions appear sharper than the Gauss–Markov estimates.

As a more practical, "real-world" exercise of the superresolution video enhancement algorithm, actual footage of an electrical engineering laboratory test bench was taken using a Sony Hi8 Handycam (Model CCD-TR930). Figure 9.17 shows the original *EELab* sequence reference picture, which was captured by panning the handheld camera across the stationary scene as a sine wave gradually appears on the oscilloscope. The sequence was digitized using a framegrabber; five progressively scanned frames were first selected from this sequence and subsequently downsampled by a factor of $r = 4$. Subpixel motion estimation was applied to the four pictures relative to the reference frame. Once again, because the motion involved a camera pan, the global scene displacements were recovered accurately. The nonlinear (Huber–Markov) estimates are definitely sharper than the linear (Gauss–Markov) estimates in this case. Note that the motion confidence parameter was reduced to $\lambda^{(i)} = 10$ in this real video example, because the single displacement estimated for each frame is only an approximation to the true motion (including translation, rotation, and perspective[6,27]) of the handheld camera.

Finally, a second "real-world" experiment was conducted using a Sony MiniDV Digital Handycam (Model DCR-TRV20) to capture DV footage. With 4:1:1 color sampling and a fixed intraframe compression ratio of 5:1 (data rate

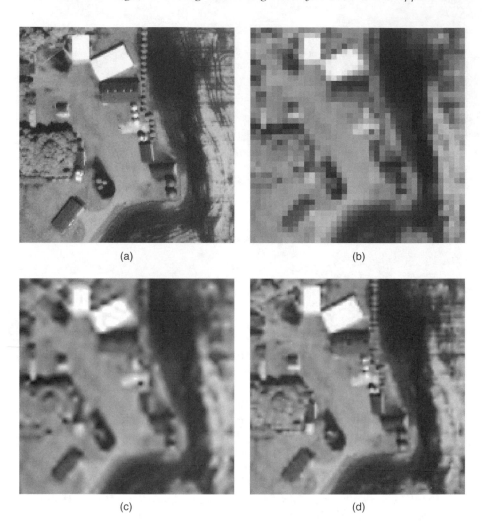

FIGURE 9.15

Superresolution video enhancement of the *IKONOS* satellite picture with an effective GSD of 4 m, using Bayesian estimation with the Huber–Markov random field prior model, $T_H = 1.5$ and $\lambda^{(i)} = 1000$. (a) Microview of the original 1-m image data. (b) 4-m image data. (c) Bayesian enhancement of the 4-m pixels using 1 frame. (d) Bayesian enhancement of the 4-m pixels using 3 frames. (e) Bayesian enhancement of the 4-m pixels using 6 frames. (f) Bayesian enhancement of the 4-m pixels using all 16 available frames. Note that these nonlinear estimates are definitely sharper in appearance when compared to the linear estimates computed using the Gauss–Markov random field model in Figure 9.14. (*Continued*)

of 3.5 MB/s), degradation of DV frames is barely noticeable.[28] Superresolution video enhancement was applied directly to a short sequence acquired by panning the MiniDV camera over a stationary scene containing a clock, with the enhancement results shown in Figure 9.18. DV frames were imported via

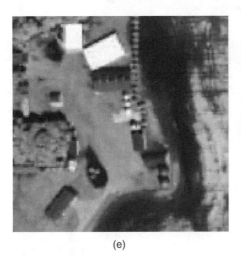

(e)   (f)

FIGURE 9.15
(*Continued*)

an IEEE 1394 (also known as FireWire) connection and the DV-NTSC standard within Adobe Premiere (720×480 pixel frames; 29.97 fps frame rate; standard NTSC video with a frame aspect ratio of 4:3 and interlacing; pixel aspect ratio of 0.9). Interlacing, which further reduces the number of useful constraints available for enhancing a particular video frame, is incorporated into the video observation model through $\hat{\mathbf{H}}^{(i)}$. In terms of appearance, the linear (Gauss–Markov) estimate in Figure 9.18 is more aesthetically pleasing than the nonlinear (Huber–Markov) estimate, probably because the jagged structures that are present due to the lack of constraints available in alternating scan lines become smoothed in the linear estimate. Note that the motion confidence parameter was further reduced to $\lambda^{(i)} = 2$ in this real video example, because of the nonuniformity of motion across the scenes. In the simulation results depicted in Figure 9.18, *no downsampling of the original data took place prior to super-resolution enhancement*. Thus, unlike most simulation results presented in the literature, this experiment is representative of real-world superresolution enhancement.

As the world moves farther in the direction of digital communications, low bit rate video compression is becoming increasingly important,[29,30] and superresolution enhancement can help restore some of the information lost due to encoding. However, the success of superresolution enhancement is predicated on the redundancy of the data found within a video sequence. When that redundancy is eliminated through intraframe and interframe compression, it becomes almost futile to extract details from an arbitrary sequence.[16] Superresolution enhancement has been applied to MPEG sequences (high variable compression ratios in excess of 10:1) in the past,[14–16] although so much redundant data is removed in the interframe coding process that the

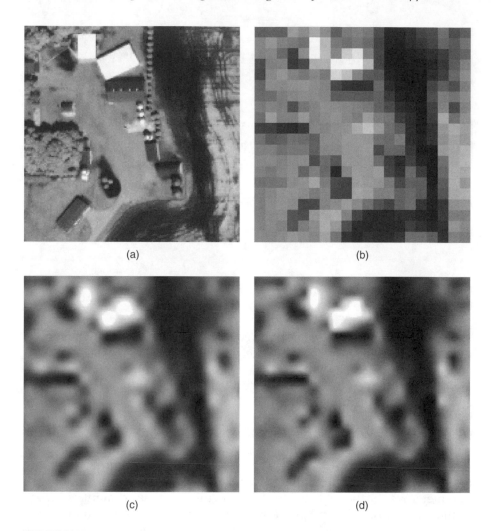

(a)   (b)

(c)   (d)

FIGURE 9.16

Bayesian superresolution video enhancement of the *IKONOS* satellite picture with an effective GSD of 8 m. (a) Microview of the original 1-m image data. (b) 8-m image data. (c) Bayesian enhancement of the 8-m pixels using 1 frame (GMRF). (d) Bayesian enhancement of the 8-m pixels using 1 frame (HMRF, $T_H = 1.5$). (e) Bayesian enhancement of the 8-m pixels using all 4 available frames (GMRF, $\lambda^{(i)} = 1000$). (f) Bayesian enhancement of the 8-m pixels using all 4 available frames (HMRF, $T_H = 1.5$ and $\lambda^{(i)} = 1000$). Note that the nonlinear (HMRF) estimates are slightly sharper in appearance when compared to the linear (GMRF) estimates. (*Continued*)

extraction of additional details from arbitrary MPEG sequences is generally not possible. Examining Figure 9.18 reveals evidence that enough redundant pixel information remains within an intra-frame coded DV sequence (low fixed compression ratio of 5:1) to warrant the application of superresolution enhancement to video coded at high bit rates.

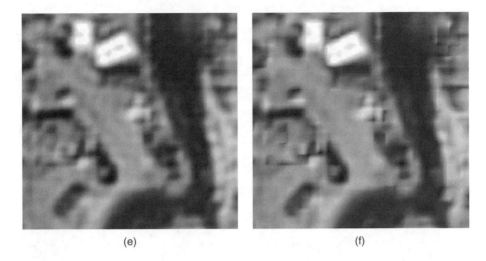

(e)   (f)

FIGURE 9.16
(*Continued*)

9.5   Summary and Future Directions

Three important image and video enhancement problems were investigated in this chapter, namely, region-of-interest image magnification, the postprocessing of block-DCT compression artifacts, and the superresolution enhancement of digital video. In all cases, Bayesian maximum *a posteriori* estimation was shown to be an effective regularization technique. Furthermore, nonlinear estimates computed using the Huber–Markov random field model were shown to be superior to linear estimates in most cases, especially with respect to edge preservation and the enhancement of discontinuities present within the observed data.

It was shown repeatedly that the number of available constraints places an upper bound on the performance of any enhancement algorithm, whether those constraints are in the form of high-frequency DCT coefficients, in the case of image coding, or multiple video frames containing shifted views of the same scene, in superresolution enhancement. Intuitively, the larger the number of available constraints, the higher the potential for enhancement accuracy. Furthermore, additional usable constraints result in a faster convergence of the gradient projection algorithm.

An obvious future extension of this research is to combine downsampling, interlacing, motion, and compression into a single video observation model, in order to postprocess compression artifacts and increase resolution simultaneously. As mentioned previously, the performance of superresolution video enhancement is predicated on the redundancy of data within a sequence.

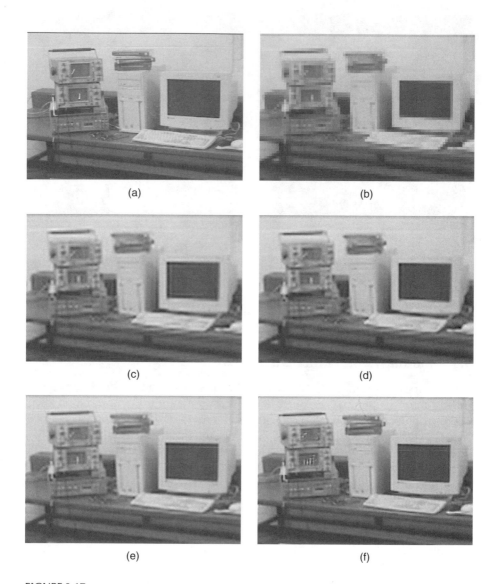

FIGURE 9.17

Bayesian superresolution video enhancement of the progressive *EELab* sequence, captured using a Sony Hi8 Handycam panning a stationary scene. (a) Original reference picture. (b) Reference picture downsampled by a factor of $r = 4$. (c) Bayesian enhancement of the downsampled pixels using 1 frame (GMRF). (d) Bayesian enhancement of the downsampled pixels using 1 frame (HMRF, $T_H = 1.5$). (e) Bayesian enhancement of the downsampled pixels using 5 frames (GMRF, $\lambda^{(i)} = 10$). (f) Bayesian enhancement of the downsampled pixels using 5 frames (HMRF, $T_H = 1.5$ and $\lambda^{(i)} = 10$). Note that the nonlinear (HMRF) estimates are sharper in appearance compared to the linear (GMRF) estimates.

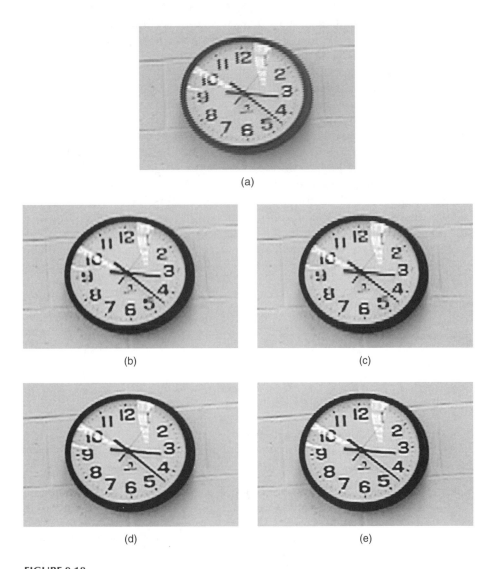

FIGURE 9.18

Bayesian superresolution video enhancement of the interlaced *Clock* sequence, captured using a Sony MiniDV camera panning a stationary scene. No downsampling of the original data took place in this example. (a) Reference picture, consisting of an upper field and a lower field. (b) Bayesian enhancement of the single lower reference field (GMRF). (c) Bayesian enhancement of the single lower reference field (HMRF, $T_H = 1.5$). (d) Bayesian enhancement of the reference field using 5 lower fields (GMRF, $\lambda^{(i)} = 2$). (e) Bayesian enhancement of the reference field using 5 lower fields (HMRF, $T_H = 1.5$ and $\lambda^{(i)} = 2$). From a subjective standpoint, the linear (GMRF) estimates are more aesthetically pleasing than the nonlinear (HMRF) estimates in this simulation.

Thus, the expectation is that compressed video will not be conducive to enhancement, because much of the data redundancy is eliminated in the coding process. However, if the encoder can be properly tuned for superresolution enhancement in the decoder, then accurate high-resolution video stills can be computed from compressed video formats.

Future research will examine real-world uses of superresolution enhancement, particularly for DV and digitized NTSC video sequences[28] of interest to forensic image analysts.[31,32] The primary challenges of applying this technology to arbitrary sequences are (1) the ability to extract extra information from a video sequence is highly dependent on the motion contained between the spatial region of interest and its temporally neighboring frames/fields; (2) superresolution enhancement is highly computational; and (3) unavoidable inaccuracies in the estimated subpixel motion vectors can cause severe degradations in visual quality within each high-resolution video still. Continuing research will be conducted to determine which video frames/fields are "integrable," i.e., to automatically detect which frames/fields are capable of contributing additional information to a spatial region of interest selected for enhancement.

References

1. R.R. Schultz and R.L. Stevenson, A Bayesian approach to image expansion for improved definition, *IEEE Trans. Image Process.*, 3(3), 233–242, 1994.
2. W.B. Pennebaker and J.L. Mitchell, *JPEG: Still Image Data Compression Standard*. Amsterdam: Kluwer Academic, 1992.
3. T.P. O'Rourke and R.L. Stevenson, Improved image decompression for reduced transform coding artifacts, *IEEE Trans. Circuits Syst. Video Technol.*, 5(6), 490–499, 1995.
4. Y. Yang, N.P. Galatsanos, and A.K. Katsaggelos, Regularized reconstruction to reduce blocking artifacts of block discrete cosine transform compressed images, *IEEE Trans. Circuits Syst. Video Technol.*, 3(6), 421–432, 1993.
5. A. Zakhor, Iterative procedures for reduction of blocking effects in transform image coding, *IEEE Trans. Circuits Syst. Video Technol.*, 2(1), 91–95, 1992.
6. S. Mann and R.W. Picard, Virtual bellows: constructing high quality stills from video, in *Proceedings of the 1994 IEEE International Conference on Image Processing*, Austin, TX, 1994, 363–367.
7. A.J. Patti, M.I. Sezan, and A.M. Tekalp, High-resolution standards conversion of low resolution video, in *Proceedings of the 1995 IEEE International Conference on Acoustics, Speech, and Signal Processing*, Detroit, MI, 1995, 2197–2200.
8. R.R. Schultz and R.L. Stevenson, Extraction of high-resolution frames from video sequences, *IEEE Trans. Image Process.*, 5(6), 996–1011, 1996.
9. R.Y. Tsai and T.S. Huang, Multiframe image restoration and registration, in *Advances in Computer Vision and Image Processing*, Vol. 1, R.Y. Tsai and T.S. Huang, Eds., JAI Press, New York, 1984, 317–339.

10. P. Cheeseman, B. Kanefsky, R. Kraft, J. Stutz, and R. Hanson, Super-resolved surface reconstruction from multiple images, in *Maximum Entropy and Bayesian Methods*, G.R. Heidbreder, Ed., Amsterdam: Kluwer Academic, 1996, 293–308.

11. R.R. Schultz, L. Meng, and R.L. Stevenson, Subpixel motion estimation for super-resolution image sequence enhancement, *J. Visual Commun. Image Representation*, 9(1), 38–50, 1998.

12. R.R. Schultz and R.L. Stevenson, Bayesian estimation of subpixel-resolution motion fields and high-resolution video stills, in *Proceedings of the 1997 IEEE International Conference on Image Processing* (on CD-ROM), Santa Barbara, CA, 1997.

13. R.R. Schultz and R.L. Stevenson, Estimation of subpixel-resolution motion fields from segmented image sequences, in *Proceedings of the SPIE—Sensor Fusion: Architectures, Algorithms, and Applications II (AeroSense'98)*, Vol. 3376, Orlando, FL, 1998, 90–101.

14. Y. Altunbasak, A.J. Patti, and R.M. Mersereau, Super-resolution still and video reconstruction from MPEG-coded video, *IEEE Trans. Circuits Syst. Video Technol.*, 12(4), 2002, 217–226.

15. D. Chen and R.R. Schultz, Extraction of high-resolution frames from MPEG image sequences, in *Proceedings of the 1998 IEEE International Conference on Image Processing* (on CD-ROM), Chicago, IL, 1998.

16. K.J. Erickson and R.R. Schultz, MPEG-1 super-resolution decoding for the analysis of video still images, in *Proceedings of the 2000 IEEE Southwest Symposium on Image Analysis and Interpretation*, Austin, TX, 2000, 13–17.

17. Y. Yang, M. Choi, and N.P. Galatsanos, New results on multichannel regularized recovery of compressed video, in *Proceedings of the 1998 IEEE International Conference on Image Processing* (on CD-ROM), Chicago, IL, 1998.

18. R.R. Schultz and R.L. Stevenson, Motion-compensated scan conversion of interlaced video sequences, in *Proceedings of the IS&T/SPIE Conference on Image and Video Processing IV*, Vol. 2666, San Jose, CA, 1996, 107–118.

19. A.M. Tekalp, *Digital Video Processing*, Upper Saddle River, NJ: Prentice-Hall, 1995.

20. S. Geman and D. Geman, Stochastic relaxation, Gibbs distributions, and the Bayesian restoration of images, *IEEE Trans. Pattern Anal. Mach. Intelligence*, 6(6), 721–741, 1984.

21. A. Blake and A. Zisserman, *Visual Reconstruction*, Cambridge, MA: MIT Press, 1987.

22. D. Geman and G. Reynolds, Constrained restoration and the recovery of discontinuities, *IEEE Trans. Pattern Anal. Mach. Intelligence*, 14(6), 367–383, 1992.

23. C.A. Bouman and K. Sauer, A generalized Gaussian image model for edge-preserving MAP estimation, *IEEE Trans. Image Process.*, 2(3), 296–310, 1993.

24. S. Baker and T. Kanade, Limits on super-resolution and how to break them, *IEEE Trans. Pattern Anal. Mach. Intelligence*, 24(9), 2002.

25. B. Furht, J. Greenburg, and R. Westwater, *Motion Estimation Algorithms for Video Compression*, Amsterdam: Kluwer Academic, 1997.

26. R.R. Schultz, Nonlinear filtering of subpixel motion vectors for improved super-resolution video frame enhancement, in *Proceedings of the 2001 IEEE-EURASIP Workshop on Nonlinear Signal and Image Processing* (on CD-ROM), Baltimore, MD, June 2001.

27. R.R. Schultz and M.G. Alford, Multiframe integration via the projective transformation with automated block matching feature point selection, in *Proceedings*

of the 1999 IEEE International Conference on Acoustics, Speech, and Signal Processing (on CD-ROM), Phoenix, AZ, 1999.

28. R.R. Schultz, Super-resolution enhancement of native digital video versus digitized NTSC sequences, in *Proceedings of the 2002 IEEE Southwest Symposium on Image Analysis and Interpretation*, Santa Fe, NM, Apr. 2002, 193–197.

29. B.G. Haskell, A. Puri, and A.N. Netravali, *Digital Video: An Introduction to MPEG-2*, New York: Chapman & Hall, 1997.

30. J.L. Mitchell, W.B. Pennebaker, C.E. Fogg, and D.J. LeGall, *MPEG Video Compression Standard*, New York: Chapman & Hall, 1996.

31. G. Fredericks, Forensic video analysis: training to be a forensic expert and getting needed equipment, *Law Enforcement Technol.*, 28(6), 34–39, 2001.

32. G. Fredericks, Forensic video analysis: understanding your frustrations with CCTV, *Law Enforcement Technol.*, 28(5), 30–34, 2001.

10

Statistical Image Modeling and Processing Using Wavelet-Domain Hidden Markov Models

Guoliang Fan and Xiang-Gen Xia

CONTENTS

10.1 Introduction

In this chapter, we study wavelet-domain hidden Markov models (HMMs) regarding both statistical image modeling and the application to various image processing problems. As prerequisites, image models often play important roles in many image processing applications. Specifically, a statistical image model regards an image as a realization of a certain probability model, and predicts a set of possible outcomes weighted by their likelihoods or probabilities. In this work, we are particularly interested in statistical image modeling and processing using the wavelet-domain HMMs proposed in Reference 1, where two major mathematical tools are involved, e.g., wavelets and HMMs.

Wavelets are powerful mathematical tools with flexible multiresolution structures and many varieties. Although the idea of multiresolution analysis goes back to the early years of this discipline, it was formally developed in the 1980s.[2] The construction of compactly supported wavelets[3] has attracted the attention of the larger scientific community and has stimulated

(a)   (b)

FIGURE 10.1
Lena (256 × 256, 8 bpp) image and its three-scale wavelet transform, where the gray level corresponds to the magnitudes of wavelet coefficients.

tremendous research activities. In particular, the discrete wavelet transform (DWT) can provide a more favorable representation that helps us develop efficient image modeling and processing techniques. For example, Figure 10.1 shows the Lena image and its three-scale DWT, which demonstrates a *compact*, *joint spatial frequency*, and *multiscale* image representation. First, the compact property of DWT indicates that most image energy can be compacted onto a few wavelet coefficients with large magnitudes. At the same time, most coefficients are very small. This compact property allows us to capture the key characteristics of an image from those large wavelet coefficients. Second, the joint spatial-frequency representation results in two evident properties of wavelet coefficient distribution, i.e., *interscale persistence* and *intrascale clustering*. These two observations inspire many researchers to develop various statistical models for image restoration, compression, and classification, etc. Third, the multiscale representation of DWT is also useful in many image processing applications, such as progressive image transmission, embedded image compression, multiresolution image analysis, etc.

Another important mathematical tool of statistical modeling discussed in this chapter is the hidden Markov model (HMM). The theory of HMMs was originally developed in the 1960s.[4] HMMs have earned popularity mainly from their successful application to speech recognition. A recent review of the HMMs theory can be found in Reference 5. An HMM model has a finite set of *states*, each of which is associated with a (generally multidimensional) probability distribution. Transitions among the states are governed by a set of

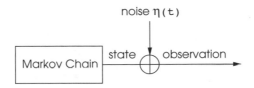

FIGURE 10.2
Illustration of an HMM.

probabilities called *transition probabilities*. In a particular state, an outcome or *observation* can be generated, according to the associated probability distribution. It is only the outcome, not the state, that is visible to an external observer, and therefore states are *hidden* to the outside—hence the name hidden markov model, as shown in Figure 10.2. HMMs with states in a finite discrete set and uncertain parameters have been widely applied in areas such as communication systems, speech processing, and biological signal processing.

There are many real-world signals that are amenable to the finite state mechanism in an HMM, e.g., speech signals. However, HMMs cannot be applied directly to image modeling in the spatial domain due to the large number of intensity levels of image pixels. Recently, a marriage of DWT and HMM produced a new, efficient mathematical tool for statistical modeling, namely, wavelet-domain HMMs, which were originally proposed in Reference 1. Wavelet-domain HMMs allow efficient statistical modeling because the decorrelation of DWT makes HMMs manipulable and useful by greatly reducing the state number in the wavelet domain, as indicated by Figure 10.1. Using probabilistic graphs, wavelet-domain HMMs can effectively characterize the joint statistics of wavelet coefficients of a given signal or image. They have been applied to image denoising,[6] image segmentation,[7] Bayesian image analysis,[8] and image retrieval.[9] In this work, we further investigate: (1) how to design and train wavelet-domain HMMs to obtain the accurate and robust statistical characterization for different applications; and (2) how to develop efficient image processing algorithms to take advantage of wavelet-domain HMMs in various applications.

The remainder of this chapter is organized into five sections as follows. We first review wavelet-domain HMMs as proposed in Reference 1, and we also show some preliminary studies. Second, image denoising is discussed, and a new wavelet-domain HMM that can outperform most state-of-the-art denoising methods is proposed. Third, multiscale Bayesian segmentation is studied, where wavelet-domain HMMs are used to obtain statistical image characterization. Fourth, texture analysis and synthesis are investigated by using an improved wavelet-domain HMM for more accurate and complete texture modeling than traditional wavelet-domain HMMs. Finally, conclusions are given.

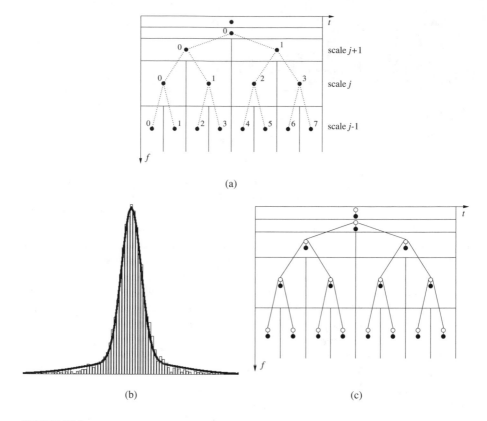

FIGURE 10.3
(a) Tiling of the time-frequency plane of DWT. The solid dot at the center corresponds to the scaling coefficients or wavelet coefficients. The tree structure is shown by the link of the dotted lines. (b) The histogram of the one-scale DWT (Daubechies-4) of the "fruit" image where a two-state zero-mean Gaussian mixture model can closely fit the real DWT data.[1] (c) Wavelet-domain HMT, where the white node represents the state variable S and the black node denotes the wavelet coefficient W.

10.2   Wavelet-Domain Hidden Markov Models

In the following, we briefly review wavelet-domain HMMs in the one-dimensional (1D) case. For more details, we refer the reader to Reference 1.

Given a bandpass wavelet function $\psi(t)$ and a lowpass scaling function $\phi(t)$, the DWT represents a signal $s(t)$ of size N in terms of shifted versions of $\phi(t)$ and shifted and dilated versions of $\psi(t)$, as shown in Figure 10.3a. The atoms of DWT are $\psi_{j,i}(t) \equiv 2^{-j/2}\psi(2^{-j}t - i)$, $\phi_{j,i}(t) \equiv 2^{-j/2}\phi(2^{-j}t - i)$,

$j, i \in \{1, \ldots, J\}$, and the wavelet representation can be written as[10]

$$s(t) = \sum_{i=0}^{N_J - 1} u_{J,i} \phi_{J,i}(t) + \sum_{j=1}^{J} \sum_{i=0}^{N_j - 1} w_{j,i} \psi_{j,i}(t), \tag{10.1}$$

where J denotes the *scale* of analysis, and scale J indicates the coarsest scale or lowest resolution of analysis. $N_j = N/2^j$ is the number of coefficients at scale j. $u_{J,i} = \int s(t) \phi_{J,i}(t) dt$ is the scaling coefficient, which measures the local mean around the time $2^J i$. $w_{j,i} = \int s(t) \psi_{j,i}(t) dt$ is the wavelet coefficient, which characterizes the local variation around the time $2^j i$ and the frequency $2^j f_0$. Because of the multiscale binary-tree structure, given a wavelet coefficient $w_{j,i}$, its parent is $w_{j+1,\lfloor i/2 \rfloor}$, where the operation $\lfloor x \rfloor$ takes the integer part of x, and its two children are $w_{j-1,2i}$ and $w_{j-1,2i+1}$, as shown in Figure 10.3a. In the following, we use **w** to denote the vector of all wavelet coefficients.

For most real-world signals and images, the set of wavelet coefficients is *sparse*. This means that the majority of the coefficients are small and only a few coefficients contain most of the signal energy. Thus, the probability density function (pdf), $f_W(w)$, of the wavelet coefficients w can be described by a peak (centered at $w = 0$) and heavy-tailed non-Gaussian density, where W stands for the random variable of w. It was presented in Reference 11 that the Gaussian mixture model (GMM) can well approximate this non-Gaussian density, as shown in Figure 10.3b. Therefore, we associate each wavelet coefficient w with a set of discrete hidden states $S = 0, 1, \ldots, M - 1$, which have probability mass functions (pmf), $p_S(m)$. Given $S = m$, the pdf of the coefficient w is Gaussian with mean μ_m and variance σ_m^2. We can parameterize an M-state GMM by $\pi = \{p_S(m), \mu_m, \sigma_m^2 | m = 0, 1, \ldots, M - 1\}$, and the overall pdf of w is determined by

$$f_W(w) = \sum_{m=0}^{M-1} p_S(m) f_{W|S}(w|S = m), \tag{10.2}$$

where

$$f_{W|S}(w|S = m) = \frac{1}{\sqrt{2\pi\sigma_m^2}} \exp \frac{-(w - \mu_m)^2}{2\sigma_m^2} \triangleq g(w; \mu_m, \sigma_m^2). \tag{10.3}$$

Although w is conditionally Gaussian given its state $S = m$, it is not Gaussian in general due to the randomness of the state variable S.

Although the orthogonal DWT can decorrelate an image with almost uncorrelated wavelet coefficients, it is widely understood that there is a considerable amount of high-order dependencies existing in **w**. This can be observed from the characteristics of the wavelet coefficient distribution, such as *intrascale clustering* and *interscale persistence*, as shown in Figure 10.1. Therefore, in Reference 1, a tree-structured hidden Markov tree (HMT) model was developed by connecting state variables of wavelet coefficients vertically across the scale, as shown in Figure 10.3c, where we can see that the HMT is able to capture the underlying interscale dependencies between parent and child state variables, which the second-order statistics cannot provide. In HMT, each coefficient $W_{j,i}$ is conditionally independent of all other random variables given its state $S_{j,i}$. Thus, an M-state HMT is parameterized by

- $p_{S_J}(m)$: The pmf of the root node S_J with $m = 0, 1, \ldots, M - 1$,
- $\epsilon_{j,j+1}^{m,n} = p_{S_j | S_{j+1}}(m | S_{j+1,\lfloor i/2 \rfloor} = n)$: The transition probability that $S_{j,i}$ is in state m given that $S_{j+1,\lfloor i/2 \rfloor}$ is in state n, $j = 1, \ldots, J - 1$ and $m, n = 0, 1, \ldots, M - 1$,
- $\mu_{j,m}$ and $\gamma_{j,m}^2$: The mean and variance, respectively, of $W_{j,i}$ given that $S_{j,i}$ is in state m, $j = 1, \ldots, J$ and $m = 0, 1, \ldots, M - 1$.

These parameters can be grouped into a model parameter vector θ as

$$\theta = \{ p_{S_J}(m), \epsilon_{j,j+1}^{m,n}, \mu_{j,m}, \gamma_{j,m}^2 | j = 1, \ldots, J; n, m = 0, \ldots, M - 1 \}. \quad (10.4)$$

The accurate estimation of HMT model parameters is essential to its practical applications, which can be effectively approached by the iterative expectation maximization (EM) algorithm.[12] This algorithm is known to numerically approximate maximum likelihood estimates for mixture-density problems. The EM algorithm has a basic structure and the implementation steps are problem dependent. The EM algorithm for HMT model training is presented briefly here, and we refer the reader to Reference 1 for more details. In the case of the HMT model training using the EM algorithm, we try to fit an M-state HMT model θ defined in Equation 10.4 to the observed J-scale tree-structured DWT, i.e., $\mathbf{w}$. The iterative structure is shown as follows:

- **Step 1. Initialization:** Set an initial model estimate θ^0, and iteration counter $l = 0$.
- **Step 2. E step:** Calculate $p(\mathbf{S} | \mathbf{w}, \theta^l)$, which is the joint pmf for the hidden state variables and is used in the maximization of $E_{\mathbf{S}}[\ln f(\mathbf{w}, \mathbf{S} | \theta) | \mathbf{w}, \theta^l]$.
- **Step 3. M step:** Set $\theta^{l+1} = \arg\max_\theta E_{\mathbf{S}}[\ln f(\mathbf{w}, \mathbf{S} | \theta) | \mathbf{w}, \theta^l]$.
- **Step 4. Iteration:** Set $l = l + 1$. If it converges, then stop; otherwise, return to Step 2.

The wavelet-domain HMMs have been applied to signal estimation, detection, and synthesis.[1,13] Specifically, an "empirical" Bayesian approach was developed to denoise a signal corrupted by additive white Gaussian noise (AWGN). It was demonstrated that signal denoising using wavelet-domain HMT outperformed other traditional wavelet-based signal denoising methods with well-preserved detailed structures. Given a noisy signal of AWGN power σ^2, the HMT model θ is first obtained via EM training, during which we can also estimate the posterior hidden-state probabilities $p(S_{j,i} | \mathbf{w}, \theta)$ for each wavelet coefficient $W_{j,i}$. Then we can obtain the conditional mean estimate for noise-free $Y_{j,i}$ by the chain rule for the conditional expectation:

$$E[Y_{j,i}|\mathbf{w}, \theta] = \sum_{m=0}^{M-1} p(S_{j,i} = m|\mathbf{w}, \theta) \times \frac{\gamma_{j,m}^2 - \sigma_\eta^2}{\gamma_{j,m}^2} \, w_{j,i}. \qquad (10.5)$$

The denoised signal is achieved by the inverse DWT (IDWT) of these estimates of wavelet coefficients.

In the following, we show two preliminary studies regarding the application of wavelet-domain HMMs to signal modeling and processing, i.e., training efficiency and modeling accuracy.[14]

10.2.1   Initialization of EM Algorithm

It was mentioned that the "intelligent" initialization of the EM algorithm may provide the fast convergence of HMT model training, but few comments on how to achieve an effective initial setting have been given.[1] We here propose an initialization scheme for the EM training algorithm that allows for an efficient HMT model learning process. Given a J-scale DWT $\mathbf{w}$, the two steps in the initialization are referred to as *horizontal scanning* and *vertical counting*. Specifically, the former step estimates the initial settings of the GMMs in different scales, and the latter step determines the initial transition probabilities $\epsilon_{j,j+1}^{m,n}$.

10.2.1.1   Horizontal Scanning

It was assumed in Reference 1 that wavelet coefficients in the same scale have the same density. Therefore, the wavelet coefficients can be grouped into different categories according to their scales, and each group can be characterized by a two-state GMM. The task of horizontal scanning is to fit a two-state GMM parameterized by $\pi_j = \{p_{S_j}(m), \mu_{j,m} = 0, \gamma_{j,m}^2|m = 0, 1\}$ to $\mathbf{w}_j$, the wavelet coefficients at scale j. We develop an EM algorithm[15] to implement the horizontal scanning. Given $\mathbf{w}_j$, we want to estimate the GMM π_j that maximizes the likelihood $E[\ln f(\mathbf{w}_j|\pi_j)|\mathbf{w}_j, \pi_j]$, where $f(\mathbf{w}_j|\pi_j)$ is

$$f(\mathbf{w}_j|\pi_j) = \prod_{i=0}^{N_j-1} f(w_{j,i}|\pi_j), \qquad (10.6)$$

$f(w_{j,i}|\pi_j)$ is given in Equation 10.3. We start the horizontal scanning from a *neutral initial setting*, π_j^0, which has equal probabilities of $p_{S_j}(0) = p_{S_j}(1) = 0.5$. $\mathbf{w}_j$ can be divided into two groups of the same number, $N_j/2$, according to their magnitudes. We use the variances of two groups as the initial values of the mixture variances, i.e., $\gamma_{j,0}^2$ and $\gamma_{j,1}^2$. After we determine π_j^0, the horizontal scanning is performed as follows.

- **Step 1. Initialization:** Set π_j^0 and the iteration counter $c = 0$.
- **Step 2. E step:** Calculate $p(S_{j,i}|w_{j,i}, \pi_j)$ that is the conditional pmf of $S_{j,i}$

$$
p(S_{j,i} = m|w_{j,i}, \pi_j) = \frac{p_{S_j}(m)g(w_{j,i}; 0, \gamma_{j,m}^2)}{\sum_{n=0}^{1} p_{S_j}(n)g(w_{j,i}; 0, \gamma_{j,n}^2)}. \tag{10.7}
$$

- **Step 3. M step:** Set $\pi_j^{c+1} = \arg\max_{\pi_j} E[\ln f(\mathbf{w}_j, \mathbf{S}_j|\pi_j)|\mathbf{w}_j, \pi_j^c]$. We update the entries of π_j^{c+1} as

$$
p_{S_j}(m) = \frac{1}{N_j} \sum_{i=0}^{N_j-1} p(S_{j,i} = m|w_{j,i}, \pi_j) \tag{10.8}
$$

$$
\gamma_{j,m}^2 = \frac{\sum_{i=0}^{N_j-1} w_{j,i}^2\, p(S_{j,i} = m|w_{j,i}, \pi_j)}{p_{S_j}(m)N_j}. \tag{10.9}
$$

- **Step 4. Iteration:** Set $c = c + 1$. If it converges, then stop; otherwise, return to Step 2.

The extra computational cost introduced by the horizontal scanning is not significant compared with the computational complexity of the HMT model training.

10.2.1.2   Vertical Counting

After estimating the initial GMM at each scale, the following vertical counting step is used to estimate the initial transition probabilities $\epsilon_{j,j+1}^{m,n}$ between two neighboring scales. Given π_j and $w_{j,i}$, we can determine the initial hidden state of $w_{j,i}$, $S_{j,i}$, based on the maximum likelihood criteria as

$$
S_{j,i} = \begin{cases} 0 & \text{if } |w_{j,i}| < T_j \\ 1 & \text{otherwise} \end{cases}. \tag{10.10}
$$

where

$$
T_j = \sqrt{\frac{\gamma_{j,0}^2\gamma_{j,1}^2\left(\ln \gamma_{j,1}^2 - \ln \gamma_{j,0}^2\right)}{\gamma_{j,1}^2 - \gamma_{j,0}^2}}.
$$

Given the initial states, we count state transition frequencies between every two neighboring scales and along the tree-structured wavelet coefficient set. Then, we set the initial values of the transition probabilities $\epsilon_{j,j+1}^{m,n}$ by the normalized transition frequencies. We average the transition probabilities of different scales to achieve a robust estimation:

$$
\epsilon_{j,j+1}^{m,n} = \frac{\sum_{j=1}^{J-1} \frac{\#(S_{j,i}=m \text{ and } S_{j+1,\lfloor i/2\rfloor}=n|i=0,....,N_j-1)}{\#(S_{j+1,\lfloor i/2\rfloor}=n|i=0,....,N_j-1)}}{J - 1}, \tag{10.11}
$$

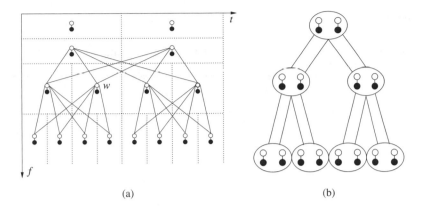

FIGURE 10.4
(a) HMT-2 model. (b) A simplified interpretation of HMT-2.

where #($A = B$) denotes the number of the event $A = B$ occurring. From our numerous simulation results, the above scale-independent initial values, by averaging the probabilities across different scales, outperform the scale-dependent ones without averaging. This averaging operation may not be necessary when we have enough data during the model training, such as in image processing.[16] The above scheme can efficiently characterize the initial marginal and joint statistics of **w**, θ^0, which allows the efficient EM model training as shown by experiments.

10.2.2   Improved HMT-2 Model

To capture more cross-correlation of wavelet coefficients between two neighboring scales, a new HMM, HMT-2, is developed and shown in Figure 10.4a, where the state of the wavelet coefficient w depends not only on the state of its parent node, but also on the state of the twin of its parent. This strategy is very popular in the graphical modeling and Bayesian network literatures.[17] Two reasons for this consideration are (1) the local stationary property of most signals and the correlation of the wavelet functions in two neighboring scales, and (2) the wavelet filter-bank decompositions, where the filter length may be long enough for wavelet coefficients to have more interscale dependencies. Usually, the analysis and training of more complicated HMMs become more difficult.[18] A simplified interpretation of HMT-2 is illustrated in Figure 10.4b, where two coefficients are integrated into one node. Actually, HMT-2 operates in the same way as the HMT model in Reference 1 except for the number of hidden states associated with each node. If we assume two hidden states for each coefficient, i.e., 0 and 1, each node of HMT-2 will have four states: 00, 01, 10, and 11. We call our new model HMT-2 instead of four-state HMT in order to distinguish it from the original M-state HMT when $M = 4$.

TABLE 10.1

Signal Denoising Results Using HMT and HMT-2, where the MSEs Were Averaged across 1000 Trials, and the EM Training Is Initialized by (a) the Neutral Setting and (b) the Proposed Two-Step Scheme

	$\left(\sigma_\eta^2 = 1.0\right)$				$\left(\sigma_\eta^2 = 2.25\right)$			
	Doppler	Bumps	Blocks	Heavisine	Doppler	Bumps	Blocks	Heavisine
HMT (a)	0.139	0.313	0.298	0.098	0.270	0.560	0.663	0.161
HMT (b)	0.129	0.306	0.290	0.089	0.262	0.553	0.518	0.154
HMT-2 (a)	0.175	0.470	0.463	0.112	0.331	0.858	0.834	0.164
HMT-2 (b)	0.120	0.292	0.273	0.080	0.241	0.569	0.529	0.142

We also developed the EM training algorithm for HMT-2 based on the one in Reference 1. It is worthwhile to note that initialization of the HMT-2 model training operates in the same way as the one for two-state HMT except for the state combination of two coefficients in one node after the horizontal scanning step.

10.2.3   Simulation Results

The proposed two-step initialization technique and the new HMT-2 model are examined here based on experiments on a set of test signals. For comparison, a neutral initial setting, where all probabilities are evenly distributed, is also studied. Our simulation is conducted on Donoho's length-1024 test signals: Bumps, Blocks, Doppler, and Heavisine.[19] The detailed experiment setup is presented in Table 10.1. A comparison is made between two different initialization schemes, the neutral initial setting and the proposed method in terms of the EM convergence rate, as shown in Figure 10.5. We can easily find that the proposed initialization scheme can accelerate the convergence rate for all test signals. The fast convergence rates of EM iterations show the effectiveness of the new initialization method, and the similar model likelihoods indicate the similar denoising performances, i.e., the mean square error (MSE).

Because the proposed HMT-2 captures more interscale dependencies of wavelet coefficients and particularly applies to the DWT with long filters, we use the Symmlet-8 DWT for four signals. In Figure 10.6 we show the HMT-2 model training results using two different initialization schemes, and we also compare the two models in terms of signal denoising with different noise powers in Table 10.1. From Figure 10.6, we find that the EM convergence rate of HMT-2 is very fast (usually several steps) with both initialization schemes. This may be because the complicated dependency structure involved in HMT-2 facilitates the wavelet-domain Bayesian inference with the more efficient EM training process. From Table 10.1, we also see that the initialization is essential to the EM training of HMT-2, i.e., the last denoising performance.

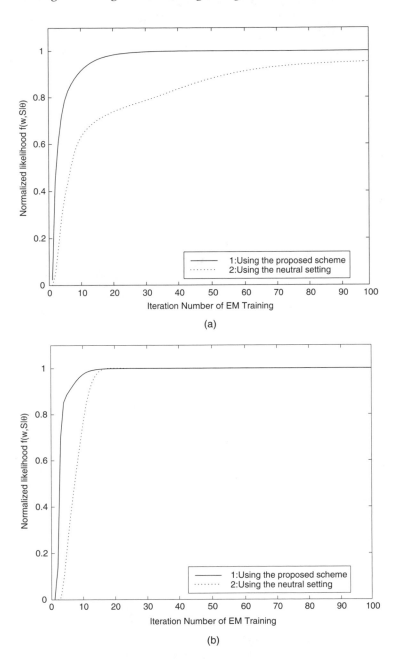

FIGURE 10.5

The plots of $\ln f(\mathbf{w}|\theta)$ with respect to the EM iteration number. Four noisy signals $\sigma_{\eta}^2 = 1.0$ are tested, where the proposed initialization scheme (solid line: 1) and the neutral initial setting (dotted line: 2) are used, respectively. (a) Doppler (1: MSE = 0.142, 2: MSE = 0.148). (b) Bumps (1: MSE = 0.261, 2: MSE = 0.261). (b) Blocks (1: MSE = 0.083, 2: MSE = 0.083). (d) Heavisine (1: MSE = 0.071, 2: MSE = 0.088). (*Continued*)

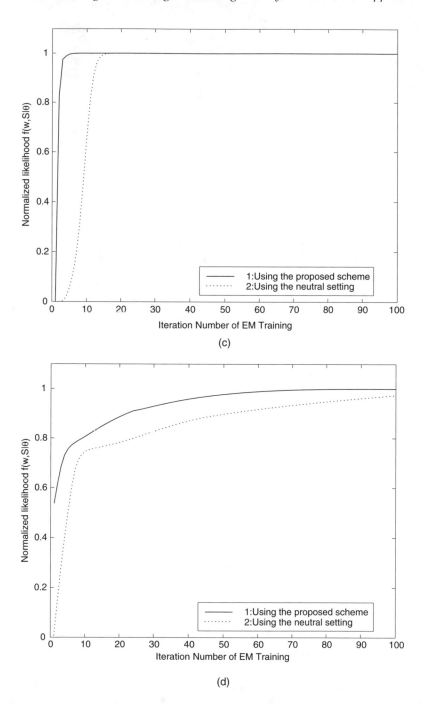

FIGURE 10.5
(*Continued*)

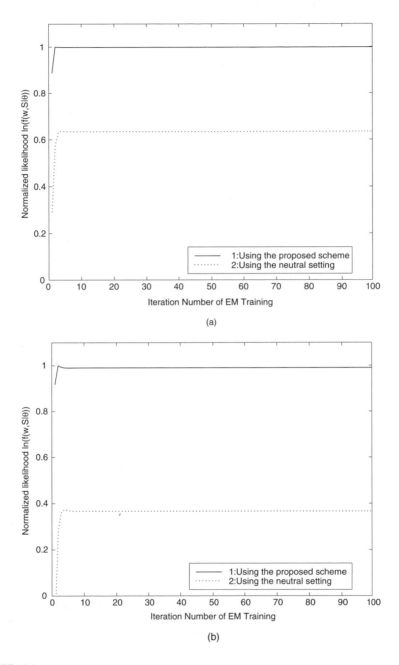

(a)

(b)

FIGURE 10.6
The plots of ln $f(\mathbf{w}|\theta)$ with respect to the EM iteration number. Four noisy signals $\sigma_\eta^2 = 1.0$ are tested, where the proposed initialization scheme (solid line: 1) and the neutral initial setting (dotted line: 2) are used, respectively. (a) Doppler (1: MSE = 0.111, 2: MSE = 0.163). (b) Bumps (1: MSE = 0.306, 2: MSE = 0.532). (b) Blocks (1: MSE = 0.256, 2: MSE = 0.468). (d) Heavisine (1: MSE = 0.070, 2: MSE = 0.095). (*Continued*)

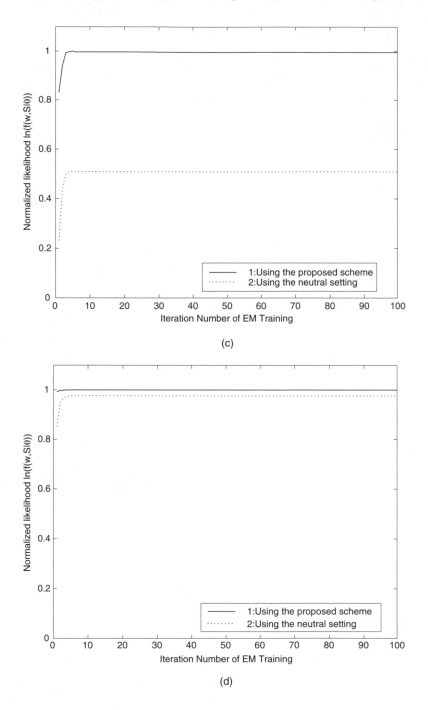

FIGURE 10.6
(*Continued*)

Compared with two-state HMT, four-state HMT-2 also improves the denoising performance for most cases.

In this section, we have briefly introduced and improved the wavelet-domain HMMs originally developed in Reference 1. This section serves the background materials of this chapter. Meanwhile, these preliminary studies will lead to more powerful wavelet-domain HMMs as well as image-processing algorithms afterward. In particular, we study image denoising, image segmentation, and texture analysis and synthesis, where wavelet-domain HMM are discussed and tailored for different applications.

10.3 Image Denoising

Wavelet-domain statistical image modeling can be roughly categorized into three groups: the interscale models,[1,6,20] the intrascale models,[21,22] and the hybrid inter- and intrascale models.[13,23–25] These models allow more accurate statistical modeling and more effective image processing, e.g., denoising and estimation, than other methods that assume wavelet coefficients to be independent. In particular,[1,6] wavelet-domain HMT imposes a tree-structured Markov chain to capture interscale dependencies of wavelet coefficients across scales. In Reference 13, both interscale and intrascale dependencies can be efficiently captured by a so-called contextual hidden Markov model (CHMM). However, the local statistics of wavelet coefficients cannot be well characterized by HMT and CHMM. In other words, neither HMT nor CHMM has the sufficient spatial adaptability that is found useful in the wavelet-domain statistical image modeling.[21] Second, although the tree structure involved in HMT captures the key characteristics of DWT along the hierarchical wavelet subtree, it also introduces undesirable denoising artifacts in denoised images due to the discontinuity of the tree structure. Third, the tree-structured EM training algorithm of HMT is computationally expensive. In this work, we propose a new wavelet-domain HMM by considering the above three issues, i.e., *spatial adaptability, reduced denoising artifacts*, and *fast model training*.[26]

10.3.1 Gaussian Mixture Field

Given the J-scale DWT of an $N \times N$ image, $w_{j,k,i}$ denotes the (k, i)th coefficient in scale j, where we omit the subband notation, $j = 1, \ldots, J$ and $k, i = 0, 1, \ldots, N_j - 1$ with $N_j = N/2^j$. $W_{j,k,i}$ and $S_{j,k,i}$ are the continuous random variable and the discrete state variable of $w_{j,k,i}$, respectively. In References 1, 6, and 13, the GMM, $\Pi_j = \{p_{S_j}(m), \sigma_{j,m}^2 | m = 0, 1\}$, is assumed for the wavelet coefficients in scale j, and S_j is the state variable associated with scale j. Sufficient data in scale j allows the robust estimation of Π_j at the loss of spatial adaptability of statistical image modeling in the wavelet domain. In this work, we propose a *Gaussian mixture field* (GMF), which can be thought

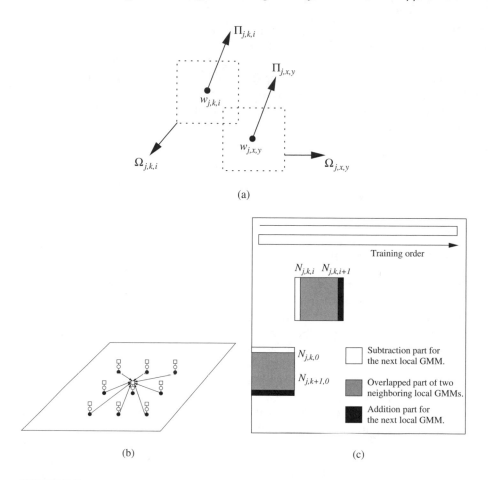

FIGURE 10.7

(a) The proposed GMF model where $w_{j,k,i}$ and $w_{j,x,y}$ are associated with two GMMs, $\Pi_{j,k,i}$ and $\Pi_{j,x,y}$, respectively. (b) Context structure in LCHMM. The white node represents the hidden state variable S. The black node denotes the continuous random variable W. The square is the context node V of w. (c) The illustration of the fast implementation of the LCHMM training, where the gray part is the overlapped part of two neighboring local models, and the black and white parts are the distinct parts.

of as an extension of the GMM. GMF assumes that each wavelet coefficient, $w_{j,k,i}$, follows a local GMM parameterized by $\Pi_{j,k,i} = \{p_{S_{j,k,i}}(m), \sigma^2_{j,k,i,m} | m = 0, 1\}$. $\Pi_{j,k,i}$ can be estimated by the neighborhood of $w_{j,k,i}$, $\Omega_{j,k,i}$, which is selected by a square window of $2C_j + 1$ and centered at $w_{j,k,i}$, as shown in Figure 10.7a, i.e., $\Omega_{j,k,i} = \{w_{j,x,y} | x = k - C_j, \ldots, k + C_j; y = i - C_j, \ldots, i + C_j\}$. GMF is a highly localized model that exploits the local statistics of the wavelet coefficients. In particular, it applies to images where the nonstationary properties are prominent.

10.3.2   Local Contextual Hidden Markov Model

In addition to GMF, we use a context model to capture intrascale dependencies of wavelet coefficients as shown in Figure 10.7b. We define the random context variable of $W_{j,k,i}$ by $V_{j,k,i}$ whose value is $v_{j,k,i} = 1$ if $\lambda_{j,k,i}^2 > \delta_j^2$, or $v_{j,k,i} = 0$ otherwise, where $\lambda_{j,k,i}^2$ is the local average energy of the eight nearest neighbors of $w_{j,k,i}$, and δ_j^2 the average energy in scale j. By conditioning on $V_{j,k,i}$ and using GMF, we develop the local contextual hidden Markov model (LCHMM) for $w_{j,k,i}$ as

$$f_{W_{j,k,i}|V_{j,k,i}}(w|v_{j,k,i} = v) = \sum_{m=0}^{1} p_{S_{j,k,i}|V_{j,k,i}}(m|v_{j,k,i} = v)g\left(w|0, \sigma_{j,k,i,m}^2\right), \quad (10.12)$$

where

$$p_{S_{j,k,i}|V_{j,k,i}}(m|v_{j,k,i} = v) = \frac{p_{S_{j,k,i}}(m)\, p_{V_{j,k,i}|S_{j,k,i}}(v|m)}{\sum_{m=0}^{1} p_{S_{j,k,i}}(m)\, p_{V_{j,k,i}|S_{j,k,i}}(v|m)}. \quad (10.13)$$

LCHMM is specified by

$$\Theta_{j,k,i} = \left\{ p_{S_{j,k,i}}(m), \sigma_{j,k,i,m}^2, p_{V_{j,k,i}|S_{j,k,i}}(v|m)|v, m = 0, 1\right\},$$

where $j = 1, \ldots, J$ and $k, i = 0, 1, \ldots, N_j - 1$. In fact, LCHMM defines a local density function for each wavelet coefficient conditioning on its context value. The EM training algorithm can be developed from that in Reference 13. Because $\Omega_{j,k,i}$ has a small number of data, one might be concerned that the estimation of $\Theta_{j,k,i}$ may not be robust. In this work, we can solve this problem by providing a good initial setting of $\Theta_{j,k,i}$ based on an idea similar to that in the previous section. Given the AWGN of variance σ_η^2, the LCHMM training is performed as follows, where $\sum_x \sum_y$ denotes $\sum_{x=k-C_j}^{k+C_j} \sum_{y=i-C_j}^{i+C_j}$.

- **Step 1. Initialization:**
  1.0 $\Pi_j^0 = \{p_{S_j}(0) = p_{S_j}(1) = 0.5, \sigma_{j,0}^2 = \sigma_\eta^2, \sigma_{j,1}^2 = 2\delta_j^2 - \sigma_\eta^2\}$ and set $p = 0$.
  1.1 E step: Given Π_j^p, calculate (Bayes rule)

$$p\left(S_{j,k,i} = m|w_{j,k,i}, \Pi_j^p\right) = \frac{p_{S_j}(m)g\left(w_{j,k,i}; 0, \sigma_{j,m}^2\right)}{\sum_{m=0}^{1} p_{S_j}(m)g\left(w_{j,k,i}; 0, \sigma_{j,m}^2\right)}.$$

$$(10.14)$$

  1.2 M step: Compute the elements of Π_j^{p+1} by

$$p_{S_j}(m) = \sum_{k=0}^{N_j-1} \sum_{i=0}^{N_j-1} p\left(S_{j,k,i} = m|w_{j,k,i}, \Pi_j^p\right), \quad (10.15)$$

$$\sigma^2_{j,m} = \frac{\sum_{k=0}^{N_j-1} \sum_{i=0}^{N_j-1} w^2_{j,k,i} \, p\left(S_{j,k,i} = m | w_{j,k,i}, \Pi^p_j\right)}{N^2_j \, p_{S_j}(m)}.$$

(10.16)

1.3 Iteration: Set $p = p + 1$. If it converges (or $p = N_p$), then go to Step 1.4; otherwise, go to Step 1.1.

1.4 Set $c = 0$ and set the elements in $\Theta^0_{j,k,i}$ by

$$p_{S_{j,k,i}}(m) = \sum_x \sum_y p(S_{j,x,y} = m | w_{j,x,y}, \Pi_j),$$

(10.17)

$$\sigma^2_{j,k,i,m} = \frac{\sum_x \sum_y w^2_{j,x,y} p(S_{j,x,y} = m | w_{j,x,y}, \Pi_j)}{(2C_j + 1)^2 p_{S_{j,k,i}}(m)},$$

(10.18)

$$p_{V_{j,k,i}|S_{j,k,i}}(v|m) = \frac{\sum_x \sum_y p(S_{j,x,y} = m | w_{j,x,y}, v_{j,x,y} = v, \Pi_j)}{p_{S_{j,k,i}}(m)}.$$

(10.19)

- **Step 2. E step:** Given $\Theta^c_{j,k,i}$, $k, i = 0, 1, \ldots, N_j - 1$, calculate (Bayes rule)

$$p_{S_{j,k,i}|V_{j,k,i},W_{j,k,i}}(m|w_{j,k,i}, v_{j,k,i} = v)$$

$$= \frac{p_{S_{j,k,i}}(m) p_{V_{j,k,i}|S_{j,k,i}}(v|m) g\left(w_{j,k,i}|0, \sigma^2_{j,k,i,m}\right)}{\sum_{m=0}^{1} p_{S_{j,k,i}}(m) p_{V_{j,k,i}|S_{j,k,i}}(v|m) g\left(w_{j,k,i}|0, \sigma^2_{j,k,i,m}\right)}.$$

(10.20)

- **Step 3. M step:** Compute the elements of $\Theta^{c+1}_{j,k,i}$, $k, i = 0, 1, \ldots, N_j-1$, by

$$p_{S_{j,k,i}}(m) = \sum_x \sum_y p_{S_{j,x,y}|V_{j,x,y},W_{j,x,y}}(m|w_{j,x,y}, v_{j,x,y}),$$

(10.21)

$$\sigma^2_{j,k,i,m} = \frac{\sum_x \sum_y w^2_{j,x,y} p_{S_{j,x,y}|V_{j,x,y},W_{j,x,y}}(m|w_{j,x,y}, v_{j,x,y})}{(2C_j + 1)^2 p_{S_{j,k,i}}(m)},$$

(10.22)

$$p_{V_{j,k,i}|S_{j,k,i}}(v|m) = \frac{\sum_x \sum_y p_{S_{j,x,y}|V_{j,x,y},W_{j,x,y}}(m|w_{j,x,y}, v_{j,x,y} = v)}{p_{S_{j,k,i}}(m)}.$$

(10.23)

- **Step 4. Iteration:** Set $c = c + 1$. If it converges (or $c = N_c$), then stop; otherwise, go to Step 2.

10.3.3 Fast EM Model Training

It seems that LCHMM training is computationally expensive because Step 3 (M step) is performed on each wavelet coefficient. As a matter of fact, it is easy to notice that there are many overlapped computations in Step 3, as shown in Figure 10.7c. The actual computational complexity of the LCHMM training is slightly higher than that of CHMM[13] and lower than those of HMMs.[1,6] Given $\Theta_{j,k,i}$, we can estimate the noise-free $Y_{j,k,i}$ from $W_{j,k,i}$ as the conditional mean as

$$E[Y_{j,k,i}|w_{j,k,i}, v_{j,k,i}] = \sum_{m=0}^{1} p_{S_{j,k,i}|V_{j,k,i},W_{j,k,i}}(m|w_{j,k,i}, v_{j,k,i}) \frac{\sigma_{j,k,i,m}^2}{\sigma_{j,k,i,m}^2 + \sigma_\eta^2} w_{j,k,i}.$$

(10.24)

The denoised image is the IDWT of the above estimates of wavelet coefficients. We expect the proposed LCHMM has the spatial adaptability, reduced denoising artifacts, and fast model training process.

10.3.4 Simplified Shift-Invariant Denoising

The lack of shift-invariant property of the orthogonal DWT results in the visually disturbing artifacts in denoised images. The "Cycle-spinning" technique has been proposed[27] to solve this problem, where signal denoising is applied to all shifts of the noisy signal, and the denoised results are then averaged. It can be shown that shift-invariant image denosing is equivalent to image denoising based on redundant wavelet transforms, such as those used in References 6, 23, and 28. In this work, we consider the 16 shifted versions that are obtained from shifting the noisy image by 1, 2, 3, and 4 pixels in each dimension, respectively. This simplification was found to be sufficient for most images in practice. We assume that the LCHMM parameters are the same as those of the 16 shifted versions. Therefore, the EM training is performed only once, and the LCHMM training results are applied to the 16 images for denoising.

10.3.5 Simulation Results

We apply LCHMM to image denoising for real images *Barbara* and *Lena* (8 bpp, 512 × 512) with AWGN of known variance σ_η^2. The experimental setting is given as follows: (1) the window size of the local GMM in GMF decreases with the increase of the scale to adapt to the higher variations of wavelet coefficients in coarser scales, and in practice, $\{C_j = 6 - j | j = 1, 2, 3, 4, 5\}$ are found both effective and efficient; (2) for simplicity, we also fix the iteration numbers of the initialization step and the EM training step to be $N_p = 20$ and $N_c = 5$; (3) we use the five-scale DWT where two wavelets, Daubechies-8 (D8) and Symmlet-8 (S8), are tested; (4) the DWT is used with two setups: the orthogonal DWT and the redundant DWT or shift-invariant (SI) techniques.

TABLE 10.2

PSNR (dB) Results from Several Recent Denoising Algorithms

Noisy Images σ_η Denoising Methods	Lena				Barbara			
	10	15	20	25	10	15	20	25
Orthogonal DWT								
Donoho's HT (D8)[19]	31.6	29.8	28.5	27.4	28.6	26.5	25.2	24.3
Wiener (MATLAB)	32.7	31.3	30.1	29.0	28.4	27.4	26.5	25.7
HMT (D8)[6]	33.9	31.8	30.4	29.5	31.9	29.4	27.8	27.1
SAWT (S8)[23]	—	31.8	30.5	29.5	—	29.2	27.6	26.5
LAWMAP (D8)[21]	34.3	32.4	31.0	30.0	32.6	30.2	28.6	27.4
AHMF (D8)[22]	34.5	**32.5**	31.1	30.1	32.7	30.3	28.7	27.5
SSM (D8)[24]	**34.8**	32.5	—	—	32.4	30.0	—	—
LCHMM (D8)	34.4	32.4	30.9	29.9	32.8	30.5	28.9	27.7
LCHMM (S8)	34.5	**32.5**	**31.2**	**30.1**	**33.1**	**30.8**	**29.2**	**28.0**
Redundant DWT								
RHMT (D8)[6]	34.6	32.6	31.2	30.1	32.8	30.3	28.6	27.7
SAWT (S8)[23]	—	**33.0**	**31.9**	**30.6**	—	30.7	28.9	27.6
SAOE[28]	34.9	**33.0**	**31.9**	**30.6**	33.3	31.1	29.4	28.2
LCHMM-SI (D8)	34.8	**33.0**	31.7	30.5	33.5	31.2	29.6	28.3
LCHMM-SI (S8)	**35.0**	**33.0**	31.7	**30.6**	**33.6**	**31.4**	**29.7**	**28.5**

The peak signal-to-noise ratio (PSNR) results are shown in Table 10.2 where several recent image denosing algorithms are compared. It is shown that LCHMM provides the excellent denoising performance for the two images, especially for the Barbara image where the nonstationarity property is prominent. LCHMM outperforms all the other methods in most cases. We also show the visual quality of image denoising (with D8 wavelet) in Figure 10.8 where LCHMM and LCHMM-SI provide better visual quality with fewer artifacts than HMT.

10.3.6 Discussions of Image Denoising

In this section, we have proposed a new wavelet-domain HMM called the local contextual hidden Markov model (LCHMM), for statistical modeling and image denoising. The simulation results show that LCHMM can achieve state-of-the-art denoising performance with three major advantages, i.e., spatial adaptability, nonstructured local-region modeling, and fast-model training. However, the main drawback of LCHMM is "overfitting" in terms of the number of model parameters, which is even larger than the number of wavelet coefficients to be modeled. This drawback may prevent LCHMM from the wider applications. Nevertheless, here LCHMM demonstrates its evident advantages in image estimation and restoration applications.

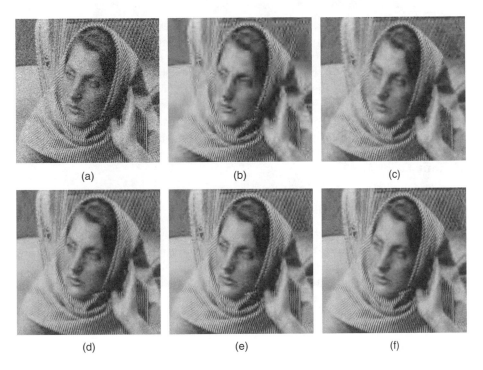

FIGURE 10.8
Partial denoising results of image Barbara. (a) Noisy image (20.34 dB). (b) Donoho's HT (24.23 dB). (c) Wiener filter (25.71 dB). (d) HMT (26.92 dB). (e) LCHMM (27.72 dB). (f) LCHMM-SI (28.43 dB).

10.4   Image Segmentation

Bayesian approaches to image segmentation have proved efficient for integrating both image features and prior contextual properties, where maximum *a posteriori* (MAP) estimation is usually involved. The Markov random field (MRF) has been developed to model the contextual behavior of image data,[29,30] and Bayesian segmentation becomes the MAP estimate of the unknown MRF from the observed data. Because the MRF model usually favors the formation of large uniformly classified regions, it may oversmooth the texture boundaries and wipe off small isolated areas. The noncausal dependence structure of MRFs typically results in high computational complexity. Recently, researchers have proposed multiscale techniques that apply contextual behavior in the coarser scale to guide the decision in the finer scale and retain the underlying MRF model in each fixed scale.[31,32] In particular, Markovian dependencies are assumed across scales to capture interscale dependencies of multiscale class labels with a causal MRF structure,[32] so that a non-iterative segmentation algorithm was developed where a sequential MAP (SMAP) estimator replaces the MAP estimator. In this section, we develop a joint multicontext and multiscale (JMCMS) approach to Bayesian segmentation, which can be formulated as a multi-objective optimization that generalizes the single-objective optimization involved in References 7, 32, and 33. To estimate the SMAP with respect to multiple context models in JMCMS, we use the heuristic multistage problem-solving technique.[34] The simulation results show that the proposed JMCMS algorithm improves the accuracy of texture classification, boundary localization, and detection at the comparable computational cost.[35]

10.4.1   Multiscale Bayesian Segmentation

We now briefly review the multiscale segmentation approaches in Reference 32. Given a random field Y, we need to accurately estimate the pixel label in X where each label specifies one of N_c possible classes. Bayesian estimators attempt to minimize the average cost of an erroneous segmentation, as shown in the following:

$$\hat{x} = \arg\max_{x} E[C(X, x)|Y = y], \tag{10.25}$$

where $C(X, x)$ is the cost of estimating the true segmentation, X. The MAP estimate is the solution of Equation 10.25, if we use the cost functional of $C_{MAP}(X, x) = 1$ whenever any pixel is incorrectly classified. This means that the MAP estimator aims at maximizing the probability that all pixels will be correctly classified. Because the MAP estimator is excessively conservative, multiscale Bayesian segmentation has been proposed,[32] where sequential MAP (SMAP) cost function, $C_{SMAP}(X, x)$, is introduced by proportionally summing together the segmentation errors from multiple scales. The SMAP estimator aims at minimizing the spatial size of errors, resulting in more desirable segmentation results with lower computational complexity than the MAP estimator. The multiscale image model proposed in Reference 32 is composed of a series of random fields at multiple scales. Each scale has a random field of image feature vectors, $Y^{(n)}$, and a random field of class labels, $X^{(n)}$. We denote an individual sample at scale n by $y_s^{(n)}$ and $x_s^{(n)}$, where s is the position in a 2D lattice $S^{(n)}$. Assuming Markovian dependencies across scales, the SMAP recursion can be computed in the fashion of coarse-to-fine as follows:

$$\hat{x}^{(n)} = \arg\max_{x^{(n)}} \{\log p_{y^{(n)}|x^{(n)}}(y|x^{(n)}) + \log p_{x^{(n)}|x^{(n+1)}}(x^{(n)}|\hat{x}^{(n+1)})\}. \tag{10.26}$$

The two terms in Equation 10.26 are the likelihood function of the image feature $y^{(n)}$ and the context-based prior knowledge from the next coarser scale, respectively. Specifically, the quadtree pyramid was developed in Reference 32 to capture interscale dependencies of multiscale class labels regarding the latter part of Equation 10.26. Thanks to the multiscale embedded structure, the quadtree model allows the efficient recursive computation of likelihood functions, but it also results in discontinuous texture boundaries because spatially adjacent samples may not have a common parent sample at the next coarser scale. Therefore, a more generalized pyramid graph model was introduced,[32] where each sample has more parent samples in the next coarser scale. However, this pyramid graph also complicates the computation of likelihood functions, and the fine-to-coarse recursion of Equation 10.26 has to be solved approximately. A trainable context model for multiscale Bayesian segmentation has been proposed,[33] where $x_s^{(n)}$ is assumed to be only dependent on $x_{\partial s}^{(n)}$, a set of neighboring samples (5×5) at the coarser scale, and $\partial s \subset S^{(n+1)}$ denotes a 5×5 window of samples at scale $n + 1$. The behavior of this simplified contextual structure can be trained off-line by providing sufficient training data including many images and their ground truth segmentations. Then, the segmentation can be accomplished efficiently via a single fine-to-coarse-to-fine iteration through the pyramid.

10.4.2   HMTseg Algorithm

A distinct context-based Bayesian segmentation algorithm has been proposed,[7] in which the context model is characterized by a context vector $v^{(n)}$ derived from a set of neighboring samples (3×3) in the coarser scale. It is assumed that, given $y_s^{(n)}$, its context vector $v_s^{(n)} = \{x_{\wp s}^{(n)}, x_{\ell s}^{(n)}\}$ can provide supplementary information regarding $x_s^{(n)}$, where $x_{\wp s}^{(n)}$ denotes the class label of the parent sample and $x_{\ell s}^{(n)}$ the dominant class label of the 3×3 samples at the coarser scale. Both $\wp s \subset S^{(n+1)}$, the position of the parent sample, and $\ell s \subset S^{(n+1)}$, a 3×3 window centered at $\wp s$, are at scale $n+1$. So given $v_s^{(n)}$, $x_s^{(n)}$ is independent with all other class labels. In particular, the contextual prior $p_{x^{(n)}|v^{(n)}}(c|u)$ is involved in the SMAP estimation, which can be estimated by maximizing the following context-based mixture model likelihood as

$$f\left(y^{(n)}|v^{(n)} = u\right) = \prod_{s \in S^{(n)}} \sum_{c=1}^{N_c} p_{x^{(n)}|v^{(n)}}\left(c|\hat{v}_s^{(n)} = u\right) f\left(y_s^{(n)}|x_s^{(n)} = c\right), \qquad (10.27)$$

where the likelihood function $f(y^{(n)}|x^{(n)} = c)$ is computed by using the wavelet-domain HMT model. An iterative EM algorithm has been developed[7] to maximize Equation 10.27, and the SMAP estimate is obtained by

$$\hat{x}^{(n)} = \arg\max_{x^{(n)}} p_{x^{(n)}|v^{(n)}, y^{(n)}}(x^{(n)}|\hat{v}^{(n)}, y^{(n)}), \qquad (10.28)$$

where

$$p_{x^{(n)}|v^{(n)}, y^{(n)}}(x^{(n)}|\hat{v}^{(n)}, y^{(n)}) = \frac{p_{x^{(n)}}(x^{(n)}) \, p_{v^{(n)}|x^{(n)}}(\hat{v}^{(n)}|x^{(n)}) \, f(y^{(n)}|x^{(n)})}{\sum_{c=1}^{N_c} p_{x^{(n)}}(c) \, p_{v^{(n)}|x^{(n)}}(\hat{v}^{(n)}|x^{(n)} = c) \, f(y^{(n)}|x^{(n)} = c)}.$$

$$\qquad (10.29)$$

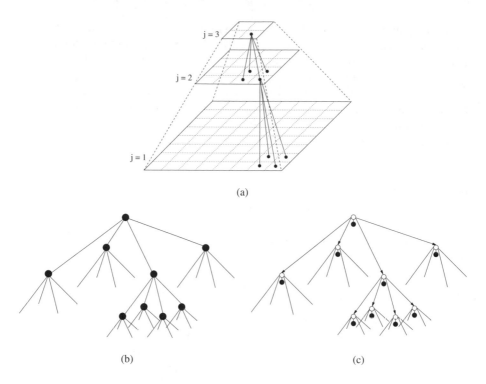

FIGURE 10.9
(a) The pyramid representation of dyadic image blocks. (b) Wavelet subtree. (c) 2D HMT.

In particular, wavelet-domain HMT was used to obtain the statistical multi-scale characterization regarding the likelihood function $f(y^{(n)}|x^{(n)} = c)$. Using the Haar DWT of the best spatial localizability, an image can be recursively divided into four subimages of the same size J times and represented in a pyramid of J scales, as shown in Figure 10.9a. We denote a dyadic block at scale n as $y^{(n)}$. Given a set of Haar wavelet coefficients $\mathbf{w}$ and a set of HMT model parameters θ, the dyadic block $y^{(n)}$ is associated with three wavelet subtrees $\{\mathcal{T}_{LH}^n, \mathcal{T}_{HL}^n, \mathcal{T}_{HH}^n\}$. The three wavelet subtrees are rooted in the tree wavelet coefficients of the same location and from three subbands at scale n. Regarding the model likelihood in Equation 10.29, the computation $f(y^n|\theta)$ is a realization of the HMT model θ and is obtained by

$$f(y^{(n)}|\theta) = f\left(\mathcal{T}_{LH}^{(n)}|\theta^{LH}\right) f\left(\mathcal{T}_{HL}^{(n)}|\theta^{HL}\right) f\left(\mathcal{T}_{HH}^{(n)}|\theta^{HH}\right), \tag{10.30}$$

where it is assumed that three wavelet subbands are independent and each component in Equation 10.30 can be computed based on the closed formula in Reference 1.

10.4.3   Joint Multicontext and Multiscale Approach

The context-based Bayesian segmentation approaches[7,32,33] have been applied to multispectral SPOT images, document images, aerial photos, etc. It was found that segmentation results in homogeneous regions, which are usually better than those around texture boundaries. This is primarily because the context models used in those approaches mainly capture interscale dependencies and encourage the formation of large, uniformly classified regions with less consideration of texture boundaries. To improve the segmentation results in both homogeneous regions and texture boundaries simultaneously, we want to discuss two questions in this work. (1) What are the characteristics of context models of different structures in terms of their segmentation results? (2) How can multiple context models of distinct advantages be integrated to implement the Bayesian segmentation? To answer the first question, we apply a set of numerical criteria to evaluate and quantify the segmentation performance, and we conduct experiments on a set of synthetic mosaics to quantitatively analyze context models. We then propose a joint multicontext and multiscale (JMCMS) approach to Bayesian segmentation, which is formulated as a multiobjective optimization problem. In particular, we use the multistage problem-solving technique to estimate SMAP of JMCMS.[34]

Given a sample $x_s^{(n)}$, its contextual information may come from some "neighbors" in the spatial and/or scale spaces. Then, we naturally have three non-overlapped contextual sources as $P = x_{\wp s}^{(n)}$, $NP = x_{\ell s}^{(n)}$, and $N = x_{\hbar s}^{(n)}$, where ℓs is the 3×3 window centered at $\wp s$ and excluding $\wp s$ at scale $n + 1$, and $\hbar s$ is the 3×3 window centered at s and excluding s at scale n. Specifically, P is the class label of $\wp s$, and PN and N are dominant class labels of ℓs and $\hbar s$, respectively. Other contextual sources could be possible, but we believe P, NP, and N are the most important because they are the nearest to $x_s^{(n)}$ in the pyramid representation, and high-order context models may introduce the context dilution problem.[36] Instead of using the majority voting scheme used in Reference 7, which may have ambiguity when $N_c > 2$, we determine the dominant class label, e.g., $x_{\hbar s}^{(n)}$, over several samples, e.g., $\hbar s$, by

$$x_{\hbar s}^{(n)} = \arg \max_{c \in \{1, \dots, N_c\}} \sum_{t \in \hbar s} p_{x^{(n)} | v^{(n)}, y^{(n)}} \left(c | \hat{v}_t^{(n)}, y_t^{(n)} \right), \qquad (10.31)$$

where we assume that each sample has the same textural contribution, which is measured by its posterior probability in Equation 10.28, to the dominant class label over several samples. $x_{\ell s}^{(n)}$ can be obtained similarly. Based on P, PN, and N, we develop five context models of different orders d as follows:

- $d = 1$: Context-1 = $\{P\}$ and Context-5 = $\{N\}$
- $d = 2$: Context-2 = $\{P, \ NP\}$ and Context-4 = $\{P, \ N\}$
- $d = 3$: Context-3 = $\{P, \ NP, \ N\}$

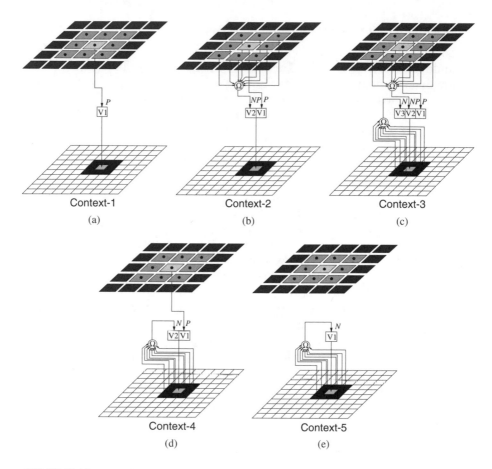

FIGURE 10.10

Five context models between two neighboring scales where the coarser scale (top) and finer scale (bottom) are shown. $\{V_1, \ldots, V_d\}$ $(d = 1, 2, 3)$ is the context vector, and Ω is defined in Equation 10.31.

The five context models are depicted in Figure 10.10, among which Context-1 and Context-2 are interscale context models, which are similar to those used to encourage the formation of large, uniformly classified regions.[7,33] Context-5 is an intrascale context model often used in the MRF literature to ensure local homogeneous labeling with high sensitivity to boundaries. Context-3 and Context-4 are hybrid inter- and intrascale context models, which have similar characteristics to those used in References 37 through 39. We anticipate that those context models have distinct effects on the segmentation results in terms of classification, boundary localization, and boundary detection. To study their characteristics, we use three numerical criteria to quantify the segmentation performance. Specifically, P_a is the percentage of pixels that are correctly classified, showing *accuracy*, P_b the percentage of

TABLE 10.3

Segmentation Results of the Five Contexts Regarding P_a, P_b, and P_c

	Context-1	Context-2	Context-3	Context-4	Context-5
$\bar{P}_a$	0.9365	**0.9728**	0.9260	0.9186	0.8327
$\bar{P}_b$	0.2098	0.3567	**0.3716**	0.3000	0.1173
$\bar{P}_c$	0.6389	0.5189	0.7071	0.7222	**0.7237**

boundaries that coincide with the true ones, showing *boundary specificity*, and P_c the percentage of true boundaries that can be detected, showing *boundary sensitivity*. We conduct segmentation experiments on 10 mosaics, as shown in Figure 10.11. For each context, we perform pixel-level segmentation on the 10 mosaics using the supervised context-based segmentation algorithm.[7] P_a, P_b, and P_c, which are averaged over 10 trials, are shown in Table 10.3.

A good segmentation requires high P_a, P_b, and P_c. Even though P_a is usually most important, high P_b and P_c provide more desirable segmentation results with high accuracy of boundary localization and detection. On the other hand, boundary localization and detection in textured regions are usually regarded as difficult issues due to abundant edges and structures around textured boundaries.[41,42] From Table 10.3, it is found that none of the five context models can work well singly in terms of the three criteria. For example, Context-2 has the best P_a but the worst P_c. This fact experimentally verifies that the context models used in References 7 and 33 are good choices in terms of P_a. Context-5 is the strongest in P_c but the weakest in P_a. Context-3 gives the highest P_b, but P_a and P_c suffer. These observations are almost completely consistent in each trial. Intuitively speaking, interscale context models, e.g., Context-1 and Context-2, favor P_a by encouraging the formation of large, uniformly classified regions across scales of the pyramid. The intrascale context model Context-5 helps P_c by being sensitive to boundaries within a scale. As a hybrid inter- and intrascale context model, Context-3 provides the best P_b by appropriately balancing both interscale and intrascale dependencies into the SMAP Bayesian estimation. Thus, a natural idea is to integrate multiple context models to achieve high P_a, P_b, and P_c simultaneously.

Generally speaking, given $\mathbf{y} = \{y^{(n)} | n = 1, 2, \ldots, L\}$ the collection of multiscale random fields of an image Y, a context model V is used to simplify the characterization of the joint statistics of $\mathbf{y}$ with local contextual modeling. Thus, given different context models, we can have different statistical characterizations of $\mathbf{y}$. Accordingly, we may have different Bayesian segmentation results. For example, the quadtree pyramid[32] and the interscale context models[7,33] emphasize the homogeneity of the labeling across scales, and the segmentation results tend to be composed of large, uniformly classified regions. However, those contexts cannot provide high accuracy of boundary

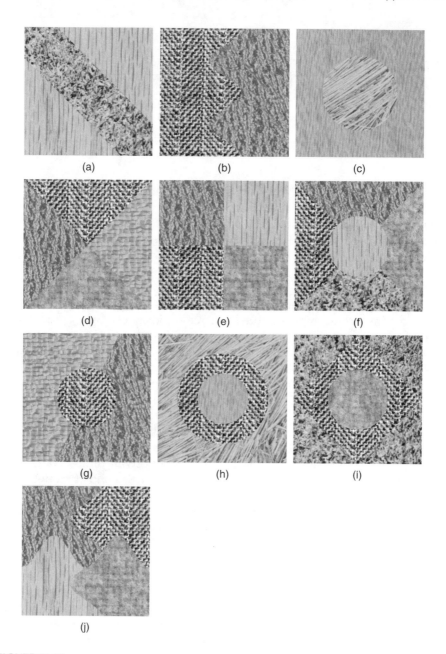

FIGURE 10.11

Ten synthetic mosaics (256 × 256, 8 bpp). (a) Mosaic1 (D9/D68), (b) Mosaic2 (D16/D24), (c) Mosaic3 (D15/D38), (d) Mosaic4 (D16/D84/D24/D19), (e) Mosaic5 (D24/D68/D16/D19), (f) Mosaic6 (D9/D16/D19/D24/D28), (g) Mosaic7 (D16/D24/D84), (h) Mosaic8 (D38/D16/D15), (i) Mosaic9 (D9/D16/D19), (j) Mosaic10 (D24/D68/D16/D19). (Mosaics from Brodatz, P., *Textures—A Photographic Album for Artists and Designers*, New York: Dover, 1966. With permission.)

localization and detection due to their limitations on boundary characterization. Similar to the multiscale image modeling,[37–39] intrascale or hybrid
interscale and intrascale context models can be used to achieve more accurate contextual modeling around boundaries, e.g., Context-3, Context-4,
and Context-5. However, those contexts may be challenged in some homogeneous regions where the homogeneity is not very good in a certain scale.
In this work, our goal is to apply multiple context models that have different
advantages for image segmentation. Hence, **y** can be represented as multiple (Z) copies and each copy is characterized by a distinct context model,
i.e., $\{\mathbf{y}_z | z = 1, 2, \ldots, Z\}$. Because different context models provide different
multiscale modeling, leading to distinct results in terms of P_a, P_b, and P_c, we
propose a joint multicontext and multiscale (JMCMS) approach to Bayesian
segmentation, which reformulates Equation 10.25 as a multiobjective optimization as

$$\hat{x} = \arg\max_{x} E[C_{\mathrm{SMAP}}(X, x) | \mathbf{Y} = \mathbf{y}_1],$$

$$\vdots \qquad\qquad\qquad (10.32)$$

$$\hat{x} = \arg\max_{x} E[C_{\mathrm{SMAP}}(X, x) | \mathbf{Y} = \mathbf{y}_Z].$$

The multiobjective optimization in Equation 10.32 is roughly analogous to
the multiple criteria of P_a, P_b, and P_c, and it can be regarded as a generalization
of the single optimization in Equation 10.25. The problem of Equation 10.32
can be approached by a heuristic algorithm, called the multistage problem-
solving technique.[34] In other words, the problem in Equation 10.32 can be bro-
ken into multiple stages, and the solution of a stage defines the constraints
on the latter stage. Thus, Equation 10.32 can be solved based on multiple
context models individually and sequentially. According to the multistage
problem-solving technique, the SMAP estimation of the posterior probabilities, as defined in Equation 10.29, is conducted for all dyadic blocks with
respect to three contexts individually and sequentially, and the SMAP decision is only made in the final step according to Equation 10.25 or 10.28.
The new JMCMS algorithm can be widely applied to different multiscale
Bayesian segmentation methods using distinct texture models or texture features. Here, in particular, we adopt the supervised segmentation algorithm[7] to
implement context-based Bayesian segmentation where the wavelet-domain
HMT is used to obtain multiscale texture characterization.

The implementation of the JMCMS approach to Bayesian segmentation is
briefly listed as follows, where Z context models are used as $\{V_1, \ldots, V_Z\}$, an
L-scale image pyramid is involved, and $n = 0$ means the pixel-level representation. An important issue that should be addressed here is the determination
of context vectors during the EM training process. Especially, the causal interscale context models, e.g., Context-1 and Context-2, have fixed context vectors
during the EM training process. Meanwhile, the noncausal intrascale or hybrid context models, e.g., Context-3, Context-4, and Context-5, require the

real-time update of context vectors during each iteration based on the results of the previous step. The JMCMS segmentation algorithm is implemented as follows.

- **Step 1.** Set $n = L - 1$, starting from the next to the coarsest scale.
- **Step 2.** Set $z = 1$, starting from the first context model V_1 in the list.
- **Step 3.** Set $p = 0$, initializing $\{p_{x^{(n)}}(c), \ p_{v^{(n)}|x^{(n)}}(u|v)\}$ and $v^{(n)}$.
- **Step 4.** Expectation (E) Step, as defined in Equation 10.29.
- **Step 5.** If context model V_z is noncausal, update $v^{(n)}$; otherwise, continue.
- **Step 6.** Maximization (M) Step, update contextual prior as

$$p_{x^{(n)}}(c) = \sum_{s \in S^{(n)}} p_{x^{(n)}|v^{(n)}, y^{(n)}} \left(x_s^{(n)} = c | \hat{v}_s^{(n)}, y_s^{(n)} \right), \tag{10.33}$$

$$p_{v^{(n)}|x^{(n)}}(u|c) = \frac{1}{p_{x^{(n)}}(c)} \sum_{\hat{v}_s^{(n)}=u} p_{x^{(n)}|v^{(n)}, y^{(n)}} \left(x_s^{(n)} = c | \hat{v}_s^{(n)}, y_s^{(n)} \right). \tag{10.34}$$

- **Step 7.** Set $p = p + 1$. If converged (or $p = N_p$), then stop; otherwise, go to Step 4.
- **Step 8.** Set $z = z + 1$. If $z > Z$, then stop; otherwise, use context X_z and go to Step 3.
- **Step 9.** Set $n = n - 1$. If $n < 0$, then stop; otherwise, go to Step 2.
- **Step 10.** $\arg\max_c \ p_{x^{(0)}|v^{(0)}, y^{(0)}}(c|\hat{v}^{(0)}, y^{(0)})$ gives the pixel-level segmentation.

In Table 10.4, we list the optimal (numerically in terms of P_a) settings of JMCMS on the 10 synthetic mosaics when $Z = 1, 2, 3$. In practice, $Z = 3$ is found sufficient for the 10 mosaics in Figure 10.11, and $Z > 3$ does not help much in terms of $\bar{P}_a$, $\bar{P}_b$, and $\bar{P}_c$. We also found that the JMCMS of Context-2-3-5, i.e., $V_1 = $ Context-2, $V_2 = $ Context-3, and $V_3 = $ Context-5, is the numerically best setting for the 10 mosaics regarding the three criteria, and it is almost completely consistent in each trial. It is interesting to note that the context ordering of the optimal JMCMS algorithms given the order Z in Table 10.4 also somehow follows a coarse-to-fine way. Three facts about JMCMS are noteworthy: (1) The JMCMS of Context-2-3-5 may not be the

TABLE 10.4

The optimal JMCMS on the 10 Mosaics with $Z = 1, 2, 3$

JMCMS	Context-2 ($Z = 1$)	Context-2-5 ($Z = 2$)	Context-2-3-5 ($Z = 3$)
P_a	0.9728	0.9893	0.9897
P_b	0.3567	0.6923	0.7259
P_c	0.5189	0.7314	0.7337

universally optimal design, and we have the flexibility to design the tailored JMCMS for a specific application. (2) As an alternative, some sophisticated methods of selecting different contexts in a spatially adaptive fashion could be developed to improve the segmentation results. However, since the contextual prior is trained by the EM algorithm, which needs sufficient data for the efficient training, the spatially adaptive context selection faces the difficulty of the robust prior estimation. The generally designed JMCMS here has good robustness and adaptability to various image data and texture boundaries. (3) The new JMCMS approach is neither preprocessing nor postprocessing on the segmentation map, as the SMAP decision is only made in the final stage. The proposed JMCMS is a new approach to driving the Bayesian estimation of posterior probabilities toward a desired solution via multiple context models, step by step.

10.4.4 Simulation Results

Here we test the proposed JMCMS approach of Context-2-3-5 ($Z = 3$) on both synthetic mosaics and remotely sensed images. At the same time, we also study the segmentation algorithm,[7] where only Context-2 ($Z = 1$) is used. For both cases, we use the wavelet-domain HMT model to obtain multiscale statistical texture characterization. We fix the total iteration numbers of the two methods to be the same, e.g., $N_p \times Z = 30$. Thus they have similar computational complexity, and the execution time is about 20 to 30 s for 256×256 images ($N_c = 2, 3, 4$) on a Pentium-II 400 computer. The average improvements on $\bar{P}_a$, $\bar{P}_b$, and $\bar{P}_c$ are about 2, 32, and 18%, respectively, across the 10 mosaics given in Figure 10.11. We also show the segmentation results of five mosaics in Figure 10.12, where the improvements on P_a, P_b, and P_c are also given.

Although Context-2 in Reference 7 provides generally good segmentation results in homogeneous regions, the texture boundaries cannot be well localized and detected, i.e., low P_b and P_c. This is the major shortcoming of most multiscale segmentation approaches, where only interscale context models are used.[7,32,33] It is shown that JMCMS can overcome this limitation. First, accuracy of boundary localization and boundary detection are significantly improved with much smoother texture boundaries, as shown by ΔP_b and ΔP_c. Second, the classification accuracy in homogeneous regions is also improved by reducing mis-classified and isolated pixels, as shown by ΔP_a. These improvements are due to multiple context models used in JMCMS, where contextual information is propagated both across scales and via multiple context models to warrant good segmentation results in both homogeneous regions and boundaries.

One may argue that some simple processing methods, such as the morphological operation, can also provide smoother boundary localization. However, there are three limitations in the morphological operation for postprocessing segmentation maps. First, it cannot deal with errors of a large size, as those that appear in Figure 10.12d. Second, it may weaken the accuracy of

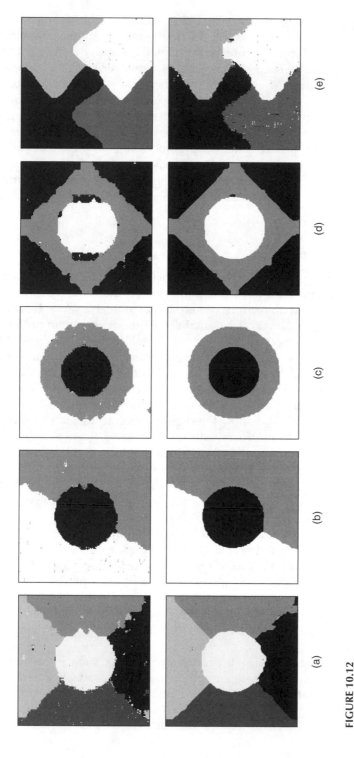

FIGURE 10.12

Segmentation results of HMTseg (top) and JMCMS (bottom). (a) Mosaic6: $\Delta P_a = 3.03\%$, $\Delta P_b = 22.12\%$, $\Delta P_c = 1.86\%$. (b) Mosaic7: $\Delta P_a = 0.98\%$, $\Delta P_b = 26.75\%$, $\Delta P_c = 20.88\%$. (c) Mosaic8: $\Delta P_a = 2.01\%$, $\Delta P_b = 39.71\%$, $\Delta P_c = 28.10\%$. (d) Mosaic9: $\Delta P_a = 3.72\%$, $\Delta P_b = 41.29\%$, $\Delta P_c = 27.48\%$. (e) Mosaic10: $\Delta P_a = 3.98\%$, $\Delta P_b = 43.44\%$, $\Delta P_c = 7.97\%$.

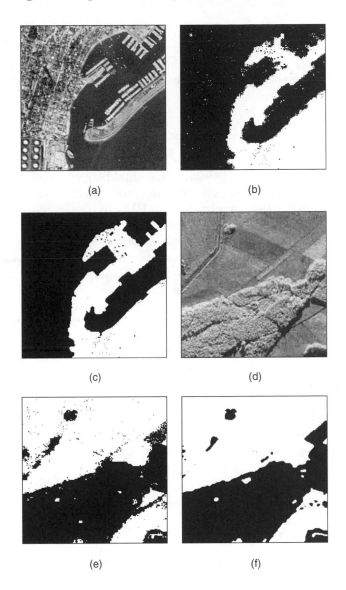

FIGURE 10.13
Segmentation of real images using HMTseg and JMCMS. (a) Aerial photo A (Sea/ground). (b) HMTseg result of A. (c) JMCMS result of A. (d) SAR image B (forest/grass). (e) HMTseg result of B. and (f) JMCMS result of B.

boundary detection by producing oversmoothed boundaries. Third, it may wipe off some small isolated targets, which are important to some applications. In the following, we conduct experiments on remotely sensed images, including an aerial photograph and a synthetic aperture radar (SAR) image, as shown in Figure 10.13. Texture models are first trained on image samples, which are manually extracted from original images (512×512, 8 bpp). We can see the improvements of JMCMS (Context-2-3-5) over HMTseg (Context-2). The accuracy of texture classification, in particular, boundary localization and detection, is improved. Meanwhile, the small targets are kept in the segmentation map.

10.4.5 Discussions of Image Segmentation

In this section, a JMCMS approach to Bayesian segmentation has been proposed. JMCMS is able to accumulate contextual behavior both across scales and via multiple context models, allowing more effective Bayesian estimation. JMCMS applies the wavelet-domain HMT to obtain multiscale texture characterization. JMCMS can be formulated as a multiobjective optimization, which can be approached by the heuristic multistage problem-solving technique. The proposed JMCMS algorithm has been applied to both synthetic mosaics and remotely sensed images. Simulation results show that JMCMS improves the accuracy of texture classification, and in particular, boundary localization and boundary detection over HMTseg. Meanwhile, small targets are kept well in the segmentation maps. We expect that the segmentation performance can be further improved by using more accurate texture models or features. The JMCMS approach can be applied to other Bayesian segmentation algorithms using different texture models or features that are suitable for characterizing the texture information in remotely sensed images. Meanwhile, all Bayesian segmentation algorithms[7,32,33] and the proposed JMCMS are supervised segmentation where the texture models are trained prior to the segmentation process. Unsupervised image segmentation using JMCMS and HMT has been studied with promising results.[43]

10.5 Texture Analysis and Synthesis

Textures play important roles in many computer-vision and image-processing applications, because images of real objects often do not exhibit regions of uniform and smooth intensities, but variations of intensities with certain repeated structures or patterns; these are referred to as visual texture. Most recent works on textures predominantly concentrate on two areas.[44] One is multichannel filtering theory, which was inspired by the multichannel filtering mechanism in neurophysiology[45] and motivated by the evident advantages of multiscale texture analysis.[31] The other area is statistical modeling, which characterizes textures as probability distributions from random fields. Statistical theories enable us to formulate and solve the problems of texture processing mathematically and systematically. Recently, texture characterization, based on the DWT, which integrates the above two aspects, has attracted much attention.

It has been found to be useful for a variety of texture analysis and synthesis applications, including texture classification, texture segmentation, and texture synthesis. These approaches have been found to be more efficient than the traditional methods, considering the characteristics of the human visual system in perceiving textures.

Wavelet-domain HMMs, e.g, HMT, are powerful in statistical signal and image modeling and processing. When HMT was applied to image processing, it was usually assumed that the three DWT subbands, i.e., *HL*, *LH*, and *HH*, were independent. This assumption is valid in modeling most real images, as real images usually carry a large amount of randomly distributed edges or structures, which weaken the cross-correlation between DWT subbands. However, we observed that, for natural textures, in particular structural textures, the regular spatial structures or patterns may result in certain dependencies across the three DWT subbands. It was also shown[46] that the dependencies across subbands are useful for wavelet-based texture characterization. Specifically, a vector wavelet-domain HMT was proposed,[9] which incorporates multivariate Gaussian densities to capture statistical dependencies across DWT subbands. The vector HMT was applied to the redundant wavelet transform to obtain rotation-invariant texture retrieval. It was demonstrated that the two-state vector HMT in Reference 9 has a moderate feature size and provides more accurate texture characterization than the two-state scalar HMT in References 1 and 6.

In this section, we propose a new wavelet-domain HMM, HMT-3S, by integrating the three DWT subbands into one tree structure. In addition to the joint DWT statistics captured by HMT, the proposed HMT-3S can also exploit statistical dependencies across DWT subbands. Differing from the vector HMM proposed in Reference 9, we still impose the single-variable Gaussian mixture densities in the wavelet domain. In HMT-3S, the state combination of three wavelet coefficients from the three DWT subbands results in an increase in the state number from 2 to 8, and the dependencies across DWT subbands can be characterized by the enlarged state transition matrices, i.e., 2×2 in HMT and 8×8 in HMT-3S. It is demonstrated that the more accurate texture characterization from HMT-3S improves the performance of texture analysis and synthesis.[47]

10.5.1   Wavelet-Domain HMT-3S

As discussed before, the two-state GMM is used to characterize the marginal statistics of wavelet coefficients. If we consider all wavelet coefficients to be independent, we obtain the so-called independence mixture model (IMM).[1] Wavelet-domain HMT was proposed mainly to capture interscale dependencies of wavelet coefficients across scales. When HMT is extended to the 2D case for image processing, the three wavelet subbands are usually considered independent.[6,7] To improve the accuracy of texture characterization by capturing dependencies across DWT subbands, we propose a new wavelet-domain

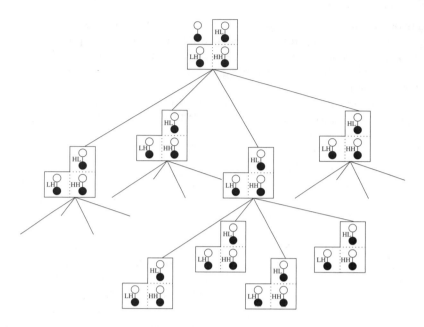

FIGURE 10.14
The simplified characterization of HMT-3S.

HMM, HMT-3S, by grouping the three DWT subbands into one quadtree structure. This grouping strategy is popular in the graphical modeling and Bayesian network literatures,[17] and as discussed before, it was used to develop an improved HMT, HMT-2, by grouping every two neighboring coefficients into one node. It was shown that the more complete statistical characterization of DWT from HMT-2 improves the signal denoising performance, as well as the model training efficiency. We show the simplified characterization of HMT-3S in Figure 10.14, where we see that HMT-3S has the same quadtree structure as HMT, except for the number of coefficients in a node, i.e., the state number. If we assume the hidden state number to be two for each wavelet coefficient, there are eight states in a node of HMT-3S.* It is worth noting that two-state GMMs are still used to characterize the DWT marginal statistics. Thus HMT-3S is parameterized by

$$\theta_{\text{HMT-3S}} = \left\{ p_J(u), \epsilon_{j,j-1}^{u,v}, \sigma_{B,j,b}^2 | B \in \mathcal{B}, j = 1, \ldots, J; u, v = 0, \ldots, 7; b = 0, 1 \right\}.$$

The EM training algorithm[1] can be straightforwardly extended to the eight-state HMT-3S. Similar to HMT, the HMT-3S model likelihood can be computed

---

* Because there are three coefficients in a node that follow three GMMs, respectively, the three-dimensional Gaussian pdf is involved in HMT-3S. For simplicity, we assume the covariances of the multivariate Gaussian pdf are zeros here.

as follows:

$$f(\mathbf{w}|\theta_{\text{HMT-3S}}) = \sum_{k,i=0}^{N_J-1} \log\left(\sum_{u=0}^{7} f_u(\mathfrak{T}_{J,k,i}|\theta_{\text{HMT-3S}}, u)\right),\tag{10.35}$$

where $\mathfrak{T}_{j,k,i}$ is the complex wavelet subtree rooted at $\mathbf{w}_{j,k,i} = \{w_{j,k,i}^{HL}, w_{j,k,i}^{LH}, w_{j,k,i}^{HH}\}$, as shown in Figure 10.14, and $f(\mathfrak{T}_{j,k,i}|\theta_{\text{HMT-3S}}, u)$ can be computed in a recursive fine-to-coarse fashion as follows:

$$f_u(\mathfrak{T}_{j,k,i}|\theta_{\text{HMT-3S}}, u)$$

$$= p_j(u)\mathbf{g}(\mathbf{w}_{j,k,i}|u)\left(\prod_{s=2k}^{2k+1}\prod_{t=2i}^{2i+1}\sum_{v=0}^{7}\left(\epsilon_{j,j-1}^{u,v}f_v(\mathfrak{T}_{j-1,s,t}|\theta_{\text{HMT}}, v)\right)\right),\tag{10.36}$$

and in the finest scale, i.e., $j = 1$, we have

$$f_v(\mathfrak{T}_{1,k,i}|\theta_{\text{HMT-3S}}, v) = p_1(v)\mathbf{g}(\mathbf{w}_{1,k,i}|v),\tag{10.37}$$

and

$$\mathbf{g}(\mathbf{w}_{j,k,i}|v) = \prod_{B\in\mathcal{B}} g\left(w_{j,k,i}^{B}|0, \sigma_{B,j,b}^{2}\right),\tag{10.38}$$

where $b = S_B \& v$, with $S_{HL} = 1$, $S_{LH} = 2$, and $S_{HH} = 4$.

Both HMT and HMT-3S have similar recursion to compute model likelihood functions. In particular, HMT-3S involves more parameters to characterize statistical dependencies across DWT subbands; i.e., the 2×2 state transition matrix in HMT, $\epsilon_{j,j-1}^{B}(m, n)$ with $m, n = 0, 1$, becomes the 8×8 one in HMT-3S, $\epsilon_{j,j-1}^{u,v}$ with $u, v = 0, 1, \ldots, 7$. We expect that HMT-3S can improve the accuracy and robustness of texture characterization by capturing more complete DWT cross-correlations, as shown in the following three texture processing applications, i.e., classification, segmentation, and synthesis.

10.5.2   Texture Classification

Texture classification is one of the most important applications of texture analysis. It involves the identification of the texture class given to a homogeneous textured region. Here, four wavelet-based texture classification methods are tested, adopting four different wavelet-domain features or models, i.e., the wavelet energy signature (WES), IMM, HMT, and HMT-3S. WES characterizes the energy distribution along the frequency axis over scales and subbands, and we adopt the Euclidean distance to measure the difference between two WESs. Moreover, we also study texture classification using the three wavelet-domain statistical models, i.e., IMM, HMT, and HMT-3S, all of which can be adapted into the maximum likelihood classifier. Given the trained model θ and the observed DWT $\mathbf{w}$, we have closed formulas to compute the model

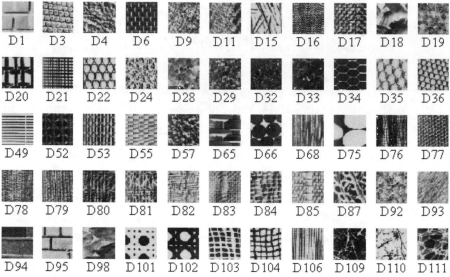

FIGURE 10.15
The sample images (64 × 64, 8 bpp) of the 55 Brodatz textures used in the experiment.

likelihood functions $f(\mathbf{w}|\theta)$ that measure how well the models θ (IMM, HMT, and HMT-3S) describe the data $\mathbf{w}$. For simplicity, we consider the case in which the prior probabilities to the texture classes are equal, and where the goal is to minimize the overall probability of classification error. The optimal decision becomes the maximum likelihood rule, which is to choose the class that makes the observed data most likely, i.e.,

$$C_{ML} = \arg \max_{c \in \{1,\dots,N_c\}} f(\mathbf{w}|\theta^c). \tag{10.39}$$

The simulation of texture classification in this work is based on 55 Brodatz textures of a size of 640 × 640 (Figure 10.15).* There were two reasons for us to choose these 55 textures for this work. One was that the overall percentage of correct classification (PCC) of the WES alone is less than 80% on this set of textures, thus creating a difficult problem in examining the IMM, HMT, and HMT-3S in terms of the accuracy of texture characterization. The other was that the classification experiment here is conducted on 64 × 64 texture samples. We excluded the textures that are nonhomogeneous in this small size, thus providing a reasonable test environment.

Prior to the classification experiment, four sets of texture features or models, i.e., WES, IMM, HMT, and HMT-3S, are obtained and stored for each texture.

---

*These textures are obtained from the Brodatz database at http://www.ux.his.no/~tranden/brodatz/.

To ensure the robustness of the model estimation, all models are trained from the whole texture image (640 × 640). It was also found that the translation of the training image has almost no effect on the model estimation. On the other hand, to test and compare the four methods with respect to their results for texture classification, we divide a texture image into 19 × 19 subimages of 64 × 64, so that horizontally and vertically neighboring subimages overlap each other with 32 columns (or rows). Here we use the four-scale Daubechies-8 DWT. The selection of an optimal wavelet filter basis for texture characterization has been studied,[48] and we are particularly interested in wavelet-domain statistical modeling, which may be further considered jointly with the results reported in Reference 48. The four texture classification methods are performed on 361 texture samples, and the PCC is recorded for each texture. Then we use the distribution, mean, and standard deviation (stdev) of 55 PCCs across 55 trials to evaluate the overall classification performance, as shown in Table 10.5. From Table 10.5, we have the following three major results:

1. **Overall Performance:** HMT-3S gives the highest overall PCC (above 95%) and the best numerical stability with the smallest Stdev of PCCs over 55 textures (below 7%). Specifically, HMT-3S can correctly identify 18 textures of 55 with 100% PCCs, and it can accurately classify most textures with more than 90% PCCs (48 of 55). There is no PCC < 70% for HMT-3S. For most textures, IMM, HMT, and HMT-3S outperform WES by significantly improving PCCs, showing that exploration of the marginal statistics and higher-order dependencies are useful for texture characterization.

2. **Structural Textures:** The advantages of HMT-3S over WES, IMM, and HMT are evident for structural textures, where dependencies across DWT subbands are relatively strong. Particularly, for 20 strong structural textures that are marked with (+) in Table 10.5, HMT-3S can outperform WES, IMM, and HMT by 17.5, 9, and 4.5% PCC improvements, respectively. This fact is consistent with the major motivation of HMT-3S.

3. **Statistical Textures:** IMM, HMT, and HMT-3S have similar performance, showing that the exploration of statistical dependencies across scales or across subbands cannot significantly improve texture characterization for statistical textures, where high-order dependencies are relatively weak. In particular, for 20 statistical textures that are marked with (*) in Table 10.5, the PCCs from IMM, HMT, and HMT-3S are very close. This observation is also consistent with the prerequisite of HMT-3S and HMT for statistical modeling.

As the main drawback, the feature size of HMT-3S is the largest. This is mainly because 8 × 8 state transition matrices are involved in HMT-3S, which characterize dependencies across both scales and subbands. Still, experiments

TABLE 10.5

Texture Classification Performance of the Four Methods in Terms of PCC (%)

Textures	WES	IMM	HMT	HMT-3S	Textures	WES	IMM	HMT	HMT-3S
D1(+)	92.2	94.2	92.5	97.8	D77	100	99.4	99.4	99.7
D3(*)	83.7	93.4	95.3	97.2	D78(*)	87.5	96.7	100	100
D4(*)	80.9	94.7	94.7	95.6	D79	95	97.8	99.2	100
D6(+)	93.9	99.4	100	100	D80(*)	70.9	85.6	93.1	95.6
D9(*)	47.6	98.1	97.8	97.5	D81(+)	72.6	89.8	84.8	92.8
D11(*)	72.3	98.3	96.7	100	D82(*)	99.7	100	99.7	100
D15(+)	74.8	83.9	83.4	91.1	D83(*)	95.3	99.2	98.1	98.6
D16(*)	99.7	100	100	100	D84(*)	95	100	100	100
D17	99.2	100	100	100	D85(*)	93.4	97.8	97	99.4
D18(*)	40.2	90	82.5	74.8	D87(+)	83.1	93.9	85	97.2
D19	85.3	86.4	75.1	90.6	D92	87.3	97.8	98.3	99.2
D20(+)	63.4	100	100	100	D93	74	74	95.3	97.5
D21	100	100	100	100	D94(+)	74.5	80.9	93.6	95.8
D22	50.4	81.2	95.8	87.3	D95(+)	47.9	92	90.6	94.5
D24(*)	89.5	92.8	92.2	92.5	D98(+)	59.8	88.1	88.1	92
D28(*)	64	96.1	91.7	91.4	D101	33	95.6	96.4	94.2
D29(*)	90.3	98.9	99.7	99.7	D102(+)	37.7	42.1	49.9	77
D32(*)	99.7	99.2	99.7	99.7	D103	80.1	99.7	98.9	98.3
D33(*)	82.8	90.3	95	93.9	D104	50.1	71.2	74.5	75.3
D34(+)	91.4	99.7	100	100	D106	98.1	9.4	100	100
D35(+)	81.4	96.7	83.7	93.6	D109(*)	57.6	80.3	83.7	85.6
D36(+)	68.4	87	91.4	93.9	D110(*)	80.6	97	99.4	100
D49	100	1.4	100	100	D111(+)	70.6	94.2	95	98.6
D52(+)	48.8	67.3	84.8	92.5	#(PCC = 100)	3	7	13	18
D53(+)	99.2	70.4	99.2	99.7	#(100 > PCC ≥ 90)	16	29	29	30
D55(+)	93.6	100	100	100	#(90 > PCC ≥ 80)	11	10	8	2
D57(*)	78.7	92.8	98.1	97.8	#(80 > PCC ≥ 70)	9	4	4	5
D65(+)	67.3	78.4	100	100	#(PCC < 70)	16	5	1	0
D66(+)	41.3	98.1	100	100	Mean of 20 PCCs (+)	72.5	86.0	90.4	**95.0**
D68	96.1	88.1	77	78.1	Mean of 20 PCCs (*)	80.5	95.1	95.7	**96.0**
D75(+)	67.9	52.4	74.8	76.5	Mean of 55 PCCs (all)	77.4	87.4	93.1	**95.2**
D76	70.6	94.7	99.4	100	Stdev of 55 PCCs (all)	18.99	20.04	9.52	**6.97**

Note: + and * signs denote a strong structural texture and a statistical texture, respectively. The feature sizes of WES, IMM, HMT, and HMT-3S are 12, 36, 45, and 199, respectively.

manifest that HMT-3S can improve the accuracy of statistical texture characterization, particularly for structural textures. This fact can be further verified by the application of texture segmentation as follows.

10.5.3 Texture Segmentation

Texture segmentation is closely related to texture classification. In particular, we want to study how segmentation performance can be improved by using a more accurate texture model, i.e., HMT-3S instead of HMT used in

Equation 10.26. It was shown that both texture modeling and contextual modeling are important to the performance of multiscale Bayesian segmentation.[49] The JMCMS algorithm described earlier improves texture segmentation by using more sophisticated contextual modeling of multiscale class labels regarding the latter part of Equation 10.26. We here attempt to improve texture segmentation by using more accurate texture models, i.e., HMT-3S instead of HMT, regarding the former part of Equation 10.26. It is worth noting that the computation of model likelihood is the only step related to the feature size and is negligible compared with the latter Bayesian estimation in JMCMS or HMTseg. Thus, we study both the HMTseg and the JMCMS algorithms by substituting HMT-3S for HMT to compute the model likelihood used in Bayesian estimation. In Figure 10.16, we show the segmentation results of two synthetic texture mosaics using both HMTseg and the JMCMS algorithms where two texture models, HMT-3S and HMT, are tested. The computational complexity of three implementations are similar for multiscale Bayesian segmentation.

We see that both JMCMS and HMT-3S can improve segmentation results over the HMTseg algorithm in terms of classification accuracy by emphasizing the two terms in Equation 10.26. Moreover, the combination of JMCMS and HMT-3S provides the best results with homogeneous regions and texture boundaries. Intuitively speaking, HMT-3S aims to generate homogeneous segmentation regions by providing robust and accurate texture characterization, and JMCMS attempts to clean misclassified pixels and smooth texture boundaries by minimizing the spatial size of errors. Although the two efforts seem to be conflicting, the integration of HMT-3S and JMCMS within the Bayesian estimation framework can provide a well-balanced segmentation result, as shown in Figure 10.16. When the number of texture types, N_c, is large, e.g., $N_c = 9$ in Figure 10.16a and e, the advantages of HMT-3S over HMT are particularly significant for segmentation. Thus, the segmentation results further verify the strengths of HMT-3S over HMT for statistical texture characterization.

10.5.4 Texture Synthesis

As the counterpart issue of texture analysis, texture synthesis attempts to generate synthetic textures that are visually indistinguishable from real textures, according to a certain parametric texture model or a set of texture features. Texture synthesis is often used in image compression and image rendering applications. Here, we want to study how to apply wavelet-domain HMMs to texture synthesis, and both HMT and HMT-3S are tested regarding their capabilities and suitability for texture synthesis. In Equation 10.35, we have closed formulas to compute likelihood functions $f(\mathbf{w}|\theta)$ given the model θ (HMT or HMT-3S) and the observed data $\mathbf{w}$. $f(\mathbf{w}|\theta)$ shows how well the model θ fits the data $\mathbf{w}$. For texture analysis, we use the EM algorithm to estimate θ by maximizing $f(\mathbf{w}|\theta)$. For texture synthesis, we need to generate

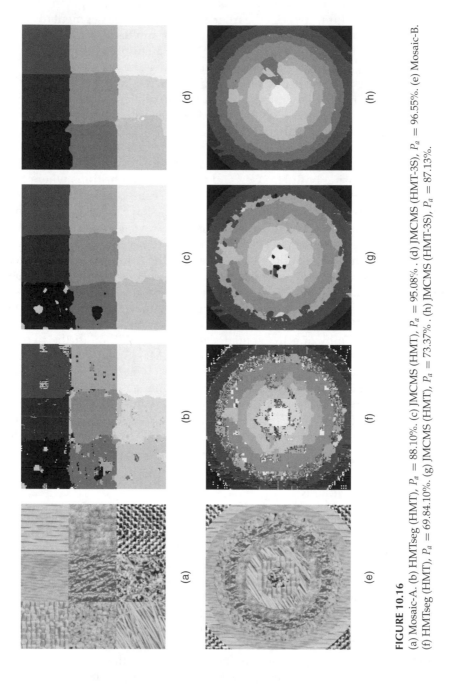

FIGURE 10.16

(a) Mosaic-A. (b) HMTseg (HMT), $P_a = 88.10\%$. (c) JMCMS (HMT), $P_a = 95.08\%$. (d) JMCMS (HMT-3S), $P_a = 96.55\%$. (e) Mosaic-B. (f) HMTseg (HMT), $P_a = 69.84.10\%$. (g) JMCMS (HMT), $P_a = 73.37\%$. (h) JMCMS (HMT-3S), $P_a = 87.13\%$.

w, which can best be parameterized by a given θ as

$$\hat{\mathbf{w}} = \arg\max_{\mathbf{w} \in \mathfrak{R}^2} f(\mathbf{w}|\theta), \tag{10.40}$$

subject to the mean and variance constraints of **w** implied by θ. Thus, the problem of texture synthesis using HMT can be formulated as a constrained optimization. Usually, to make it easier, we can change a constrained optimization into an unconstrained one by using the penalty function technique.[50] On the other hand, because it is hard to compute the overall **w** by a single operation, we adopt the multiscale scheme in Reference 46 to update **w** in a coarse-to-fine fashion, as described in the following, where the problem is discussed in the 1D form and applies to the 2D case.

Given a 1D HMT θ, and $\mathbf{w}_j$, the set of N_j wavelet coefficients in scale j, the local model likelihood of two adjacent scales, scale j and scale $j + 1$, $f(\mathbf{w}_j, \mathbf{w}_{j+1}|\theta)$ can be computed as

$$f(\mathbf{w}_j, \mathbf{w}_{j+1}|\theta) = \sum_k \log \left(\sum_{m=0}^1 \alpha_{j+1,k}(m)\beta_{j+1,k}(m) \right), \tag{10.41}$$

where

$$\alpha_{j+1,k}(m) = p_{j+1}(m)g\left(w_{j+1,k}; 0, \sigma_{j+1,m}^2\right), \tag{10.42}$$

$$\beta_{j+1,k}(m) = \prod_{i=0}^1 \left(\sum_{n=0}^1 \epsilon_{j,j+1}^{n,m} g\left(w_{j,2k+i}; 0, \sigma_{j,n}^2\right) \right). \tag{10.43}$$

Because we use the coarse-to-fine scheme, we fix $\mathbf{w}_{j+1}$ to find $\mathbf{w}_j$, which can maximize $f(\mathbf{w}_j, \mathbf{w}_{j+1}|\theta)$, i.e., $f(\mathbf{w}_j|\theta, \mathbf{w}_{j+1})$, subject to the mean and variance constraints of expected $\mathbf{w}_j$. As stated earlier, we can define a new objective function $h(\mathbf{w}_j|\theta)$ by introducing two penalty functions as follows:

$$h(\mathbf{w}_j|\theta, \mathbf{w}_{j+1}) = f(\mathbf{w}_j|\theta, \mathbf{w}_{j+1}) - K_1 \left(\frac{\sum_k w_{j,k}}{N_j} - \eta_j \right)^2 - K_2 \left(\frac{\sum_k w_{j,k}^2}{N_j} - \delta_j^2 \right)^2,$$

$$= f(\mathbf{w}_j|\theta, \mathbf{w}_{j+1}) - K_1(\Delta e_1)^2 - K_2(\Delta e_2)^2, \tag{10.44}$$

where $\eta_j = 0$ and $\delta_j^2 = \sum_{m=0}^1 p_j(m)\sigma_{j,m}^2$ are the expected mean and variance implied by θ, respectively. Δe_1 and Δe_2 are the errors between the mean and variance of estimated $\mathbf{w}_j$ and the expected ones, respectively. K_1 and K_2 are two positive constants whose values are set empirically to balance the effects of three terms in $h(\mathbf{w}_j|\theta, \mathbf{w}_{j+1})$ appropriately. Thus, the constrained maximum likelihood–based texture synthesis in Equation 10.40 is changed into an unconstrained optimization as

$$\hat{\mathbf{w}}_j = \arg\max_{\mathbf{w}_j \in \mathfrak{R}} h(\mathbf{w}_j|\theta, \mathbf{w}_{j+1}). \tag{10.45}$$

The solution to Equation 10.45 can be iteratively obtained by the steepest ascent algorithm that updates $\mathbf{w}_j$ in the direction of maximizing $h(\mathbf{w}_j | \theta, \mathbf{w}_{j+1})$ as

$$\mathbf{w}'_j = \mathbf{w}_j + \lambda_j \nabla(h(\mathbf{w}_j | \theta, \mathbf{w}_{j+1})), \qquad (10.46)$$

where λ_j is the step size, which is tuned to warrant the convergence of the algorithm, and

$$\mathbf{w}_j = \begin{bmatrix} w_{j,0} \\ w_{j,1} \\ \vdots \\ w_{j,L} \end{bmatrix}, \qquad \nabla(h_j(\mathbf{w}_j|\theta)) = \begin{bmatrix} \frac{\partial h}{\partial w_{j,0}} \\ \frac{\partial h}{\partial w_{j,0}} \\ \vdots \\ \frac{\partial h}{\partial w_{j,L}} \end{bmatrix},$$

where $L = N_j - 1$, and

$$\frac{\partial h}{w_{j,k}} =$$

$$\frac{\sum_{m=0}^{1} \alpha_{j+1,l}(m) \overbrace{\left(\sum_{n=0}^{1} \epsilon_{j,j+1}^{n,m} g\left(w_{j,i}; 0, \sigma_{j,n}^2\right) \right)}^{\Lambda_p} \overbrace{\left(\sum_{n=0}^{1} \epsilon_{j,j+1}^{n,m} g'\left(w_{j,k}|0, \sigma_{j,n}^2\right) \right)}^{\Lambda_o}}{\sum_{m=0}^{1} \alpha_{j+1,l}(m) \beta_{j+1,l}(m)}$$

$$\underbrace{- \frac{2K_0}{N_j} \Delta e_1 - \frac{4K_1}{N_j} \Delta e_2 w_{j,k}}_{\Lambda_e}, \qquad (10.47)$$

where $k = 0, \ldots, L$, $l = \lfloor k/2 \rfloor$, $i = 2l$ $(2l + 1)$ if k is odd (even), and g' is the derivative of the Gaussian function g. Actually, $w_{j,i}$ and $w_{j,k}$ are two neighbors sharing the same parent $w_{j+1,l}$ in the wavelet subtree. From Equation 10.47, we see the update of $w_{j,k}$ is dependent on four terms, i.e., Λ_p from its parent, Λ_n from its neighbor, Λ_o from its own, and Λ_e from the errors of mean and variance, so that $w_{j,k}$ can be updated along the direction of $\nabla(h_j(\mathbf{w}_j|\theta))$ subject to those constraints. In particular, the cross-correlation constraint from HMT is mainly represented by the $\epsilon_{j,j+1}^{n,m}$ in Λ_n and Λ_o. Equations 10.41 to 10.47 can be straightforwardly extended to the 2D HMT and HMT-3S because 1D HMT, 2D HMT, and 2D HMT-3S share a similar tree structure and have the similar recursive computation of model likelihood defined in Equation 10.35. The major changes to Equation 10.47 will be on Λ_n and Λ_o when the 2D HMT or HMT-3S is used. Thanks to the integrated characterization of the joint DWT statistics from wavelet-domain HMT and HMT-3S, the problem of texture synthesis can be formulated mathematically and solved systematically, as shown from Equation 10.40 to Equation 10.47. It is expected that HMT and

HMT-3S will allow us to impose statistical constraints efficiently for texture analysis.

Because HMT mainly captures cross-correlations of the DWT, autocorrelations are also needed here to represent the periodicity and globally oriented structures in textures.[46,51] The statistical constraints from autocorrelations can be imposed by zero-phase 2D linear filtering.[46] In addition, we adopt the histogram specification algorithm to impose the statistics of gray-level texture pixels.[52] Using a similar framework to the one in Reference 46, we develop a new texture synthesis algorithm, as shown in Figure 10.17, where only a two-scale DWT is used for simplicity.

The synthesis process begins with an image containing samples of Gaussian white noise. In the wavelet domain, a recursive coarse-to-fine procedure imposes the statistical constraints through HMT and the autocorrelations, while simultaneously reconstructing a lowpass image, until we obtain the synthesized image after histogram specification in the spatial domain. The entire process is repeated until convergence (visual). In parallel with most synthesis algorithms, we cannot guarantee convergence. In practice, the proposed algorithm has a fast visual convergence rate with appropriately tuned constants, i.e., λ_j in Equation 10.46, and K_1 and K_2 in Equation 10.44. The current implementation requires roughly 10 min to synthesize a 256×256 texture (10 iterations) on a 400 MHz Pentium-II computer.

In this experiment, we adopt the four-level Daubechies-8 wavelet for texture synthesis. We have tried other wavelets, such as Daubechies-4,6,10 and Symmlet-4,5,6, etc. It seems that the wavelet with longer filter banks may provide slightly better texture synthesis results than one with shorter filter banks regarding the convergence rate and visual quality. This may be because the longer filter bank can provide a more compact DWT representation that allows more accurate statistical texture modeling. Both HMT and HMT-3S are tested here. Even though we found that HMT-3S allows faster visual convergence of the synthesized textures than HMT, eventually they are quite close. We guess that autocorrelation constraints of the reconstructed lowpass images can partially compensate for the loss of the characterization of dependencies across DWT subbands. We show four synthesized textures with five iterations in Figure 10.18. It is shown that HMT-3S provides perceptually more favorable results than HMT, and spatial structures or patterns can be generally replicated with fewer artifacts and distortions, e.g., D17, D68, and D87. Two models work similarly well for the textures dominated by random structures, which may weaken dependencies across subbands, e.g, D57. This fact is consistent with the texture classification results in Section 10.5.2. More texture synthesis results can be found in Figure 10.19. However, we also note that both HMT and HMT-3S cannot effectively capture the spatial periodicity and regularity in textures, as shown in Figure 10.20.

Generally speaking, there are three major observations from the above experiment regarding the texture synthesis results:

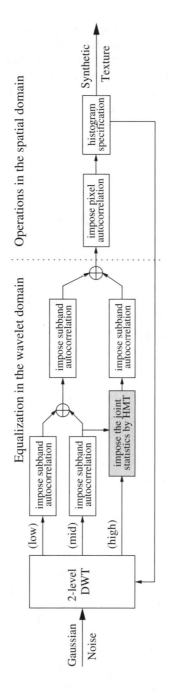

FIGURE 10.17

The texture synthesis algorithm using HMT where a two-scale DWT is adopted.

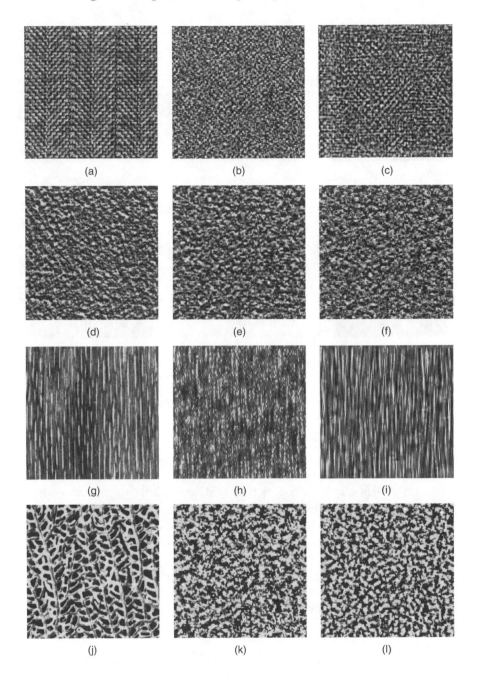

FIGURE 10.18

Texture synthesis using HMT and HMT-3S. (a) D17 texture. (b) Synthetic D17 (HMT). (c) Synthetic D17 (HMT-3S). (d) D57 texture. (e) Synthetic D57 (HMT). (f) Synthetic D57 (HMT-3S). (g) D68 texture. (h) Synthetic D68 (HMT). (i) Synthetic D68 (HMT-3S). (j) D87 texture. (k) Synthetic D87 (HMT). (l) Synthetic D87 (HMT-3S).

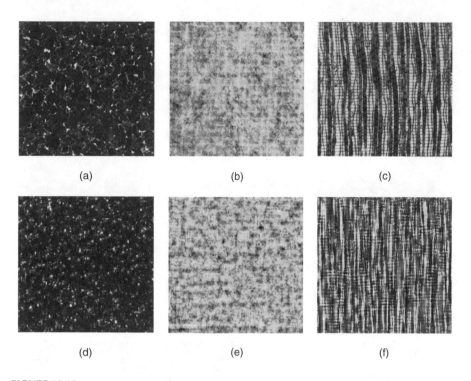

FIGURE 10.19
Texture synthesis results using HMT-3S. (a) D33 texture. (b) D19 texture. (c) D76 texture. (d) Synthetic D33. (e) Synthetic D19. (f) Synthetic D76.

1. For structural textures with irregular patterns, HMT-3S outperforms HMT by providing smoother and shaper structures, such as D68, D87, D76, etc. This shows that HMT-3S is able to reproduce the sharpness and smoothness of the irregular structures by characterizing statistical dependencies across DWT subbands.

2. For structural textures with periodic or regular patterns, both HMT-3S and HMT fail to reproduce the regular or periodic structures, such as D20, D36, D75, etc. This is because both HMT and HMT-3S are statistical models, which cannot handle regular and periodic structures or patterns. Second, the orthogonal/nonredundant DWT has a poor shift-invariant property and cannot preserve regularity or periodicity well in the wavelet domain.[48]

3. For statistical texture, the texture synthesis results are very close for HMT and HMT-3S; for example, D57, D33, D19, etc. That is because the high-order dependencies across scales or subbands are relatively weak for those textures where HMT and HMT-3S offer similar statistical characterization.

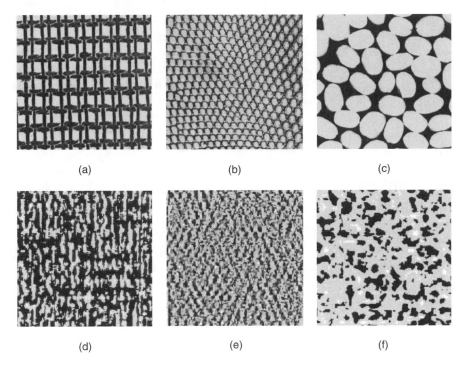

(a)   (b)   (c)

(d)   (e)   (f)

FIGURE 10.20
Texture synthesis failures. (a) D20 texture. (b) D36 texture. (c) D75 texture. (d) Synthetic D20.
(e) Synthetic D36. (f) Synthetic D75.

10.5.5   Discussions of Texture Analysis and Synthesis

In this section, we have studied texture analysis and synthesis using wavelet-domain HMMs; in particular, we have proposed a new wavelet-domain HMT-3S model to capture statistical dependencies across both subbands and scales. The basic idea of HMT-3S is that a more complete statistical characterization of DWT can be implemented by more sophisticated graphical structures for Bayesian inference in the wavelet domain. The proposed HMT-3S has been applied to texture analysis, including classification and segmentation, as well as texture synthesis. The simulation results show that the proposed HMT-3S outperforms HMT by improving the PCC of texture classification, as well as the segmentation accuracy and the visual similarity of synthetic textures. This work demonstrates the capabilities and suitability of wavelet-domain HMMs for texture analysis and synthesis applications, and shows that wavelet-domain statistical image modeling plays an important role in texture characterization. One limitation of this work is the proposed HMT-3S established in the nonredundant DWT, which is inferior to the redundant DWT for statistical image modeling and processing. However, this work can be useful for the applications where the nonredundant DWT is preferred, e.g., compression-domain image segmentation and feature extraction, etc. Moreover, the proposed maximum likelihood–based texture synthesis algorithm can also be applied to other HMM in the redundant DWT that has a moderate feature size, such as the HMT of complex wavelet transform[53] and the vector HMM of steerable wavelet transform.[9]

10.6   Conclusions

In this chapter, we have studied wavelet-domain statistical image modeling and processing. In particular, we have investigated wavelet-domain hidden Markov models (HMMs), which were originally proposed in Reference 1 for statistical signal/image processing. We first improved wavelet-domain HMMs in terms of their training efficiency and modeling accuracy by developing several new techniques that further inspire our studies toward four applications: image denoising, image segmentation, and texture analysis and synthesis. With these techniques we can obtain state-of-the-art performance or promising results by developing new wavelet-domain HMMs as well as efficient image processing algorithms.

- We show that training efficiency and modeling accuracy of wavelet-domain HMMs are important in their applications to practical signal and image processing problems. Specifically, an efficient EM initialization scheme can improve the training performance of HMMs, especially for those newly developed HMMs, i.e., HMT-2, LCHMM, and HMT-3S. Meanwhile, the graphical grouping and classification schemes have been found efficient for obtaining more accurate statistical modeling.

- We suggest that spatial adaptability and nonstructured local regional modeling are essential to the application of wavelet-domain HMMs to image denoising. Thus, a new local contextual hidden Markov model (LCHMM) was proposed, one that provides state-of-the-art denoising performance with improved visual quality at low computational complexity.

- We argue that the performance of multiscale Bayesian segmentation can be improved by strengthening two factors: contextual modeling and texture characterization. A new joint multicontext and multiscale (JMCMS) approach to Bayesian segmentation was developed to consider the first factor. In JMCMS, contextual behavior can be accumulated both across scales and via multiple-context models. On the other hand, a new wavelet-domain HMM, HMT-3S, was proposed to emphasize the second factor by providing more accurate texture characterization than HMT. It was shown that the combination of JMCMS and HMT-3S provides the best segmentation results among all tested methods, as measured by the three numerical criteria.

- We point out that wavelet-domain HMMs are useful in texture analysis and texture synthesis applications. Unlike with image denoising, hierarchical tree-structured HMMs such as HMT or HMT-3S are desired; they regard the whole wavelet subtree as one instance of the statistical model. Meanwhile, efficient texture processing algorithms are also very important to the applications of wavelet-domain HMMs for texture-related processing. This may include maximum likelihood–based texture classification and maximum likelihood–based texture synthesis.

References

1. M.S. Crouse, R.D. Nowak, and R.G. Baraniuk, Wavelet-based statistical signal processing using hidden Markov models, *IEEE Trans. Signal Process.*, 46, 886–902, April 1998.
2. S. Mallat, A theory for multiresolution signal decomposition: the wavelet representation, *IEEE Trans. PAMI*, 11, 674–693, 1989.
3. I. Daubechies, Orthonormal bases of compactly supported wavelets, *Commun. Pure Appl. Math.*, 41, 909–996, Nov. 1988.
4. L.E. Baum and T. Petrie, Statistical inference for probabilistic functions of finite state Markov chains, *Ann. Math. Stat.*, 37, 360–363, 1967.
5. L.R. Rabiner, A tutorial on hidden Markov models and selected applications in speech recognition, *Proc. IEEE*, 77(2), 257–286, 1989.
6. J.K. Romberg, H. Choi, and R.G. Baraniuk, Bayesian tree-structured image modeling using wavelet-domain hidden Markov models, *IEEE Trans. Image Process.*, 10, 1056–1068, July 2001.
7. H. Choi and R. Baraniuk, Multiscale image segmentation using wavelet-domain hidden Markov models, *IEEE Trans. Image Process.*, 10, 1309–1321, Sept. 2001.
8. R.D. Nowak, Multiscale hidden Markov models for Bayesian image analysis, in *Bayesian Inference in Wavelet Based Models*, P. Müller and B. Vidakovic, Eds., New York: Springer-Verlag, 1999, 243–266.
9. M.N. Do and M. Vetterli, Rotation invariant texture characterization and retrieval using steerable wavelet-domain hidden Markov models, *IEEE Trans. Multimedia*, 4(4), 517–527, 2002.
10. I. Daubechies, *Ten Lectures on Wavelets*, Philadelphia: SIAM, 1992.
11. H. Chipman, E. Kolaczyk, and R. McCulloch, Adaptive Bayesian wavelet shrinkage, *J. Am. Stat. Assn.*, 440, 1413–1421, Dec. 1997.
12. A.P. Dempster, N.M. Laird, and D.B. Rubin, Maximum likelihood from incomplete data via the EM algorithm, *J. R. Stat. Soc.*, 39, 1–38, 1977.

13. M.S. Crouse and R.G. Baraniuk, Contextual hidden Markov models for wavelet-domain signal processing, in *Proc. 31th Asilomar Conf. Signals, Systems, and Computers*, Pacific Grove, CA, Nov. 1997.

14. G. Fan and X.-G. Xia, Improved hidden Markov models in the wavelet-domain, *IEEE Trans. Signal Process.*, 49, 115–120, Jan. 2001.

15. B. Everitt, Ed., *Finite Mixture Distribution*, New York: Chapman & Hall, 1981.

16. G. Fan and X.-G. Xia, Wavelet-based statistical image processing using hidden Markov tree model, in *Proc. 34th Annual Conf. on Information Sciences and Systems*, Princeton, NJ, Mar. 2000.

17. B. Frey, *Graphical models for Machine Learning and Digital Communication*, Cambridge, MA: MIT Press, 1998.

18. H. Lucke, Which stochastic models allow Baum-Welch training, *IEEE Trans. Signal Process.*, 44, 2746–2756, Nov. 1994.

19. D. Donoho and I. Johnstone, Ideal spatial adaptation via wavelet shrinkage, *Biometrika*, 81, 1994.

20. J.M. Shapiro, Embedded image coding using zerotrees of wavelet coefficients, *IEEE Trans. Signal Process.*, 41(12), 3445–3663, 1993.

21. M.K. Mihcak, I. Kozintsev, and K. Ramchandran, Low-complexity image denoising based on statistical modeling of wavelet coefficients, *IEEE Signal Process. Lett.*, 6, 300–303, Dec. 1999.

22. S. Xiao, I. Kozintsev, and K. Ramchandran, Stochastic wavelet-based image modeling using factor graphs and its application to denoising, in *Proc. SPIE Image and Video Communications and Processing*, Vol. 3974, San Jose, CA, 2000.

23. S.G. Chang, B. Yu, and M. Vetterli, Spatially adaptive wavelet thresholding with context modeling for image denoising, in *Proc. IEEE Int. Conf. on Image Processing*, Chicago, IL, Oct. 1998.

24. J. Liu and P. Moulin, Image denoising based on scale-space mixture modeling of wavelet coefficients, in *Proc. IEEE Int. Conf. on Image Processing*, Kobe, Japan, Oct. 1999.

25. E.P. Simoncelli, Statistical models for images: compression, restoration and synthesis, in *Proc. 31st Asilomar Conf. on Signals, Systems and Computers*, Pacific Grove, CA, Nov. 1997, 673–678.

26. G. Fan and X.-G. Xia, Image denoising using local contextual hidden Markov model in the wavelet-domain, *IEEE Signal Process. Lett.*, 8, 125–128, May 2001.

27. R.R. Coifman and D.L. Donoho, *Translation-Invariant De-Noising*, New York: Springer-Verlag, 1995.

28. X. Li and M. Orchard, Spatially adaptive denoising under overcomplete expansion, in *Proc. IEEE Int. Conf. on Image Processing*, Vancouver, Canada, Sept. 2000.

29. J. Besag, Spatial interaction and statistical analysis of lattice systems, *J. R. Stat. Soc. B*, 36(2), 192–236, 1974.

30. H. Derin and H. Elliott, Modeling and segmentation of noisy and textured images using Gibbs random fields, *IEEE Trans. Pattern Anal. Mach. Intelligence*, 9, 39–55, Jan. 1987.

31. C.A. Bouman and B. Liu, Multiple resolution segmentation of textured images, *IEEE Trans. Pattern Anal. Mach. Intelligence*, 13, 99–113, Feb. 1991.

32. C.A. Bouman and M. Shapiro, A multiscale random field model for Bayesian image segmentation, *IEEE Trans. Image Process.*, 3, 162–177, Mar. 1994.

33. H. Cheng and C.A. Bouman, Multiscale Bayesian segmentation using a trainable context model, *IEEE Trans. Image Process.*, 10, 511–525, Apr. 2001.

34. P. Dasgupta, P.P. Chakrakbarti, and S.C. DeSarkar, *Multiobjective Heuristic Search*, Wiesbaden, Germany: Friedrich Vieweg & Sohn, 1999.

35. G. Fan and X.-G. Xia, A joint multi-context and multiscale approach to Bayesian image segmentation, *IEEE Trans. Geosci. Remote Sensing*, 39, 2680–2688, Dec. 2001.

36. X. Wu, Low complexity high-order context modeling of embedded wavelet bit streams, in *Proc. IEEE Data Compression Conference*, Snowbird, UT, March 1999, 112–120.

37. J. Li, R.M. Gray, and R.A. Olshen, Multiresolution image classification by hierarchical modeling with two dimensional hidden Markov models, *IEEE Trans. Inf. Theor.*, 46, 1826–1841, Aug. 2000.

38. Z. Kato, M. Berthod, and J. Zerubia, Parallel image classification using multiscale Markov random fields, in *Proc. IEEE Int. Conf. Acoust. Speech Signal Process.*, 5, 137–140, Apr. 1993.

39. M.L. Comer and E.J. Delp, Segmentation of textured images using a multiresolution Gaussian autoregressive model, *IEEE Trans. Image Process.*, 8, 408–420, Mar. 1999.

40. P. Brodatz, *Textures—A Photographic Album for Artists and Designers*, New York: Dover, 1966.

41. S.R. Yhann and T.Y. Young, Boundary localization in texture segmentation, *IEEE Trans. Image Process.*, 4, 849–855, June 1995.

42. P.L. Palmer and M. Petrou, Locating boundaries of textured regions, *IEEE Trans. Geosci. Remote Sensing*, 35, 1367–1371, Sept. 1997.

43. X. Song and G. Fan, A study of supervised, semi-supervised, and unsupervised Bayesian image segmentation, in *Proc. IEEE Int. Midwest Symposium on Circuits and Systems*, Tulsa, OK, Aug. 2002.

44. S.C. Zhu, Y. Wu, and D. Mumford, Filters, random fields and maximum entroy, *Int. J. Comput. Vision*, 27, 1–20, Mar./Apr. 1998.

45. M.S. Silverman, D.H. Crosof, R.L.D. Valois, and S.D. Elfar, Spatial-frequency organization in primate strate cortex, *Proc. Natl. Acad. Sci. U.S.A.*, 86, 1989.

46. J. Portilla and E.P. Simoncelli, A parametric texture model based on joint statistics of complex wavelet coefficients, *Int. J. Comput. Vision*, 40(1), 2000.

47. G. Fan and X.-G. Xia, Wavelet-based texture analysis and synthesis using hidden Markov models, *IEEE Trans. Circuits Syst.*, Part I, 50(1), 106–120, 2003.

48. A. Mojsilovic, M.V. Popovic, and D.M. Rackov, On the selection of an optimal wavelet basis for texture characterization, *IEEE Trans. Image Process.*, 9, 2043–2050, Dec. 2000.

49. G. Fan and X. Song, A study of contextual modeling and texture characterization for multiscale Bayesian segmentation, in *Proc. IEEE Int. Conf. on Image Processing*, Rochester, NY, Sept. 2002.

50. B.S. Gottfried, *Introduction to Optimization Theory*, Englewood Cliffs, NJ: Prentice-Hall, 1973.

51. J.M. Francos, A.Z. Meiri, and B. Porat, A unified texture model based on a 2-d Wold-like decomposition, *IEEE Trans. Signal Process.*, 41(8), 2665–2678, 1993.

52. A.K. Jain, *Fundamentals of Digital Image Processing*. Englewood Cliffs, NJ: Prentice-Hall, 1989.

53. H. Choi, J.K. Romberg, R.G. Baraniuk, and N.G. Kingsbury, Hidden Markov tree modeling of complex wavelet transforms, in *Proc. IEEE Int. Conf. Acoust. Speech, Signal Processing*, Istanbul, Turkey, June 2000.

11

Self-Organizing Maps and Their Applications in Image Processing, Information Organization, and Retrieval

Constantine Kotropoulos and Ioannis Pitas

CONTENTS

11.1 Introduction

Neural networks (NNs) are able to learn from their environment so that their performance is improved. In several NN categories, learning is provided by a desirable input–output mapping that the NN approximates. This is called *supervised learning*. Typical NNs where supervised learning is employed are called multilayer perceptron or radial-basis functions networks. Another principle is the unsupervised learning or self-organized learning that aims at identifying the important features in the input data without a supervisor. Unsupervised learning algorithms are equipped with a set of rules that locally update the synaptic weights of the network. The topologies of NNs that are trained using unsupervised learning are more similar to neurobiological structures than are those of NNs that are trained using supervised learning. The basic topologies of self-organizing NNs are as follows:

1. NNs that consist of an input layer and output layer where the neurons of the input layer are connected to the neurons of the output layer with feedforward connections, and the neurons of the output layer are connected with lateral connections

2. NNs of multiple layers in which the self-organization proceeds from one layer to another

There are two self-organized learning methods:

1. *Hebbian learning* that yields NNs that extract the principal components[1,2]

2. *Competitive learning* that yields K-means clustering[3]

Self-organized learning is essentially a repetitive updating of NN synaptic weights as a response to input patterns, according to a set of prescribed rules, until a final configuration is obtained.[2] A number of observations have motivated the research toward self-organized learning. It is worth noting that in 1952 Turing stated that "global ordering can arise from local interactions," and von der Malsburg observed that self-organization is achieved through self-amplification, competition, and cooperation of the synaptic weights of the NN (see Reference 2, and references therein). In this chapter, we focus on competitive learning and on self-organizing maps (SOM) in particular. The latter can be viewed as a computational procedure for finding a discrete approximation of principal curves.[4] Principal curves could be conceived of as a nonlinear principal component analysis method.[1]

Self-organizing maps are based on competitive learning. That is, the output neurons of the network compete among themselves to be activated, and accordingly only one output neuron or one neuron per group is active at each time instant. The neuron that wins the competition is called a *winner-takes-all* neuron. In an SOM, the neurons are placed at the nodes of a lattice. Although high-dimensional lattices could also be employed, a one- or two-dimensional map is frequently used, because such maps facilitate data visualization. During the training that implements competitive learning, the neurons are tuned selectively to various input patterns or classes of input patterns. In addition, the coordinates of the neurons become ordered so that a meaningful coordinate system for the different intrinsic statistical properties of the input data, the so-called *features*, is created over the lattice. The seminal work of Kohonen dominates this research field.[5–7]

The outline of the chapter is as follows. The basic feature-mapping techniques are reviewed in Section 11.2. Kohonen's SOM is described in Section 11.3. The convergence analysis of Kohonen's SOM is treated in Section 11.4. SOM properties are outlined in Section 11.5. Variants of SOMs based on robust statistics are described in Section 11.6, which also discusses their applications to color image quantization and document organization and retrieval. A class of split-merge SOMs that incorporate outlier rejection and cluster validity tests are analyzed in Section 11.7; applications of split-merge

SOMs to color image quantization and image segmentation are also presented in this section. Concluding remarks are given in Section 11.8.

11.2 Basic Feature-Mapping Models

The research in self-organizing maps was motivated by similar studies at the neurophysiological and biological level. For example, the different sensory inputs (motor, somatosensory, visual, auditory, etc.) are mapped onto corresponding areas of the cerebral cortex of the human brain. Two feature-mapping models have been proposed:

1. The model of Willshaw–von der Malsburg
2. The model of Kohonen

We will briefly describe the first model and then study the second in more detail. The model of Willshaw–von der Malsburg is older than that of Kohonen.

The model of Willshaw–von der Malsburg uses two distinct two-dimensional (2D) neuron grids that are connected so that the one grid is projected onto the other.[9] The first grid represents the presynaptic neurons (input) and the second one represents the postsynaptic neurons (output). Let $\mathbf{a} = (a_1, a_2)^T$ be the position vector of one neuron that lies in the presynaptic grid $\mathcal{X}$ and $\mathbf{b} = (b_1, b_2)^T$ be the corresponding position vector of a postsynaptic neuron in the output grid $\mathcal{Y}$. Each neuron, e.g., the one indexed by $\mathbf{a}$, computes a feature vector $f_\mathbf{a}$ and its activation value $x_\mathbf{a}$. The postsynaptic neurons are connected with the presynaptic neurons through interlayer synaptic weights $J_\mathbf{ba}$ that are called *dynamic links*. For the interlayer synaptic weights we have $J_\mathbf{ba} > 0$. The weights of the links that abut to the postsynaptic neuron indexed by $\mathbf{b}$ are normalized, that is, $\sum_\mathbf{a} J_\mathbf{ba} = 1$. Let us denote by $T_\mathbf{ba}$ the similarity function between the postsynaptic and presynaptic neurons. Such a function measures the similarity of the local features in the two neurons, i.e.,

$$T_\mathbf{ba} = \psi(f_\mathbf{b}, f_\mathbf{a}), \tag{11.1}$$

where $\psi(\cdot)$ is any valid similarity function, e.g., autocorrelation. Obviously, a plausible initial value of the dynamic link is

$$J_\mathbf{ba} = \frac{T_\mathbf{ba}}{\sum_{\mathbf{a}'} T_{\mathbf{ba}'}}. \tag{11.2}$$

Self-organization is achieved through an iterative algorithm whose principle is described subsequently. Besides the dynamic links $J_\mathbf{ba}$ inherent in layers $\mathcal{X}$ and $\mathcal{Y}$ are static homogeneous intralayer connections that are described by an interaction kernel[9]

$$\kappa(\xi) = \gamma \exp\left(-\frac{\xi^2}{2s^2}\right) - \beta, \tag{11.3}$$

where the first term corresponds to an excitatory short-range mechanism with width s and the second term corresponds to a generic inhibitory mechanism of relative strength β. It can be shown that such an interaction kernel activates only one connected region at the equilibrium, the *active bubble* (described in more detail below).[8]

At each iteration of the algorithm, the active bubbles are simultaneously developed in layers X and Y according to a procedure described by the set of coupled differential equations:[9]

$$\dot{x}_a = -\alpha x_a + (\kappa * X)_a + I_a^{(x)}, \quad X_a = \varphi(x_a) \tag{11.4}$$

$$\dot{y}_b = -\alpha x_b + (\kappa * Y)_b + I_b^{(y)}, \quad Y_b = \varphi(y_b) \tag{11.5}$$

$$x_a(0) = y_b(0) = 0,$$

where $\dot{x}$ denotes a time derivative, $\varphi(\xi) = 1/1 + \exp(-\lambda\xi)$ is the sigmoidal function, and $*$ is the 2D spatial convolution operator. In the presynaptic layer, $I_a^{(x)}$ is a noise random variable that is uncorrelated for two different neurons. It represents a random excitation of the presynaptic neurons that is slowly varying compared to the dynamics of X and Y. In the postsynaptic layer, we have

$$I_b^{(y)} = \upsilon \sum_a J_{ba} T_{ba} X_a, \tag{11.6}$$

where υ is the coupling coefficient between the two layers. Equation 11.6 implies that the activity flow from one active presynaptic layer to the postsynaptic layers $\mathbf{b} \in Y$ is proportional to the weight of the dynamic link J_{ba} and the similarity function T_{ba} of the local features. When the activation in the two layers is led to equilibrium, then the dynamic links are amplified according to

$$\Delta J_{ba} = \upsilon J_{ba} T_{ba} Y_b X_a. \tag{11.7}$$

The constraint $\sum_a (J_{ba} + \Delta J_{ba}) = 1, \forall \mathbf{b} \in Y$ is externally enforced and the iteration concludes by resetting the active zones, i.e., $x_a = y_b = 0, \forall \mathbf{a, b}$.

The next iteration begins with a new noise amplitude $I_a^{(x)}$ that produces a different active bubble in the presynaptic layer. The mechanism of active bubble formation in the postsynaptic layer is affected by the previous iterations, as these are reflected in the present state of the dynamic link weights. A property of this NN is that a presynaptic neuron $\mathbf{a}$ is not activated in isolation, but always together with some of its neighbors. If the neighbors of $\mathbf{a}$ have strong links with a region in the postsynaptic layer Y, the dynamic links that emanate from neuron $\mathbf{a}$ and abut on this region will be further amplified.

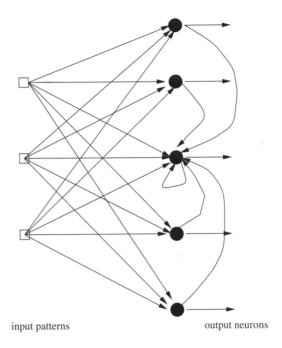

input patterns   output neurons

FIGURE 11.1
One-dimensional lattice of neurons with feedforward and feedback connections.

The training algorithm of the self-organizing network of Willshaw–von der Malsburg is called *fast dynamic link matching*.[9] This algorithm is a neural network implementation of the older idea of the group of von der Malsburg, the dynamic link architecture.[10]

In the following, we study the self-organization model proposed by Kohonen.[2] The analysis starts assuming neurons that pass a linear combination of the input signals through a thresholding function. We demonstrate how we can simplify this usual neuron model and under which assumptions we can confine ourselves only to distance comparisons between the vectors of synaptic weights and the input pattern.

Let us consider the NN of Figure 11.1. There are feedforward connections from the input, self-feedback connections, and lateral feedback connections. The weighted sum of the input signals at each neuron performs feature detection. The feedback connections produce the excitatory or inhibitory effects depending on the distance between the neurons.

Let $x_1, x_2, \ldots, x_p$ be the signals applied to the input nodes. Let $w_{j1}, w_{j2}, \ldots, w_{jp}$ be the synaptic weights of the jth neuron. Let us denote by $c_{j,-K}, \ldots, c_{j,-1}, c_{j,0}, c_{j,1}, \ldots, c_{j,K}$ the lateral feedback weights connected to jth neuron where K is the radius of the lateral interaction. The weight of the lateral feedback connection with respect to the distance from the neuron j, $j - |k|$,

$k = -K, \ldots, -1, 0, 1, \ldots, K$, usually forms a "Mexican hat"–shaped function. Let $y_1, y_2, \ldots, y_N$ be the signals in the N output neurons. The output of the jth neuron is given by[11]

$$y_j = \theta \left(I_j + \sum_{k=-K}^{K} c_{j,k} y_{j+k} \right), \quad j = 1, 2, \ldots, N, \tag{11.8}$$

where $\theta(\cdot)$ is a nonlinear function that limits the output of the jth neuron and

$$I_j = \sum_{l=1}^{p} w_{j,l} x_l \tag{11.9}$$

is the stimulus of the jth neuron. The solution of the nonlinear Equation 11.8 can be found iteratively using a relaxation technique. If we reformulate Equation 11.8 as a difference equation, we have

$$y_j(n+1) = \theta \left(I_j + \beta \sum_{k=-K}^{K} c_{jk} \, y_{j+k}(n) \right), \quad j = 1, 2, \ldots, N, \tag{11.10}$$

where n denotes discrete time. The parameter β controls the rate of convergence of the relaxation technique and plays the role of feedback factor. Depending on the sign of the weights $c_{j,k}$, both positive and negative feedback can be included in the system. The limiting action of the function $\theta(\cdot)$ enforces the spatial response $y_j(n)$ to stabilize depending on the value assigned to parameter β. If β is large enough, then at the final state corresponding to $n \to \infty$, the values of y_j tend to concentrate inside a spatially bounded cluster, the *activity bubble*, as shown in Figure 11.2.

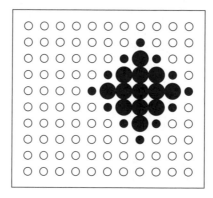

FIGURE 11.2
Shape of activity bubble in a 2D self-organizing network. The circles correspond to neurons. Their radius corresponds to their activation. The active neurons are shown as filled-in black circles.

The bubble is centered at the point where the initial response $y_j(0)$ is maximized with respect to I_j. The width of the bubble depends on the ratio of excitatory to inhibitory lateral connections. If the positive feedback is strengthened, the bubble becomes wider. If the negative feedback is amplified, the bubble becomes thinner.

If we consider the piecewise-linear function $\theta(\cdot)$

$$\theta(x) = \begin{cases} a & \text{if } x \geq a \\ x & \text{if } 0 \leq x < a \\ 0 & \text{otherwise,} \end{cases} \tag{11.11}$$

then the jth neuron output can be written as

$$y_j = \begin{cases} a & \text{if neuron } j \text{ is inside the bubble} \\ 0 & \text{otherwise.} \end{cases} \tag{11.12}$$

Accordingly, we can exploit the bubble formation as a computational shortcut that emulates the effect of lateral feedback connections. Therefore, it is sufficient to introduce the topological neighborhood of active neurons that corresponds to the activity bubble. Henceforth, in Kohonen's model we are interested in:

1. Finding the winner neuron
2. Defining the topological neighborhood of the winner
3. Updating the weights of the neurons that fall inside this neighborhood

The winner neuron is the one that corresponds to the maximum stimulus $I_j = \mathbf{w}_j^T \mathbf{x}$, where $\mathbf{w}_j = (w_{j1}, w_{j2}, \ldots, w_{jp})^T$, $j = 1, 2, \ldots, N$, is the weight vector of the jth neuron and $\mathbf{x} = (x_1, x_2, \ldots, x_p)^T$ is the input pattern. If we assume that all weight vectors are normalized to a constant norm, then the neuron that maximizes the inner product is simply the neuron that minimizes the Euclidean distance between the weight vectors and the input pattern, $d(\mathbf{w}_j, \mathbf{x})$, because

$$d^2(\mathbf{w}_j, \mathbf{x}) = \|\mathbf{w}_j - \mathbf{x}\|^2 = \|\mathbf{w}_j\|^2 - 2\mathbf{w}_j^T \mathbf{x} + \|\mathbf{x}\|^2. \tag{11.13}$$

Equation 11.13 indicates that, under the assumptions of a piecewise-linear function (Equation 11.11) and weight vectors normalized to a constant norm, the self-organizing model of Kohonen performs like a K-means clustering algorithm. In the following, we will resort to this simplification.

11.3   Kohonen's Self-Organizing Map

Let $d(\mathbf{x}, \mathbf{w}_i)$ denote a generic distance measure between $\mathbf{x}$ and $\mathbf{w}_i$. The index of the best matching neuron to input pattern $\mathbf{x}$ is given by

$$c(\mathbf{x}) = \arg\min_i\{d(\mathbf{x}, \mathbf{w}_i)\}, \qquad i = 1, 2, \ldots, N. \tag{11.14}$$

The neuron that satisfies Equation 11.14 is the so-called *winner* for the input pattern $\mathbf{x}$. By using Equation 11.14, the continuous space of input patterns is mapped into a set of discrete neurons. Depending on the application, the response of the network could be either the index of the winner neuron or the vector of the synaptic weights of the winner neuron. Figure 11.3a depicts the topology of the NN under study.

Our aim is to determine $\mathbf{w}_i$ so that an ordered mapping is obtained, one that describes the distribution of patterns $\mathbf{x}$. One possible way is to determine $\mathbf{w}_i$ according to the principles of *vector quantization*.[12] Let $f(\mathbf{x})$ be the joint probability density function (pdf) of input patterns and $\mathbf{w}_c$ the closest weight vector to $\mathbf{x}$ with respect to Equation 11.14. The elements of $\mathbf{w}_c$ are chosen so that the expected average quantization error

$$\varepsilon = \int_X g(d(\mathbf{x}, \mathbf{w}_c)) f(\mathbf{x}) d\mathbf{x} \tag{11.15}$$

is minimized,[6] where $g(\cdot)$ is a monotonically increasing function of the distance $d(\mathbf{x}, \mathbf{w}_c)$. One can observe that the index c is also a function of $\mathbf{x}$ as well as of all $\mathbf{w}_i$. Accordingly, the integrand in Equation 11.15 is not continuously differentiable. The minimization of Equation 11.15 is, in general, difficult and does not guarantee that all neurons are indexed with one particular way, i.e., that an ordered mapping results. However, if ε is modified so that the quantization error is smoothed locally by a kernel h_{ci}, which is a function of the distance between the neurons c and i, the minimization of the new criterion

$$\varepsilon' = \int \sum_i h_{ci} g(d(\mathbf{x}, \mathbf{w}_i)) f(\mathbf{x}) d\mathbf{x} \tag{11.16}$$

results in ordered weight vectors. The minimization of Equation 11.16 could be done by using stochastic approximation methods, such as the Robbins–Monro algorithm.[13] Let $\mathbf{x} = \mathbf{x}(n)$ be the input pattern at discrete time instant n. Let $\mathbf{w}_i(n)$ be the approximation of $\mathbf{w}_i$ at the same time instant. Let us consider the sample function $\varepsilon''(n)$ defined by[7]

$$\varepsilon''(n) = \sum_i h_{ci} g(d(\mathbf{x}(n), \mathbf{w}_i(n))). \tag{11.17}$$

The Robbins–Monro algorithm for the minimization of Equation 11.17 yields

$$\mathbf{w}_i(n+1) = \mathbf{w}_i(n) - \frac{1}{2}\beta(n)\frac{\partial \varepsilon''(n)}{\partial \mathbf{w}_i(n)}, \tag{11.18}$$

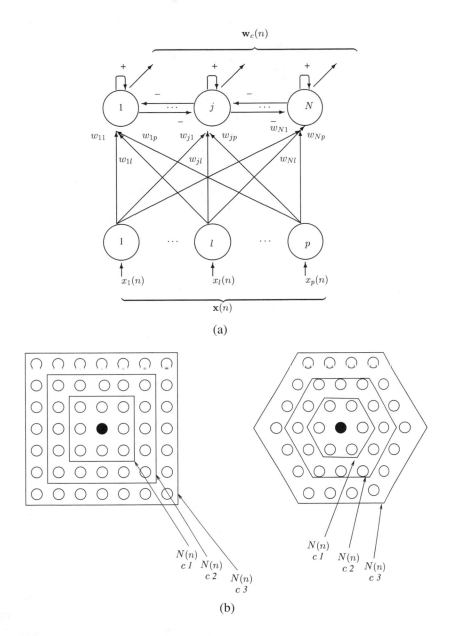

FIGURE 11.3

(a) Topology of the self-organizing feature map. (b) Shrinking square and hexagonal topological neighborhoods around the winner neuron, which is marked by a black circle, for $n_1 < n_2 < n_3$.

where $\beta(n)$ is a time-varying adaptation step. The selection of an adaptation step sequence that satisfies the following conditions:[13,14]

$$\lim_{n\to\infty} \beta(n) = 0, \qquad \sum_{n=0}^{\infty} \beta(n) = \infty, \qquad \sum_{n=0}^{\infty} \beta^2(n) < \infty, \qquad (11.19)$$

forces the covariance matrix of the weight errors to tend to a zero matrix. If $d(\cdot)$ is the Euclidean distance and $g(d)$ is chosen to be $g(d) = d^2$, then Equation 11.18 is rewritten as

$$\mathbf{w}_i(n+1) = \mathbf{w}_i(n) + \tilde{h}_{ci}(n)[\mathbf{x} - \mathbf{w}_i(n)], \qquad (11.20)$$

where the adaptation step and the kernel function were merged in the term $\tilde{h}_{ci}(n)$. The simplest choice for $\tilde{h}_{ci}(n)$ is

$$\tilde{h}_{ci}(n) = \begin{cases} \alpha(n) & \text{if } i \in \mathcal{N}_c(n) \\ 0 & \text{otherwise} \end{cases}, \qquad (11.21)$$

where $\mathcal{N}_c(n)$ is the topological neighborhood of neuron $c = c(\mathbf{x})$ at time instant n. By using Equation 11.21, the recursion (Equation 11.20) is rewritten as

$$\mathbf{w}_i(n+1) = \begin{cases} \mathbf{w}_i(n) + \alpha(n)[\mathbf{x}(n) - \mathbf{w}_i(n)] & \text{if } i \in \mathcal{N}_c(n) \\ \mathbf{w}_i(n) & \text{otherwise.} \end{cases} \qquad (11.22)$$

A plethora of simulations has demonstrated that the topological neighborhood $\mathcal{N}_c(n)$ has to be selected to be wide enough at the beginning and to shrink gradually in the subsequent iterations[6] (Figure 11.3b).

11.4   Convergence Analysis of Self-Organizing Maps

The convergence properties of self-organizing maps (or topology preserving maps) have been studied by Kohonen himself[15] as well as by a number of other researchers.[14,16–19] Cottrell and Fort[16] presented a formal proof of the self-organizing procedure that is based on the assumption that the training of the network is a Markovian process, and demonstrated that the states of the network that correspond to a topologically correct mapping are *absorbing states*. Ritter and Shulten[14,17] described the training procedure at the equilibrium of the neural network in terms of an equivalent *Fokker–Planck differential equation*. They formulated the conditions that should be satisfied by the sequence of adaptation steps so that convergence is achieved. Tolat[18] presented an analysis of the Kohonen algorithm using a set of energy equations that are defined for each neuron. Such equations can be used so that the synaptic weights at equilibrium are determined, as well as bounds on the adaptation step or on the learning rate.[18] Lo et al.[19] followed a different procedure to prove the convergence of the NN under study. Their approach extends the notion of the energy for the entire NN and proves directly the convergence using Glasdshev's theorem. A different consideration, also based on the assumption that the weight updating is a Markovian chain, aims at finding the set of states whose probability to remain in the same class is 1. Such sets of states are known as *ergodic classes* of the Markovian procedure. Their derivation answers the question *when* self-organization is achieved. Kohonen was the first who presented analytical results for an 1D neural network that maps the 1D input space to the interval $(0,1)$.[5] Extensions of this work in the 2D case for uniformly distributed input patterns as well as for several neighborhood functions are included in Reference 19. For simplicity, we qualitatively describe the results of Kohonen's work. Let us denote by $w_1, w_2, \ldots, w_N$ the N scalar synaptic weights. Self-organization is achieved if and only if the weights are ordered in either strictly increasing order ($w_1 < w_2 < \cdots < w_N$) or strictly decreasing order ($w_1 > w_2 > \cdots > w_N$). When such an ordering is set up, subsequently the ordering is preserved. Moreover, the pdf of w_i approximates that of the input patterns, $f(x)$, as is explained next. In the following, we describe in detail the proof of convergence that is based on the assumption that training is a Markovian process.

In its generality, the NN maps a pattern space X with elements $x(n)$, $n = 1, 2, \ldots$ into a lattice of neurons A, where a neuron is indexed by the vector $\mathbf{r}$ that represents its spatial coordinates on the lattice. That is, we have a mapping $\Psi : X \to A$. Each $x(n)$ is applied as input to all neurons $\mathbf{r}$ of A. A vector of synaptic weights $\mathbf{w_r}(n) \in X$ is attached to each neuron that determines the response of neuron $\mathbf{r}$ when it is excited by the pattern $x(n)$. The response of the neuron is given by a smooth real-valued function that attains its maximum at the origin and resembles a Gaussian function. The union of all patterns in X that are closer to $\mathbf{w_r}(n)$ than to any other neuron $\mathbf{w_s}(n)$, $\mathbf{s} \neq \mathbf{r}$, forms the receptor $V_\mathbf{r}$ of neuron $\mathbf{r}$. Accordingly, the receptor of a neuron is its Voronoi neighborhood.[20] We always shall have the maximal response for that neuron $\mathbf{r}$ for which $x(n) \in V_\mathbf{r}$. Initially, the vectors $\mathbf{w_r}(0)$ as well as the receptors of all neurons are randomly distributed in X. For each incoming pattern $x(n) \in X$, $n = 1, 2, \ldots$:

1. We select the neuron $\mathbf{r}$ that admits the maximum response for this pattern.
2. We update the weight vector of neuron $\mathbf{r}$ and those of neighboring neurons $\mathbf{s}$ according to

$$\mathbf{w_s}(n+1) = \mathbf{w_s}(n) + \alpha h(\mathbf{r} - \mathbf{s})(x(n) - \mathbf{w_s}(n))$$
$$= \mathbf{w_s}(n) + \alpha h_{\mathbf{rs}}(x(n) - \mathbf{w_s}(n)), \tag{11.23}$$

where $h_{\mathbf{rs}}$ is the *neighboring function*.

The aforementioned steps mathematically constitute a Markovian process whose states are the vectors of synaptic weights $\mathbf{w_s}$. The transitions between the states are defined in terms of the pdf of input patterns $f(\mathbf{x})$. In other words, each pattern $\mathbf{x} \in X$ is mapped into the position $\mathbf{r}$ of the lattice and such a discrete mapping of X into A is determined by the set of weight vectors $\mathbf{w_s}$ that gradually evolves in this Markovian process. Kohonen proved that by decreasing both the adaptation step α and the amplitude of h_{rs} the weight vectors $\mathbf{w_s}$ converge asymptotically to equilibrium values that represent a useful mapping Ψ.[5]

Let us confine ourselves to the case where h_{rs} is a Kronecker delta function, i.e., when only the weight vector of the winner neuron is updated. We shall denote the weight vectors by $\mathbf{w}_1, \mathbf{w}_2, \ldots, \mathbf{w}_N$. The expected state of the network is given by

$$\overline{\mathbf{w}}_i = E[\mathbf{w}_i] = \frac{\int_{V_i(\overline{\mathbf{W}})} \mathbf{x} f(\mathbf{x}) d\mathbf{x}}{\int_{V_i(\overline{\mathbf{W}})} f(\mathbf{x}) d\mathbf{x}} \qquad i = 1, 2, \ldots, N \quad \overline{\mathbf{W}} = \left(\overline{\mathbf{w}}_1^T \mid \ldots \mid \overline{\mathbf{w}}_N^T\right)^T,$$

(11.24)

where $V_i(\overline{\mathbf{W}})$ is the Voronoi neighborhood of the ith neuron. It is evident that Equation 11.24 is an implicit definition of the stationary state of the neural network. The nonlinear Equation 11.24 can be solved by any iterative method, e.g., Newton's method. Let us assume that the stationary weight vectors $\overline{\mathbf{w}}_i$ are known. The following study, which is a variant of the analysis presented in Reference 24, focuses on the derivation of the convergence rate toward the stationary solution. Let $\mathbf{u}_i(t)$ denote the weight error vector at time instant t, i.e.,

$$\mathbf{u}_i(t) = \mathbf{w}_i - \overline{\mathbf{w}}_i. \tag{11.25}$$

The expected weight error vector is given by[14,21]

$$E[\mathbf{U}(t)] = \mathbf{Y}(t)E[\mathbf{U}(0)]; \quad \mathbf{U}(t) = \left(\mathbf{u}_1^T(t) \mid \ldots \mid \mathbf{u}_N^T(t)\right)^T, \tag{11.26}$$

where the expected value is computed with respect to the distribution of the divergence of synaptic weights from the stationary state. Let us define the $(Np \times Np)$ matrix $\mathbf{Y}(t)$

$$\mathbf{Y}(t) = \exp\left(-\mathbf{B}\int_0^t \alpha(\zeta)d\zeta\right), \tag{11.27}$$

where $\mathbf{B}$ is a $(Np \times Np)$ coefficient-matrix that can be partitioned as follows:

$$\mathbf{B} = \begin{bmatrix} \mathbf{B}_{11} & \mathbf{B}_{12} & \cdots & \mathbf{B}_{1N} \\ \mathbf{B}_{21} & \mathbf{B}_{22} & \cdots & \mathbf{B}_{2N} \\ \vdots & & \ddots & \vdots \\ \mathbf{B}_{N1} & \mathbf{B}_{N2} & \cdots & \mathbf{B}_{NN} \end{bmatrix}. \tag{11.28}$$

Each $\mathbf{B}_{kl}, k, l = 1, \ldots, N$ is a $(p \times p)$ square submatrix, whose mn-element is given by

$$[\mathbf{B}_{kl}(\overline{\mathbf{W}})]_{mn} = \left[w_{km} \frac{\partial}{\partial w_{ln}} \widehat{F}_k(\mathbf{W}) + \widehat{F}_k(\mathbf{W})\delta(k - l, m - n) \right.$$
$$\left. - \frac{\partial}{\partial w_{ln}} \int_{\mathcal{V}_k(\mathbf{W})} x_m f(\mathbf{x})d\mathbf{x} \right]_{\mathbf{W} = \overline{\mathbf{W}}}, \qquad (11.29)$$

where

$$\widehat{F}_k(\mathbf{W}) = \int_{\mathcal{V}_k(\mathbf{W})} f(\mathbf{x})d\mathbf{x} \qquad (11.30)$$

and $\delta(k - l, m - n)$ is the 2D Kronecker delta function, i.e.,

$$\delta(k - l, m - n) = \begin{cases} 1 & k = l \text{ and } m = n \\ 0 & \text{otherwise.} \end{cases} \qquad (11.31)$$

The coefficient-matrix $\mathbf{B}$ does not have any physical interpretation for stochastic processes that are described by multivariate Fokker–Planck differential equations, as is the training procedure of the self-organizing NN under study. However, it is well known that for stochastic processes that are described by single variable Fokker–Planck differential equations (also known as *diffusion processes*), the matrix $\mathbf{B}$ degenerates to a vector called the *drift vector*.[22] Let us assume that the (real) matrix $\mathbf{B}$ is symmetric, so that it is diagonalizable. It is worth noting that there is no guarantee that $\mathbf{B}$ is symmetric in the general case of a stochastic process described by a multivariate Fokker–Planck differential equation.[21] Under this condition, it can be shown that the necessary and sufficient condition for convergence in the mean is that the symmetric matrix $\mathbf{B}$ is positive-definite. In other words, the eigenvalues of matrix $\mathbf{B}$ denoted by $\lambda_i, i = 1, 2, \ldots, Np$, should be positive.[24] The careful reader would add the additional condition $\lim_{t \to \infty} \int_0^t \alpha(\zeta)d\zeta = \infty$. The necessity for a symmetric matrix $\mathbf{B}$ can be alleviated if the convergence analysis is described in terms of the trace of matrix $\mathbf{Y}(t)$, as has been shown in Reference 25, and is explained subsequently. Clearly, the convergence is of the form of a negative exponential for a constant adaptation step and it is hyperbolic when $\alpha(t) = 1/t$. We elaborate the case of a constant adaptation step. In this case, $\mathbf{Y}(t)$ is a square matrix of negative exponentials. Accordingly, each weight converges toward the stationary solution (Equation 11.24) as a weighted sum of negative exponentials of the form $\exp(-\lambda_i \alpha t)$. The time constant needed for each term to reach the $1/e$ of its initial value is given by

$$\tau_i = \frac{1}{\alpha \lambda_i}. \qquad (11.32)$$

However, the total time constant τ_a, defined as the time needed so that any expected synaptic weight decays to $1/e$ of its initial value, cannot be expressed with a simple closed-form expression like Equation 11.32. For the same reasons that hold in the adaptive filter theory,[23] the total time constant τ_a for any expected synaptic weight can be bounded by

$$\frac{1}{\alpha \lambda_{max}} \le \tau_a \le \frac{1}{\alpha \lambda_{min}}, \tag{11.33}$$

where λ_{min} and λ_{max} denote the smallest and the largest eigenvalue of $\mathbf{B}$.

The study of the convergence in the mean square involves the correlation matrix of the synaptic weight error vectors $\mathbf{C}(t)$. The matrix $\mathbf{C}(t)$ is of dimensions $Np \times Np$ and has the structure of Equation 11.28, where the $(p \times p)$ square submatrices $\mathbf{C}_{kl}(t)$, $k, l = 1, 2, \ldots, N$ are defined by $\mathbf{C}_{kl}(t) = E[\mathbf{u}_k(t)\mathbf{u}_l^T(t)]$. $\mathbf{C}(t)$ can be computed by[21]

$$\mathbf{C}(t) = \mathbf{Y}(t)\left[\mathbf{C}(0) + \int_0^t \alpha^2(\zeta)\mathbf{Y}(\zeta)^{-1}\mathbf{D}(\mathbf{Y}(\zeta)^{-1})^T d\zeta\right]\mathbf{Y}(t)^T, \tag{11.34}$$

where $\mathbf{C}(0)$ is the initial correlation matrix and $\mathbf{D}$ is an $(Np \times Np)$ diagonal matrix that has the structure of Equation 11.28 with the diagonal submatrices being only nonzero (i.e., $\mathbf{D}_{kl} = \mathbf{0}_{p \times p}, k \ne l$). The mn-element of the submatrix $\mathbf{D}_{kk}$ is given by

$$[\mathbf{D}_{kk}(\overline{\mathbf{W}})]_{mn} = \left[w_{km}w_{kn}\hat{F}_k(\mathbf{W}) - w_{km}\int_{\mathcal{V}_k(\mathbf{W})} x_n f(\mathbf{x})d\mathbf{x} \right.$$
$$\left. - w_{kn}\int_{\mathcal{V}_k(\mathbf{W})} x_m f(\mathbf{x})d\mathbf{x} + \int_{\mathcal{V}_k(\mathbf{W})} x_m x_n f(\mathbf{x})d\mathbf{x} \right]_{\mathbf{W}=\overline{\mathbf{W}}}. \tag{11.35}$$

One may observe that the coefficient-matrix $\mathbf{D}$ is a symmetric matrix by definition. Moreover, it can be shown that $\mathbf{D}$ is positive semidefinite in any case.[22] For 1D stochastic processes that are described by the Fokker–Planck differential equation, the matrix $\mathbf{D}$ is known as diffusion matrix.[22] For $\alpha(t) = \alpha$ it can be shown, following lines similar to that of Reference 24, that the trace of the correlation matrix $\mathbf{C}(t)$, $J(t)$, can be bounded as follows:

$$\left(J(0) - \frac{\alpha \text{tr}[\mathbf{D}]}{2\lambda_{max}}\right) \exp(-2\alpha \lambda_{max}t) + \frac{\alpha \text{tr}[\mathbf{D}]}{2\lambda_{max}} \le J(t)$$
$$\le \left(J(0) - \frac{\alpha \text{tr}[\mathbf{D}]}{2\lambda_{min}}\right) \exp(-2\alpha \lambda_{min}t) + \frac{\alpha \text{tr}[\mathbf{D}]}{2\lambda_{min}}, \tag{11.36}$$

where $J(0) = \text{tr}[\mathbf{C}(0)]$. In this case, we can choose α so that $J(t) < J_b$ if

$$0 < \alpha < \frac{J_b \lambda_{min}}{2\text{tr}[\mathbf{D}]}. \tag{11.37}$$

In general, the following theorem can be proved.

THEOREM 1[25]
Let μ_i be the eigenvalues of the matrix $(\mathbf{B} + \mathbf{B}^T)$, $\mathbf{Q}$ be the modal matrix that has as columns the eigenvectors that correspond to the eigenvalues μ_i, and $\mathbf{M} = \mathrm{diag}[\mu_1, \mu_2, \ldots, \mu_{Np}]$. We define also the following terms:

$$e_{ii} = [\mathbf{Q}^T \mathbf{C}(0)\mathbf{Q}]_{ii} \quad q_{ii} = [\mathbf{Q}^T \mathbf{D}\mathbf{Q}]_{ii}, \tag{11.38}$$

where $[\cdot]_{ii}$ denotes the ii-element of the matrix inside brackets. The following statements hold:

(i)

$$\|\mathbf{Y}(t)\| \le \exp\left\{ -\frac{1}{2} \min_i \mu_i \int_0^t \alpha(\zeta)d\zeta \right\}. \tag{11.39}$$

(ii) If the adaptation step is constant, $\alpha(t) = \alpha$, then

$$J(t) = \alpha \sum_i \left\{ \frac{q_{ii}}{\mu_i} + \left(e_{ii} - \alpha \frac{q_{ii}}{\mu_i} \right) \exp\{-\mu_i \alpha t\} \right\}. \tag{11.40}$$

(iii) If $(\mathbf{B} + \mathbf{B}^T)$ is positive definite and $\int_0^\infty \alpha(\zeta)d\zeta = \infty$, then

$$\lim_{t \to \infty} J(t) = \left(\sum_i \frac{q_{ii}}{\mu_i} \right) \lim_{t \to \infty} \alpha(t). \tag{11.41}$$

One can easily verify that, for a constant adaptation step, Equation 11.40 is consistent with Equation 11.36. For $\alpha(t) = 1/t$, we have $\lim_{t \to \infty} J(t) = 0$.

Kosmatopoulos and Christodoulou[27] derived another proof of convergence. They applied a time-coordinate transformation similar to that analyzed in Reference 26. For a neighborhood function chosen to be Kronecker delta, they transformed Equation 11.20 into a linear time-varying stochastic difference equation and applied Lyapunov stochastic stability arguments.

11.5 Self-Organizing Map Properties

When the training algorithm has led to convergence, the feature map computed by the algorithm depicts important statistical characteristics of the space of input patterns. We have already said that the map computed by the neural network is essentially a nonlinear transformation Ψ that maps the input space X into the output space A, $\Phi : X \to A$. From this point of view, we have the following.

PROPERTY 1[2]
The feature map Ψ represented by the set of weight vectors $\{\mathbf{w}_j \in \mathcal{A} \mid j = 1, 2, \ldots, N\}$ provides an approximation of the input space $\mathcal{X}$.

We have already explained in Section 11.3 that the theoretical basis for analyzing the approximation of the input space lies in the theory of vector quantization.[12] The origins of vector quantization can be traced to multivariate statistical analysis[37,38] and in particular to cluster analysis.[39] We can easily recognize that Equation 11.14 is the encoding rule of input vector $\mathbf{x}$. We claim that this rule is actually a *nearest neighbor rule*. Given the winner neuron, the reconstruction vector is given by the mean vector of all patterns that fall into the Voronoi neighborhood (or receptor) of the winner neuron (Equation 11.24). This is a *minimum mean distortion rule*. The training algorithm of the NN given by Equation 11.22 is an online algorithm for the minimization of the mean distortion (Equation 11.16). On the contrary, the commonly used Linde–Buzo–Gray (LBG) algorithm (also known as generalized Lloyd algorithm, or GLA) in vector quantizer design is a batch algorithm.[28] For completeness, we note that a batch SOM algorithm was also proposed by Kohonen.[29] Let us assume that the weight vectors $\mathbf{w}_i, i = 1, 2, \ldots, N$, are randomly initialized. For each input pattern $\mathbf{x}(t)$ we determine the index $i = c(\mathbf{x})$ of the winner neuron $\mathbf{w}_i$. Let $\mathcal{V}_i$ denote the Voronoi set of the winner neuron, i.e., the set of all input patterns that have $\mathbf{w}_i$ as their closest weight vector, and z_i be the cardinality of $\mathcal{V}_i$. Batch SOM computes first the mean vector of all $\mathbf{x}(t) \in \mathcal{V}_i$

$$\bar{\mathbf{x}}_i = \frac{\sum_{\mathbf{x}(t) \in \mathcal{V}_i} \mathbf{x}(t)}{z_i}, \qquad i = 1, 2, \ldots, N \tag{11.42}$$

and updates the reference vector $\mathbf{w}_i$ as follows:

$$\mathbf{w}_i = \frac{\sum_{j=1}^{N} z_j h_{ji} \bar{\mathbf{x}}_j}{\sum_{j=1}^{N} z_j h_{ji}}. \tag{11.43}$$

It becomes evident that the adaptive capability of the SOM algorithm is lost when the batch version is used.

Closely related to the aforementioned discussion is the following property:

PROPERTY 2[30]
The training procedure of the SOM algorithm is the vector quantizer that minimizes the L_2 norm of the distortion, when the index of the winner neuron is distorted due to noise.

The next two properties deal with the topological properties of the map.

PROPERTY 3[2]
The feature map Ψ computed by the SOM algorithm is topologically ordered in the sense that the spatial location of a neuron in the lattice corresponds to a particular domain of the input space.

We may therefore visualize the feature map Ψ as an elastic net with the topology of a 1D or 2D lattice in the output space whose nodes have weights as coordinates in the input space $\mathcal{X}$. Let us assume that the dimension of the input space $\mathcal{X}$ equals that of the output space $\mathcal{A}$. If we represent the neurons by lattice nodes, whose coordinates are given by the weights of the neurons, and connect each neuron with its neighbors with edges corresponding to the lattice topology, then the map Ψ approximates the distribution of input patterns $\mathbf{x} \in \mathcal{X}$. If the output space is 1D and the distribution of input patterns is 2D, then $\mathcal{A}$ forms a Peano curve in order to approximate the input space.[6]

PROPERTY 4[31]
The SOM is an unsupervised nonlinear projection method from space $\mathcal{X}$ to $\mathcal{A}$.

Another well-known unsupervised nonlinear projection algorithm is Sammon's algorithm,[34] whereas unsupervised linear projections are the Karhunen–Loeve transform (or principal component analysis) and the projection-pursuit method. The projection methods usually preserve the distance between the patterns in the input space $\mathcal{X}$ and in the space $\mathcal{A}$ that has smaller dimensionality.[32,33] The SOM algorithm does not define a distance measure between the neurons in the output space $\mathcal{A}$.[31] To alleviate this drawback, Kraaijveld et al.[31] defined the distance between the neuron i and the neuron j, d_{ij}, as follows:

- The distance d_{ij}^* in the input space, if neurons i and j are 8-connected neighbors, or
- The minimum sum of distances between neighboring neurons from the set of all possible paths from neuron i to neuron j, where all intermediate nodes are 8-connected

Moreover, the distance between two patterns in $\mathcal{A}$ was defined as the distance between the corresponding winner neurons.

The following property defines quantitatively the approximation capabilities of the map Ψ.

PROPERTY 5[5]
The density of the neurons approximates the pdf of the input patterns.

The map Ψ depicts the variations of the statistical properties of the input distribution. More neurons are assigned in $\mathcal{A}$ to describe the regions in the input space where patterns appear with a high probability. Let $m(\mathbf{x})$ be the magnification factor defined as the number of neurons in an infinitesimal volume $d\mathbf{x}$ in $\mathcal{X}$, i.e., the density of neurons. If N is the number of neurons, then for the 1D case and for nearest-neighbor encoding it has been proved:[35]

$$m(\mathbf{x}) \propto f^{2/3}(\mathbf{x}). \tag{11.44}$$

A choice for the neighborhood function $h_{ci}(n)$ that is grounded on neurophysiological results is proposed below.

PROPERTY 6[36]
The neighborhood function $h_{ci}(n)$ can be chosen as follows:

$$h_{ci}(n) = a_1 + a_2 \exp\left(-\frac{a_3 d_{ic}^2}{2\sigma^2}\right), \tag{11.45}$$

where a_1, a_2, a_3 are constants, and σ denotes the effective width of the topological neighborhood. Frequently, a time-varying effective width is assumed, i.e.,

$$\sigma(n) = \sigma(0) \exp\left(-\frac{n}{\tau_1}\right), \tag{11.46}$$

where σ_0 is a constant and τ_1 is the time constant of the exponentially decaying function (Equation 11.46).

As a concluding remark of this section, we would like to point out that the updating equation of neuron weights that fall into the Voronoi neighborhood of the winner neuron is actually a stochastic-gradient-descent algorithm (e.g., the Robbins–Monro algorithm[13]), if the adaptation step-size sequence is chosen so that it satisfies the conditions

$$\lim_{t\to\infty} \alpha(t) = 0 \quad \lim_{t\to\infty} \int_0^t \alpha(\zeta)d\zeta = \infty. \tag{11.47}$$

In the special case where a constant step size is chosen, it follows that the synaptic weights are updated according to the least-mean squares (LMS) algorithm.[23] Tools from the convergence analysis of adaptive filters were exploited in Section 11.4.

11.6   Variants of Self-Organizing Maps Based on Robust Statistics

Let us denote by $f_i(\mathbf{x})$, $i = 1, 2, \ldots, K$, the pdfs of the various data classes. If sample data from these classes are mixed to form the sample set with mix percentages (i.e., *a priori* probabilities) ϵ_i, $i = 1, 2, \ldots, K$, such that $\sum_{i=1}^{K} \epsilon_i = 1$,

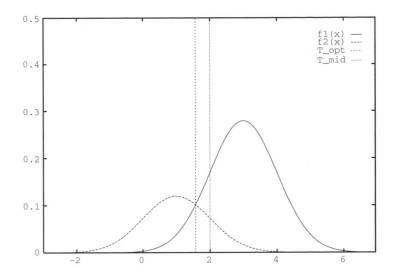

FIGURE 11.4
One-dimensional Gaussian mixture model.

the sample set distribution has the form

$$f(\mathbf{x}) = \sum_{i=1}^{K} \epsilon_i f_i(\mathbf{x}). \tag{11.48}$$

Nearest mean reclassification algorithms, such as the K-means, may have a serious shortcoming, particularly when a mixture distribution of the form of Equation 11.48 consists of several overlapping distributions.[40] In the following we confine ourselves to a 1D Gaussian mixture to maintain simplicity, i.e.,

$$f(x) = \epsilon \mathbb{N}(x; m_1, \sigma) + (1 - \epsilon)\mathbb{N}(x; m_2, \sigma), \tag{11.49}$$

where $\mathbb{N}(x; m, \sigma)$ denotes a 1D Gaussian pdf having mean m and standard deviation σ. The pdf of such a Gaussian mixture is plotted in Figure 11.4. An important goal is to decompose a mixture into several Gaussian-like distributions. However, the clustering procedures decompose the mixture by using a properly defined threshold. For example, the nearest mean reclassification algorithm with piecewise quadratic boundary applied to Equation 11.49 employs a threshold T_{opt} defined by[40]

$$\epsilon \mathbb{N}(T_{\text{opt}}; m_1, \sigma) = (1 - \epsilon)\mathbb{N}(T_{\text{opt}}; m_2, \sigma). \tag{11.50}$$

As a result, the distribution of class 1 includes the tail of the distribution $\mathbb{N}(x; m_2, \sigma)$ and does not include the tail of the distribution $\mathbb{N}(x; m_1, \sigma)$. Accordingly, the estimated mean values from the "truncated" distributions could be significantly different from the true ones. The same applies for the SOM whose threshold is simply the midpoint between the stationary weight vectors given by Equation 11.24. Moreover, it is not difficult to show that the weight vectors determined by SOM are linear combinations of the input vectors.[41] Accordingly, SOM employs a linear estimation of location whose robustness properties are poor in the presence of outliers that are inevitable in any clustering problem, as we have just argued.

To overcome these drawbacks, a novel class of SOMs has been proposed that is based on order statistics.[41–46] It is well known that operators based on order statistics[47] (e.g., the median) have very good robustness properties.[48] In the case of SOM, we rely on multivariate order statistics.

The notion of data ordering cannot be extended in a straightforward way in the case of multivariate data. There are several ways to order multivariate data. There is no unambiguous, universally agreed total ordering of N p-dimensional patterns $\mathbf{x}_1, \mathbf{x}_2, \ldots, \mathbf{x}_N$, where $\mathbf{x}_i = (x_{1i}, x_{2i}, \ldots, x_{pi})^T$, $i = 1, \ldots, N$. The following so-called subordering principles are discussed in References 49 and 50: *marginal ordering, reduced (aggregate) ordering, partial ordering,* and *conditional (sequential) ordering.* In marginal ordering, the multivariate samples are ordered along each of the p-dimensions

$$x_{j(1)} \leq x_{j(2)} \leq \cdots \leq x_{j(N)} \qquad j = 1, \ldots, p, \tag{11.51}$$

i.e., the sorting is performed in each channel of the multichannel signal independently. The ith marginal order statistic is the vector $\mathbf{x}_{(i)} = (x_{1(i)}, x_{2(i)}, \ldots, x_{p(i)})^T$. Let us assume that N is odd, i.e., $N = 2\nu + 1$. Accordingly, the marginal median is the vector $\mathbf{x}_{(\nu+1)}$.

Based on multivariate data-ordering principles, we introduce the marginal median SOM (MMSOM) in Section 11.6.1. Another definition of the multichannel median based on reduced ordering principles is the so-called *vector median* proposed in Reference 51. Following similar lines, extensions to vector median SOM[42] as well as weighted median SOM[45] and SOM that employs the Wilcoxon test [46] have also been proposed. The expected stationary states of the SOM and the MMSOM are derived for Equation 11.49 in Section 11.6.2. A modified asymptotic relative efficiency is also introduced in this section. Applications are presented in Section 11.6.3.

11.6.1   Marginal Median SOM

The marginal median is defined as follows:

$$\mathbf{x}_{\text{med}} = \text{marginal median}\{\mathbf{x}_1, \mathbf{x}_2, \ldots, \mathbf{x}_N\}$$

$$= \begin{cases} (x_{1(\nu+1)}, x_{2(\nu+1)}, \ldots, x_{p(\nu+1)})^T & \text{for } N = 2\nu + 1 \\ \left(\frac{x_{1(\nu)} + x_{1(\nu+1)}}{2}, \frac{x_{2(\nu)} + x_{2(\nu+1)}}{2}, \ldots, \frac{x_{p(\nu)} + x_{p(\nu+1)}}{2} \right)^T & \text{for } N = 2\nu. \end{cases} \tag{11.52}$$

It can be used in the following way to define the MMSOM. Let us denote by $X_i(n)$ the set of the vector-valued patterns assigned to class i, $i = 1, 2, \ldots, K$, until time $n - 1$. At time n, we find the winner vector $\mathbf{w}_c(n)$ that minimizes $\|\mathbf{x}(n) - \mathbf{w}_i(n)\|$, $i = 1, 2, \ldots, K$. The MMSOM updates the winner vector as follows:

$$\mathbf{w}_c(n+1) = \text{marginal median}\{\mathbf{x}(n) \cup X_c(n)\}, \tag{11.53}$$

where the marginal median operator is defined by Equation 11.52. Thus, all past class assignment sets $X_i(n)$, $i = 1, 2, \ldots, K$ are needed for MMSOM. MMSOM keeps track of all its history and past data samples contribute in exactly the same way as the new ones. This poses certain problems when sample statistics change with time. In this case, the marginal median operator can be applied to a predefined number of the most recent observations. MMSOM requires the calculation of the median of data sets of ever-increasing size, as can be seen from Equation 11.53. This may pose severe computational problems for relatively large n. A modification of the running median algorithm[52] can be employed to speed computations. Another useful shortcut is to quantize the discrete feature values that comprise the input patterns to a limited number of quantization levels.

11.6.2 First- and Second-Order Statistical Analysis of MMSOM

In this section, the expected stationary state of the MMSOM is derived and compared to the expected stationary state of the SOM given by Equation 11.24. Our goal is to compare quantitatively the bias that SOM and MMSOM introduce in estimating the true cluster means. Subsequently, a modified relative efficiency is introduced to cope with the biased nature of the cluster mean estimates provided by the weight vectors of the SOM and MMSOM.

Let $\tilde{f}_i(\mathbf{x})$ be the conditional pdf of $\mathbf{x}$ when $\mathbf{x}$ is restricted within the Voronoi neighborhood of class i. It is given by[53]

$$\tilde{f}_i(\mathbf{x}) = \frac{f(\mathbf{x})}{\int_{V_i(\overline{\mathbf{W}})} f(\mathbf{x})d\mathbf{x}'}, \quad \mathbf{x} \in V_i, \tag{11.54}$$

where $\overline{\mathbf{W}}$ is as in Equation 11.24. Let us denote by $\tilde{f}_{ij}(x_j)$ the marginal pdfs of $\tilde{f}_i(\mathbf{x})$ along each dimension j, $j = 1, 2, \ldots, p$, i.e.,

$$\tilde{f}_{ij}(x_j) = \int_{V_i(\overline{\mathbf{W}})} \tilde{f}_i(\mathbf{x})d\mathbf{x}_j, \tag{11.55}$$

where $\mathbf{x}_j = (x_1, \ldots, x_{j-1}, x_{j+1}, \ldots, x_p)^T$. The stationary state of the ith neuron of MMSOM is $\overline{\mathbf{w}}_{Mi} = (\overline{w}_{Mi1}, \ldots, \overline{w}_{Mip})^T$ where $\overline{w}_{Mij}$ is the population median of the marginal distribution $\tilde{f}_{ij}(x_j)$, i.e.,

$$\int_{\phi_a}^{\overline{w}_{Mij}} \tilde{f}_{ij}(x_j)dx_j = \int_{\overline{w}_{Mij}}^{\phi_b} \tilde{f}_{ij}(x_j)dx_j \tag{11.56}$$

with $[\phi_a, \phi_b]$ the domain of $\tilde{f}_{ij}(x_j)$. Equations 11.24 and 11.54 give an implicit definition of the stationary state of SOM and MMSOM, respectively. To maintain simplicity, we confine the analysis to the 1D case (Equation 11.49). Without loss of generality, we assume that $m_1 < m_2$. Obviously, to decide if an observation x belongs to the class $\mathcal{C}_i$ described by the pdf $\mathbb{N}(x; m_i, \sigma), i = 1, 2$, we need a threshold T, such that if $x \leq T, x \in \mathcal{C}_1$; otherwise, $x \in \mathcal{C}_2$. As has already been said, the nearest mean reclassification algorithm would yield the threshold given by Equation 11.50. It can be shown that

$$T_{opt} = \frac{m_1 + m_2}{2} - \frac{\sigma^2}{m_1 - m_2} \ln\left(\frac{\epsilon}{1-\epsilon}\right), \tag{11.57}$$

where the subscript "opt" implies that T_{opt} minimizes the probability of false classification. From Equation 11.57, it is seen that only for $\epsilon = 0.5$,

$$T_{opt} = \frac{m_1 + m_2}{2} = T_{mid}. \tag{11.58}$$

The SOM yields the following Voronoi neighborhoods:

$$\mathcal{V}_1(\overline{\mathbf{W}}) = \{x \leq T_{SOM}\} \qquad \mathcal{V}_2(\overline{\mathbf{W}}) = \{x > T_{SOM}\}. \tag{11.59}$$

Let us assume that $T_{SOM} = T$ is known. Then, by using Equation 11.24, we obtain

$$\begin{aligned}
\overline{w}_1 = \frac{1}{F(T)} \Bigg\{ &\frac{1}{2}[\epsilon m_1 + (1-\epsilon)m_2] \\
&+ \left[\epsilon m_1 \operatorname{erf}\left(\frac{T - m_1}{\sigma}\right) + (1-\epsilon)m_2 \operatorname{erf}\left(\frac{T - m_2}{\sigma}\right)\right] - \frac{\sigma}{\sqrt{2\pi}} \\
&\times \left[\epsilon \exp\left[-\frac{1}{2}\left(\frac{T - m_1}{\sigma}\right)^2\right] + (1-\epsilon)\exp\left[-\frac{1}{2}\left(\frac{T - m_2}{\sigma}\right)^2\right]\right] \Bigg\},
\end{aligned} \tag{11.60}$$

where

$$F(T) = \frac{1}{2} + \epsilon \operatorname{erf}\left(\frac{T - m_1}{\sigma}\right) + (1-\epsilon)\operatorname{erf}\left(\frac{T - m_2}{\sigma}\right) \tag{11.61}$$

and $\operatorname{erf}(a)$ is the error function defined as $\operatorname{erf}(a) = 1/\sqrt{2\pi} \int_0^a \exp(-t^2/2)dt$. For the second neuron, we have

$$\overline{w}_2 = \frac{[\epsilon m_1 + (1-\epsilon)m_2] - F(T)\overline{w}_1}{1 - F(T)}. \tag{11.62}$$

From Equations 11.60 and 11.62, it is seen that $\overline{w}_1 \simeq m_1$ and $\overline{w}_2 \simeq m_2$, if and only if $(m_2 - m_1) \gg \sigma$. Let us define the following variables:

$$\mathrm{erf}\left(\frac{T - m_1}{\sigma}\right) = \zeta_1, \qquad \exp\left[-\frac{1}{2}\left(\frac{T - m_1}{\sigma}\right)^2\right] = \eta_1$$

$$\tag{11.63}$$

$$\mathrm{erf}\left(\frac{T - m_2}{\sigma}\right) = \zeta_2, \qquad \exp\left[-\frac{1}{2}\left(\frac{T - m_2}{\sigma}\right)^2\right] = \eta_2.$$

Then, at equilibrium, the threshold T determined by the SOM is

$$T_{\mathrm{SOM}} = \frac{\overline{w}_1 + \overline{w}_2}{2}$$

$$= \frac{1}{1/4 - [\epsilon\zeta_1 + (1 - \epsilon)\zeta_2]^2} \left\{ \frac{\epsilon m_1 + (1 - \epsilon)m_2}{4} - [\epsilon\zeta_1 + (1 - \epsilon)\zeta_2] \right.$$

$$\left. \times \left[[\epsilon m_1\zeta_1 + (1 - \epsilon)m_2\zeta_2] - \frac{\sigma}{\sqrt{2\pi}}[\epsilon\eta_1 + (1 - \epsilon)\eta_2] \right] \right\}. \tag{11.64}$$

It can be easily verified that the pair of values $\epsilon = 0.5$ and $T_{\mathrm{SOM}} = T_{\mathrm{mid}}$, is a solution of Equation 11.64. For $\epsilon \neq 0.5$, we have to solve the set of Equations 11.60 and 11.62 that define implicitly the stationary weights of SOM by an iterative algorithm because, at equilibrium, the threshold T_{SOM} should be the midpoint between the stationary weights. The most straightforward approach is to employ Newton's algorithm. However, Newton's algorithm involves the evaluation of a Jacobian matrix and its inversion. The above-mentioned set of equations can be solved more easily by employing the generalized Lloyd algorithm (also known as the Linde-Buzo-Gray, or LBG, algorithm).[12,28]

ALGORITHM 1

Step 1: Begin with arbitrary weights $w_1(0)$ and $w_2(0)$.

Step 2: Calculate their midpoint $T_{\mathrm{SOM}} = (w_1(i) + w_2(i))/2$.

Step 3: Reevaluate $w_1(i)$ and $w_2(i)$ for the specific T_{SOM} of Step 2.

Step 4: Calculate the total absolute error in w_1 and w_2 between two successive iterations:

$$\mathcal{E}(i) = \sum_{k=1}^{2} |w_k(i) - w_k(i - 1)|. \tag{11.65}$$

Step 5: If $(\mathcal{E}(i) - \mathcal{E}(i - 1))/\mathcal{E}(i) > 10^{-5}$ go to Step 2; otherwise,

$$\overline{w}_1 = w_1(i - 1), \qquad \overline{w}_2 = w_2(i - 1). \tag{11.66}$$

By using Equation 11.56, it is found that the stationary weights of the MMSOM are implicitly defined as follows:

$$\frac{1}{F(T)} \left[\frac{1}{2} + \epsilon \operatorname{erf}\left(\frac{\overline{w}_{M1} - m_1}{\sigma} \right) + (1 - \epsilon)\operatorname{erf}\left(\frac{\overline{w}_{M1} - m_2}{\sigma} \right) \right] = 0.5 \quad (11.67)$$

$$\frac{1}{1 - F(T)} \left[\frac{1}{2} - \epsilon \operatorname{erf}\left(\frac{\overline{w}_{M2} - m_1}{\sigma} \right) - (1 - \epsilon)\operatorname{erf}\left(\frac{\overline{w}_{M2} - m_2}{\sigma} \right) \right] = 0.5. \quad (11.68)$$

It can be seen that $\overline{w}_{M1}$ and $\overline{w}_{M2}$ are defined implicitly not only in terms of T, but also due to the definition of the population median. Obviously, the left-hand side of Equations 11.67 and 11.68 are simply the values of the conditional cumulative density functions (cdf) $\tilde{F}_1$ at $\overline{w}_{M1}$, $\tilde{F}_1(\overline{w}_{M1})$, and $\tilde{F}_2$ at $\overline{w}_{M2}$, $\tilde{F}_2(\overline{w}_{M2})$, respectively. If $(m_2 - m_1) \gg \sigma$, then

$$\operatorname{erf}\left(\frac{m_2 - m_1}{\sigma} \right) = -\operatorname{erf}\left(\frac{m_1 - m_2}{\sigma} \right) = 0.5.$$

In this case, it can be verified that $\overline{w}_{M1} = m_1$ and $\overline{w}_{M2} = m_2$, because $\tilde{F}_1(m_1) = \tilde{F}_2(m_2) = 0.5$. If $\epsilon = 0.5$ and a real ξ is chosen so that

$$\operatorname{erf}\left(\frac{m_1 - m_2 + 2\xi}{2\sigma} \right) - \operatorname{erf}\left(\frac{m_1 - m_2 - 2\xi}{2\sigma} \right) = 0.5, \quad (11.69)$$

the stationary weights of the MMSOM are given by

$$\overline{w}_{M1} = T_{\text{mid}} - \xi, \quad \overline{w}_{M2} = T_{\text{mid}} + \xi, \quad (11.70)$$

which implies that $T_{\text{MMSOM}} = T_{\text{mid}}$. Indeed, in order for Equation 11.70 to describe the stationary state of MMSOM, $\overline{w}_{M1}$ and $\overline{w}_{M2}$ should satisfy $\tilde{F}_1(\overline{w}_{M1}) = 0.5$ and $\tilde{F}_2(\overline{w}_{M2}) = 0.5$, respectively. By substituting Equation 11.70 into Equations 11.67 and 11.68, Equation 11.69 results. Equation 11.70 can be interpreted as follows. For $\epsilon = 0.5$, there always exists one real ξ (i.e., the solution of Equation 11.69) so that $T_{\text{MMSOM}} = T_{\text{mid}}$. For $\epsilon \neq 0.5$, we have to solve the set of Equations 11.67 and 11.68. Algorithm 1 described above for solving the set of Equations 11.60 and 11.62 can easily be modified to solve the set of equations that define implicitly the stationary state of MMSOM as well.

The thresholds determined by SOM and MMSOM are compared to T_{opt} for the model (Equation 11.49) with $m_1 = 5$, $m_2 = 10$, and $\sigma = 1, 3$. T_{SOM}, T_{MMSOM}, and T_{opt} are plotted vs. the contamination percentage $\epsilon \in [0.2, 0.8]$ for $\sigma = 1$ in Figure 11.5a and b for $\sigma = 3$.

It is seen that T_{SOM} and T_{opt} are related as follows:

$$\epsilon < 0.5: \quad T_{\text{opt}} < T_{\text{mid}} < T_{\text{SOM}}$$

$$\epsilon = 0.5: \quad T_{\text{opt}} = T_{\text{SOM}} = T_{\text{mid}} \quad (11.71)$$

$$\epsilon > 0.5: \quad T_{\text{SOM}} < T_{\text{mid}} < T_{\text{opt}}.$$

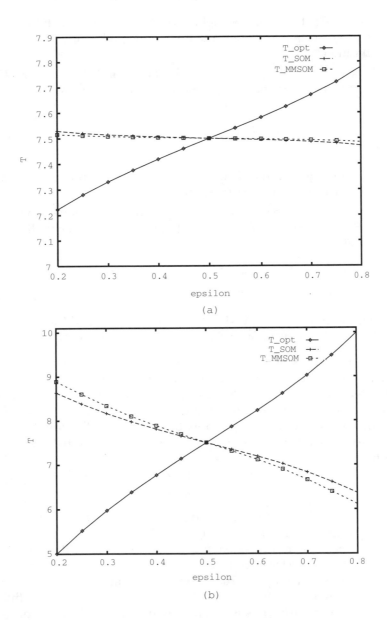

FIGURE 11.5
Thresholds predicted by SOM and MMSOM compared to T_{opt} for several values of the contamination percentage ϵ. (a) $\sigma = 1$; (b) $\sigma = 3$.

We can also observe for T_{MMSOM} in relation to T_{SOM} and T_{opt} that:

1. For $\epsilon = 0.5$, the MMSOM yields the same threshold with the SOM. In this case, both methods attain the same probability of false classification.

2. When $(m_2 - m_1) \gg \sigma$, e.g., $\sigma = 1$ for $m_1 = 5$ and $m_2 = 10$, MMSOM introduces a correction to the threshold toward the direction of T_{opt}, i.e.,

$$
\begin{aligned}
\epsilon < 0.5: \quad & T_{opt} < T_{mid} < T_{MMSOM} < T_{SOM} \\
\epsilon > 0.5: \quad & T_{SOM} < T_{MMSOM} < T_{mid} < T_{opt}.
\end{aligned}
\tag{11.72}
$$

As a consequence, the MMSOM attains a smaller probability of false classification than the SOM for the same ϵ.

3. When the overlap between the cluster pdfs progressively increases (e.g., $\sigma = 3$ for $m_1 = 5$ and $m_2 = 10$), the following inequalities are satisfied:

$$
\begin{aligned}
\epsilon < 0.5: \quad & T_{opt} < T_{mid} < T_{SOM} < T_{MMSOM} \\
\epsilon > 0.5: \quad & T_{MMSOM} < T_{SOM} < T_{mid} < T_{opt}
\end{aligned}
\tag{11.73}
$$

which imply that the MMSOM is inferior to the SOM with respect to the probability of false classification for the same ϵ.

The bias introduced by the SOM and the MMSOM in estimating the true cluster means of the Gaussian mixture model (Equation 11.49) is studied in Figure 11.6a and b. We have also included the conditional means that correspond to the decision regions predicted by T_{opt}. The bias introduced in estimating the true mean for the dominating cluster by all methods is plotted in Figure 11.6a. In other words, in Figure 11.6a we have plotted the following quantities:

$$
\begin{aligned}
|\overline{w}_2 - m_2| & \quad \text{for } \epsilon \leq 0.5 \\
|\overline{w}_1 - m_1| & \quad \text{for } \epsilon > 0.5
\end{aligned}
\tag{11.74}
$$

vs. ϵ for $\sigma = 3$. The sum of biases introduced in estimating both the true means is plotted vs. ϵ for $\sigma = 3$ in Figure 11.6b. For $\epsilon > 0.55$ or $\epsilon < 0.45$ the conditional mean for the decision region determined by T_{opt} that corresponds to the dominating cluster introduces the smallest bias. MMSOM introduces smaller bias in estimating the true mean of the dominating cluster than the SOM in all cases. For $\epsilon = 0.5$, T_{opt} and T_{SOM} yield the same bias. On the other hand, for $0.45 < \epsilon < 0.55$, it is worth noting that MMSOM outperforms both the choice T_{opt} and the SOM with respect to the bias. From the preceding analysis, we conclude that:

1. The conditional mean for the decision region defined by T_{opt} enables a more accurate representation of the patterns that belong to the dominating cluster. Such an approach favors the majority.

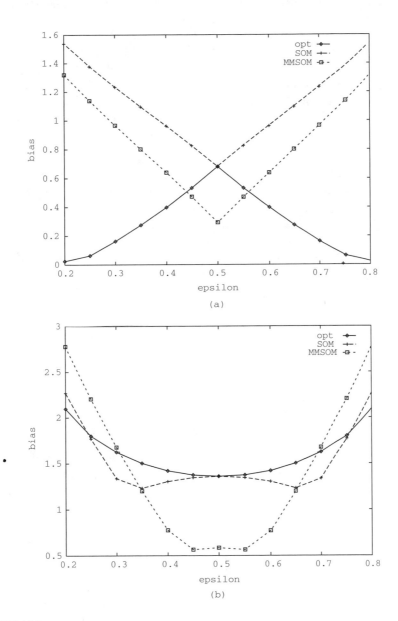

FIGURE 11.6
Bias introduced by the SOM, MMSOM, and the nearest mean reclassification algorithm (T_{opt}) in estimating (a) the true mean for the dominating cluster; (b) the true means for both clusters vs. the contamination percentage ϵ for $\sigma = 3$.

2. The SOM enables a more accurate representation of the patterns that are in the minority at the expense of the representation of the patterns that belong to the dominating cluster. In other words, the SOM favors the minority due to the vulnerability of linear estimators to outliers.

3. The marginal median SOM outperforms the SOM with respect to the bias in any case due to the threshold shift when $m_2 - m_1 \gg \sigma$, and the replacement of the conditional means by the conditional medians.

It is evident that the SOM and the MMSOM converge asymptotically to the conditional mean and to the conditional marginal median, respectively, of the pdf $\tilde{f}_i(\mathbf{x})$ defined by Equation 11.54, and not to the unconditional mean of the pdf $f_i(\mathbf{x})$ for each data class. In terms of the 1D Gaussian mixture model under study (Equation 11.49), the asymptotic value $\overline{w}_i(F)$ of each SOM weight $\{w_i(n); n \geq 1\}$ at F as well as the asymptotic value $\overline{w}_{Mi}(F)$ of each MMSOM weight $\{w_{Mi}(n); n \geq 1\}$ at F, are not Fisher consistent.[54] Accordingly, the *asymptotic variance* $V(\overline{w}, F)$ of the functional $\overline{w}$ for $\overline{w} = \overline{w}_i, \overline{w}_{Mi}$ defined by

$$V(\overline{w}, F) = \int \mathrm{IF}(x; \overline{w}, F)^2 f(x) dx, \qquad (11.75)$$

where $\mathrm{IF}(x; \overline{w}, F)$ is the influence function of $\overline{w}$ at F[54], does not take into account the bias introduced by the SOM or MMSOM weight. Indeed, Equation 11.75 is simply the variance of the random variable $\sqrt{n}(\omega(n) - \overline{w}(F))$ that is normally distributed with zero mean as $n \to \infty$.[55] Therefore, the *asymptotic relative efficiency* (ARE) of SOM and MMSOM, defined by

$$\mathrm{ARE(MMSOM, SOM)} = \frac{V(\mathrm{SOM}, F)}{V(\mathrm{MMSOM}, F)} = \frac{V(\overline{w}_1, F) + V(\overline{w}_2, F)}{V(\overline{w}_{M1}, F) + V(\overline{w}_{M2}, F)},$$

$$(11.76)$$

is not appropriate for performance comparisons of the two estimators. In our case, we are interested in evaluating, approximately, the mean-squared estimation error $E[(\omega(n) - m_i)^2]$ for $\omega = w_i, w_{Mi}$, where m_i is the true cluster mean. If we assume that both SOM and MMSOM operate in the vicinity of the equilibrium, we may write that

$$E[(\omega(n) - m_i)^2] \approx (\overline{w}(F) - m_i)^2 + \frac{V(\overline{w}, F)}{n}. \qquad (11.77)$$

In the right-hand side of Equation 11.77 the first term is the squared bias (or dc-error) and the second term is the output variance. It should be emphasized that, for small n, Equation 11.77 is a rough approximation and, in addition, it holds if and only if both SOM and MMSOM operate in the vicinity of the equilibrium. Accordingly, we propose the following modified ARE:

$$\widetilde{\text{ARE}}(\text{MMSOM}, \text{SOM}) = \frac{\sum_{i=1}^{2} E[(w_i(n) - m_i)^2]}{\sum_{i=1}^{2} E[(w_{Mi}(n) - m_i)^2]}. \tag{11.78}$$

From Equations 11.77 and 11.78, it is seen that the performance of SOM and MMSOM with respect to the mean-squared estimation error is different for small and large n. For large n, the dc-error dominates, and therefore the output variance does not play any role. Therefore, all the conclusions drawn from the analysis of bias are still valid: i.e., the MMSOM outperforms the linear SOM. This is not the case for small n. To demonstrate the role of output variance, we evaluate Equation 11.78 for $n = 1$.

First, let us consider the influence functions of the SOM weights. They are given by

$$\text{IF}(x; \overline{w}_1, F) = x[1 - u_s(x - T_{\text{SOM}})] - E[x|x \in \mathcal{V}_1(\overline{\mathbf{W}})] \tag{11.79}$$

$$\text{IF}(x; \overline{w}_2, F) = x u_s(x - T_{\text{SOM}}) - E[x|x \in \mathcal{V}_2(\overline{\mathbf{W}})], \tag{11.80}$$

where $u_s(x)$ is the unit-step function

$$u_s(x) = \begin{cases} 1 & \text{if } x \geq 0 \\ 0 & \text{otherwise.} \end{cases} \tag{11.81}$$

By applying Equation 11.75, we find

$$V(\overline{w}_i, F) = E[x^2|x \in \mathcal{V}_i(\overline{\mathbf{W}})] - \overline{w}_i^2 \qquad i = 1, 2. \tag{11.82}$$

In the case of the Gaussian mixture model, for the first SOM weight and $T = T_{\text{SOM}}$, we obtain

$$E[x^2|x \in \mathcal{V}_1(\overline{\mathbf{W}})]$$

$$= \frac{1}{F(T)} \left\{ \epsilon \left[\frac{1}{2} + \text{erf}\left(\frac{T - m_1}{\sigma}\right) \right] (m_1^2 + \sigma^2) + (1 - \epsilon) \right.$$

$$\times \left[\frac{1}{2} + \text{erf}\left(\frac{T - m_2}{\sigma}\right) \right] (m_2^2 + \sigma^2) - \frac{\sigma}{\sqrt{2}} \left[\epsilon(m_1 + T) \exp\left[-\frac{1}{2}\left(\frac{T - m_1}{\sigma}\right)^2 \right] \right.$$

$$\left. + (1 - \epsilon)(m_2 + T) \exp\left[-\frac{1}{2}\left(\frac{T - m_2}{\sigma}\right)^2 \right] \right] \right\}. \tag{11.83}$$

For the second SOM weight, we have

$$E[x^2|x \in \mathcal{V}_2(\overline{\mathbf{W}})] = \frac{1}{1 - F(T)} \{ \sigma^2 + \epsilon m_1^2 + (1 - \epsilon)m_2^2 - F(T)E[x^2|x \in \mathcal{V}_1(\overline{\mathbf{W}})] \}. \tag{11.84}$$

The influence functions of the MMSOM weights are

$$\text{IF}(x; \overline{w}_{M1}, F) = \frac{1}{2\tilde{f}_1(\overline{w}_{M1})} \text{sgn}(x - \overline{w}_{M1}) [1 - u_s(x - T_{\text{MMSOM}})] \quad (11.85)$$

$$\text{IF}(x; \overline{w}_{M2}, F) = \frac{1}{2\tilde{f}_2(\overline{w}_{M2})} \text{sgn}(x - \overline{w}_{M2}) u_s(x - T_{\text{MMSOM}}), \quad (11.86)$$

where $\text{sgn}(x)$ is the signum function. Accordingly,

$$V(\overline{w}_{Mi}, F) = \frac{1}{4\tilde{f}_i^2(\overline{w}_{Mi})}, \quad i = 1, 2. \quad (11.87)$$

The modified ARE (Equation 11.78) for the Gaussian mixture model are plotted in Figure 11.7a for several values of ϵ and σ. It is seen that for $n = 1$, the SOM outperforms MMSOM with respect to the mean-squared error. The performance of MMSOM is improved, as σ increases. However, even in this case, the SOM is still better than the MMSOM. If the mixture comprised Laplacian distributions, i.e.,

$$f(x) = \epsilon f_1(x) + (1 - \epsilon) f_2(x), \quad (11.88)$$

where

$$f_i(x) = \frac{1}{\sigma_i \sqrt{2}} \exp\left[-\sqrt{2}\frac{|x - m_i|}{\sigma_i}\right] \quad i - 1, 2, \quad (11.89)$$

with $m_1 = 5$ and $m_2 = 10$, the same analysis would reveal that the MMSOM always outperforms the SOM.[44] The corresponding modified ARE is plotted in Figure 11.7b.

11.6.3   Applications

Two applications of the variants of SOM that are based on robust statistics are described. The first application is in color image quantization and the second is in document organization and retrieval.

11.6.3.1   *Color Image Quantization*

In practice, 8 bits are used for representing each of the R, G, B components in color images. That is, 24 bits are needed for each pixel in total. Color image quantization aims at reducing the number of RGB triplets (which are 2^{24} at most) to a predefined number (e.g., 16 to 256) of codevectors so that 8 bits or less suffice to encode all the RGB.[56] Color image quantization is a useful preprocessing step in histogram-based algorithms for color image retrieval.[57]

A set of experiments has been conducted to assess the performance of MMSOM in color image quantization and to compare it to that of well-known vector quantization (VQ) methods such as the LBG algorithm[28] and the SOM. We have also included noisy color images as inputs to the learning phase.

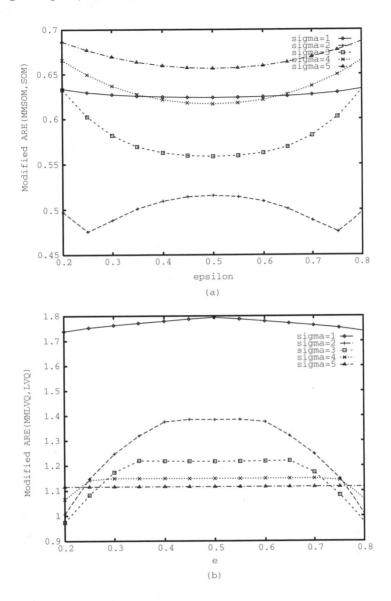

FIGURE 11.7
Modified asymptotic relative efficiency of the MMSOM against the SOM vs. the contamination percentage ϵ for several σ. (a) Gaussian mixture model; (b) Laplacian mixture model.

Let us first define when we declare that the learning procedure in the VQ techniques included in our study has converged. During the learning phase, each VQ algorithm is applied to the training set several times. Each presentation of the training set is called training session hereafter. At the end of each training session k, the mean-squared error (MSE) between the quantized and the original training patterns (i.e., RGB triplets) is evaluated as follows:

$$\mathcal{E}(k) = \frac{1}{\text{card}(\mathcal{X}_t)} \sum_{\mathbf{x}(i,j) \in \mathcal{X}_t} \|\mathbf{x}(i, j) - \tilde{\mathbf{x}}^{(k)}(i, j)\|^2, \tag{11.90}$$

where $\mathcal{X}_t$ denotes the training set, $\text{card}(\mathcal{X}_t)$ stands for the cardinality of the training set, $\mathbf{x}(i, j) = (x_R(i, j), x_G(i, j), x_B(i, j))^T$ represents the original training pattern and $\tilde{\mathbf{x}}(i, j)$ is the quantized pattern. The training patterns can be obtained from the input color image, e.g., by subsampling. In steepest-descent algorithms (e.g., SOM), the MSE is a nonmonotonically decreasing function of the iteration index. Consequently, unifying terminating rules, such as

$$\frac{\mathcal{E}(k - 1) - \mathcal{E}(k)}{\mathcal{E}(k)} \leq \vartheta \tag{11.91}$$

may not work, because there is no guarantee that $\mathcal{E}(k)$ is always less than $\mathcal{E}(k-1)$. We have decided to tolerate the same maximum number of violations of the terminating rule (Equation 11.91) in all the algorithms to be compared. We decide that the learning procedure has converged, if Equation 11.91 is satisfied for $\vartheta = 10^{-4}$. In our comparative study, we have used as figure of merit the MSE at the end of the recall phase that is defined similarly to Equation 11.90, i.e.,

$$\text{MSE} = \frac{1}{M^2} \sum_{i=1}^{M} \sum_{j=1}^{M} \|\mathbf{x}(i, j) - \tilde{\mathbf{x}}(i, j)\|^2, \tag{11.92}$$

where M is the number of image rows/columns.

Let us consider the case of a codebook of 32 RGB triplets that is learned from a training set of 4096 patterns, extracted from a noisy frame of color image sequence "Trevor White," shown in Figure 11.8a. It is applied to quantize several original frames of the same color image sequence. Figure 11.8a shows the first frame of "Trevor White" corrupted by adding mixed white zero-mean Gaussian noise having standard deviation $\sigma = 20$ and impulsive noise with probability of impulses 7% independently to each R, G, B component. The weight vectors determined at the end of the learning procedure on the training set have been applied to quantize the 10th, 50th, 100th, 120th, and 129th frame. The MSE achieved at the end of the recall phase of both SOM and MMSOM is shown in Table 11.1. It is seen that the color palette determined by the MMSOM yields the smallest MSE in all cases. The quantized images produced by the SOM and the MMSOM when frame 50 is considered are shown in Figure 11.8b and c. For comparison purposes, the original frame 50 is shown in Figure 11.8d. It can be verified that the visual quality of the quantized output of MMSOM is higher than that of the SOM.

11.6.3.2 Document Organization and Retrieval

An architecture based on the SOM algorithm, which employs the vector space model[58] and is capable of clustering documents according to their semantic

FIGURE 11.8
Application of SOM and MMSOM in quantizing the 50th frame of the color image sequence "Trevor White" in the presence of mixed additive Gaussian and impulsive noise (Codebook size = 32). (a) First noisy frame of the image sequence "Trevor White" used in the learning phase. (b) Quantized image produced by the recall phase of the SOM on the 50th noise-free frame of "Trevor White." (c) Quantized image produced by the recall phase of the MMSOM on the 50th noise-free frame of "Trevor White." (d) Original 50th noise-free frame of "Trevor White."

similarities, is the so-called WEBSOM architecture.[29] The frequencies of bi-grams (i.e., word pairs) are used to encode the documents. The WEBSOM consists of two distinct layers where the SOM algorithm is applied. The first layer is used to cluster the words appearing in the documents of the training corpus into semantically related classes. The second layer, which is activated

TABLE 11.1

Performance of the LBG, SOM, and MMSOM
in Quantizing Several Frames of Color Image
Sequence "Trevor White"[a]

Frame No.	Recall MSE		
	LBG	SOM	MMSOM
10	289.65	313.46	240.32
50	283.27	298.84	222.38
100	295.43	312.01	242.52
120	285.27	308.36	226.46
129	282.22	307.14	224.05

[a] When the learning procedure was applied on the
first frame in the presence of mixed additive Gaussian
and impulsive noise, Codebook size = 32.

after the completion of the first layer, clusters the documents of the training corpus into classes that contain relevant documents with respect to their context. Accordingly, the WEBSOM architecture is a prominent candidate for document organization and retrieval. The MMSOM has replaced the SOM in the aforementioned architecture.[41] To assess the quality of the document organization (i.e., segmentation) that results by the SOM and the MMSOM, the average recall-precision curve[59] has been derived for a test (unseen) document query of known categorization. Figure 11.9 demonstrates that the MMSOM improves the precision rate for a range of recall rates. The frequencies of bigrams can be sorted into descending order and signed ranks can be assigned to them. A novel metric that incorporates these signed ranks has been proposed for evaluating the context similarity between document pairs. A novel SOM variant, the *Wilcoxon SOM*, is built based on this metric.[46] Figure 11.10 depicts the average recall-precision curve for the SOM and the Wilcoxon SOM for the "ACQ" category of the Reuters-21578 corpus.[60]

11.7 A Class of Split-Merge Self-Organizing Maps

One common feature of both SOM and LBG algorithms is that they rely on the assumption that the number of output neurons or the size of codebook N is known in advance or is preset to a desired value, usually a power of 2. Although several VQ design techniques, such as the pairwise nearest neighbor algorithm[39,61] or the tree-structured VQ,[62] employ splitting criteria that might be useful in determining how many output neurons or codevectors adequately approximate the training set provided in the input of VQ, the availability of splitting criteria is not sufficient on its own. Because a quantization in terms of $N + 1$ output neurons always is expected to yield a lower MSE than a quantization in terms of N output neurons,[63] the issue of deciding what

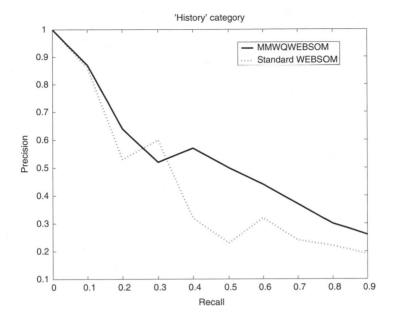

FIGURE 11.9
Average recall-precision curves for the SOM and the MMSOM in document retrieval from the
Hypergeo corpus. The sample test document was classified into the "history" category.

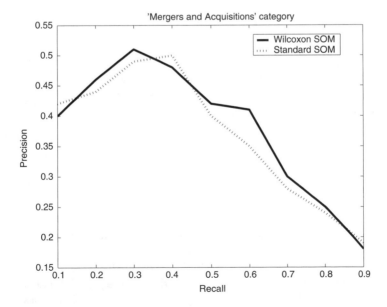

FIGURE 11.10
Average recall-precision curve for the SOM and the Wilcoxon SOM for the "Mergers and Acqui-
sitions (ACQ)" category.

constitutes a statistically significant improvement in MSE should also be addressed. The answer to this question could be found in the literature available on cluster validity.[33,38] For example, there are several indices (Davies–Bouldin index, the Hubert's Γ statistic, etc.) that can be used to assess cluster quality.[33]

Another common feature of both SOM and LBG algorithms is the unconditional inclusion of each input training vector (i.e., pattern) in the cluster of the winner neuron and the nearest-neighbor codevector, respectively, without taking into account whether the incoming training vector is statistically similar or not to the training vectors that were previously included in the same cluster. Therefore, efficient means for rejecting outliers in the formation of minimum distortion partition of a VQ should be developed.

It has been pointed out above that VQ design is a computationally intensive task whether it is a batch algorithm (like LBG) or an online algorithm (like SOM). If the number of output neurons is N and the dimension of training vectors is p, given a training vector, Np multiplications, $N(p-1)$ additions, and $N-1$ binary comparisons are required. One approach followed in the VQ design literature aims at reducing the computational load per each training vector. However, reduction of the computation time can also be achieved if training parallelism is devised by splitting the training set into subsets. Then, a second layer VQ has to be equipped with appropriate criteria for merging the partial codevectors provided by a number of first-layer VQs that operate independently and in parallel on each training subset. In the following, we confine ourselves to the design of SOM and all the remaining discussion refers to this VQ algorithm. Motivated by the three open questions just described, namely,

1. The optimal number of output neurons
2. The outlier rejection in the formation of minimum distortion partition
3. The implementation of training parallelism

a novel class of *split-merge SOM* algorithms is described that successfully addresses the just-posed questions. First, a modified SOM algorithm is proposed that incorporates statistical hypothesis testing on the mean vectors as well as additional tests that are used to determine if cluster splitting is statistically significant. Testing statistical hypotheses on mean vectors[37,38] may serve as an outlier detection mechanism. Cluster splitting can be performed by applying tests on the sum of squared-errors (or average distortion criterion). These tests determine if the reduction in the just-mentioned criterion is statistically significant.[38,63] Cluster merging is performed on the grounds of statistical tests that decide if two mean vectors are statistically equal.[37,38] The latter tests will be called *homogeneity tests*. The proposed split-merge SOM belongs to the family of dynamic neural network models.[64] Other algorithms of this family are the *growing cell structures*,[65] the *neural gas algorithm*,[66] and the *incremental grid growing*.[67]

A two-layer SOM architecture is also described that exploits the aforementioned homogeneity tests in its training phase and allows parallelism by splitting the training set into disjoint subsets. The training vectors of the first-layer SOMs are the input patterns. The training patterns of the second-layer SOM are the weight vectors of the first-layer SOMs after their convergence. The second layer SOM clusters the weight vectors provided by the first-layer SOMs. Some of them have been trained by patterns extracted from the same population; therefore they should be merged. Some others are reference vectors associated with distinct populations; consequently they should be preserved. A similar approach is adopted in Reference 68. Moreover, simplified versions of the homogeneity tests are also proposed to be used in a parallel implementation of the two-layer SOM. Applications of the split-merge SOMs to color image quantization and image segmentation are presented as well.

The proposed split-merge SOM algorithm embodies in its training phase criteria for:

1. Merging the current input pattern $x(n)$ into the cluster of patterns associated with the winner neuron $w_c(n)$, or equivalently, criteria for detecting if $x(n)$ is outlier to the cluster of patterns represented by the winner neuron $w_c(n)$
2. Testing if cluster splitting is statistically significant

In general, when the whole training set is presented in the input of SOM for the first time, many wrong decisions are expected, because the winner vectors fail to adequately approximate the true cluster means. Therefore, a need for further testing the correctness of the classification of the input training patterns to the cluster represented by the winner is recognized. Similarly, when a training pattern moves from one cluster to another, an additional test is needed to approve the correctness of such a decision. In the latter case, the cluster where the input training pattern was formerly classified may be considered unstable, because it has been affected by outliers. Consequently, during the session when a pattern removal has occurred, we have decided to check further whether the classification of input training patterns to this unstable cluster on the basis of the Euclidean distance metric is still correct. The outline of the split-merge SOM learning algorithm is as follows.

ALGORITHM 2

1. For each pattern presentation $x(n)$:
   a. Find the winner $w_c(n)$.
   b. Test if $x(n)$ is outlier to the patterns that are represented by $w_c(n)$ for
      (i) Pattern presentations during the first session
      (ii) Patterns that are moved from one cluster to another

(iii) Patterns of a cluster where a removal of a pattern has occurred during the session when this modification took place

c. If $\mathbf{x}(n)$ is not an outlier, proceed as in standard SOM.

d. If $\mathbf{x}(n)$ is an outlier, examine whether the cluster represented by the winner can be split in two subclusters, and test possible inclusion of $\mathbf{x}(n)$ in one of the resulting subclusters. Otherwise, create a new cluster having seed $\mathbf{x}(n)$.

2. When the training set is exhausted, test the integrity of the cluster associated with each output neuron.

3. Repeat steps 1 and 2, until convergence is attained.

Step 1.b is analyzed in detail in Section 11.7.1 and step 1.d is described in Section 11.7.2. To begin with, let us describe our motivation.

SOM, as well as the majority of VQ designs, relies on the Euclidean distance metric. However, the sole use of the Euclidean distance metric is not appropriate in many practical cases. When the training patterns provided as input to SOM are highly correlated or have different variances, outliers may lie near the winner vector $\mathbf{w}_c(n)$ in a Euclidean sense. For highly correlated observations or observations having different variances, it is well known that they must be linearly transformed into observations that are uncorrelated and of equal variances[38,69] in order for the Euclidean distance to be a most effective measure. Such a linear transformation yields the so-called Mahalanobis distance,[37,38] i.e.,

$$\|\mathbf{x}(n) - \mathbf{w}_c(n)\|_M = \left[(\mathbf{x}(n) - \mathbf{w}_c(n))^T \mathbf{S}_c^{-1}(n)(\mathbf{x}(n) - \mathbf{w}_c(n)) \right]^{1/2}, \qquad (11.93)$$

where $\mathbf{S}_c(n)$ is the sample dispersion matrix of the the winner's cluster. It will be shown subsequently that a quadratic form similar to Equation 11.93 needs to be evaluated and compared to a threshold when employing tests on the mean vectors as an outlier rejection mechanism.

11.7.1 Criteria for Detecting Outliers

Let us assume that an arbitrary number of output neurons exists in the network topology of Figure 11.3a. Let $N(0)$ denote the initial number of output neurons. Because all the statistical tests that are employed subsequently rely on first- and second-order statistics, each output neuron evaluates the sample mean vector $\mathbf{m}_j$ and the sample dispersion matrix $\mathbf{S}_j$ associated with the cluster it represents. We also assume that a counter z_j is associated with each neuron that counts the number of patterns that belong to the cluster represented by that neuron. During the learning phase, the training set is presented several times to the input of the NN. Each presentation of the whole training set constitutes a training session. Let k be the index that counts the training sessions. Accordingly, $\mathbf{m}_j^{(k)}(n)$, $\mathbf{S}_j^{(k)}(n)$, and $z_j^{(k)}(n)$ denote the sample mean, the sample dispersion matrix, and the number of patterns of the jth neuron cluster at the presentation of the nth training pattern during the kth training session. At the beginning, $\mathbf{m}_j$, $\mathbf{S}_j$, and n_j are appropriately initialized. For each pattern presentation $\mathbf{x}(n)$, $n = 1, 2, \ldots, M$, during the kth session, the winner neuron is found:

$$\left\| \mathbf{x}(n) - \mathbf{w}_c^{(k)}(n) \right\| = \min_{j=1}^{N^{(k)}(n)} \left\{ \left\| \mathbf{x}(n) - \mathbf{w}_j^{(k)}(n) \right\| \right\}. \tag{11.94}$$

In Equation 11.94, $N^{(k)}(n)$ is the number of output neurons, when the nth training vector is presented to the input of SOM during the kth training session. For the sake of simplicity, let us first describe the test for detecting outliers in the case of the first training session ($k = 1$), i.e., in case (i) of Algorithm 2.[70,71] More specifically, we decide that $\mathbf{x}(n)$ is not an outlier and, therefore, can be merged with the patterns already represented by the winner neuron if

$$\left(\frac{z_c^{(k)}(n-1) - p}{p} \right) \mathbf{d}^T \left[\mathbf{S}_c^{(k)}(n-1) \right]^{-1} \mathbf{d} \leq \mathcal{F}_{p, z_c^{(k)}(n-1)-p; 0.05}, \tag{11.95}$$

where $\mathcal{F}_{p, z-p; 0.05}$ denotes the upper 5% level of significance for the F-distribution with p and $z - p$ degrees of freedom, and $\mathbf{d}$ is given by

$$\mathbf{d} = -\frac{1}{z_c^{(k)}(n-1) + 1} \left(\mathbf{x}(n) - \mathbf{m}_c^{(k)}(n-1) \right). \tag{11.96}$$

It should be noted that the test (Equation 11.95) cannot be applied when $z_c^{(k)}(n-1) < p$. Therefore, the input pattern $\mathbf{x}(n)$ is unconditionally merged with the cluster of the training vectors represented by $\mathbf{w}_c^{(k)}(n)$. The same analysis is also applied to case ii outlined in Algorithm 2 (i.e., when $c^{(k)}(n) \neq c^{(k-1)}(n)$, with $c^{(k)}(n)$ denoting the index of the winner neuron at the presentation of the nth pattern during kth session, $k \geq 2$).

If Equation 11.95 is satisfied, the winner vector is updated as SOM suggests:[6]

$$\mathbf{w}_c^{(k)}(n+1) = \mathbf{w}_c^{(k)}(n) + \alpha(k) \left[\mathbf{x}(n) - \mathbf{w}_c^{(k)}(n) \right]. \tag{11.97}$$

Furthermore, the number of patterns, the sample mean, and the sample dispersion matrix of the cluster associated with either the winner $c^{(k)}(n)$ in case i or the previous winner $c^{(k-1)}(n)$ in case ii are updated. Next, we consider case iii of Algorithm 2. It refers to the remaining patterns of a cluster, which has been modified due to a removal of another pattern. Because $c^{(k)}(n) = c^{(k-1)}(n)$, for a moment we exclude the pattern $\mathbf{x}(n)$ from the cluster of patterns that is represented by the winner, and we verify whether its inclusion to this cluster is still valid by applying a test similar to Equation 11.95. For the remaining neurons, all the corresponding parameters are left intact. In the following, we consider what happens when $\mathbf{x}(n)$ is found to be an outlier. It is reasonable then to examine whether the cluster represented by the winner neuron can be split into two subclusters.

11.7.2   Splitting Criteria

Let us denote by $\mathcal{C}_c^{(k)}(n-1)$ the cluster that can be tentatively split into two subclusters. We use two statistics from the field of cluster analysis[38,63] that rely on the sum of squared errors $J_e(g)$, $g = 1, 2$, to test the validity of the following possibilities:

1. Cluster $\mathcal{C}_c^{(k)}(n-1)$ is kept united ($g = 1$).
2. Cluster $\mathcal{C}_c^{(k)}(n-1)$ is subdivided into two clusters ($g = 2$), say, $\mathcal{C}_\zeta^{(k)}(n-1)$ and $\mathcal{C}_\eta^{(k)}(n-1)$.

First let us define the sum of squared errors in cases (1) and (2) outlined above. We have

$$
J_e(g) = \begin{cases} \sum_{\mathbf{x}_j \in \mathcal{C}_c^{(k)}(n-1)} \left\| \mathbf{x}_j - \mathbf{m}_c^{(k)}(n-1) \right\|^2 & \text{for } g = 1 \\ \sum_{\gamma \in \{\zeta, \eta\}} \sum_{\mathbf{x}_j \in \mathcal{C}_\gamma^{(k)}(n-1)} \left\| \mathbf{x}_j - \mathbf{m}_\gamma^{(k)}(n-1) \right\|^2 & \text{for } g = 2, \end{cases} \tag{11.98}
$$

where $\mathbf{m}_\zeta^{(k)}(n-1)$ and $\mathbf{m}_\eta^{(k)}(n-1)$ denote the sample mean vectors of the resulting subclusters. In the sequel, we describe how a tentative splitting is performed.

We determine the direction in which cluster $\mathcal{C}_c^{(k)}(n-1)$ variation is greatest. This is equivalent to finding the first principal component of the sample dispersion matrix (i.e., the eigenvector that corresponds to the largest eigenvalue of $\mathbf{S}_c^{(k)}(n-1)$). Let us denote by $\mathbf{e}_c^{(k)}(n-1)$ the first normalized principal eigenvector of $\mathbf{S}_c^{(k)}(n-1)$. Having determined $\mathbf{e}_c^{(k)}(n-1)$, we examine the splitting of cluster $\mathcal{C}_c^{(k)}(n-1)$ with a hyperplane that is perpendicular to the direction of $\mathbf{e}_c^{(i)}(n-1)$ and passes through the sample mean $\mathbf{m}_c^{(k)}(n-1)$. Therefore, all patterns in $\mathcal{C}_c^{(k)}(n-1)$ are sorted into sets $\mathcal{C}_\zeta^{(k)}(n-1)$ and $\mathcal{C}_\eta^{(k)}(n-1)$ as follows:

$$
\mathcal{C}_\zeta^{(k)}(n-1)
$$
$$
= \left\{ \mathbf{x} \in \mathcal{C}_c^{(k)}(n-1) : \mathbf{e}_c^{(k)}(n-1)^T \mathbf{x} \le \mathbf{e}_c^{(k)}(n-1)^T \mathbf{m}_c^{(k)}(n-1) \right\}
$$

$$
\mathcal{C}_\eta^{(k)}(n-1) \tag{11.99}
$$
$$
= \left\{ \mathbf{x} \in \mathcal{C}_c^{(k)}(n-1) : \mathbf{e}_c^{(k)}(n-1)^T \mathbf{x} > \mathbf{e}_c^{(k)}(n-1)^T \mathbf{m}_c^{(k)}(n-1) \right\}.
$$

As mentioned earlier, splitting any cluster into two subclusters will result in a lower sum of squared errors, i.e., $J_e(2) < J_e(1)$. We decide to consider as valid any splitting that yields a statistically significant improvement (i.e., decrease) in the sum of squared errors. To this end, a binary hypothesis-testing problem is formulated as follows.[63]

Under the null hypothesis we assume that there is exactly one cluster present. Furthermore, it is assumed that all $z_c^{(k)}(n-1)$ patterns come from a multivariate distribution with mean $\boldsymbol{\mu}$ and covariance matrix $\sigma^2 \mathbf{I}$. In other words, the observations that form the training vectors are assumed to be uncorrelated and to have the same variance. Under the null hypothesis, $J_e(1)$ is argued, which is approximately normal with mean $z_c^{(k)}(n-1)p\sigma^2$ and variance $2z_c^{(k)}(n-1)p\sigma^4$.[63] Next, the sampling distribution for $J_e(2)$ is computed under the null hypothesis. This distribution expresses what kind of apparent improvement is to be expected when the one cluster partition is actually correct. For the splitting provided by a hyperplane through the sample mean and for large $z_c^{(k)}(n-1)$, $J_e(2)$ is again approximately normal with mean $z_c^{(k)}(n-1)(p-2/\pi)\sigma^2$ and variance $2z_c^{(k)}(n-1)(p-8/\pi^2)\sigma^4$.[63]

The null hypothesis is rejected (therefore, splitting is accepted) at the ρ-percentage significance level, if[63]

$$\frac{J_e(2)}{J_e(1)} < 1 - \frac{2}{\pi p} - \beta \sqrt{\frac{2\left(1 - \frac{8}{\pi^2 p}\right)}{z_c^{(k)}(n-1)p}}, \tag{11.100}$$

where β is determined by

$$\rho = 100 \int_\beta^\infty \frac{1}{\sqrt{2\pi}} \exp\left[-\frac{u^2}{2}\right] du. \tag{11.101}$$

Another possibility for testing, if splitting is statistically significant, is to use Beale's F-statistic.[38]

If cluster splitting is accepted, we proceed to the evaluation of the sample dispersion matrices for $\mathcal{C}_\zeta^{(k)}(n-1)$ and $\mathcal{C}_\eta^{(k)}(n-1)$. Next, we examine whether the current training pattern $\mathbf{x}(n)$ can be merged with one of the subclusters $\mathcal{C}_\zeta^{(k)}(n-1)$ or $\mathcal{C}_\eta^{(k)}(n-1)$. The following two cases are considered:

1. If $\mathbf{e}_c^{(k)}(n-1)^T\mathbf{x}(n) \le \mathbf{e}_c^{(k)}(n-1)^T\mathbf{m}_c^{(k)}(n-1)$, possible inclusion of $\mathbf{x}(n)$ in $\mathcal{C}_\zeta^{(k)}(n-1)$ is tested by applying the statistic described in Section 11.7.1.

2. Otherwise, possible inclusion of $\mathbf{x}(n)$ in $\mathcal{C}_\eta^{(k)}(n-1)$ is tested by applying the statistic described in Section 11.7.1.

In either case, if the null hypothesis of Section 11.7.1 is accepted, the number of patterns, the sample mean and the sample dispersion matrix of the subcluster, where $\mathbf{x}(n)$ is merged, are appropriately updated. The corresponding parameters of the other cluster are left intact. Moreover, the winner neuron is replaced by the two newly created ones. Their weight vectors are set equal to the sample mean vectors $\mathbf{m}_\zeta^{(k)}(n)$ and $\mathbf{m}_\eta^{(k)}(n)$.

If $\mathbf{x}(n)$ cannot be merged with any of the subclusters created by splitting $\mathcal{C}_c^{(k)}(n-1)$, a third subcluster is formed having seed $\mathbf{x}(n)$. In this case, the winner neuron is replaced by three new neurons, i.e., the two products of cluster splitting and a third neuron whose weight vector is set to $\mathbf{x}(n)$. When cluster splitting is not accepted, i.e., when Equation 11.100 does not hold, the winner neuron is kept united and an additional neuron is formed corresponding to a distinct cluster having seed $\mathbf{x}(n)$.

The procedure described so far is applied for each training pattern presentation. When the training set is exhausted, the integrity of the cluster associated with each output neuron can be tested once more by applying the splitting criterion described above. Having completed the latter test, we compute the average distortion (i.e., MSE) at the end of session k as follows:

$$\mathcal{E}(k) = \frac{1}{M} \sum_{n=1}^{M} \left\| \mathbf{x}(n) - \mathbf{w}_{c(n)}^{(k)}(M) \right\|^2. \tag{11.102}$$

If $(\mathcal{E}(k-1) - \mathcal{E}(k))/\mathcal{E}(k) > 10^{-3}$, we proceed to an additional training session.

11.7.3   Two-Layer Self-Organizing Map

In this section, we discuss the design of a novel two-layer SOM architecture, which incorporates second-order statistics in its training phase and allows training parallelism by splitting patterns into groups.[72] The proposed two-layer SOM architecture is comprised of L SOMs that are trained independently in the first-layer and a single SOM network in the second layer. The training patterns of the first-layer SOMs are input patterns. The training patterns of the second-layer SOM are the weight vectors of the first-layer SOMs after their convergence. The second layer classifies the weight vectors provided by the L maps of the first layer. Let us suppose that the L SOMs of the first layer classify p-dimensional data into N-many classes; then the second layer SOM has p input nodes and $L \times N$ output nodes at most. Some of the weight vectors of the first-layer SOMs have been trained by patterns extracted from the same population. Therefore, they must be merged. Some others are reference vectors associated with different populations. Therefore, they must be preserved. The incorporation of homogeneity and proximity statistical tests based on second-order statistics in the second-layer SOM learning algorithm is proposed so that the second-layer SOM can group the partial results provided by the first-layer SOMs to provide the final weight vectors.[72] The final weight vectors are the reference vectors that represent the whole training set. Furthermore, the proposed two-layer SOM architecture is easily parallelized.

In the following, the learning and recall procedures of the first- and second-layer SOMs are described. The learning procedure of each first-layer SOM may be the one that has been described in either Sections 11.7.1 and 11.7.2 or even the typical learning procedure of a multiple-winner SOM.[6,73] The recall procedure of each SOM network in the first-layer is applied only to the patterns used for the training of this network. It also provides the necessary information about the sample mean vector and the sample dispersion matrix of the classes produced at the output of the network. Let $\mathbf{v}_l = (v_{1l}, \ldots, v_{pl})^T$, $l = 1, 2, \ldots, N$, be the weight vectors for a first-layer SOM. The *recall procedure* of a network in the first layer has the following steps:

1. Initialize the sample mean vector $\mathbf{m}_j$, the sample dispersion matrix $\mathbf{S}_j$, and the number of patterns z_j associated with each class.
2. Determine the class $\mathcal{C}_c$ represented by the weight vector $\mathbf{v}_c$ with which the training pattern $\mathbf{x}(n)$ is most closely associated.
3. Increment the number of patterns belonging to $\mathcal{C}_c$ by one and update the sample mean vector and the sample dispersion matrix of this class. For the remaining classes, $j = 1, 2, \ldots, p$ with $j \neq c$, the number of patterns, the sample mean, and the sample dispersion matrix are not altered.

The SOM network of the second layer is used to find the weight vectors provided by the first-layer SOMs that are candidates for merging. As has already been discussed, the criterion of minimum Euclidean distance metric used in the SOM is not sufficient for the above-described task, because it does not take into account the presence of outliers. Consequently, additional tests must be implemented to test the similarity between the weight vector provided by the first-layer SOMs and the winner vector determined by the second-layer SOM. The following *learning algorithm* for the second-layer SOM is proposed:

ALGORITHM 3

1. Initialize randomly all the weight vectors $\mathbf{w}_l = (w_{1l}, w_{2l}, \ldots, w_{pl})^T$, $l = 1, 2, \ldots, q$, where $N \leq q < LN$. Set $k = 1$.
2. For each weight vector provided by the first-layer SOMs $\mathbf{v}(n) = (v_1(n), v_2(n), \ldots, v_p(n))^T$:
   a. Find the closest weight vector of the second-layer SOM, i.e., the final winner vector $\mathbf{w}_c(n)$ by using

$$\|\mathbf{v}(n) - \mathbf{w}_c(n)\| = \min_{j=1}^{q}\{\|\mathbf{v}(n) - \mathbf{w}_j(n)\|\}. \qquad (11.103)$$

   b. If there is an output node of the second-layer SOM that has not been activated yet, i.e., if there is a free class:
      (i) Test the similarity between $\mathbf{w}_c(n)$ and $\mathbf{v}(n)$ (to be described subsequently).
      (ii) If $\mathbf{w}_c(n)$ and $\mathbf{v}(n)$ are proved similar, then merge them. The final winner is updated as SOM suggests:

$$\mathbf{w}_c(n+1) = \mathbf{w}_c(n) + \alpha(k)[\mathbf{v}(n) - \mathbf{w}_c(n)]. \qquad (11.104)$$

   To achieve a faster rate of convergence, the sequence of step-size parameters is chosen to be $\alpha(k) = 1/(k+1)$. Let $z_c(n)$,

$\mathbf{m}_c(n)$, and $\mathbf{S}_c(n)$ denote the cardinality, the sample mean vector, and the sample dispersion matrix associated with the class of $\mathbf{w}_c(n)$, respectively. Let also z_v, $\mathbf{m}_v$ and $\mathbf{S}_v$ be the corresponding quantities associated with the class of $\mathbf{v}(n)$. Subsequently, modify the sample mean vector and the sample dispersion matrix.

(iii) Otherwise, assign $\mathbf{v}(n)$ to a free class, say $\mathcal{C}_f$. Update $\mathbf{w}_f(n)$ by using Equation 11.104. Set the sample mean vector $\mathbf{m}_f(n)$ and the sample dispersion matrix $\mathbf{S}_f(n)$ of the free class equal to $\mathbf{m}_v$ and $\mathbf{S}_v$, respectively.

c. If there is no free class, merge unconditionally $\mathbf{v}(n)$ and $\mathbf{w}_c(n)$. Update the winner vector and modify the sample mean vector and the sample dispersion matrix associated with the class $\mathcal{C}_c$.

3. Repeat step 2 for $k = 2, 3, \ldots$ until convergence is attained.

The *recall procedure* of the second-layer SOM is used for the classification of input patterns that have either been taken from the training set or not. It is used to determine the class $\mathcal{C}_c$ represented by $\mathbf{w}_c$, with which the input pattern $\mathbf{x}(n)$ is most closely associated, i.e.,

$$\mathbf{x}(n) \in \mathcal{C}_c \quad \text{if} \quad \|\mathbf{x}(n) - \overline{\mathbf{w}}_c\| = \min_{j=1}^{q}\{\|\mathbf{x}(n) - \overline{\mathbf{w}}_j\|\}, \tag{11.105}$$

where $\overline{\mathbf{w}}_c$ denotes the weight vector of the winner neuron after the convergence of the learning algorithm.

The homogeneity of the winner vectors evaluated by the SOM in the second layer and the input weight vectors provided by the SOMs of first layer can be tested (step 2.b.(i)) by employing statistical tests on the mean vectors and on the covariance matrices. Let $\boldsymbol{\mu}_c(n)$ and $\boldsymbol{\Sigma}_c(n)$ denote the mean vector and the covariance matrix associated with the class of $\mathbf{w}_c$, respectively. Let also $\boldsymbol{\mu}_v(n)$ and $\boldsymbol{\Sigma}_v(n)$ be the corresponding quantities associated with the class of $\mathbf{v}(n)$. The homogeneity of the covariance matrices $\boldsymbol{\Sigma}_c$ and $\boldsymbol{\Sigma}_v$ is tested by using the statistic[38]

$$T_1 = z_c \ln |\mathbf{S}_c^{-1}\mathbf{S}| + z_v \ln |\mathbf{S}_v^{-1}\mathbf{S}|, \tag{11.106}$$

where

$$\mathbf{S} = \frac{1}{z_c + z_v}(z_c\mathbf{S}_c + z_v\mathbf{S}_v) \tag{11.107}$$

and $|\cdot|$ denotes the determinant of a matrix. The statistic T_1 is distributed as $\chi^2_{p(p+1)/2}$. To test the homogeneity of the mean vectors, the following two cases are considered:

1. *Inhomogeneous covariance matrices*: A test statistic for the hypothesis that μ_c and μ_v are homogeneous is given by[38]

$$T_2 = (\mathbf{m}_c - \mathbf{m}_v)^T \left(\frac{1}{z_c} \mathbf{S}_c + \frac{1}{z_v} \mathbf{S}_v \right)^{-1} (\mathbf{m}_c - \mathbf{m}_v) \leq k_\rho. \qquad (11.108)$$

The threshold k_ρ in Equation 11.108 can be approximately expressed in terms of the upper $\rho\%$ percentile of the χ_p^2 distribution.[38]

2. *Homogeneous covariance matrices*: A test statistic for the hypothesis that μ_c and μ_v are homogeneous is the following:[37]

$$T_3 = \left| \mathbf{I}_{p \times p} + \mathbf{S}_s^{-1} \mathbf{B} \right|^{-1}, \qquad (11.109)$$

where

$$\mathbf{S}_s = z_c \mathbf{S}_c + z_v \mathbf{S}_v$$
$$\mathbf{B} = \sum_{i=c,v} z_i (\mathbf{m}_i - \mathbf{m}) (\mathbf{m}_i - \mathbf{m})^T$$
$$\mathbf{m} = \frac{z_c \mathbf{m}_c + z_v \mathbf{m}_v}{z_c + z_v}. \qquad (11.110)$$

The statistic T_3 is approximately distributed according to the Wilks distribution $\Lambda(q, z_c + z_v - 2, 1)$.[37]

In many practical cases, the above-described rigorous procedure is computationally demanding because it requires matrix inversion and the calculation of matrix determinants. In addition, the number of matrices to be handled may be extraordinarily large (e.g., in the case of color image quantization, to be discussed in the Section 11.7.4). These problems can be alleviated by testing if there is any intersection between the hyperellipsoids associated with the winner vector of the second-layer SOM and the weight vectors provided by the first-layer SOMs. A simple proximity test of the form:

$$\frac{|w_{ic} - v_i(n)|}{\sqrt{S_{c_{ii}}} + \sqrt{S_{v_{ii}}}} \leq 1, \qquad (11.111)$$

where $S_{c_{ii}}$ and $S_{v_{ii}}$ are the ii diagonal elements of the sample dispersion matrices of the corresponding classes can be used in step 2.b.(i). Inequality 11.111 implies that the hyperellipsoids are approximated by hyperparallelepipeds and simply tests if there is overlap along the ith dimension. If such an overlap exists along any dimension, it is inferred that $\mathbf{v}(n)$ and $\mathbf{w}_c$ are similar.

11.7.4 Applications

Two applications of the proposed split-merge SOM algorithms are discussed, namely, color image quantization and image segmentation.

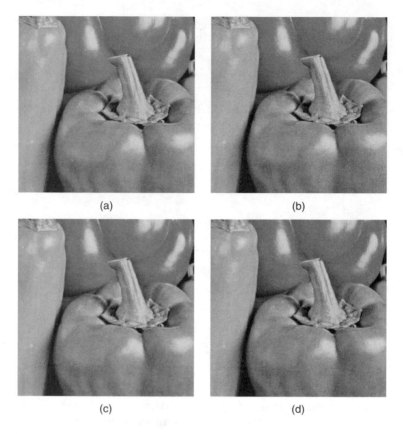

FIGURE 11.11
Color image quantization. (a) Original image. (b) Quantized color image by the split-merge SOM.
(c) Quantized color image by the two-layer SOM. (d) Quantized color image by the standard SOM
that is initialized as the LBG algorithm.

11.7.4.1 Color Image Quantization

Figure 11.11a shows an RGB color image of dimensions 256×256 with 24
bits per pixel. We compare the performance of single-layer and two-layer
split-merge SOM algorithms to that of the standard SOM in quantizing the
color image of Figure 11.11a with approximately 256 quantization levels. In
our study we use as a figure of merit the MSE given by Equation 11.92, or
equivalently the peak signal-to-noise ratio (PSNR), at the end of the recall
phase. PSNR is defined as follows:

$$\text{PSNR} = 10 \log \left(\frac{255^2}{\text{MSE}} \right) \quad \text{(in dB)}. \tag{11.112}$$

We also quote the number of iterations for each algorithm. Although the
number of iterations is not an objective measure, it is used as a rough estimate

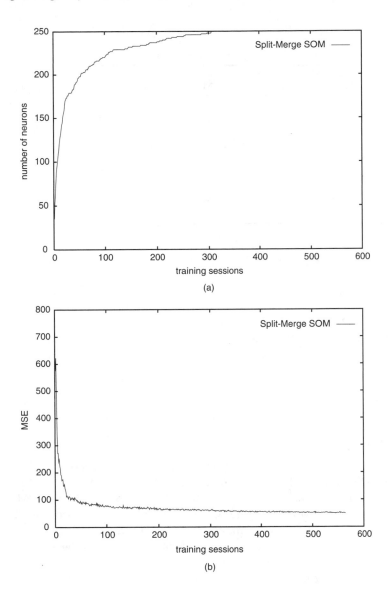

FIGURE 11.12
(a) Number of activated neurons at the end of each training session for the single-layer split-merge SOM. (b) MSE at the end of each training session for the single-layer split-merge SOM.

of the computational requirements of each algorithm. First, let us consider the performance of the single-layer split-merge SOM algorithm. A SOM with 16 output neurons initially was used to quantize the original color image shown in Figure 11.11a. The training set is formed by subsampling the original image by a factor of 2. At the end of the learning phase, 256 neurons have been created. The learning phase has lasted 562 sessions. Figure 11.12a shows the

TABLE 11.2

Figures of Merit for Color Image Quantization

Neural Network	No. of Neurons	Learning MSE	Recall MSE	PSNR	Iterations
Standard SOM random initialization	256	146.369	91.35	28.523	560
Standard SOM initialization of LBG	256	51.866	44.863	31.611	572
Single-layer split-merge SOM	256	48.795	44.137	31.682	562
Two-layer split-merge SOM	241	73.987	52.183	30.955	529 (1st FL SOM) 546 (2nd FL SOM) 18 (SL SOM)

FL-first layer; SL-second layer.

number of neurons created at the end of each training session. The MSE at the end of each training session vs. the training session index is plotted in Figure 11.12b. The MSE at the end of the learning phase as well as at the end of the recall phase is listed in Table 11.2. The result of quantization is shown in Figure 11.11b. It is seen that the quantized image is almost identical to the original image.

Subsequently, we have applied the two-layer split-merge SOM to the same task. Two single-layer split-merge SOMs with 16 output neurons initially were used in the first layer. Each of them has been trained on the half training set. The first split-merge SOM results in 112 neurons at the end of its learning phase. The number of created neurons vs. the training session index is plotted in Figure 11.13a. The MSE at the end of the learning phase for that NN is found to be 80.942. The plot of the MSE at the end of each training session is shown in Figure 11.13b. The second first-layer split-merge SOM results in 139 output neurons at the end of its training phase. Figures 11.13c and d show the evolution of the number of output neurons and the MSE vs. the training session index, respectively. The MSE at the end of the learning phase is 67.288 for that neuron. The second-layer split-merge SOM algorithm is fed with the 251 weight vectors created by the SOMs in the first layer. It merges 10 clusters represented by 10 output neurons created by the first split-merge SOM with another 10 clusters resulting from the second SOM in the first layer. Therefore, 241 output neurons are activated in the output of the second layer. The resulting MSE at the end of the learning phase for the split-merge SOM in the second layer is found to be 73.987. The quantized image created at the end of the recall phase is shown in Figure 11.11c. The quantized image is still of high quality.

For comparison purposes, we have included the corresponding figures of merit for the standard SOM algorithm in Table 11.2. The number of output neurons for the standard SOM is set to 256. The performance of the standard

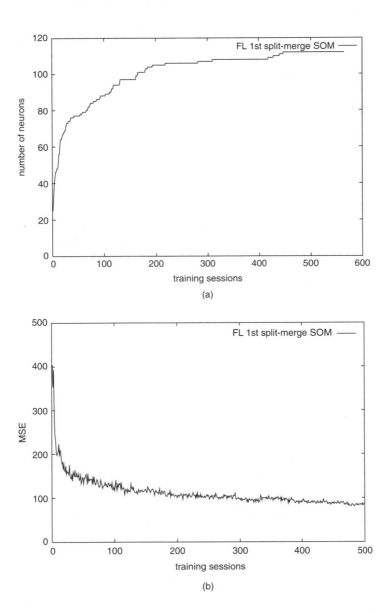

FIGURE 11.13
(a) Number of activated neurons at the end of each training session for the first split-merge SOM in the first layer of the two-layer SOM architecture. (b) MSE at the end of each training session for the first split-merge SOM in the first layer of the two-layer SOM architecture. (c) Number of activated neurons at the end of each training session for the second split-merge SOM in the first layer of the two-layer SOM architecture. (d) MSE at the end of each training session for the second split-merge SOM in the first layer of the two-layer SOM architecture.

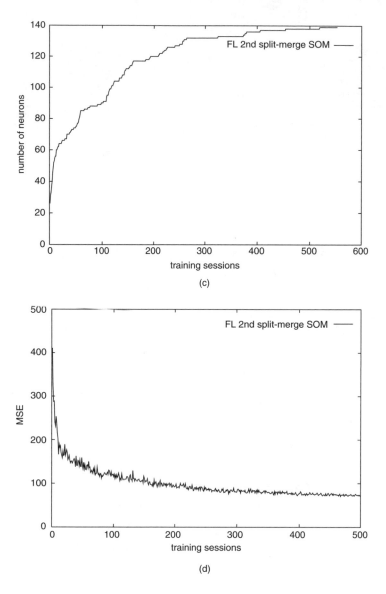

FIGURE 11.13
(*Continued*)

SOM algorithm depends strongly on the initialization procedure that is employed. Two different initialization procedures have been used: (a) random initialization and (b) the initialization of LBG algorithm.[28] The quantized images created at the end of the recall phase for a standard SOM that is initialized by the initialization procedure of LBG is shown in Figure 11.11d. By inspecting Table 11.2, one can see that the proposed single-layer split-merge SOM achieves a slightly better performance than the standard SOM that uses the same initialization with LBG, but with one fundamental difference: The split-merge SOM algorithm *has found* the number of clusters present in the input training set, while the standard SOM has been initialized in an optimal way for 256 output neurons. The two-layer split-merge SOM architecture is expected to give the best result when there is strong overlap between the training subsets. It seems that this is not the case in our experiment. If the two single-layer split-merge SOMs in the first layer were trained in parallel, then the two-layer split-merge SOM architecture would provide almost identical results with a (single-layer) split-merge SOM, but in half the computation time.

11.7.4.2 Image Segmentation

Segmentation algorithms are usually employed to generate indices for image retrieval. For example, a hierarchical tree-structured SOM is used for indexing in the PicSOM system.[74] The evaluation of image segmentation algorithms is a difficult task, mainly due to the absence of a clearly defined criterion of "success." It also affects the evaluation of retrieval systems. In this section we compare the performance of the proposed split-merge SOM algorithm in image segmentation to that of the adaptive iterated conditional modes (ICM) algorithm.[75,76] The split-merge SOM is applied to five-dimensional input vectors $\mathbf{x}(i, j) = (x_R(i, j), x_G(i, j), x_B(i, j), i, j)^T$. The training data set is obtained by sampling the entire image with a step of 10 pixels per row, every 10th row. All statistical tests were performed at 95% significance level. The adaptation step size $\alpha = 0.2$ and $\vartheta = 0.01$. The algorithm starts with one cluster. The ICM algorithm is applied on the luminance component of the image. The ICM algorithm starts with four regions.

The nonzero value of the Pott's potential function was set to 2.5. Six images from the collection of paintings of the Bridgeman Art Library were used in the study.[77] The ground truth for the test images is given as a quantitative description:

Image 107931— five objects: sky, mountains, people, foreground, water

Image 122100— four objects: figure, reflection, hammerhead cloud, the rest

Image 134975— five objects: sky, buildings, trees, people, pavement

Image 47794— three objects: sky, people, hat

Image 72398— four objects: sky, trees, waters, sails

Image 98640— three objects: horses-sledges-people, snow, sky

Using the aforementioned verbal ground truth, the number of the obtained regions by both algorithms is presented in Table 11.3.

TABLE 11.3

Number of Regions in Test Images Segmented by
the ICM and the Split-Merge SOM Algorithms

Image	Ground Truth	ICM	Split-Merge SOM
107931	5	7	7
122100	4	10	8
134975	5	13	8
47794	3	14	8
72398	4	15	11
98640	3	5	8

Figure 11.14 depicts the original image 98640 and the segmentation maps produced by the ICM and the split-merge SOM. ICM is found to yield spatially continuous regions of relatively large size that have well-defined boundaries. The split-merge SOM yields small, spatially connected regions. However, more regions describe the same object enabling a more accurate description of the object shape in the latter algorithm.

To evaluate the performance of both algorithms, a ranking scheme of the segmented images based on visual examination by a human observer is employed. Let us define by $\mathcal{O} = \{O_1, O_2, \ldots, O_K\}$ the set of objects implied by the qualitative description of the ground truth. Let $\mathcal{C} = \{C_1, C_2, \ldots, C_N\}$ be the set of regions with a unique label derived by the segmentation algorithm. Three cases associate the segmentation outcome with the ground truth:

Case 1, Best match: Best match is obtained when there is one-to-one correspondence between the segmented regions and the ground truth objects.

Case 2, Reasonable match: Reasonable match is declared when there is a one-to-many correspondence between a ground truth object and the segmented regions.

Case 3, Mismatch: Mismatch occurs when there is no correspondence between the ground truth objects and the segmented regions.

Let i denote the case number. For the ground truth object O_j, $j = 1, 2, \ldots, K$, the three cases are defined formally as follows:

$$\text{card}\{O_j \cap \mathcal{C}\} = \begin{cases} 1 & \text{if } i = 1 \\ l > 1 & \text{if } i = 2 \\ 0 & \text{if } i = 3. \end{cases} \tag{11.113}$$

Let r_{ij} be a binary index that indicates the detection of object O_j in the ith case. The index r_{ij} is determined for each object by a visual examination of

(a)

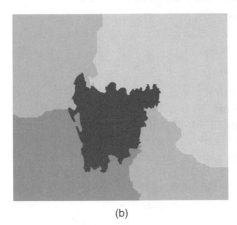

(b)

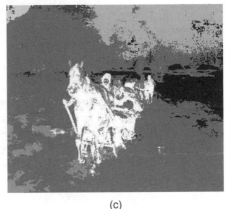

(c)

FIGURE 11.14
(a) Original image 98640 from the Bridgeman Art Library collection of paintings. (b) Segmented image by the ICM algorithm. (c) Segmented image by the split-merge SOM algorithm.

the correspondence between the segmented image and the ground truth. The total rank of the entire segmented image for each case is calculated as

$$R_i = \frac{1}{K} \sum_{j=1}^{K} r_{ij}. \tag{11.114}$$

The ranking of the segmented images by the ICM and the split-merge SOM are presented in Table 11.4. It is seen that the split-merge SOM tends to capture the ground truth and gives a more detailed segmentation. However, the ICM is the best for images that contain less details and more homogeneous regions.

TABLE 11.4

Total Ranks of the Segmentations Obtained by Using the ICM and the Split-Merge SOM Algorithms under the Best Match (BM), the Reasonable Match (RM), and the Mismatch (MM) Cases

	ICM			Split-Merge SOM		
Image	**BM**	**RM**	**MM**	**BM**	**RM**	**MM**
107931	0	0.75	0.25	0.5	0.5	0
122100	0	0.75	0.25	0.5	0.5	0
134975	0.4	0.4	0.2	0.2	0.8	0
47794	0	1	0	0	1	0
72398	0.5	0.25	0.25	0.25	0.75	0
98640	0.33	0	0.66	0	1	0

11.8   Conclusions

The SOM is one of the most popular neural networks. A total of 5384 papers related to SOM are cited at http://www.cis.hut.fi/research/som-bibl/. In this chapter, we have reviewed how the SOM algorithm results from simple neurons that compute the inner product between their synaptic weight vector and the input pattern and subsequently apply a threshold to the product. We have derived the updating equations of the SOM and we have studied its convergence properties for the winner-takes-all case. It is worth mentioning that there is no general proof of convergence in the multiple-winner case. Self-organization has been proved only for specific input distributions, e.g., the uniform one.[19] Accordingly, there is still space for theoretical contributions in understanding SOM properties. We have also described in detail our research results on nonlinear versions of SOM that aim at increasing its robustness to outliers, and have presented applications where the proposed SOM variants have been shown to be beneficial.

References

1. K.I. Diamantaras and S.Y. Kung, *Principal Component Neural Networks*, New York: John Wiley & Sons, 1996.
2. S. Haykin, *Neural Networks: A Comprehensive Foundation*, 2nd ed., Upper River Saddle, NJ: Prentice-Hall, 1999.
3. J. MacQueen, Some methods for classification and analysis of multivariate observation, in *Proc. 5th Berkeley Symposium on Mathematical Statistics and Probability*, Vol. 1, L.M. LeCun and J. Neyman, Eds., Berkeley: University of California Press, 1967, 281–297.

4. V. Cherkassky and F. Mulier, Self-organization as an iterative kernel smoothing process, *Neural Computation*, 7, 1165–1177, 1995.

5. T. Kohonen, *Self-Organizating Maps*, 2nd ed., Berlin: Springer-Verlag, 1997.

6. T. Kohonen, The self-organizing map, *Proc. IEEE*, 78(9), 1464–1480, Sept. 1990.

7. T. Kohonen, E. Oja, O. Simula, A. Visa, and J. Kangas, Engineering applications of the self-organizing map, *Proc. IEEE*, 84(10), 1355–1357, Oct. 1996.

8. S. Amari, Dynamical stability of formation of cortical maps, in *Dynamic Interactions in Neural Networks: Models and Data*, M.A. Arbib and S. Amari, Eds., New York: Springer, 1989.

9. W.K. Konen, T. Maurer, and C. von der Malsburg, A fast dynamic link matching algorithm for invariant pattern recognition, *Neural Networks*, 7(6/7), 1019–1030, 1994.

10. M. Lades, J.C. Vorbrüggen, J. Buhmann, J. Lange, C.v.d. Malsburg, R.P. Würtz, and W. Konen, Distortion invarinat object recognition in the dynamic link architecture, *IEEE Trans. Comput.*, 42(3), 300–311, Mar. 1993.

11. T. Kohonen, The neural phonetic typewriter, *IEEE Comput.*, 21(3), 11–22, Mar. 1988.

12. A. Gersho and R.M. Gray, *Vector Quantization and Signal Compression*, Dordrecht: Kluwer, 1992.

13. H. Robbins and S. Monro, A stochastic approximation method, *Ann. Math. Stat.*, 22, 400–407, 1951.

14. H. Ritter and K. Schulten, Convergence properties of Kohonen's topology conserving maps: fluctuations, stability and dimension selection, *Biol. Cybern.*, 60, 59–71, 1989.

15. T. Kohonen, Analysis of a simple self-organizing process, *Biol. Cybern.*, 44, 135–140, 1982.

16. M. Cottrell and J.C. Fort, A stochastic model for retinotopy: a self-organizing process, *Biol. Cybern.*, 53, 405–411, 1986.

17. H. Ritter and K. Schulten, On the stationary state of Kohonen's self-organizing sensory mapping, *Biol. Cybern.*, 54, 99–106, 1986.

18. V.V. Tolat, An analysis of Kohonen's self-organizing maps using a system of energy functions, *Biol. Cybern.*, 64, 155–164, 1990.

19. Z.-P. Lo, Y. Yu, and B. Bavarian, Analysis of the convergence properties of topology preserving neural networks, *IEEE Trans. Neural Networks*, 4(2), 207–220, Mar. 1993.

20. G. Voronoi, Recherches sur les paralleloedres primitives, *J. Reine Angew. Math.*, 134, 198–287, 1908.

21. N.G. Van Kampen, *Stochastic Processes in Physics and Chemistry*, Amsterdam: North Holland, 1981.

22. C.W. Gardiner, *Handbook of Stochastic Methods*, 2nd ed., Berlin: Springer-Verlag, 1985.

23. S. Haykin, *Adaptive Filter Theory*, Englewood Cliffs, NJ: Prentice-Hall, 1986.

24. C. Kotropoulos, X. Magnisalis, I. Pitas, and M.G. Strintzis, Nonlinear ultrasonic image processing based on signal-adaptive filters and self-organizing neural networks, *IEEE Trans. Image Process.*, 3(1), 65–77, Jan. 1994.

25. M.G. Strintzis, Convergence analysis of variants of the learning vector quantizer, personal communication.

26. D.M. Clark and K. Ravishankar, A convergence theorem for Grossberg learning, *Neural Networks*, 3, 87–92, 1990.

27. E.B. Kosmatopoulos and M.A. Christodoulou, Convergence properties of a class of learning vector quantization algorithms, *IEEE Trans. Image Process.*, 5(2), 361–368, Feb. 1996.

28. Y. Linde, A. Buzo, and R.M. Gray, An algorithm for vector quantizer design, *IEEE Trans. Commun.*, 28(1), 84–95, Jan. 1980.

29. T. Kohonen, S. Kaski, K. Lagus, J. Salojärvi, J. Honkela, V. Paatero, and A. Saarela, Self-organization of a massive document collection, *IEEE Trans. Neural Networks*, 11(3), 574–585, May 2000.

30. S.P. Luttrell, Derivation of a class of training algorithms, *IEEE Trans. Neural Networks*, 1(2), 229–232, June 1990.

31. M.A. Kraaijveld, J. Mao, and A.K. Jain, A nonlinear projection method based on Kohonen's topology preserving map, *IEEE Trans. Neural Networks*, 6(3), 548–559, May 1995.

32. E. Backer, *Computer-Assisted Reasoning in Cluster Analysis*, Englewood Cliffs, NJ: Prentice-Hall, 1995.

33. A.K. Jain and R.C. Dubes, *Algorithms for Clustering Data*, Englewood Cliffs, NJ: Prentice-Hall, 1988.

34. J.W. Sammon, A nonlinear mapping for data structure analysis, *IEEE Trans. Comput.*, C-18, 401–409, 1969.

35. H. Ritter, Asymptotic level density for a class of vector quantization processes, *IEEE Trans. Neural Networks*, 2, 173–175, 1991.

36. Z.-P. Lo and B. Bavarian, On the rate of convergence in topology preserving neural networks, *Biol. Cybern.*, 65, 55–63, 1991.

37. K.V. Mardia, J.T. Kent, and J.M. Bibby, *Multivariate Analysis*, London: Academic Press, 1979.

38. G.A.F. Seber, *Multivariate Observations*, New York: John Wiley & Sons, 1984.

39. R.L. Bottemiller, Comments on "a new vector quantization clustering algorithm," *IEEE Trans. Signal Process.*, 40(2), 455–456, Feb. 1992.

40. K. Fukunaga, *Introduction to Statistical Pattern Recognition*, 2nd ed., San Diego, CA: Academic Press, 1990.

41. A. Georgakis, C. Kotropoulos, A. Xafopoulos, and I. Pitas, MM-WEBSOM: a variant of WEBSOM based on order statistics, in *CD-ROM Proc. IEEE-EURASIP Workshop Nonlinear Signal and Image Processing*, Baltimore, June 2001.

42. I. Pitas, C. Kotropoulos, N. Nikolaidis, R. Yang, and M. Gabbouj, Order statistics learning vector quantizer, *IEEE Trans. Image Process.*, 5(6), 1048–1053, June 1996.

43. C. Kotropoulos, N. Nikolaidis, A.G. Borş, and I. Pitas, Robust and adaptive techniques in self-organizing neural networks, *Int. J. Comput. Math.*, 7, 183–200, 1998.

44. C. Kotropoulos, I. Pitas, and M. Gabbouj, Marginal median learning vector quantizer, in *Signal Processing VII: Theories and Applications*, M. Holt, C. Cowan, P. Grant, and W. Sandham, Eds., European Association for Signal Processing, 1994, 1496–1499.

45. R. Yang, M. Gabbouj, I. Pitas, and C. Kotropoulos, A class of robust learning vector quantizers, in *Signal Processing VII: Theories and Applications*, M. Holt, C. Cowan, P. Grant, and W. Sandham, Eds., European Association for Signal Processing, 1994, 564–567.

46. A. Georgakis, C. Kotropoulos, and I. Pitas, A SOM variant based on the Wilcoxon test for document organization and retrieval, in *Proc. 2002 IEEE Int. Conf. on Artificial Neural Networks*, Barcelona, 2002.

47. I. Pitas and A.N. Venetsanopoulos, *Nonlinear Digital Filters: Principles and Applications*. Hingham, MA: Kluwer, 1990.
48. F. Hampel, E. Ronchetti, P. Rousseeuw, and W. Stahel, *Robust Statistics*, New York: John Wiley & Sons, 1986.
49. V. Barnett, The ordering of multivariate data, *J. R. Stat. Soc. A*, 139(3), 318–354, 1976.
50. I. Pitas and P. Tsakalides, Multivariate ordering in color image restoration, *IEEE Trans. Circuits Syst. Video Technol.*, 1(3), 247–259, Sept. 1991.
51. J. Astola, P. Haavisto, and Y. Neuvo, Vector median filters, *Proc. IEEE*, 78(4), 678–689, April 1990.
52. T.S. Huang, G.J. Yang, and G.Y. Tang, A fast two-dimensional median filtering algorithm, *IEEE Trans. Acoust. Speech Signal Process.*, ASSP-27, 13–18, 1979.
53. A. Papoulis, *Probability, Random Variables and Stochastic Processes*, New York: McGraw-Hill, 1984.
54. P.J. Huber, *Robust Statistics*, New York: John Wiley & Sons, 1981.
55. E.L. Lehman, *Theory of Point Estimation*, New York: John Wiley & Sons, 1983.
56. M.T. Orchard and C.A. Bouman, Color quantization of images, *IEEE Trans. Signal Process.*, 39(12), 2677–2690, Dec. 1991.
57. A. Del Bimbo, *Visual Information Retrieval*, San Francisco: Morgan Kaufmann, 1999.
58. G. Salton and M.J. McGill, *Introduction to Modern Information Retrieval*, New York: McGraw-Hill, 1983.
59. R.R. Korfhage, *Information Storage and Retrieval*, New York: John Wiley & Sons, 1997.
60. D.D. Lewis, Reuters-21578 text categorization test collection, Distribution 1.0, 1997. Available at http://kdd.ics.uci.edu/databases/reuters21578/reuters21578.html.
61. W.H. Equitz, A new vector quantization clustering algorithm, *IEEE Trans. Acoust. Speech Signal Process.*, 37(10), 1568–1575, Oct. 1989.
62. J. Makhoul, S. Roucos, and H. Gish, Vector quantization in speech coding, *Proc. IEEE*, 73(11), 1551–1588, Nov. 1985.
63. R.O. Duda and P.E. Hart, *Pattern Classification and Scene Analysis*, New York: John Wiley & Sons, 1973.
64. D. Alahakoon, S.K. Halgamuge, and B. Srinivasan, Dynamic self-organizing maps with controlled growth for knowledge discovery, *IEEE Trans. Neural Networks*, 11(3), 601–614, May 2000.
65. B. Fritzke, Growing cell structure: a self-organizing network for supervised and unsupervised learning, *Neural Networks*, 7, 1441–1460, 1994.
66. T. Martinetz and K. Schulten, A neural-gas network learns topologies, in *Artificial Neural Networks*, T. Kohonen, K. Mäkisara, O. Simula, and J. Kangas, Eds., Amsterdam: Elsevier, 1991, 397–402.
67. J. Blackmore and R. Miikkulainen, Visualizing high-dimensional structure with the incremental grid growing neural network, in *Proc. Twelfth Int. Conf. Machine Learning*, 55–63, San Francisco: Morgan Kaufmann, 1995.
68. J. Vesanto and E. Alhoniemi, Clustering of the self-organizing map, *IEEE Trans. Neural Networks*, 11(3), 586–600, May 2000.
69. R.C. Hardie and G. Arce, Ranking in $\mathcal{R}^p$ and its use in multivariate image estimation, *IEEE Trans. Circuits Syst. Video Technol.*, 1(2), 197–209, June 1991.

70. C. Kotropoulos and I. Pitas, A variant of learning vector quantizer based on split-merge statistical tests, in *Lecture Notes in Computer Science: Computer Analysis of Images and Patterns*, D. Chetverikov and W.G. Kropatsch, Eds., New York: Springer-Verlag, 1993, 822–829.

71. C. Kotropoulos and I. Pitas, Split-merge learning vector quantizer algorithm, in *Proc. of the European Conference on Circuit Theory and Design (ECCTD '93)*, Davos, Switzerland, 1993, 465–468.

72. C. Kotropoulos, E. Augé, and I. Pitas, Two-layer learning vector quantizer for color image quantization, in *Signal Processing VI: Theories and Applications*, Amsterdam: Elsevier, 1992, 1177–1180.

73. P.K. Simpson, *Artificial Neural Systems*, Oxford: Pergamon Press, 1990.

74. J. Laaksonen, M. Koskela, and E. Oja, PicSOM–Self-organizing image retrieval with MPEG-7 content descriptors, *IEEE Trans. Neural Networks*, 13(4), 841–853, July 2002. Working demonstration available at http://www.cis.hut.fi/picsom/.

75. J. Besag, On the statistical analysis of dirty pictures, *J. R. Stat. Soc. B*, 48(3), 259–302, 1986.

76. T. Pappas, An adaptive clustering algorithm for image segmentation, *IEEE Trans. Signal Process.*, 40(4), 901–914, April 1992.

77. E. Pranckeviciene, C. Kotropoulos, and I. Pitas, Still image segmentation by using iterative conditional modes and split-merge self-organizing maps, in preparation.

12

Nonlinear Techniques for Color
Image Processing

Bogdan Smolka, Konstantinos N. Plataniotis,
and Anastasios N. Venetsanopoulos

CONTENTS

12.1 Introduction

The perception of color is of paramount importance to humans because they routinely use color features to sense the environment, recognize objects, and convey information. That is the reason why it is necessary to use color information for computer vision, because in many practical cases location of scene objects can be obtained only when color information is considered.[1]

Noise filtering is one of the most important tasks in many image analysis and computer vision applications. Its goal is the removal of unprofitable information that may corrupt any of the subsequent image processing steps.

The reduction of noise in digital images without degradation of the underlying image structures has attracted much interest in the last years.[2–8] Recently, increasing attention has been given to nonlinear processing of vector-valued signals. Many of the techniques used for color noise reduction are direct implementations of the methods used for gray-scale imaging. The independent processing of color image channels is, however, inappropriate and leads to strong artifacts. To overcome this problem, the standard techniques developed for monochrome images have to be extended in a way that exploits correlation among the image channels.

The acquisition or transmission of digital images through sensors or communication channels is often inferred by mixed impulsive and Gaussian noise. In many applications it is indispensable to remove the corrupted pixels to facilitate subsequent image processing operations, such as edge detection, image segmentation, and pattern recognition.

Numerous filtering techniques have been proposed to date for color image processing. Nonlinear filters applied to color images are required to preserve edges and image details and to remove different kinds of noise. Edge information is especially important for human perception. Therefore, its preservation and possible enhancement are very important subjective features of the performance of nonlinear image filters.

12.1.1   Noise in Color Images

Noise introduces random variations into sensor readings, making them different from the real values, and thus introducing errors and undesirable side effects in subsequent stages of image processing. Faulty sensors, optic imperfectness, electronics interference, data transmission errors, or aging of the storage material may introduce noise to digital images. In considering the signal-to-noise ratio (SNR) over practical communication media, such as microwave or satellite links, there can be degradation in quality, due to low power of the received signal. Image quality degradation can be also a result of processing techniques, such as de-mosaicking or aperture correction, that introduce various noise-like artifacts.

The noise encountered in digital image processing applications cannot always be described by the commonly assumed Gaussian model. Very often it has to be characterized in terms of impulsive sequences, which occur in the form of short-duration, high-energy spikes attaining large amplitudes with probability higher than predicted by the Gaussian density model. Thus, image filters should be robust to impulsive or generally heavy-tailed noise. In addition, when color images are processed, care must be taken to preserve image chromaticity, edges, and fine image structures.

12.1.1.1   Impulsive Noise Models

In many practical applications, images are corrupted by noise caused either by faulty image sensors or by transmission corruption resulting from anthropogenic phenomena, such as ignition transients in the vicinity of the receivers or even natural phenomena such as lightning in the atmosphere.

Transmission noise, also known as *salt and pepper* noise in gray-scale imaging, is modeled by an impulsive distribution. However, one of the problems encountered in the research of noise effects on image quality is the lack of commonly accepted multivariate impulsive noise models.

A number of simplified models have been introduced to assist performance evaluation of the different color image filters. The impulsive noise model considered in this chapter is as follows:[5,9,10]

$$
\mathbf{F}_I = \begin{cases}
(F_1, F_2, F_3) & \text{with probability } (1 - p) \\
(d, F_2, F_3) & \text{with probability } p_1 \cdot p \\
(F_1, d, F_3) & \text{with probability } p_2 \cdot p \\
(F_1, F_2, d) & \text{with probability } p_3 \cdot p \\
(d, d, d)^T & \text{with probability } p_4 \cdot p
\end{cases}
\tag{12.1}
$$

where $\mathbf{F}_I$ denotes the noisy signal, $\mathbf{F} = (F_1, F_2, F_3)$ is the noise-free color vector, and d is the impulse value, $p_1 + p_2 + p_3 + p_4 = 1$. Impulse d can have either positive or negative values and we assume that when an impulse is introduced, forcing the pixel value outside the $[0, 255]$ range, clipping is applied to push the corrupted noise value into the integer range specified by the 8-bit arithmetic.

12.1.1.2 Mixed Noise

In many practical situations, an image is often corrupted by both additive Gaussian noise due to sensors (thermal noise) and impulsive transmission noise introduced by environmental interference or faulty communication channels. An image can therefore be thought of as being corrupted by mixed noise according to the following model:

$$
\mathbf{F}_M = \begin{cases}
\mathbf{F} + \mathbf{F}_G & \text{with probability } (1 - p), \\
\mathbf{F}_I & \text{otherwise,}
\end{cases}
\tag{12.2}
$$

where $\mathbf{F}$ is the noise-free color signal, the additive noise $\mathbf{F}_G$ is modeled as zero mean, white Gaussian noise, and $\mathbf{F}_I$ is the transmission noise modeled as multivariate impulsive noise.[5]

This chapter is organized as follows. In the second section a short introduction to the adaptive techniques of noise removal in gray-scale images is presented. In Section 12.3 the anisotropic diffusion approach is described and its relation to the adaptive smoothing presented in Section 12.2 is discussed. In Section 12.4 a brief survey of the noise attenuation techniques applied in color image processing is presented. Section 12.5 is devoted to the new technique of noise reduction based on the concept of digital paths. In the last section the effectiveness of the new filtering framework is evaluated, a comparison between the new filter class and some of the filters presented in Section 12.4 is provided, and the relation of the new filter class to the anisotropic diffusion presented in Section 12.3 is shown.

12.2   Adaptive Noise Reduction Filtering

In this section we examine some adaptive techniques used for the reduction of noise in gray-scale images. Some of the presented concepts can be redefined, so that they can be used to suppress noise in the multidimensional case.

The most frequently used noise reduction transformations are the linear filters, which are based on the convolution of the image with the filter kernel of constant coefficients. This kind of filtering replaces the central pixel value F_0 from the set of pixels $F_0, F_1, \ldots, F_n$ (Figure 12.1), belonging to the filter mask W, with the weighted average of the gray-scale values of the central pixel F_0 and its n neighbors $F_1, \ldots, F_n$.[11,12] The result of the convolution F_0^* of the kernel H with the pixels in W is

$$F_0^* = \frac{1}{Z} \sum_{k=0}^{n} H_k F_k, \quad Z = \sum_{k=0}^{n} H_k. \tag{12.3}$$

Linear filters are simple and fast, especially when they are separable, but their major drawback is that they cause blurring of the edges. This effect can be diminished by choosing an appropriate adaptive nonlinear filter kernel, which performs the averaging in a selected neighborhood. The term *adaptive*[13,14] means that the filter kernel coefficients change their values according to the image structure, which is to be smoothed. Adaptive smoothing can be seen as a nonlinear process, in which noise is removed while important image features are preserved.

Different kinds of edge- and structure-preserving filter kernels have been proposed in the literature.[7,11,15] One of the simplest nonlinear schemes works with a filter kernel of the form $H_k = 1 - |F_0 - F_k|$,

$$F_0^* = \frac{1}{Z} \sum_{k=0}^{n} [1 - |F_0 - F_k|] \cdot F_k, \quad Z = \sum_{k=0}^{n} [1 - |F_0 - F_k|], \quad F_k \in [0, 1]. \tag{12.4}$$

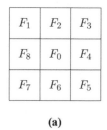

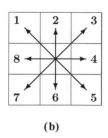

(a)   (b)

FIGURE 12.1

The filtering mask of size 3×3 with the pixel F_0 in the center (a) and the directions between the central pixel and its neighbors (b).

With greater weighting coefficients, this filter takes those pixels of the neighborhood, whose intensity are close to the intensity of the central pixel F_0, and does not take into consideration the value of F_0, when defined as[16-20]

$$F_0^* = \frac{1}{Z} \sum_{k=1}^{n} [1 - |F_0 - F_k|] \cdot F_k, \quad Z = \sum_{k=1}^{n} [1 - |F_0 - F_k|], \quad (12.5)$$

which leads to a more robust filter performance. Similar structure has the gradient inverse weighted operator, which forms a weighted mean of the pixels belonging to a filter window. Again, the weighting coefficients depend on the difference of the gray-scale values between the central pixel and its neighbors:[17,19]

$$F_0^* = \frac{1}{Z} \sum_{k=0}^{n} \frac{F_k}{\max\{\gamma, |F_0 - F_k|\}},$$

$$Z = \sum_{k=0}^{n} \frac{1}{\max\{\gamma, |F_0 - F_k|\}}, \quad (\text{in Reference 17 } \gamma = 0.5). \quad (12.6)$$

Lee's local statistics filter[18,21,22] estimates the local mean and variance of the intensities of pixels belonging to a specified filter window W and assigns to the pixel F_0 the value $F_0^* = F_0 + (1 - \alpha)\hat{F}$, where $\hat{F}$ is the arithmetic mean of the image pixels belonging to the filter window and α is estimated as $\alpha = \max\{0, (\sigma_0^2 - \sigma^2)/\sigma_0^2\}$, where σ_0^2 is the local variance calculated for the samples in the filter window and σ^2 is the variance calculated over the whole image. If $\sigma_0 \gg \sigma$, then $\alpha \approx 1$ and no changes are introduced. When $\sigma_0 \ll \sigma$, then $\alpha = 0$ and the central pixel is replaced with the local mean. In this way, the filter smoothes with a local mean when the noise is not very intensive and leaves the pixel value unchanged when strong signal activity is detected.

In Reference 23 and 24 a powerful adaptive smoothing technique related to anisotropic diffusion (which will be discussed in the next section) was proposed. In this approach, the central pixel F_0 is replaced by a weighted sum of all the pixels contained in the filtering mask:

$$F_0^* = \frac{1}{Z} \sum_{k=0}^{n} w_k F_k, \quad \text{with} \quad w_k = \exp\left\{-\frac{|G_k|^2}{\beta^2}\right\}, \quad Z = \sum_{k=0}^{n} w_k, \quad (12.7)$$

where $|G_k|$ is the magnitude of the gradient calculated in the local neighborhood of the pixel F_k and β is a smoothing parameter.

In Reference 25 another efficient adaptive technique was proposed:

$$F_0^* = \frac{1}{Z} \sum_{k=1}^{N} \exp\left\{-\frac{\rho_k^2}{\beta_1^2}\right\} \exp\left\{-\frac{|F_k - F_0|^2}{\beta_2^2}\right\} \cdot F_k, \quad (12.8)$$

where ρ_k denotes the topological distance between the central pixel F_0 and the pixels F_k ($k = 1, 2, \ldots, N$) of the filtering mask, β_1, β_2, and N (number of neighbors of F_0 in W) are filter parameters. The concept of combining the topological distance between pixels with their intensity similarities has been further developed in so-called bilateral filtering,[26–28] which can be seen as a generalization of the adaptive smoothing proposed in References 23 through 25 and 29 through 31.

Good results of noise reduction can usually be obtained by performing the σ-filtering.[7,22,32] This procedure computes a weighted average over the filter window, but only those pixels whose gray values do not deviate too much from the value of the center pixel are permitted into the averaging process. This procedure computes a weighted mean over the filter window, but only those pixels whose values lie within $\kappa \cdot \sigma$ of the central pixel value are taken into the average. This filter attempts to estimate a new pixel value with only those neighbors whose values do not deviate too much from the value of F_0:

$$F_0^* = \frac{1}{Z} \sum_k H_k F_k, \quad \{k : |F_k - F_0| \le \kappa \sigma\}, \tag{12.9}$$

where Z is the normalizing factor, κ is a parameter (typically $\kappa = 2$), σ is the standard deviation of all pixels belonging to W or the value of the standard deviation estimated from the whole image, and H_k values are filter parameters.

Another adaptive scheme, called *k-nearest neighbor filter*, suggested in Reference 33, replaces the gray level of the central pixel F_0 by the average of its k neighbors whose intensities are closest to that of F_0 ($k = 6$ and a window of size 3×3 was recommended in Reference 20). The image noise also can be reduced by applying a filter that substitutes the gray-scale value of the central pixel with a gray tone from the neighborhood that is closest to the average of all points in the filter window W (*nearest neighbor filter*). In this way $F_0^* = F_q$, where $q = \arg \{\min\{|F_k - \hat{F}|\}\}$.

Another class of filters divides the filter masks into a set of regions, in which the variance of the pixel intensities is calculated. The aim of these filters is to find clusters of pixels that are similar to the central pixel of the filtering mask. Their output is defined as a mean value of the pixel values belonging to the subwindow in which the variance reaches the minimum. The Kuwahara filter[34–36] divides the 5×5 filtering mask into four subwindows as depicted in Figure 12.2a. In each of the subwindows, the mean and the variance are calculated and the output of the filter is the mean value of the pixels from that subwindow whose pixels have the smallest variance. This filtering scheme, based on searching for pixel clusters with similar intensities, was further extended by introducing new regions in which the variance was measured[37–39] (Figures 12.2b, c, and d).

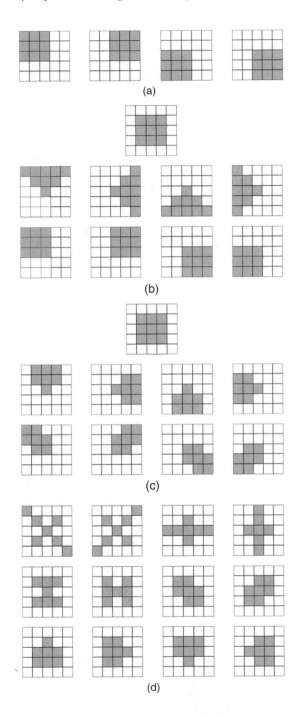

FIGURE 12.2
Different subwindow structures used in the filtering framework proposed in References 34 and 37 (a), in References 37 and 38 (b, c), and in Reference 39 (d).

This approach is in some way similar to the technique we propose in Section 12.5, in which the filters based on digital path are introduced. In the new approach, instead of looking for subwindows with similar pixels, we investigate digital paths linking the central pixel with pixels belonging to the filter window.

Another class of adaptive algorithms is based on the rank transformations, defined using an ordering operator whose goal is the transformation of the set of pixels lying in a given filtering window W into a monotonically increasing sequence $\{F_0, F_1, \ldots, F_n\} \rightarrow \{F_{(0)}, F_{(1)}, \ldots, F_{(n)}\}$, with the property: $F_{(k)} \leq F_{(k+1)}, k = 0, \ldots, n-1$. In this way the rank operator is defined on the ordered values from the set $\{F_{(0)}, \ldots, F_{(n)}\}$ and has the form

$$F_0^* = \frac{1}{Z} \sum_{k=0}^{n} \varrho_{(k)} F_{(k)}, \quad Z = \sum_{k=0}^{n} \varrho_{(k)}, \tag{12.10}$$

where ϱ_k are nonzero weighting (ranking) coefficients. Taking appropriate ranking coefficients allows the definition of a variety of useful operators. The sequence

- $\{1, 1, \ldots, 1\}$ corresponds to the moving average operator.
- $\{0, \ldots, 0, \varrho_m = 1, 0, \ldots, 0\}$, $m = n/2$, generates the median (for even number of neighbors n).
- $\{0, \ldots, 0, \varrho_{m-\alpha} = 1 = \cdots = \varrho_m = \cdots = \varrho_{m+\alpha} = 1, 0, \ldots, 0\}$, $0 \leq \alpha \leq m$ defines the α-trimmed mean, which is a compromise between the median ($\alpha = 0$) and the moving average ($\alpha = m$).
- $\{1, 0, \ldots, 0, 1\}$ determines the so-called mid-range filter.

The standard median exploits the rank-order information (order statistics) to eliminate impulsive noise. This filter substitutes the corrupted pixel with the middle-position element (median) of the ordered input samples. Since its introduction, it has been extensively studied and extended to the weighted median and its special case center-weighted median filter.

The median filter is one of the most commonly used nonlinear filters. It has the ability to attenuate strong impulse noise while preserving image edges. Its major drawback, however, is that it wipes out structures that are of the size of the filter window, and this effect causes the texture of a filtered image to be strongly distorted. Another drawback of the standard median is that it inevitably alters the details of the image not distorted by the noise process; because the standard median cannot distinguish between the corrupted and original pixels, and whether a pixel is corrupted or not, it is replaced by the local median within a filtering window. Therefore, a trade-off between the suppression of noise and the preservation of fine image details and edges has to be found. This can be accomplished in different ways; the goal, however, is always to diminish the filtering effect in image regions not affected by the noise process.[40–50]

12.3 Anisotropic Diffusion

A powerful filtering technique, called anisotropic diffusion (AD), has been introduced by Perona and Malik (P-M)[29,53] in order to selectively enhance image contrast and reduce noise using a modified heat diffusion equation and the concepts of scale space.[54]

The main concept of *AD* is based on modification of the isotropic diffusion equation (Equation 12.12), with the aim of inhibiting the smoothing across image edges. This modification is done by introducing a conductivity function that encourages intraregion smoothing over interregion smoothing.

Since the introduction of the P-M method, a wide variety of techniques have been elaborated including multiscale approaches, extensions to vector-valued imaging,[55,56] multigrid methods,[57] mathematical morphology–inspired techniques, and many others.[56,58–65]

Diffusion is a transport process that tends to level out concentration differences, and in this way it leads to equalization of the spatial concentration differences. The elementary law of diffusion states that flux density $\Im$ is directed against the gradient of concentration F in a given medium $\Im = -c\nabla F$, where c is the diffusion coefficient. If we use the continuity equation

$$\frac{\partial F}{\partial t} + \nabla \Im = 0, \quad \text{we obtain} \quad \frac{\partial F}{\partial t} = \nabla\,[c\,\nabla F]\,. \tag{12.11}$$

If $F(x, y, t)$ denotes a real-valued function representing the digital image, the equation of linear and isotropic diffusion is

$$\frac{\partial F(x, y, t)}{\partial t} = c\left[\frac{\partial^2 F(x, y, t)}{\partial x^2} + \frac{\partial^2 F(x, y, t)}{\partial y^2}\right], \tag{12.12}$$

where x, y are the image coordinates, t denotes time, and c is the conductivity coefficient.

Perona and Malik suggested that conductivity coefficient c should be dependent on the image structure, and therefore they proposed the following partial derivative equation (PDE):

$$\frac{\partial F(x, y, t)}{\partial t} = \nabla\,[c(x, y, t)\nabla F(x, y, t)]\,. \tag{12.13}$$

The conductivity coefficient $c(x, y, t)$ is a monotonically decreasing function of the image gradient magnitude and usually contains a free parameter K,

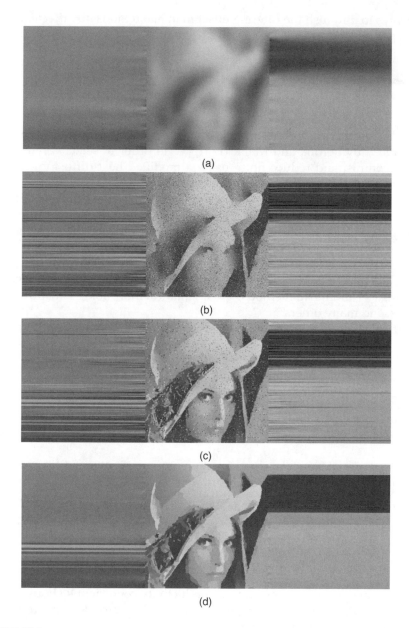

(a)

(b)

(c)

(d)

FIGURE 12.3

Illustrations of the the development of the anisotropic diffusion process. The central part of the images shows the result obtained after 300 iterations. Left and right parts show the evolution of the column 25 and 325 of the 350×350 color *LENA* image distorted by mixed impulsive and Gaussian noise. (a) Isotropic diffusion process (Equation 12.12), (b) P-M AD with c_1 (Equation 12.14), (c) regularized AD of Catté,[51,52] (d) new filter DPAF introduced in Section 12.5.

which determines the amount of smoothing introduced by the nonlinear diffusion process. Different functions of $c(x, y, t)$ have been suggested in the literature.[57,66–71] The most popular are those introduced in Reference 29:

$$c_1 = \exp\left(-\frac{|\nabla F(x, y, t)|^2}{2K^2}\right), \quad c_2 = \left(1 + \frac{|\nabla F(x, y, t)|^2}{K^2}\right)^{-1}. \quad (12.14)$$

The conductivity function $c(x, y, t)$ is time and space varying; it is chosen to be large in homogeneous regions to encourage smoothing and small at edges to preserve image structures.

The discrete version of Equation 12.13 is

$$F_0^{t+1} = F_0^t + \lambda \sum_{k=0}^{n} c_k^t [F_k^t - F_0^t], \quad \text{for stability} \quad \lambda \leq \lambda_0 = \frac{1}{n}, \quad (12.15)$$

where t denotes discrete time (iteration number), c_k^t are the diffusion coefficients in n directions (Figure 12.1 b), F_0^t denotes the central pixel of the filtering window at time t, F_k^t are its neighbors, and λ_0 is the largest value of λ, which guarantees the stability of the diffusion process.

It is quite easy to notice[28] that this equation is quite similar to the adaptive smoothing scheme proposed in References 23, 24, and 72. Equation 12.7, formulated in an iterative way

$$F_0^{t+1} = \sum_{k=0}^{n} w_k F_k^t \bigg/ \sum_{k=0}^{n} w_k, \quad (12.16)$$

can be written as

$$F_0^{t+1} = F_0^t + \frac{\sum_{k=0}^{n} w_k F_k^t - F_0^t \sum_{k=0}^{n} w_k}{\sum_{k=0}^{n} w_k} = F_0^t + \frac{\sum_{k=0}^{n} w_k (F_k^t - F_0^t)}{\sum_{k=0}^{n} w_k}$$

$$= F_0^t + \sum_{k=0}^{n} w_k^* (F_k^t - F_0^t), \quad (12.17)$$

where w_k^* are the normalized weighting coefficients. In this way, every adaptive smoothing scheme based on averaging with weighting coefficients can be seen as a special realization of the general nonlinear diffusion scheme.

The equation of AD (Equation 12.15) can be written as

$$F_0^{t+1} = F_0^t \left[1 - \lambda \sum_{k=0}^{n} c_k^t\right] + \lambda \sum_{k=0}^{n} c_k^t F_k^t, \quad \lambda \leq \lambda_0 = \frac{1}{n}. \quad (12.18)$$

If we set $[1 - \lambda \sum_{k=1}^{n} c_k^t] = 0$, then we can switch off to some extent the influence of the central pixel F_0 in the iteration process. This requires, however, that in each iteration step the λ value is a variable, dependent on time and image structure, equal to $\lambda^t = [\sum_{k=0}^{n} c_k^t]^{-1}$. The effect of diminishing the influence of the central pixel can, however, be achieved in a more natural way. Introducing the normalized conductivity coefficients C_k^t

$$C_k^t = \frac{c_k^t}{\sum_{k=0}^{n} c_k^t}, \qquad \sum_{k=0}^{n} C_k^t = 1, \qquad (12.19)$$

Equation 12.18 takes the form

$$F_0^{t+1} = F_0^t (1 - \lambda^*) + \lambda^* \sum_{k=0}^{n} C_k^t F_k^t, \quad \lambda^* = \lambda \sum_{k=0}^{n} c_k^t, \quad \lambda^* \in [0, 1], \qquad (12.20)$$

which has the nice property that for $\lambda^* = 0$ no filtering takes place—$F_0^{t+1} = F_0^t$—and for $\lambda^* = 1$, the central pixel is not taken into the weighted average and the anisotropic smoothing scheme reduces to a nonlinear, weighted average of the neighbors of F_0

$$F_0^{t+1} = \sum_{k=1}^{n} C_k^t F_k^t. \qquad (12.21)$$

In this way, the central pixel is replaced by a weighted average of its neighbors and the weights correspond to the similarity measure of the central pixel and its neighbors.

This scheme is very similar to the iterative approach proposed by Wang et al.[17] (Equation 12.6), who recommended a gradient-inverse weighted noise smoothing algorithm

$$F_0^{t+1} = c_0 F_0^t + \sum_{k=0}^{n} c_k F_k^t \quad \text{with} \quad c_k = \frac{\max\{\gamma, |F_k - F_0)|}{\sum_{k=0}^{n} \max\{\gamma, |F_k - F_0|)\}}. \qquad (12.22)$$

This is also quite similar to the approach of Lee[22] and to the algorithm of Smith and Brady[25] (Equation 12.8)

$$F_0^{t+1} = \frac{1}{Z} \sum_{k=1}^{n} c_k \cdot F_k^t, \quad c_k = \exp\left\{-\frac{\rho_k^2}{\beta_1^2}\right\} \exp\left\{-\frac{|F_k - F_0|^2}{\beta_2^2}\right\}, \quad k = 1, \ldots, n,$$

$$\qquad (12.23)$$

which corresponds to the case of $\lambda^* = 1$ in Equation 12.20. The robustness of this scheme is achieved by rejecting the central pixel value of the filter mask when calculating the filter output. This scheme is especially efficient when the image is corrupted by heavy impulsive noise process.

Setting $\lambda^* = 1$ in Equation 12.20 is similar to taking the largest possible value of λ in Equation 12.18, $\lambda_0 = 1/n$, which ensures the stability of the anisotropic diffusion process.[67] The good performance of an anisotropic diffusion scheme with $\lambda^* = 1$ is confirmed by Figure 12.4, which depicts the dependence of the

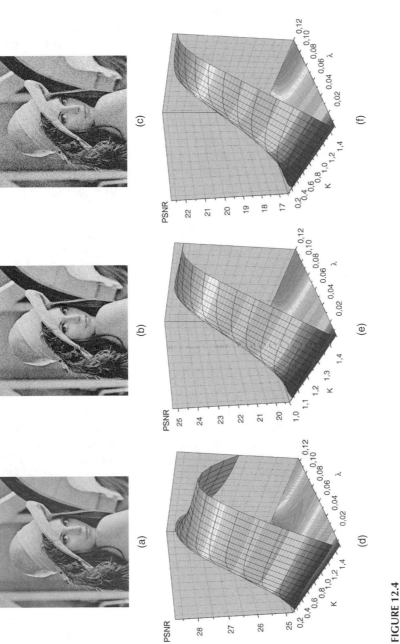

FIGURE 12.4

Dependence of the efficiency of the P-M scheme in terms of PSNR using the c_1 conductivity function on the λ and K parameters (Equations 12.14 and 12.15). The test gray-scale image *LENA* contaminated with Gaussian noise of (a) $\sigma = 10$, (b) $\sigma = 20$, (c) $\sigma = 30$ are shown, and the respective plots of the noise-reduction efficiency in terms of PSNR after three iterations are presented (d–f).

efficiency of the P-M approach using the c_1 conductivity function on the K and λ parameters for the gray-scale *LENA* image distorted by Gaussian noise of different intensity. In this figure, it is clearly visible that the best filter performance in terms of peak SNR (PSNR) is achieved for λ close to $\lambda_0 = \frac{1}{8}$ (3×3 mask), especially in the case of images distorted by Gaussian noise process of high σ. Such a setting of λ enables the diminishing of the influence of the central pixel, which ensures the suppression of the outliers injected by the noise process.

One of the major drawbacks of the anisotropic approach is that the optimal values of the parameters K and λ are unknown. Although K can be calculated using some *a priori* knowledge or can be estimated using some heuristic rules, the algorithm is very slow and needs many iterations to achieve the desired solution. Also, some stopping criterion is needed to finish the iteration process, before the image converges to the trivial solution (the average value of the image pixels).[61,73]

Another disadvantage of the P-M approach is that this algorithm is not able to cope with impulsive noise and, as a result, the noisy image goes through the diffusion process without perceptible improvement (Figure 12.3). The only way to force the diffusion to smooth out the impulsive noise is to increase the K value in Equation 12.14, which results, however, in a higher blurring.

To improve the efficiency of the original scheme, a regularized version was proposed, in which the conductance coefficient is a function of the gradient convolved with the Gaussian linear filter[51,52]

$$\frac{\partial F(x, y, t)}{\partial t} = \text{div}[\tilde{c}(x, y, t)\nabla F(x, y, t)], \qquad (12.24)$$

where $\tilde{c}(x, y, t) = f(|\nabla \mathcal{G}_o * F(x, y, t)|)$, $\mathcal{G}$ denotes the Gaussian kernel with standard deviation σ, $*$ denotes the convolution, and f is a decreasing function. The advantage of this formulation is that it is mathematically well posed, contrary to the P-M scheme. However, the drawback of this approach is that the image discontinuities tend to be blurred and the whole scheme leads to higher computational complexity of the AD process (Figure 12.3).

Another solution to the impulsive noise problem is the introduction of robust conductivity functions. In Reference 66 robust statistic norms were chosen to design the AD process. However, these conductivity functions do not help increase the efficiency of the filtering in the case of strong Gaussian or impulsive noise.

12.3.1   Anisotropic Diffusion Applied to Color Images

Let $\mathbf{F}(x, y, t) = [F_r(x, y, t), F_g(x, y, t), F_b(x, y, t)]$ denote a color image pixel at position (x, y), where $F_r(x, y, t), F_g(x, y, t), F_b(x, y, t)$ are the red, green, and blue channels, respectively. The PDE equation (Equation 12.13) can be written for the multichannel case as

$$\frac{\partial \mathbf{F}(x, y, t)}{\partial t} = \nabla \left[c(x, y, t) \nabla \mathbf{F}(x, y, t) \right],$$

$$\mathbf{F}(x, y) = \begin{bmatrix} F_r(x, y) \\ F_g(x, y) \\ F_b(x, y) \end{bmatrix}, \quad \frac{\partial \mathbf{F}(x, y)}{\partial t} = \begin{bmatrix} \frac{\partial F_r(x, y)}{\partial t} \\ \frac{\partial F_g(x, y)}{\partial t} \\ \frac{\partial F_b(x, y)}{\partial t} \end{bmatrix}, \quad (12.25)$$

where $c(x, y, t) = f(\|\mathbf{G}\|)$ is a conductivity function, which couples the three color image channels.[56,74-77] The conductivity function is the same for all the image channels and is a function of the local gradient vector $\mathbf{G}(x, y)$

$$\begin{bmatrix} \frac{\partial F_r(x, y, t)}{\partial t} \\ \frac{\partial F_g(x, y, t)}{\partial t} \\ \frac{\partial F_b(x, y, t)}{\partial t} \end{bmatrix} = \begin{bmatrix} \nabla \left[c(x, y, t) \nabla F_r(x, y, t) \right] \\ \nabla \left[c(x, y, t) \nabla F_g(x, y, t) \right] \\ \nabla \left[c(x, y, t) \nabla F_b(x, y, t) \right] \end{bmatrix},$$

$$\mathbf{G}(x, y) = \begin{bmatrix} \frac{\partial \mathbf{F}(x, y)}{\partial x} \\ \frac{\partial \mathbf{F}(x, y)}{\partial y} \end{bmatrix} = \begin{bmatrix} \frac{\partial F_r(x, y)}{\partial x}, & \frac{\partial F_g(x, y)}{\partial x}, & \frac{\partial F_b(x, y)}{\partial x}, \\ \frac{\partial F_r(x, y)}{\partial y}, & \frac{\partial F_g(x, y)}{\partial y}, & \frac{\partial F_b(x, y)}{\partial y}, \end{bmatrix}. \quad (12.26)$$

Estimating the local multichannel image gradient is one of the most important tasks in designing an AD scheme. Many of the approaches devised for color images are based on the *vector gradient norm* introduced by Di Zenzo.[78] Local variations of the color image $\|d\mathbf{F}\|^2$ are expressed as

$$\|d\mathbf{F}\|^2 = \begin{bmatrix} dx \\ dy \end{bmatrix}^T \begin{bmatrix} g_{11} & g_{12} \\ g_{21} & g_{22} \end{bmatrix} \begin{bmatrix} dx \\ dy \end{bmatrix}, \quad (12.27)$$

where

$$\begin{cases} g_{11} = \left(\frac{\partial F_r(x, y)}{\partial x} \right)^2 + \left(\frac{\partial F_g(x, y)}{\partial x} \right)^2 + \left(\frac{\partial F_b(x, y)}{\partial x} \right)^2 \\ g_{22} = \left(\frac{\partial F_r(x, y)}{\partial y} \right)^2 + \left(\frac{\partial F_g(x, y)}{\partial y} \right)^2 + \left(\frac{\partial F_b(x, y)}{\partial y} \right)^2 \\ g_{12} = \left(\frac{\partial F_r(x, y)}{\partial x} \right) \left(\frac{\partial F_r(x, y)}{\partial y} \right) + \left(\frac{\partial F_g(x, y)}{\partial x} \right) \left(\frac{\partial F_g(x, y)}{\partial y} \right) + \left(\frac{\partial F_b(x, y)}{\partial x} \right) \left(\frac{\partial F_b(x, y)}{\partial y} \right) \end{cases}.$$

$$(12.28)$$

The eigenvalues of the matrix $[g_{i,j}]$, $i = 1, 2$,

$$\lambda_+ = \frac{g_{11} + g_{22} + \sqrt{(g_{11} - g_{22})^2 + 4g_{12}^2}}{2}, \quad \lambda_- = \frac{g_{11} + g_{22} - \sqrt{(g_{11} - g_{22})^2 + 4g_{12}^2}}{2},$$

$$(12.29)$$

are the extremum of $\|d\mathbf{F}\|^2$ and the orthogonal eigenvectors determine the corresponding variation directions η and ξ

$$\eta = \frac{1}{2} \arctan \frac{2g_{12}}{g_{11} - g_{22}} , \quad \xi = \eta + \frac{\pi}{2} . \tag{12.30}$$

Based on the eigenvalues, different gradient norms leading to various PDE schemes can be developed.[55,65,68,79–81]

12.4 Noise Reduction Filters for Color Image Processing

Several nonlinear techniques for color image processing have been proposed over the years. Among them are linear processing methods, whose mathematical simplicity and the existence of a unifying theory make their design and implementation easy. However, not all filtering problems can be efficiently solved using linear techniques. For example, conventional linear techniques cannot cope with nonlinearities of the image formation model and fail to preserve edges and image details.

To this end, nonlinear color image processing techniques are introduced. Nonlinear techniques, to some extent, are able to suppress non-Gaussian noise and preserve important image elements, such as edges, corners, and fine details, and eliminate degradations occurring during image formation and transmission through noisy channels.

12.4.1 Order-Statistics Filters

One of the most popular families of nonlinear filters for impulsive noise removal are order-statistics filters.[4,82–86,136] These filters utilize algebraic ordering of a windowed set of data to compute the output signal.

The early approaches to color image processing usually comprised extensions of the scalar filters to color images. Ordering of scalar data, such as the values of pixels in gray-scale images, is well defined and has been extensively studied.[4] However, the concept of input ordering, initially applied to scalar quantities, is not easily extended to multichannel data, as there is no universal way to define ordering in vector spaces. A number of different ways to order multivariate data have been proposed. These techniques are generally classified into the following:[86–89]

- *Marginal ordering* (M-ordering), where the multivariate samples are ordered independently along each dimension
- *Reduced* or *aggregated ordering* (R-ordering), where each multivariate observation is reduced to a scalar value according to a distance metric

- *Partial ordering* (P-ordering), where the input data are partitioned into smaller groups, which are then ordered
- *Conditional ordering* (C-ordering), where multivariate samples are ordered conditionally on one of their marginal sets of observations

12.4.1.1   R-ordering Filters

Let $\mathbf{F}(\mathbf{x})$ be a multichannel image and let W be a window of finite size $n+1$ (filter length). The noisy image vectors inside the filtering window W will be denoted as $\mathbf{F}_j$, $j = 0, 1, \ldots, n$. If the distance between two vectors $\mathbf{F}_i, \mathbf{F}_j$ is denoted as $\rho(\mathbf{F}_i, \mathbf{F}_j)$, then the scalar quantity

$$R_i = \sum_{j=0}^{n} \rho(\mathbf{F}_i, \mathbf{F}_j) \tag{12.31}$$

is the aggregated distance associated with the noisy vector $\mathbf{F}_i$ inside the processing window. Assuming a reduced ordering of the $R_i - R_{(0)} \leq R_{(1)} \leq \ldots \leq R_{(\tau)} \leq \ldots, \leq R_{(n)}$—implies the same ordering of the corresponding vectors $\mathbf{F}_i : \mathbf{F}_{(0)}; \mathbf{F}_{(1)}; \ldots; \mathbf{F}_{(\tau)}; \ldots; \mathbf{F}_{(n)}$. Nonlinear ranked-type multichannel filters define the vector $\mathbf{F}_{(0)}$ as the output of the filtering operation. This selection is due to the fact that vectors that diverge greatly from the data population usually appear in higher-indexed locations in the ordered sequence.[90,91]

12.4.1.2   Vector Median Filter

The best-known member of the family of the ranked-type multichannel filters is the *vector median filter* (VMF).[9,10,92–99] The definition of the multichannel median is a direct extension of the ordinary scalar median definition with the L_1 or L_2 norm utilized to order vectors according to their relative magnitude differences.[92] The output of the VMF is the pixel $\mathbf{F}^* \in W$ for which the following condition is satisfied:

$$\sum_{j=0}^{n} \rho(\mathbf{F}^*, \mathbf{F}_j) \leq \sum_{j=0}^{n} \rho(\mathbf{F}_i, \mathbf{F}_j), \quad i = 0, \ldots, n. \tag{12.32}$$

It has been observed through experimentation that the VMF discards impulses and preserves edges and details in the image.[92] However, its performance in the suppression of additive white Gaussian noise, which is frequently encountered in image processing, is inferior to that of the arithmetic mean filter (AMF). If a color image is corrupted by both additive Gaussian noise and impulsive noise, an effective filtering scheme should make an appropriate compromise between the AMF and the VMF.

12.4.1.3   Extended Vector Median Filter

The VMF concept may be combined with linear filtering when the vector median is inadequate for filtering out noise (such as in the case of additive

Gaussian noise). The filter based on this idea, the *extended vector median filter* (EVMF) has been presented in Reference 92. If the output of the AMF is denoted as F_{AMF} then

$$\mathbf{F}^* = \begin{cases} \mathbf{F}_{AMF} & \text{if} \quad \sum_{j=0}^{n} ||\mathbf{F}_{AMF} - \mathbf{F}_j|| < \sum_{j=0}^{n} ||\mathbf{F}_{VMF} - \mathbf{F}_j|| \\ \mathbf{F}_{VMF} & \text{otherwise} \end{cases}. \tag{12.33}$$

12.4.1.4 *α-Trimmed Vector Median Filter*

In this filter, the $1 + \alpha$ samples closest to the vector median are selected as inputs to an average type of filter (see page 452). The output of the α-trimmed VMF (VMF$^\alpha$) can be defined as follows:[9,88]

$$\mathbf{F}^* = \sum_{i=0}^{\alpha} \frac{1}{1+\alpha} \mathbf{F}_{(i)}. \tag{12.34}$$

The trimming operation guarantees good performance in the presence of long-tailed or impulsive noise and helps in the preservation of sharp edges. On the other hand, the averaging operation causes the filter to perform well in the presence of short-tailed noise.

12.4.1.5 *Crossing Level Median Mean Filter*

On the basis of the vector ordering, another efficient technique combining the idea of the VMF and the AMF can be proposed, the *crossing level median mean filter* (CLMMF). Let w_i be a weight associated with ith element of the ordered vectors $\mathbf{F}_{(0)}; \mathbf{F}_{(1)}; \ldots; \mathbf{F}_{(n)}$; then the filter output is declared as $\mathbf{F}_0^* = \sum_{i=0}^{n} w_{(i)} \cdot \mathbf{F}_{(i)}$. One of the simplest possibilities of weight selection is

$$w_{(i)} = \begin{cases} 1 - \dfrac{n}{\sqrt{(n+1)(n+1+\gamma)}} & \text{for } i = 0 \\ \dfrac{1}{\sqrt{(n+1)(n+1+\gamma)}} & \text{for } i = 1, \ldots, n, \end{cases} \tag{12.35}$$

where γ is the filter parameter. For $\gamma \to \infty$ we obtain the standard vector median filter, and for $\gamma = 0$ this filter reduces to the arithmetic mean (AMF).

12.4.1.6 *Weighted Vector Median Filter*

In [References 9, 49, and 99] the vector median concept has been generalized and the *weighted vector median filter* (WVMF) has been proposed. By using the weighted vector median approach, the filter output is the vector $\mathbf{F}^*$, for which the following condition holds:

$$\sum_{j=0}^{n} w_j \, \rho(\mathbf{F}^*, \mathbf{F}_j) \leq \sum_{j=0}^{n} w_j \, \rho(\mathbf{F}_i, \mathbf{F}_j), \quad i = 0, \ldots, n. \tag{12.36}$$

12.4.1.7 Basic Vector Directional Filter

Within the framework of ranked-type nonlinear filters, the orientation difference between color vectors can also be used to remove vectors with atypical directions. The *basic vector directional filter* (BVDF) is a rank-ordered filter, similar to the VMF, which uses the angle between two color vectors as the distance criterion. This criterion is defined using the scalar measure

$$A_i = \sum_{j=0}^{n} \alpha(\mathbf{F}_i, \mathbf{F}_j), \quad \text{with} \quad \alpha(\mathbf{F}_i, \mathbf{F}_j) = \cos^{-1}\left(\frac{\mathbf{F}_i \cdot \mathbf{F}_j}{|\mathbf{F}_i|\,|\mathbf{F}_j|}\right). \tag{12.37}$$

As in the case of VMF, the ordering of the A_i implies the same ordering of the corresponding vectors $\mathbf{F}_i$. The BVDF outputs the vector $\mathbf{F}_{(0)}$ that minimizes the sum of angles with all the other vectors within the processing window. Since the BVDF uses only information about vector directions, it cannot remove achromatic noisy pixels.

12.4.1.8 Generalized Vector Directional Filter

To overcome the deficiencies of the BVDF, the *generalized vector directional filter* (GVDF) was introduced.[100] The GVDF generalizes BVDF in the sense that its output is a superset of the single BVDF output. The first vector in the ordered sequence constitutes the output of the BVDF, whereas the first τ vectors constitute the output of the GVDF:

$$\text{BVDF}\{\mathbf{F}_0, \mathbf{F}_1, \ldots, \mathbf{F}_n\} = \mathbf{F}_0,$$
$$\text{GVDF}\{\mathbf{F}_0, \mathbf{F}_1, \ldots, \mathbf{F}_n\} = \{\mathbf{F}_0, \mathbf{F}_1, \ldots, \mathbf{F}_\tau\}, \quad 1 \le \tau \le n. \tag{12.38}$$

The output of GVDF is subsequently passed through an additional filter to produce a single output vector. In this step the designer can consider only the magnitudes of the vectors $\mathbf{F}_0, \mathbf{F}_1, \ldots, \mathbf{F}_\tau$ because they have approximately the same direction in the vector space. As a result the GVDF separates the processing of color vectors into directional processing and then magnitude processing (the vector's direction signifies its chromaticity, while its magnitude is a measure of its brightness). The resulting cascade of filters is usually complex and the implementations may be slow because they operate in two steps.[101,102]

12.4.1.9 Directional Distance Filter

To overcome the deficiencies of the directional filters, another method called *directional-distance filter* (DDF) was proposed.[103] DDF constitutes a combination of VMF and BVDF and is derived by simultaneous minimization of their defining functions. Specifically, in the case of the DDF the accumulated distance inside the processing window is defined as

$$B_i = \left(\sum_{j=0}^{n} \alpha(\mathbf{F}_i, \mathbf{F}_j)\right)^{\varsigma} \left(\sum_{j=0}^{n} \rho(\mathbf{F}_i, \mathbf{F}_j)\right)^{1-\varsigma}, \tag{12.39}$$

where $\alpha(\mathbf{F}_i, \mathbf{F}_j)$ is the directional (angular) distance defined in Equation 12.37 and distance $\rho(\mathbf{F}_i, \mathbf{F}_j)$ could be calculated using the Minkowski L_p norm. The parameter ς regulates the influence of angle and distance components. As for any other ranked-order filter, an ordering of the B_i implies the same ordering of the corresponding vectors $\mathbf{F}_i$. Thus, DDF defines the $\mathbf{F}_{(0)}$ vector as its output: $\mathbf{F}_{\text{DDF}} = \mathbf{F}_0$. For $\varsigma = 0$ we obtain the VMF and for $\varsigma = 1$ the BVDF. The DDF is defined for $\varsigma = 0.5$ and its usefulness stems from the fact that it combines both the criteria used in BVDF and VMF.[100,104]

12.4.1.10 Hybrid Directional Filter

Another efficient rank-ordered operation called *hybrid directional filter* (HDF) has been proposed.[95] This filter operates independently on the direction and magnitude of the color vectors and then combines them to produce a final output. This hybrid filter, which can be viewed as a nonlinear combination of the VMF and BVDF filters, produces an output according to the following rule:

$$\mathbf{F}^* = \begin{cases} \mathbf{F}_{\text{VMF}} & \text{if } \mathbf{F}_{\text{VMF}} = \mathbf{F}_{\text{BVDF}} \\ \dfrac{\|\mathbf{F}_{\text{VMF}}\|}{\|\mathbf{F}_{\text{BVDF}}\|} \mathbf{F}_{\text{BVDF}} & \text{otherwise} \end{cases}, \tag{12.40}$$

where $\mathbf{F}_{\text{BVDF}}$ is the output of the BVDF filter, $\mathbf{F}_{\text{VMF}}$ is the output of the VMF and $\| \cdot \|$ denotes the vector norm.

12.4.2 Fuzzy Adaptive Filters

The performance of the different nonlinear filters based on order statistics depends heavily on the problem under consideration. The types of noise that are present in an image considerably affect the filter performance. To overcome difficulties associated with the uncertainty that comes with the data, adaptive designs based on local statistics have been introduced.[105–109,134] Such filters utilize data-dependent coefficients to adapt to local image characteristics. The weights of the adaptive filters are determined by fuzzy transformations based on features from local data. The general form of the fuzzy adaptive filters is given as a nonlinear transformation of a weighted average of the input vectors inside the processing window:

$$\mathbf{F}^* = f\left(\sum_{i=0}^{n} w_i^* \mathbf{F}_i \right) = f\left(\sum_{i=0}^{n} w_i \mathbf{F}_i \bigg/ \sum_{i=0}^{n} w_i \right), \tag{12.41}$$

where $f(\cdot)$ is a nonlinear function that operates over the weighted average of the input set. The relationship between the pixel under consideration and each pixel in the window should be reflected in the decision for the filter's weights. In the adaptive design, the weights provide the degree to which an input vector contributes to the output of the filter. They are determined adaptively using fuzzy transformations of a distance criterion at each image position.

In this framework the weights are determined by fuzzy transformations based on features from local data. The fuzzy module extracts information without any *a priori* knowledge about noise characteristics. The weighting coefficients are transformations of the distance between the vector under consideration (center of the processing window W) and all other vector samples inside the processing window W. This transformation can be considered as a membership function with respect to a specific window component. The adaptive algorithm evaluates a membership function based on a given vector signal and then uses the membership values to calculate the filter output. Adaptive fuzzy algorithms utilize features extracted from local data, here in the form of a sum of distances, as inputs to the fuzzy weights. In this case, the distance functions are not used to order input vectors. Instead, they provide selected features in reduced space, features used as inputs for the fuzzy membership function.

Several candidate functions, such as triangular, trapezoidal, piecewise linear, or Gaussian-like functions, can be used as a membership function. If the distance criterion described by Equation 12.37 is used as a distance measure, a sigmoidal membership function can be selected[5,110]

$$w_i = \beta(1 + \exp\{A_i\})^{-r},\qquad(12.42)$$

where A_i is a cumulative distance from Equation 12.37, while β and r are parameters to be determined. The r value is used to adjust the weighting effect of the membership function and β is a weight-scale threshold. If the Minkowski L_p metric is used as the distance function, the fuzzy membership function with exponential form gives good results

$$w_i = \exp\left(-\frac{R_i^r}{\beta}\right),\qquad(12.43)$$

where R_i is a cumulative distance associated with the ith vector in the processing window W using generalized Minkowski norm, r is a positive constant, and β is a distance threshold.

Within the general fuzzy adaptive filter framework, numerous filters may be constructed by changing the form of the nonlinear function $f(\cdot)$, as well as the way the fuzzy weights are calculated. The choice of these two parameters determines the filter characteristics.

12.4.2.1 *Fuzzy Weighted Average Filter*

The first class of filters derived from the general nonlinear fuzzy algorithm is the *fuzzy weighted average filters* (FWAF). In this case, the output of the filter is a fuzzy weighted sum of the input set. The form of the filter is given as

$$\mathbf{F}_0^* = \frac{1}{Z}\sum_{i=0}^{n} w_i \mathbf{F}_i,\qquad Z = \sum_{i=0}^{n} w_i.\qquad(12.44)$$

This filter provides a vector-valued signal, which is not included in the original set of inputs. The weighted average form of the filter provides a compromise between a nonlinear order-statistics filter and an adaptive filter with data-dependent coefficients. Depending on the form of the distance criterion and the corresponding fuzzy transformation, different fuzzy filters can be designed. If the distance criterion selected is the sum of vector angles, the *fuzzy vector directional filter* (FVDF) is obtained. If an L_1 norm is used as the distance criterion, a fuzzy generalization of the VMF is constructed.

12.4.2.2　Maximum Fuzzy Vector Directional Filters

Another possible choice of the nonlinear function $f(\cdot)$ is the maximum selector. In this case, the output of the nonlinear function is the input vector that corresponds to the maximum fuzzy weight. Using the maximum selector concept, the output of the filter is a part of the original input set. The form of this filter is

$$\mathbf{F}_0^* = \mathbf{F}_i \quad \text{with} \quad i = \arg\max\ w_i,\ i = 0, \ldots, n. \qquad (12.45)$$

In other words, as an output the input vector associated with the maximum fuzzy weight is selected. It must be emphasized that through the fuzzy membership function, the maximum fuzzy weight corresponds to the minimum distance. If the vector angle criterion is used to calculate distances, the fuzzy filter delivers the same output as the BVDF.[5,110] If the L_1 or L_2 is adopted as the distance criterion, the filter provides the same output as the VMF. By utilizing the appropriate distance function, different filters can be obtained. Thus, filters such as VMF or BVDF can be seen as special cases of this specific class of fuzzy filters.

12.4.2.3　Fuzzy Ordered Vector Directional Filters

In many cases it is favorable not to use all the inputs inside the operational window to produce the final output of the nonlinear filter. Instead, only a part of the vector-valued input signals can be used. The input vectors are ordered according to their respective fuzzy membership strengths. The form of the fuzzy ordered vector directional filter is given as

$$\mathbf{F}^* = \frac{1}{Z} \sum_{i=0}^{\tau} w_{(i)} \mathbf{F}_{(i)}, \qquad Z = \sum_{i=0}^{\tau} w_{(i)}, \qquad (12.46)$$

where $w_{(i)}$ represents the ith ordered fuzzy membership function and $w_{(\tau)} \leq w_{(\tau-1)} \leq \ldots \leq w_{(0)}$, with $w_{(0)}$ the fuzzy coefficient with the largest membership strength.

The above form of the filter constitutes a fuzzy generalization of the α-trimmed filters (Equation 12.34).[4] Through the fuzzy transformation, the weights to be sorted are scalar values. In this way the nonlinear ordering process does not introduce any significant computational burden. Depending on the distance criterion and the associate fuzzy chosen by the designer, a number of different α-trimmed filters can be obtained.

The fuzzy transformations of Equations 12.42 and 12.43 are not the only way in which the adaptive weights can be constructed. In addition to fuzzy membership functions, other design concepts can be utilized for the task. One such design is the *nearest neighbor rule*,[137] in which the value of the weight w_i in Equation 12.41 is calculated according to the following formula:

$$w_i = \frac{D_{(n)} - D_{(i)}}{D_{(n)} - D_{(0)}}, \tag{12.47}$$

where $D_{(n)}$ is the maximum distance in the filtering window, measured using an appropriate distance criterion, and $D_{(0)}$ is the minimum distance, which is associated with the centermost vector inside the window. As in the case of the fuzzy membership function, the value of the weight in Equation 12.47 expresses the degree to which the vector $\mathbf{F}_i$ is close to the centermost vector, and far away from the worst value, the outer rank.

In Reference 137 an adaptive vector processing filter named *adaptive nearest neighbor filter* (ANNF) was devised utilizing the general framework of Equation 12.41. The weights in ANNF were calculated by using the formula of Equation 12.47 with the angular distance as a measure of dissimilarity between the color vectors.

It is evident that the outcome of such an adaptive vector processing filter depends on the choice of the distance criterion selected as a measure of dissimilarity among vectors. As before, the L_p norm or the angular distance (sum of angles) between the color vectors can be used to remove vector signals with atypical directions. However, both these distance metrics utilize only part of the information carried by the color image vectors. As in the case of DDF, it is anticipated that an adaptive vector processing filter based on an ordering criterion, which utilizes both vector features, namely, magnitude and direction, will provide a robust solution whenever the noise characteristics are unknown.

In Reference 111 a distance measure for the noisy vectors was introduced:

$$J_i = \sum_{j=0}^{n} [1 - S(\mathbf{F}_i, \mathbf{F}_j)], \quad \text{with}$$

$$S(\mathbf{F}_i, \mathbf{F}_j) = \left(\frac{\mathbf{F}_i \cdot \mathbf{F}_j}{|\mathbf{F}_i||\mathbf{F}_j|} \right) \left(1 - \frac{|\, \|\mathbf{F}_i\| - \|\mathbf{F}_j\| \,|}{\max(\|\mathbf{F}_i\|, \|\mathbf{F}_j\|)} \right). \tag{12.48}$$

As can be seen, the similarity measure of Equation 12.48 takes into consideration both the direction and the magnitude of the vector inputs. The first part of the measure S is equivalent to the angular distance (*vector angle criterion*) and the second part is related to the normalized difference in magnitude. Thus, if the two vectors under consideration have the same length, the second part of $S(\mathbf{F}_i, \mathbf{F}_j)$ equals to one and only the directional information is

used in Equation 12.48. On the other hand, if the vectors under consideration have the same direction in the vector space (collinear vectors), the first
part of $S(\mathbf{F}_i, \mathbf{F}_j)$ (directional information) equals to one and the similarity
measure of Equation 12.48 is based only on the magnitude of the difference
part.

Utilizing this similarity measure, an adaptive vector processing filter based
on the general framework of Equation 12.41 and the weighting formula of
Equation 12.48 was devised in Reference 111. The *adaptive nearest neighbor
multichannel filter* (ANNMF) belongs to the adaptive vector processing filter family defined through Equation 12.41. However, ANNMF combines
the weighting formula of Equation 12.47 with the new distance measure of
Equation 12.48 to evaluate its weights.

12.4.3   Nonparametric Adaptive Multichannel Filter

Consider the following model for the color image degradation process:

$$\mathbf{F}_j = \mathbf{X}_j + \mathbf{G}_j, \tag{12.49}$$

where $\mathbf{X}_j$ is a *three-dimensional* uncorrupted image vector, $\mathbf{F}_j$ is the corresponding noisy vector to be filtered, and $\mathbf{G}_j$ is an additive noise vector. In
our analysis, it is assumed that the color image vectors are unknown and that
the noise vectors are uncorrelated at the different image locations and are
signal independent.

Let us denote with $\Phi(\mathbf{F})$ the minimum variance estimator of the color
vector $\mathbf{X}$, given the noisy measurement vector $\mathbf{F}$. The expected square error
of the filter, when the image vectors are corrupted by additive noise as in
Equation 12.49, can be written as

$$V = \int \int [\mathbf{X} - \Phi(\mathbf{F})][\mathbf{X} - \Phi(\mathbf{F})]^T f(\mathbf{X}|\mathbf{F}) f(\mathbf{F}) \, d\mathbf{X} \, d\mathbf{F}, \tag{12.50}$$

$$V = \int_{-\infty}^{\infty} \left[\int_{-\infty}^{\infty} [\mathbf{X} - \Phi(\mathbf{F})][\mathbf{X} - \Phi(\mathbf{F})]^T f(\mathbf{X}|\mathbf{F}) \, d\mathbf{X} \right] f(\mathbf{F}) \, d\mathbf{F}, \tag{12.51}$$

where z^T denotes the transpose of z. Because $\Phi(\mathbf{F})$ does not enter into
the outer integral and $f(\mathbf{F})$ is always positive, it is sufficient for the optimal
minimum variance estimator to minimize the expected value of the estimation
cost (conditional Bayesian risk), given the observation $\mathbf{F}$. Thus, it is sufficient
to minimize the quantity

$$V_{BR} = \int_{-\infty}^{\infty} [\mathbf{X} - \Phi(\mathbf{F})][\mathbf{X} - \Phi(\mathbf{F})]^T f(\mathbf{X}|\mathbf{F}) \, d\mathbf{X}. \tag{12.52}$$

The minimum variance estimator, which minimizes the above cost, is then known to be

$$\Phi(\mathbf{F})_{MV} = \int_{-\infty}^{\infty} \mathbf{X} f(\mathbf{X}|\mathbf{F}) \, d\mathbf{X} = \int_{-\infty}^{\infty} \frac{\mathbf{X} f(\mathbf{X}, \mathbf{F})}{f(\mathbf{F})} \, d\mathbf{X}, \qquad (12.53)$$

with

$$f(\mathbf{F}) = \int_{-\infty}^{\infty} f(\mathbf{X}, \mathbf{F}) f(\mathbf{X}) \, d\mathbf{X}. \qquad (12.54)$$

If the densities in Equation 12.52 are known and a training record of the sample pairs $(\mathbf{X}, \mathbf{F})$ is available, the minimum variance estimator can be derived. Unfortunately, in a realistic image processing scenarios, no *a priori* knowledge about the noise process or the image itself is available. Thus, a nonparametric estimator must be utilized to approximate the probability density functions (pdf) in Equation 12.52.

Let us assume a window of finite length n centered around a noisy vector $\{\mathbf{F}\}_0$. Through this window, a set of multivariate noisy samples $W = (\mathbf{F}_0, \mathbf{F}_1, \ldots, \mathbf{F}_n)$ becomes available. Based on the samples from the filtering window W, an adaptive, data-dependent multivariate kernel estimator can be devised to approximate the densities in Equation 12.52. The form of the adaptive kernel estimator selected is as follows:

$$\hat{f}(\mathbf{X}, \mathbf{F}) = \frac{1}{N} \sum_{i=0}^{n} \frac{1}{h_i^L} K\left(\frac{\mathbf{F} - \mathbf{F}_i}{h_i}\right), \qquad N = n + 1, \qquad (12.55)$$

where $\mathbf{F}_i$ is the ith training vector, with $i = 0, 1, \ldots, n$, $L = 3$ is the dimensionality of the measurement space, and h_i is the data-dependent smoothing parameter, which regulates the shape of the kernel. The variable kernel density estimator exhibits local smoothing, which depends both on the point at which the density is evaluated and also on the information on the local neighborhood in W.

The h_i can be any function of the sample size N.[112] The bandwidths h_i (smoothing factors) can be defined as a function of the aggregated distance between the local observation under consideration and all the other vectors inside the W window. Thus,

$$h_i = N^{-\frac{k}{L}} A_i = N^{-\frac{k}{L}} \sum_{k=0}^{n} \|\mathbf{F}_i - \mathbf{F}_k\|, \qquad (12.56)$$

where k is a design parameter. The choice of the kernel function in Equation 12.55 is not nearly as important as the bandwidth (smoothing factor). For the applications, the multivariate extension of the exponential kernel $K(z) = \exp(-|z|)$ or the Gaussian kernel $K(z) = \exp(-|z^T z|/2)$ can be selected.[112]

Given Equations 12.52 through 12.55, the nonparametric estimator can be defined as

$$\Phi(\mathbf{F})_{NP} = \int_{-\infty}^{\infty} \frac{\mathbf{X} \hat{f}(\mathbf{X}, \mathbf{F})}{\hat{f}(\mathbf{F})} \, d\mathbf{X} = \sum_{i=0}^{n} \mathbf{X}_i \left(\frac{(N^{-1}) h_i^{-L} K \left(\frac{\mathbf{F} - \{\mathbf{F}\}_i}{h_i} \right)}{\sum_{i=0}^{n} (N^{-1}) h_i^{-L} K \left(\frac{\mathbf{F} - \mathbf{F}_i}{h_i} \right)} \right) \qquad (12.57)$$

$$\Phi(\mathbf{F})_{NP} = \sum_{l=0}^{n} \mathbf{X}_i \left(\frac{h_i^{-L} K \left(\frac{\mathbf{F} - \mathbf{F}_i}{h_i} \right)}{\sum_{i=0}^{n} h_i^{-L} K \left(\frac{\mathbf{F} - \mathbf{F}_i}{h_i} \right)} \right) = \sum_{i=0}^{n} w_i^* \mathbf{X}_i \qquad (12.58)$$

where w_i^* is a weighting function defined in the interval $[0,1]$.

To obtain the required estimate we must assume that, in the absence of noise, discrete sample vectors $\mathbf{X}_i$ are available. This is not a severe restriction, because in many cases such samples may be obtained by a calibration procedure in a controlled environment, perhaps at a very high SNR. In a real-time image processing application, however, that is not the case. Therefore, alternative suboptimal solutions are introduced. In a first approach, we substitute the vectors $\mathbf{X}_i$ in Equation 12.57 with their noisy measurements. The resulting *adaptive nonparametric multichannel filter* (ANMF) is based solely on the available noisy vectors and the form of the minimum variance estimator. Thus, the form of the ANMF is

$$\Phi_1(\mathbf{F})_{ANMF} = \sum_{i=0}^{n} \mathbf{F}_i \left(\frac{h_i^{-L} K \left(\frac{\mathbf{F} - \mathbf{F}_i}{h_i} \right)}{\sum_{i=0}^{n} h_i^{-L} K \left(\frac{\mathbf{F} - \mathbf{F}_i}{h_i} \right)} \right). \qquad (12.59)$$

A different form of the adaptive nonparametric estimator can be obtained if a reference vector is used instead of the actual noisy measurement. The ideal reference vector is, of course, the actual value of the multidimensional signal in the specific location under consideration. However, since the $\mathbf{X}_0$ vector is not available, a robust estimate, usually evaluated in a small subset of the input vector set, is utilized instead. Usually the vector median $\mathbf{X}^{VM}$ is the preferable choice because it smoothes out impulsive noise and preserves the edges to some extent. The median-based ANMF then has the following form:

$$\Phi_2(\mathbf{F})_{ANMF} = \sum_{i=0}^{n} \mathbf{X}_i^{VM} \left(\frac{h_i^{-L} K \left(\frac{\mathbf{F} - \mathbf{F}_i}{h_i} \right)}{\sum_{i=0}^{n} h_i^{-L} K \left(\frac{\mathbf{F} - \mathbf{F}_l}{h_l} \right)} \right). \qquad (12.60)$$

This filter can be viewed as a double-window, two-stage estimator. First, the original image is filtered by a multichannel median filter in a small processing window in order to reject possible outliers, and then an adaptive nonlinear filter with data-dependent coefficients defined in Equation 12.57 is utilized to provide the final filtered output.

12.5 Digital Paths Approach to Color Image Filtering

In this section a novel approach to color image filtering is proposed. Instead of using a fixed window, the new method exploits connections between image pixels using the concept of digital paths. According to the proposed methodology, image pixels are grouped together, forming paths that reveal the underlying structural dynamics of the image (Figure 12.5 and Figure 12.6). Depending on the design principles and the computational constraints, the new filter framework allows the paths to be considered on the entire image or restricted to a predefined search area.[113,114] The new approach focuses on the latter case.

To facilitate comparisons with existing ranked-type operations and to illustrate the computational efficiency of the proposed framework, the path-searching area is allowed to match the window W used by the ranked-type filters. However, instead of the indiscriminate use of the window pixels—an approach advocated by the majority of existing multichannel filters—the proposed framework here allows for the formation of a number of digital path models, which in turn are used to determine the coefficients of a weighted average type of filtering operation.

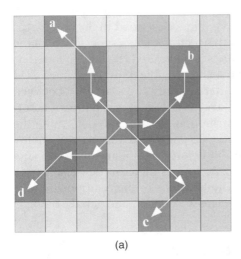

221	134	240	244	190	238	221
228	219	129	215	224	124	226
224	205	115	214	200	106	240
221	221	210	116	106	206	239
235	110	121	214	116	225	206
122	180	229	212	211	112	225
224	235	229	235	130	210	207

(a)   (b)

FIGURE 12.5
Illustration of the concept of digital paths and connection cost. The pixels a, b, c, d are connected with the central pixel along paths whose connection costs are minimal.

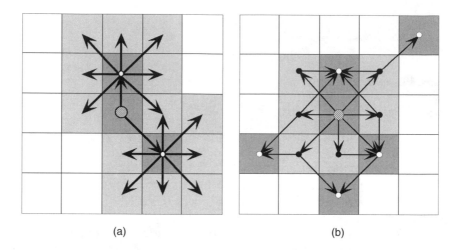

(a)   (b)

FIGURE 12.6
In the digital path approach to first and last (DPAF and DPAL) filters, the weights are assigned to the pixels surrounding the central pixel and are determined in different ways. In the DPAF approach (a), the weights in Equation 12.74 are calculated exploring all digital paths starting from the central pixel and crossing its nearest neighbors; then a weighted average of the nearest neighbors of the central pixel is calculated (Equation 12.75). In the DPAL approach, the weights are obtained by exploring all digital paths leading from the central pixel to the pixels contained in the filtering window (b) and then a weighted average of all pixels from that window is calculated (Equation 12.81).

The new filter class based on digital paths and connection cost can be seen as a powerful generalization of the multichannel AD presented in Section 12.3, and as an extension of the fuzzy adaptive filters described in Section 12.4.2. The filters discussed there are shown in this section to be a special case of the new filtering scheme, when a digital path is degenerated to a step of length 1.

The path connection costs evaluated over all possible digital paths are used to derive fuzzy membership functions that quantify the similarity between vectorial inputs. The proposed filtering structure then uses the function outputs to appropriately weight input contributions in order to determine the filtering result. The proposed filtering schemes parallelize the structure of the adaptive multichannel filter introduced in Reference 115, and they can successfully eliminate Gaussian, impulsive, as well as mixed-type noise. However, thanks to the introduction of the digital paths in their supporting element, the new filters not only preserve edges and fine image details, but also can act as image-sharpening operators.

12.5.1 Connection Cost Defined over Digital Paths

To perform operations based on the distances, we first need to precisely define the notion of a topological distance. The concept of a topological distance between image points is of extreme importance in many applications based on the distance transformation, which is one of the fundamental operations of mathematical morphology.[116–119]

Let $\mathcal{B}$ be a non-empty set. We can measure distances between points in $\mathcal{B}$, which amounts to defining a real valued function on the Cartesian product $\mathcal{B} \times \mathcal{B}$ of $\mathcal{B}$ with itself. Let the function $\rho : \mathcal{B} \times \mathcal{B} \to \mathbb{R}$ be called a distance if it is positive definite: $\rho(x, y) \geq 0$, with $\rho(x, y) = 0$, when $x = y$, and symmetric: $\rho(x, y) = \rho(y, x)$, for all $x, y \in \mathcal{B} \times \mathcal{B}$. A distance is called a metric if, additionally, it satisfies the triangle inequality:[120] $\rho(x, z) \leq \rho(x, y) + \rho(y, z)$, for all $x, y, z \in \mathcal{B} \times \mathcal{B}$.

In digital image processing, three basic distance functions are usually applied. If $p = (p_1, p_2)$ and $q = (q_1, q_2)$ denote two image points ($p, q \in Z^2$), then we define the *city-block distance*: $\rho_4(p, q) = |p_1 - q_1| + |p_2 - q_2|$; the *chessboard distance*: $\rho_8(p, q) = \max\{|p_1 - q_1|, |p_2 - q_2|\}$; and the *Euclidean distance*: $\rho_E(p, q) = [(p_1 - q_1)^2 + (p_2 - q_2)^2]^{1/2}$. Using the city-block and chessboard distances we are able to define the two basic types of neighborhoods, 4-neighborhood $\mathcal{N}_4(x) = \{y : \rho_4(x, y) = 1\}$ and 8-neighborhood $\mathcal{N}_8(x) = \{y : \rho_8(x, y) = 1\}$.

Let $\omega \in \{4, 8\}$. Two points $p, q \in Z^2$ are said to be in $\mathcal{N}_\omega$-neighborhood relation (denoted as $\sim$), or to be $\mathcal{N}_\omega$-adjacent if $q \in \mathcal{N}_\omega(p)$ or equivalently $p \in \mathcal{N}_\omega(q)$. This $\mathcal{N}_\omega$ adjacency relation defines a graph structure on the image domain, called an $\mathcal{N}_\omega$-adjacency graph. On the graph, a finite $\mathcal{N}_\omega$-path can be defined as a sequence of points $(p_0, p_1, \ldots, p_\eta)$ such that for $i \in \{1, 2, \ldots, \eta\}$ the point p_{i-1} is $\mathcal{N}_\omega$ adjacent to p_i. A path is called simple if $i \neq j$ implies that $p_i \neq p_j$. This is a very important property of a path, as it means that a path does not intersect itself or, in other words, it is self-avoiding.[121,122]

Using the distances between neighboring points, which are called prime distances,[123] we are able to define a distance between any two image points by following all admissible paths linking those points and then taking the minimum of the total length over all possible paths, which is the sum of the prime distances between the nodes of the paths. In this way, the distance between two image points is the length of the path for which the sum of the prime distances between the path nodes is minimal. For the city-block distance the admissible paths consist of horizontal and vertical moves only, whereas for the chessboard distance the diagonal moves are also allowed. The prime distances for the two kinds of neighborhood are declared in this work to be equal to 1.

Let us now introduce the definition of a geodesic distance. Let us assume that $\mathbb{R}^2$ is the Euclidean space, S is a planar subset of $\mathbb{R}^2$, and x, y are points belonging to set S. A path from x to y is a continuous mapping $\Pi : [a, b] \to S$, such that $\Pi(a) = x$ and $\Pi(b) = y$. The point x is considered the starting point, while y is the ending point on the path Π.[117]

An increasing polygonal line P on the path Π is any polygonal line such that $P = \{\Pi(\lambda_i)\}_{i=0}^\eta$, $a = \lambda_0 < \cdots < \lambda_\eta = b$. The length of the polygonal line P is considered to be the total sum of its constitutive line segments $L(P) = \sum_{i=1}^\eta \rho(\Pi(\lambda_{i-1}), \Pi(\lambda_i))$, where $\rho(x, y)$ is the distance between the points x and y, when a specific metric is adopted. A path Π from x to y is called rectifiable, if and only if $L(P)$, where P is an increasing polygonal line, is bounded. Its upper bound is called the length of the path Π.

The geodesic distance $\rho^S(x, y)$ between points x and y is the lower bound of the length of all paths leading from x to y, which are totally included in S. If such paths do not exist, then the value of the geodesic distance is set to ∞. In general $\rho^S(x, y) \geq \rho(x, y)$. However, if the set S is convex, meaning that there are no points on the line between x and y that are not members of S, the geodesic distance verifies $\rho^S(x, y) = \rho(x, y)$.

The notion of a path can be extended to a lattice, which is a set of discrete points on the plane, in our case the spatial locations of the image pixels. Let a digital lattice $\mathcal{H} = (\mathbf{F}, \mathcal{N})$ be defined by $\mathbf{F}$, which is the set of all points of the plane (pixels of a color image) and a neighborhood relation $\mathcal{N}$ between the lattice points.[124]

A digital path $P = \{p_i\}_{i=0}^{\eta}$ defined on the lattice $\mathcal{H}$ is a sequence of neighboring points $(p_{i-1}, p_i) \in \mathcal{N}$. The length $L(P)$ of the digital path $P\{p_i\}_{i=0}^{\eta}$ is simply $\sum_{i=1}^{\eta} \rho^{\mathcal{H}}(p_{i-1}, p_i)$, where $\rho^{\mathcal{H}}$ denotes the distance between two neighboring points of the lattice $\mathcal{H}$ and the geodesic distance between p_0 and p_η is the minimal length of $L(P)$.

Constraining the paths to be totally included in a predefined set W yields the digital geodesic distance ρ^W. In this work $\mathcal{N}_\omega$-neighborhood system ($\omega = 4$ or $\omega = 8$) is considered, with a topological distance of 1 assigned to any neighboring points and the set W is the supporting window of appropriate size. All paths considered in this chapter are included in the filtering window W (Figure 12.7).

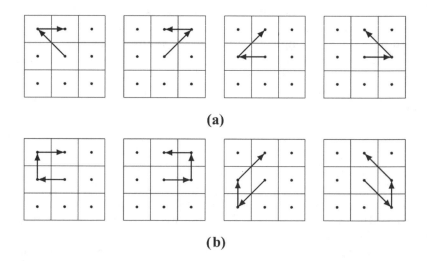

(a)

(b)

FIGURE 12.7
Digital paths of (a) length 2 and (b) length 3, connecting two neighboring points within a predefined window W of size 3×3, when the 8-neighborhood system is applied.

Let us now adopt the following notation, which will help us define the distance functions defined over geodesic paths. The starting point of a path is denoted as $p_0 = (x_0, y_0)$. Its neighbors are denoted as $p_1 = (x_{u_1}, y_{v_1})$, which means that the neighbors are the second points of all digital paths originating at p_0. Then the third point of a digital path starting at p_0 is $p_2 = (x_{u_2}, y_{v_2})$ and so on, until the path reaches in η steps the ending point $p_\eta = (x_{u_\eta}, y_{v_\eta})$. In this way the sequences $x_{u_1}, \ldots, x_{u_\eta}$ and $y_{v_1}, \ldots, y_{v_\eta}$ uniquely define the digital path starting at x_0, y_0 and ending at x_{u_η}, y_{v_η}. The set of all possible digital paths contained in W joining two points $x, y \in W$ is denoted as $\Psi^W(x, y)$.

Two pixels x and y are called connected (hereafter denoted as $x \Leftrightarrow y$), if there exists a digital path $P^W(x, y)$ contained in the set W starting from x and ending at y. If two pixels p_0 and p_η are connected by a digital path $P^{W,\eta}$ $\{p_0, p_1, \ldots, p_\eta\}$ of length η, then let $\Lambda^{W,\eta}\{p_0, p_1, \ldots, p_\eta\}$ be a measure of the connection cost defined over the digital path linking the starting point p_0 and ending point p_η (f is a nonnegative scalar function of η vector variables)

$$\Lambda^{W,\eta}\{p_0, \ldots, p_\eta\} = f\{\mathbf{F}(p_0), \ldots, \mathbf{F}(p_\eta)\}$$
$$= f\{\mathbf{F}(x_0, y_0), \mathbf{F}(x_{u_1}, y_{v_1}), \ldots, \mathbf{F}(x_{u_\eta}, y_{v_\eta})\}. \quad (12.61)$$

The connection cost over the digital path $\Lambda^{W,\eta}$ can be seen as a measure of dissimilarity between color image pixels at points $p_0, p_1, \ldots, p_\eta$ forming a specific path linking p_0 and p_η.[119,125,126] If a path joining two distinct points x, y, such that $\mathbf{F}(x) = \mathbf{F}(y)$ consists of the pixels of the same channel values, then the connection cost should be zero; otherwise $\Lambda^{W,\eta} > 0$.

Let us now define a generalized connection cost function, based on the distance transform on the curved space (DTOCS),[119,125] introduced by Toivanen for the gray-scale images. For two given points $p_i = (x_{u_i}, y_{v_i})$ and $p_{i-1} = (x_{u_{i-1}}, y_{v_{i-1}})$, $i = 1, 2, \ldots, \eta$, which are in neighborhood relation, let the generalized distance between the two points be called the connection cost defined on the hybrid spatial-color space discussed in Reference 127 and 128: $\Lambda^{W,1}\{p_{i-1}, p_i\} = \|\mathbf{F}(p_i) - \mathbf{F}(p_{i-1})\| + \xi \cdot \rho^W(p_i, p_{i-1})$, where ξ establishes a proper weighting in the hybrid spatial-color space. The connection cost of a whole digital path $p_0, p_1, \ldots, p_\eta$ will be then

$$\Lambda^{W,\eta}\{p_0, p_1, \ldots, p_\eta\} = \sum_{i=1}^{\eta} \left[\|\mathbf{F}(p_i) - \mathbf{F}(p_{i-1})\| + \xi \cdot \rho^W(p_i, p_{i-1}) \right]. \quad (12.62)$$

As we work with a small filtering window, we focus on the color space only, by setting $\xi = 0$.

Similarly to the gray-scale case, we call the minimal connection cost $\Gamma^{W,\eta}(x, y)$ of a path of length η linking two points $x, y \in W$, the η-geodesic between x and y: $\Gamma^{W,\eta}(x, y) = \min\{\Lambda(\gamma), \gamma \in \Psi^{W,\eta}\}$.

In this way the η-geodesic is defined as the path of length η, which gives the minimal connection cost between two points linked by a digital path. If we take the minimum of the connection costs generated by all possible paths joining two points x and $y \in W$, then we obtain the generalized multichannel geodesic distance between these points:

$$\Gamma^W(x, y) = \min_{\eta}\{\Gamma^{W,\eta}(x, y)\} = \min\{\Lambda(\pi), \pi \in P^{W,\eta}(x, y), \eta \in \mathbb{N}\}.$$

$\Gamma^W(x, y)$ defines the multidimensional distance transform, which is a generalization of DTOCS.[125]

In general, two distinct pixels' locations on the image lattice can be connected by many paths. Moreover, the number of possible geodesic paths of certain length η connecting two distinct points depends on their locations, length of the path, and the neighborhood system used (Figure 12.7).

12.5.2   General Filter Framework

In this work, the fuzzy filtering structure proposed in References 105, 106, and 110 is used. The general form of the fuzzy adaptive filters presented here is defined as a weighted average of input vectors inside W

$$\mathbf{F}_0^* = \sum_{i=0}^{n} w_i^* \mathbf{F}_i = \frac{\sum_{i=0}^{n} w_i \mathbf{F}_i}{\sum_{i=0}^{n} w_i}. \tag{12.63}$$

The relationship between the pixel under consideration $\mathbf{F}_0$ and each pixel in the window should be reflected in the decision how to define the filter weights. In our case, the weights are determined using the similarity functions calculated over digital paths included in the processing window W.

On the basis of the connection cost-function concept, it is possible to define different classes of similarity functions. Choosing a specific form of a similarity function yields different filters of specific properties, which can be applied for a wide range of low-level vision tasks.

12.5.3   Digital Path Approach Filter Class

Let us now define a similarity function, analogous to a membership function used in fuzzy systems, between two pixels connected through all possible digital paths leading from x to y

$$w^{W,\eta}(x, y) = \sum_{m=1}^{\omega} f\{\Lambda_m^{W,\eta}(x, y)\}, \tag{12.64}$$

where ω is the number of all paths connecting x and y, $\Lambda_m^{W,\eta}(x, y)$ is a dissimilarity value along a specific path m from the set of all ω possible paths leading from x to y, and $f(\cdot)$ is smooth function of $\Lambda_m^{W,\eta}$. By definition $w^{W,\eta}(x, y)$ returns a value evaluated over all possible routes linking the starting point

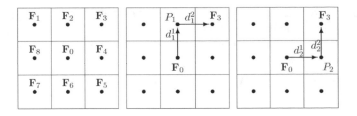

FIGURE 12.8
Digital paths of length $n = 2$ connecting points F_0 and F_3.

x with the end point y. The smooth function $f : (0; \infty) \to \mathbb{R}$ should satisfy following conditions: f is decreasing in $(0; \infty)$, f is convex in $(0; \infty)$, $f(0) = 1$, $f(\chi) \to 0$, when $\chi \to \infty$. Several functions satisfying the above conditions have been proposed in the literature.[5,75,115,129,133] However, for impulsive noise removal good results are obtained using the exponential form of the function f.[130] Therefore,

$$w^{W,\eta}(x, y) = \sum_{m=1}^{\omega} \exp\left[-\beta \cdot \Lambda_m^{W,\eta}(x, y)\right], \tag{12.65}$$

where β is the filter design parameter.

For $\eta = 1$ and a square (3×3) window W, the similarity function w is defined according to Equation 12.62 as $w^{W,1}(x, y) = \exp\{-\beta\|\mathbf{F}(x) - \mathbf{F}(y)\|\}$, and then if $\mathbf{F}(x) = \mathbf{F}(y)$, $\Lambda^{W,n}(x, y) = 0$, $w(x, y) = 1$, and for $\|\mathbf{F}(x) - \mathbf{F}(y)\| \to \infty$ then $w \to 0$.[110]

Figure 12.8 illustrates the calculation of the similarity function between two points connected by two geodesic paths of length $\eta = 2$. In this case, the cost functions related to paths P_1 and P_2 are

$$\Lambda_1^{W,2}(x, y) = d_1^1 + d_1^2, \quad \Lambda_2^{W,2}(x, y) = d_2^1 + d_2^2, \tag{12.66}$$

where d_1^1 and d_1^2 are connection costs between neighboring points on the path P_1 defined according to Equation 12.62, while d_2^1, d_2^2 are connection costs defined on path P_2. The total similarity value can be expressed as

$$w^{W,2} = \exp\left(-\beta \cdot \Lambda_1^{W,2}\right) + \exp\left(-\beta \cdot \Lambda_2^{W,2}\right). \tag{12.67}$$

A normalized form of the similarity function is defined as

$$w^*(x, y) = \frac{w^{W,n}(x, y)}{\sum_{z \Leftrightarrow x} w^{W,n}(x, z)}, \tag{12.68}$$

where $y \Leftrightarrow x$ denotes all points y connected by digital paths with x contained in W.

Assuming that the pixel $\mathbf{F}_x$ is the pixel under consideration, with $\mathbf{F}_y$ representing the pixel included in the supporting element W, which is connected to $\mathbf{F}_x$ via a digital path, the filter output $\mathbf{F}_x^*$ is given as

$$\mathbf{F}_x^* = \sum_{y \Leftrightarrow x} w^*(x, y) \cdot \mathbf{F}_y, \quad w^*(x, y) = \frac{w^{W,\eta}(x, y)}{\sum_{z \Leftrightarrow x} w^{W,\eta}(x, z)}. \tag{12.69}$$

As can be easily noticed, $\mathbf{F}_x^*$ is the weighted average of all points $\mathbf{F}_y^*$ connected by digital paths with the pixel $\mathbf{F}_x^*$. The pixel $\mathbf{F}_y$ is the ending point of a path leading from x and therefore this filter structure is called digital path-approach last (DPAL), as y is the last point on the path (see Figure 12.6b).

Another possible filtering scheme takes into account the similarity between the starting point $x = p_0$ and point $y = p_1$ crossed by a digital path connecting pixel p_0 and its neighbor p_1 with all possible points $p_\eta \in W$, which can be reached in η steps from p_0 (digital path approach first, DPAF).

The aim of taking into account the points $p_2, \ldots, p_\eta$ when calculating the filter output is to explore not only the direct neighborhood of p_0 but also to use the information on the local image structure. This can be done by acquiring the information on the local image features investigating the connection costs of digital paths originating at p_0, passing p_1, and then visiting successive points, until the path reaches length η. In this case the new similarity function takes the form

$$w^{W,\eta}(x, y) = w^{W,\eta}(p_0, p_1) = \sum_{\{p_2^*, p_3^*, \ldots, p_\eta^*\}} f\left(\Lambda^{W,\eta}\{p_0, p_1, p_2^*, p_3^*, \ldots, p_\eta^*\}\right), \tag{12.70}$$

where $\{p_0, p_1, p_2^*, \ldots, p_\eta^*\}$ denotes all paths originating at $x = p_0$ crossing $y = p_1$ end, ending at p_η^*, which are totally included in W, $f(\cdot)$ is a smooth function of $\Lambda^{W,\eta}$.

If the exponential function is used, then the similarity function takes the form

$$w^{W,\eta}(x, y) = w^{W,\eta}(p_0, p_1) = \sum_{\{p_2^*, p_3^*, \ldots, p_\eta^*\}} \exp\left[-\beta \cdot \Lambda^{W,\eta}\{p_0, p_1, p_2^*, \ldots, p_\eta^*\}\right], \tag{12.71}$$

where β is the smoothing parameter. A normalized form of the similarity function can be defined as follows:

$$w^*(x, y) = w^*(p_0, p_1) = \frac{\sum_{\{p_2^*, p_3^*, \ldots, p_\eta^*\}} \exp\left[-\beta \cdot \Lambda^{W,\eta}\{p_0, p_1, p_2^*, \ldots, p_\eta^*\}\right]}{\sum_{\{p_1^*, p_2^*, \ldots, p_\eta^*\}} \exp\left[-\beta \cdot \Lambda^{W,\eta}\{p_0, p_1^*, p_2^*, \ldots, p_\eta^*\}\right]}, \tag{12.72}$$

where $\{p_0, p_1, p_2^*, \ldots, p_\eta^*\}$ denotes a path joining $x = p_0$ and p_η, crossing $y = p_1$, whereas $\{p_0, p_1^*, p_2^*, \ldots, p_\eta^*\}$ do not necessarily cross $y = p_1$ when joining p_0 and p_η.

Assuming that the pixel $\mathbf{F}_x$ at the position $x = p_0$ is the pixel under consideration, with $\mathbf{F}_y$ representing the pixel at $y = p_1$, the filter output $\mathbf{F}_x^*$ is given as

$$\mathbf{F}_x^* = \mathbf{F}_{p_0}^* = \sum_{y \Leftrightarrow x} w^*(x, y) \cdot \mathbf{F}_y = \sum_{y \sim x} w^*(x, y) \cdot \mathbf{F}_y = \sum_{p_1^* \sim p_0} w^*(p_0, p_1^*) \cdot \mathbf{F}_{p_1^*},$$

(12.73)

and combining this with Equation 12.72 gives

$$\begin{aligned} \mathbf{F}_x^* = \mathbf{F}_{p_0}^* &= \sum_{p_1^* \sim p_0} \frac{\sum_{\{p_2^*, p_3^*, \dots, p_n^*\}} \exp\left[-\beta \cdot \Lambda^{W, \eta}\left\{p_0, p_1^*, p_2^*, \dots, p_n^*\right\}\right]}{\sum_{\{p_1^*, p_2^*, \dots, p_n^*\}} \exp\left[-\beta \cdot \Lambda^{W, \eta}\left\{p_0, p_1^*, p_2^*, \dots, p_n^*\right\}\right]} \cdot \mathbf{F}_{p_1^*} \\ &= \sum_{p_1^* \sim p_0} w^*(p_0, p_1^*) \cdot \mathbf{F}_{p_1^*}. \end{aligned}$$

(12.74)

Using the notation from Sections 12.3 and 12.2 and Equation 12.20, we can formulate Equation 12.74 as

$$\mathbf{F}_0^* = \sum_{k=1}^{n} w_k^* \mathbf{F}_k,$$

(12.75)

where w_k^*, the normalized weighting coefficients, play the role of the generalized conductivity coefficients from Section 12.3, and $\mathbf{F}_k$ are the neighbors of $\mathbf{F}_0$, which is the central pixel in the filter mask W.

The general form of the AD scheme based on the digital paths can be written as

$$\mathbf{F}_0^* = (1 - \lambda^*)\mathbf{F}_0^* + \lambda^* \sum_{k=1}^{n} w_k^* \mathbf{F}_k,$$

(12.76)

or using the iterative notation, as

$$\mathbf{F}_0^{t+1} = (1 - \lambda^*)\mathbf{F}_0^t + \lambda^* \sum_{k=1}^{n} w_k^* \mathbf{F}_k^t.$$

(12.77)

By using the relation $\lambda^* = \lambda \sum_{k=0}^{n} c_k$ (Equation 12.20), it is possible to obtain the classical form of the AD scheme defined by Equation 12.15.

Figure 12.9 shows the dependence of PSNR on the λ^* and K values for the color *LENA* image contaminated by impulsive and mixed noise for the classic multichannel AD scheme and the new DPAF filter defined by Equation 12.75. Especially interesting is the behavior of the plots as a function of λ^*. As can be seen, for images contaminated by a noise process of high intensity, the maximum of PSNR is obtained for λ^* very close to 1, which means that it is favorable to omit the central pixel while calculating the weighted average in

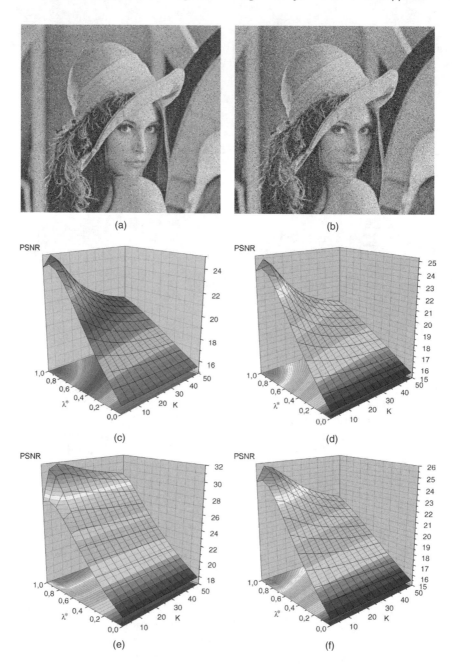

FIGURE 12.9
Dependence of the efficiency of the P-M AD filter and the DPAF on the λ^* parameter: (a) color image *LENA* contaminated with impulsive noise ($p = 0.12$, $p_1 = p_2 = p_3 = 0.3$), (b) test image corrupted by mixed noise ($\sigma = 30$, $p = 0.12$, $p_1 = p_2 = p_3 = 0.3$), (c and d) results obtained with the P-M AD filter, (e and f) results obtained with the DPAF ($\eta = 2$). As expected the maximum of PSNR is achieved for λ^* close to 1.

Equation 12.20 and also in Equation 12.75. This was already noticed in the scheme of Smith and Brady[25] (Equation 12.8), who did not take the central pixel into the averaging process, which is equivalent to setting $\lambda^* = 1$. That is why we set $\lambda^* = 1$ in Equation 12.75 to define the new DPAF filter Equations 12.74 and 12.75.

The superiority of this approach over the classic scheme is clearly seen in Figure 12.9, where especially for highly corrupted images, the difference in terms of PSNR is quite significant (see also Tables 12.4 and 12.5 in Section 12.6.2).

In a similar way, the DPAL filter can be defined as

$$\mathbf{F}_x^* = \mathbf{F}_{p_0}^* = \frac{\sum_{\{p_1^*, p_2^*, p_3^*, \dots, p_\eta^*\}} \exp\left[-\beta \cdot \Lambda^{W,\eta}\{p_0, p_1^*, p_2^*, \dots, p_\eta^*\}\right] \cdot \mathbf{F}_{p_\eta^*}}{\sum_{\{p_1^*, p_2^*, \dots, p_\eta^*\}} \exp\left[-\beta \cdot \Lambda^{W,\eta}\{p_0, p_1^*, p_2^*, \dots, p_\eta^*\}\right]}$$

$$= \sum_{p_\eta^*} w^*(p_0, p_\eta^*) \cdot \mathbf{F}_{p_\eta^*}, \qquad (12.78)$$

which can be written as

$$\mathbf{F}_0^* = \sum_{k=1}^{N} w_k^* \mathbf{F}_k, \qquad (12.79)$$

where N denotes the number of pixels surrounding $\mathbf{F}_0$ in the filtering window. Analogously to Equation 12.76, we can introduce the general form of DPAL defined by Equation 12.78

$$\mathbf{F}_0^* = (1 - \lambda^*)\mathbf{F}_0^* + \lambda^* \sum_{k=1}^{N} w_k^* \mathbf{F}_k, \qquad (12.80)$$

and its iterative version

$$\mathbf{F}_0^{t+1} = (1 - \lambda^*)\mathbf{F}_0^t + \lambda^* \sum_{k=1}^{N} w_k^* \mathbf{F}_k^t, \qquad (12.81)$$

where w_k^* are the normalized weighting coefficients from Equation 12.78.

The concept of the DPAF and DPAL filters is presented in Figure 12.6. The weights assigned to the pixels surrounding the central pixel $\mathbf{F}_0$ are determined in different ways. In the DPAF approach, the weights in Equation 12.74 are calculated exploring all digital paths starting from the central pixel and crossing its neighbors (Figure 12.6a), and then a weighted average of the nearest neighbors of the central pixel is calculated (Equation 12.75).

In the DPAL approach, the weights are obtained by exploring all digital paths leading from the central pixel to any of the pixels in the filtering window (Figure 12.6b), and then a weighted average of all pixels contained in that window is calculated (Equation 12.81).

Although both schemes work on supporting windows of the same size, determined by the number of steps η and the kind of neighborhood relation $\sim$, the DPAL has more powerful smoothing properties, as it involves all the N pixels from the filtering window into the averaging process, whereas the DPAF determines the weighted output using only its nearest neighbors. The efficiency of the new class of filters DPAF and DPAL is evaluated and compared with some of the standard filtering techniques in Section 12.6.

The computational complexity of the DPA filters depends on the path length η and the number of paths that can be constructed in the supporting window W of size $(k \times k)$. It is not difficult to see that for large k, which may be required in certain applications, the computational complexity of the filters makes them inapplicable. To decrease the computational burden, another filter structure is introduced. In the fast digital paths approach (FDPA), the size of the supporting window W is set to (3×3) independently of the digital path lengths η.

It is possible to construct both fast DPAF and fast DPAL filters; however, their properties are quite similar and therefore only the filtering approach based on DPAL (denoted as FDPA) is investigated. Using the FDPA formulation a number of interesting properties of the proposed filtering structure can be observed. For example, let us assume that parameter β used in Equation 12.65 is very small, $\beta \to 0$. Then the weights in Equation 12.69 reduce to $w^*(x, y) = \omega(x, y)/\Omega$, where $\omega(x, y)$ is the number of digital paths of length η connecting points x and y, and Ω denotes the number of all possible digital paths starting from x, which are totally included in W.

Examination of the convolution masks obtained in this way reveals their similarity to the masks obtained through Gaussian kernels. Therefore, the FDPA can be viewed as a nonlinear generalization of the Gaussian kernel-based schemes, which are widely used in many image processing tasks.

The parameter β in Equations 12.65 and 12.71 regulates the smoothness of the similarity function. Because the filtering structure of Equation 12.63 is a regression estimator, which enables a smooth interpolation among the observed, noise-corrupted image vectors, the parameter β provides the required balance between smoothing and detail preservation. Therefore, it is not surprising that the best results are obtained when the smoothing operators $\mathbf{F}^*$ in Equations 12.69 and 12.73 are applied in an iterative way. Starting with a low value of β enables the smoothing of the image noise components. At each iteration step the parameter β can be increased, following a scheme similar to that used in simulated annealing applications. In particular, β can be increased exponentially: $\beta(\kappa) = \beta(\kappa - 1) \cdot \alpha$, $\kappa \in \mathbb{N}$, where κ is the iteration number and α is a design parameter. Increasing the β parameter has the result that, after a few iterations, no further changes are introduced to the image, as for high β the filter output is that pixel, which lies on the geodesic digital path in the color space. The influence of α on the performance of the DPAL and FDPA filters is shown in Figure 12.10. The value of α is not critical for the efficiency of the new filter class, and taking α

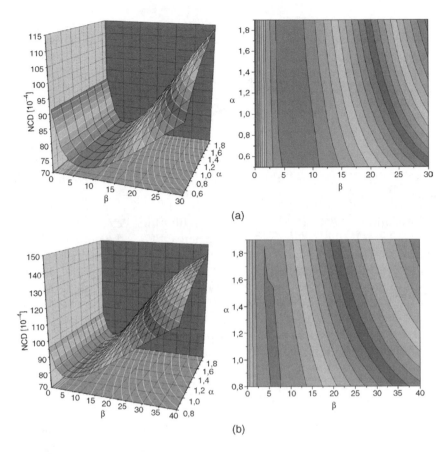

FIGURE 12.10
Efficiency of the (a) DPAL and (b) FDPA in terms of normalized color difference (NCD) and their dependence on α and β for *LENA* image corrupted by impulsive ($p = 0.12$, $p_1 = p_2 = p_3 = 0.3$) and Gaussian noise ($\sigma = 30$) ($n = 2$, third iteration).

from the interval [1, 2] guarantees fast filter convergence and good filtering results.

12.5.4 Computational Complexity and Fast Filter Design

Apart from the numerical behavior of any proposed algorithm, its computational complexity is a realistic measure of its practicality and usefulness because it determines the required computing power and processing (execution) time. A general framework to evaluate the computational requirements of image filtering algorithms based on a fixed processing window is given in References 93 and 88. The requirement of this approach is that the filter window W is symmetric ($k \times k$) and contains k^2 vector samples of dimension L.

TABLE 12.1

Number of Possible Simple Digital
Paths Ω in Dependence on Path
Length η

n	1	2	3	4
DPA	8	56	368	2336
FDPA	8	24	56	69

In most image processing applications a value $k = 3$ is considered, while for color RGB images $L = 3$.

The computational complexity of a specific filter is given in terms of the total execution time needed for a complete filtering cycle. The total time is calculated as: TIME $= \sum w_{\text{OPER}} \cdot \text{OPER}$, where OPER is the number of particular operations required for a complete cycle, and w_{OPER} is the relative weight of this operation. In the following analysis the several operations are used: ADDS (additions), MULTS (multiplications), DIVS (divisions), SQRTS (square roots), COMPS (comparisons), ARCCOS (arc cosines), and EXPS (exponents). Most of the time w_{ADDS} is assumed to be 1, while other w_{OPER} values depend on the computing platform. The determination of the weights of different operations is beyond the scope of this chapter.

Because the structure of the new filters is not based on a fixed window, the methodology presented in References 93 and 88 cannot be directly applied to evaluate the new filter class complexity. The computational burden of the proposed filters depends mostly on the number of possible digital paths, which in turn depends on the path length. For a given path of length η, the number of simple paths Ω can be easily computed. Table 12.1 depicts the number of possible paths corresponding to the DPA and FDPA filters.[114, 131–133]

The complexity of the DPA and FDPA filters can be determined as follows:[131, 132]

1. Filtering of 1 pixel requires computation of all weights w^* (see point 2), $L(\Omega - 1)$ additions, and $L \cdot \Omega$ multiplications.

2. Computation of all weights w^* requires computation of all similarity functions $w^{W,\eta}$ (see point 3), Ω divisions, and $(\Omega - 1)$ additions.

3. Computation of all similarity functions $w^{W,\eta}$ requires Ω computations of distance $\Lambda_m^{W,\eta}$ (see point 4), $(\Omega - 1)$ additions, Ω multiplications, and Ω computations of an exponent.

4. Computation of one distance $\Lambda_m^{W,\eta}$ along path m requires n computations of Euclidean distance (if the L_2 metric is used) and $(\eta - 1)$ additions.

5. Computation of one particular Euclidean distance requires L multiplications, $2L$ additions, and 1 square root.

TABLE 12.2

Number of Elementary Operations for a Complete Processing Cycle

Filter	ADDS	MULTS	DIVS	SQRTS	EXPS	COMPS	ARCCOS	Total
DPA_2	947	228	56	112	56	—	—	1399
DPA_3	8827	1478	368	1104	368	—	—	12145
$FDPA_2$	403	100	24	48	24	—	—	599
$FDPA_3$	1139	230	56	168	56	—	—	1649
$FDPA_2^*$	169	22	24	9	24	—	—	**248**
$FDPA_3^*$	721	24	56	9	56	—	—	866
$VMF_{3\times3}$	186	63	—	21	—	8	—	**278**
$VMF_{5\times5}$	855	330	—	110	—	24	—	1319
$BVDF_{3\times3}$	375	210	21	21	—	8	21	656
$BVDF_{5\times5}$	1970	1100	110	110	—	24	110	3424
$DDF_{3\times3}$	540	282	21	42	—	8	21	914
$DDF_{5\times5}$	2785	1455	110	220	—	24	110	4704

Thus, the total number of operations needed to implement the filters is

$$(2\eta L\Omega + \Omega p + L\Omega - L - 2) \cdot \text{ADDS} + (\Omega + L\Omega + 2\eta)$$
$$\times \text{MULTS} + \Omega \cdot \text{DIVS} + \Omega\eta \cdot \text{SQRTS} + \Omega \cdot \text{EXPS}. \qquad (12.82)$$

Using the framework of Reference 93 and assuming that the size of the processing window is $(k \times k)$, the computational complexity for the VMF, BVDF, and DDF can be evaluated (Table 12.2).

It should be emphasized at this point that the computational complexity analysis of the new filter was based on straightforward application of the described algorithms without any consideration of a particular implementation. However, it is possible to significantly reduce the computational complexity of the proposed filters. To illustrate this, the FDPA filter is considered. The analysis of the filtering structure reveals that the L_2 distance should be evaluated η times for each path of length η. If the total number of paths in the supporting window is Ω, the number of L_2 norm evaluations is $(\Omega \cdot \eta)$. However, most of these calculations are unnecessary because values already computed for other paths can be used. For example, in a (3×3) window there are only 20 possible distances to be calculated. These values can be computed and stored in order to be used to determine the path-related weights for a neighboring pixel. Furthermore, other techniques used to improve the performance of the VMF presented in Reference 93 can be applied in the DPA or FDPA filter design.

Table 12.2 summarizes the total number of operations for different filters, with DPA_η denoting the basic DPA filter of length η, $FDPA_\eta$ denoting straightforward application of FDPA algorithms, and $FDPA_\eta^*$ the optimized version of FDPA. As can be seen, the fast implementation of the proposed filter is computationally more attractive than the VMF and it significantly outperforms filters based on angular distances.

TABLE 12.3

Filters Taken for Comparison with the Proposed Noise
Reduction Techniques

Notation	Method	Ref.
AMF	Arithmetic mean filter	5
VMF	Vector median filter	92
BVDF	Basic vector directional filter	138, 139
GVDF	Generalized vector directional filter	100
DDF	Directional-distance filter	103
HDF	Hybrid directional filter	95
AHDF	Adaptive hybrid directional filter	95
FVDF	Fuzzy vector directional filter	110
ANNF	Adaptive nearest neighbor filter	111, 137
ANP-E	Adaptive nonparametric (exponential) filter	5, 105
ANP-G	Adaptive nonparametric (Gaussian) filter	5, 105
ANP-D	Adaptive nonparametric (Directional) filter	5, 105
VBAMMF	Vector Bayesian adaptive median/mean filter	5, 105
AD	Perona–Malik anisotropic diffusion filter with c_1	29, 53
GD-PDE	Geometric diffusion PDE	79

12.6   Efficiency of the New Filter Class

In this section the performance of the new filter class is evaluated comparing
the results with some of the noise reduction techniques listed in Table 12.3,
using synthetic and natural color images corrupted by Gaussian and mixed
Gaussian and impulsive noise.

12.6.1   Simulations Performed on Artificial Images

The use of nonlinear filters in color image processing is motivated primarily
by the good performance of the filters near edges and other sharp signal
transitions. Edges are basic image features that carry valuable information,
useful in image analysis and object classification. Therefore, any nonlinear
noise reduction operator is required to preserve edges and smooth out noise
without altering sharp signal transitions.

   In this section some examples of the efficiency of the new filter class are presented to illustrate its excellent noise-reduction properties. To quantitatively
evaluate the behavior of the proposed algorithms, two color synthetic images
were prepared. To examine the performance of the new filters in case of an
artificial step edge, a three-channel image called *SQUARE* of size (60×60)
containing a square of size (30×30) was generated (Figure 12.11a). Further,
for the evaluation of the filter performance in case of a ramp edge, a synthetic

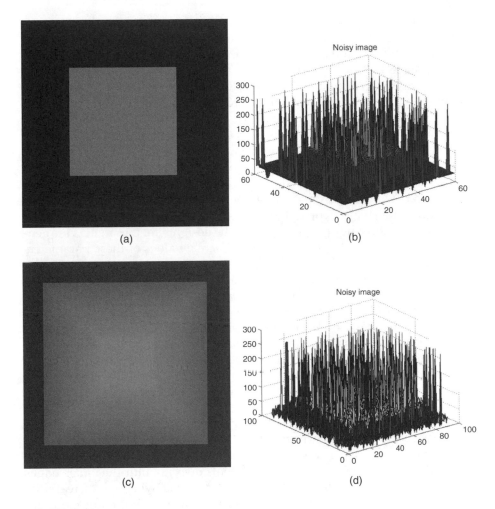

FIGURE 12.11
(a) Test image *SQUARE*, (b) *SQUARE* image corrupted by impulsive noise (green channel),
(c) test image *PYRAMID*, (d) *PYRAMID* image corrupted by mixed impulsive and Gaussian
noise, (green channel).

test image called *PYRAMID* was constructed. The three-channel image of size
(90×90) contains a top-cut pyramid, which is used to simulate a "ramp-edge"
scenario (Figure 12.11c).

The test image *SQUARE* was corrupted by multivariate impulsive noise
following the model given by Equation 12.1 in Section 12.1, with the degree
of contamination $p = 0.1$ and $p_1 = p_2 = p_3 = 0.25$ (Figure 12.11b). The test
image *PYRAMID* was corrupted by mixed impulsive noise with $p = 0.1$ and
$p_1 = p_2 = p_3 = 0.25$ and $\sigma = 20$ (Figure 12.11d).

The new techniques based on the DPA (DPAF, DPAL) and the FDPA algorithms were compared in terms of objective quality criteria with the VMF, with the AMF, with the classic P-M AD, and with other filtering techniques listed in Table 12.3.

In the DPAF, DPAL, and FDPA filters, the paths of length $\eta = 2$ with design parameters set at $\beta = 20$ and $\alpha = 1.2$ were used. The AMF and VMF operated on a filtering window of size (3×3). The anisotropic diffusion filter used in the experiments denoted as AD is a vector implementation of the P-M AD, which utilizes the conductivity function c_1 (Equation 12.14).[29,56] For the AD filter the parameters that gave the best results in terms of PSNR were used.

It should be pointed out that the parameters used for the FDPA, DPAF, and DPAL filters were not optimal and in the majority of cases better results can be obtained for images corrupted by a specific noise process. However, in practical situations the optimal values of the design filter parameters are generally unknown, and therefore the experimental values of these parameters were used.

In case of images corrupted with Gaussian noise, the AMF as expected gave better results than the VMF, especially in the flat homogeneous regions, but it heavily blurred the image edges. Classic P-M AD gives good results for images corrupted with Gaussian noise of low intensity, but it requires many iterations to smooth the image before its performance can be comparable with the new filter class in terms of objective quality criteria. In case of images distorted by a Gaussian noise process with high σ, the P-M approach is not able to suppress the spikes, which leads to a poor overall performance.

The experimentations with images corrupted by mixed Gaussian and impulsive noise revealed as expected that the AMF filter introduces extensive smoothing into the image and impulses are still visible as blurred "bumps." AD with parameters used in the experiments does not blur the image edges but it leaves impulses almost unchanged (of course, when we increase the threshold parameter K in Equation 12.14 we can smooth the noise out but then the AD will also destroy the image edges). The VMF efficiently reduces the noise component but tends to blur the edges and produces color blotches in flat image regions. The results obtained using the DPAF, DPAL, and FDPA filters confirm their excellent properties in the case of images corrupted by both impulsive and Gaussian noise.

The new filtering structure gives excellent results both in flat regions and also at the edges (see Figure 12.12 through Figure 12.14). The results obtained with anisotropic diffusion and with filters proposed in this chapter are quite similar in the case of images corrupted by low-intensity Gaussian noise. Both the schemes provide efficient smoothing in homogeneous image regions and achieve excellent edge preservation. However, the new filters achieve the goal much faster and work efficiently even when the intensity of the Gaussian noise is high (Figure 12.15). For images corrupted with mixed

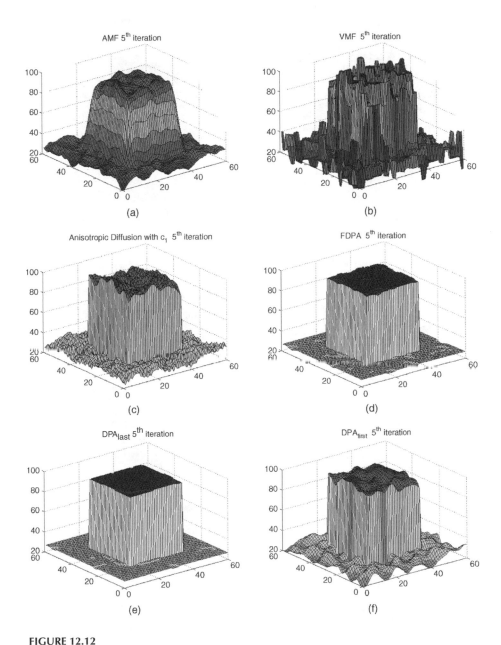

FIGURE 12.12
Three-dimensional representation of the results of noise attenuation in the green channel of the *SQUARE* image corrupted by impulsive noise, using the standard and new techniques: (a) AMF, (b) VMF, (c) AD, (d) FDPA, (e) DPAL, and (f) DPAF (five iterations, $\eta = 2$).

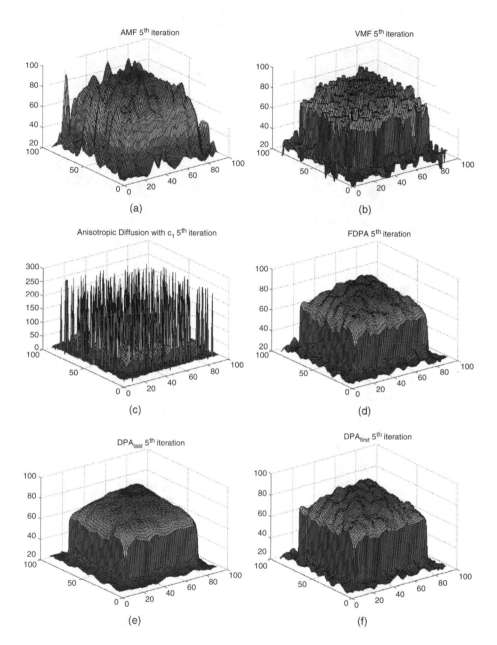

FIGURE 12.13
Three-dimensional representation of the results of noise attenuation in the the green channel of the *PYRAMID* image corrupted by mixed Gaussian and impulsive noise using the standard and new techniques: (a) AMF, (b) VMF, (c) AD, (d) FDPA, (e) DPAL, and (f) DPAF (five iterations, $\eta = 2$).

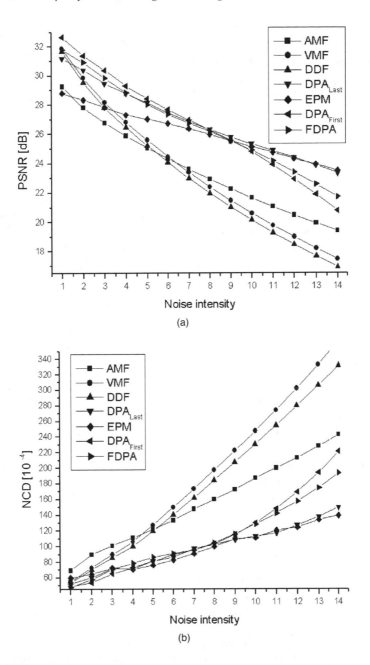

FIGURE 12.14
Comparison of the efficiency of the standard filter's efficiency with the new filter class in terms of (a) PSNR and (b) NCD for different amounts of noise (mixed Gaussian and impulsive noise intensities, $p = 0.01$ to 0.12, $p_1 = p_2 = p_3 = 0.3$), (c). The escaping particle model (EPM) denotes a path in which, with every step, the distance between the current point and the origin increases. (*Continued*)

Noise intensity	1	2	3	4	5	6	7	8	9	10	11	12	13	14
Gaussian σ	5	10	15	20	25	30	35	40	45	50	55	60	65	70
Impulsive [%]	1	2	3	4	5	6	7	8	9	10	11	12	13	14

(c)

FIGURE 12.14
(*Continued*)

Gaussian and impulsive noise, neither the VMF nor the AMF provide acceptable results. While the anisotropic diffusion filter smoothes out only the Gaussian noise component and AMF introduces blurring, the DPAF, DPAL, and FDPA filter's performance is excellent. The new filters remove outliers introduced by impulsive noise, and smooth flat noisy regions leaving the edges of the objects almost unchanged. The simulations performed on the synthetic images revealed the following:

- The VMF performs poorly in the presence of Gaussian noise.
- The AMF works well in homogeneous regions with additive Gaussian noise.
- The classic P-M AD scheme performs well in images corrupted by low-intensity Gaussian noise, but fails in the presence of impulsive noise.
- The proposed filtering class is able to suppress Gaussian as well as mixed Gaussian and impulsive noise in homogeneous regions and also near edges. The obtained results confirm the much better performance of the new filters when compared to the AMF, VMF, and P-M AD scheme.

12.6.2 Filter Performance for Natural Color Images

The noise attenuation properties of different filters are examined using the color test image *LENA*, which has been contaminated by Gaussian and mixed Gaussian and impulsive noise in order to compare the new filters with the filtering techniques listed in Table 12.3. The test images were contaminated by additive Gaussian noise of $\sigma = 30$ and also by mixed impulsive ($p = 0.12$, $p_1 = p_2 = p_3 = 0.3$) and Gaussian noise of $\sigma = 30$. As the results for *LENA* and *PEPPERS* are consistent, only the results obtained with the *LENA* image are reported here.

The root mean squared error (RMSE), SNR, PSNR, normalized mean square error (NMSE), and the normalized color difference (NCD)[5] are used for the

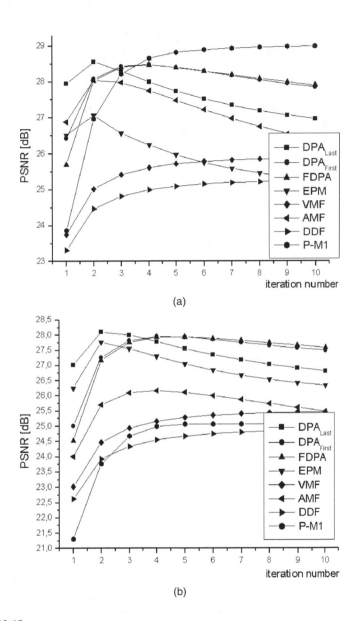

FIGURE 12.15
Plots of PSNR in subsequent iterations for various filters applied to color *LENA* image contaminated with Gaussian, $\sigma = 30$ (a) and mixed impulsive and Gaussian noise, $\sigma = 30$, $p = 0.12$, $p_1 = p_2 = p_3 = 0.3$ (b).

analysis. The objective quality measures are defined by the following formulas:

$$\text{RMSE} = \sqrt{\frac{1}{NML} \sum_{i=1}^{N} \sum_{j=1}^{M} \sum_{l=1}^{L} (F^l(i, j) - \hat{F}^l(i, j))^2},$$

$$\text{NMSE} = \frac{\sum_{i=1}^{N} \sum_{j=1}^{M} \sum_{l=1}^{L} (F^l(i, j) - \hat{F}^l(i, j))^2}{\sum_{i=1}^{N} \sum_{j=1}^{M} \sum_{l=1}^{L} \hat{F}^l(i, j)^2},$$

(12.83)

$$\text{SNR} = 10 \log \left[\frac{\sum_{i=1}^{N} \sum_{j=1}^{M} \sum_{l=1}^{L} F^l(i, j)^2}{\sum_{i=1}^{N} \sum_{j=1}^{M} \sum_{l=1}^{L} (F^l(i, j) - \hat{F}^l(i, j))^2} \right],$$

$$\text{PSNR} = 20 \log \left(\frac{255}{\text{RMSE}} \right),$$

(12.84)

where M, N are the image dimensions, and $F^l(i, j)$ and $\hat{F}^l(i, j)$ denote the lth component of the original image vector and its estimation at pixel position (i, j), respectively. The NCD perceptual measure is evaluated over the uniform $L^*u^*v^*$ color space. The difference measure NCD is defined as

$$\text{NCD} = \frac{\sum_{i=1}^{N} \sum_{j=1}^{M} \Delta E}{\sum_{i=1}^{N} \sum_{j=1}^{M} E^*}, \quad \Delta E = [(\Delta L^*)^2 + (\Delta u^*)^2 + (\Delta v^*)^2]^{1/2},$$

$$E^* = [(L^*)^2 + (u^*)^2 + (v^*)^2]^{1/2},$$

(12.85)

where ΔE is the perceptual color error and E^* is the *norm* or *magnitude* of the uncorrupted original color image pixel in the $L^*u^*v^*$ space.

Results obtained using the new filtering techniques are compared with the filtering algorithms from Table 12.3 in Table 12.4 and Table 12.5. For the denoising of both contaminated *LENA* images with the new filtering techniques, predefined parameter values were used: path length $\eta = 2$, $\beta = 13$, $\alpha = 1.2$. For all evaluated filters, 10 iterations were performed, and the best result in terms of PSNR is presented in Table 12.4 and Table 12.5.

Figure 12.10 depicts the efficiency of the proposed algorithms (DPAL and FDPA) in terms of NCD quality measure, as a function of the design parameters α and β. It can be easily noticed that both algorithms yield comparable results with a flat minimum of NCD, which ensures their robustness to optimal parameter settings. The parameter α ensures quick convergence of the proposed filters to a stable state, and as can be seen in Figure 12.10, good results can be obtained for any α in the range [1, 2].

Figure 12.16 presents the efficiency of the DPAL filter applied to a scanned road map. The new filtering technique was able to remove the raster structure, while image details such as roads, names, etc. were preserved and

TABLE 12.4

Comparison of the Efficiency of the New Algorithms with Different Techniques (Table 12.3), Using the *LENA* Standard Color Image Corrupted by Gaussian Noise of $\sigma = 30$

Filter	NMSE [10^{-3}]	RMSE	SNR [dB]	PSNR [dB]	NCD [10^{-4}]
NONE	420.55	29.075	13.762	18.860	250.090
AMF	66.452	11.558	21.775	26.873	95.347
VMF	87.314	13.248	20.589	25.688	117.170
BVDF	279.54	23.705	15.536	20.634	117.400
GVDF	76.713	12.418	21.151	26.250	84.876
DDF	100.50	14.213	19.979	25.077	108.960
HDF	66.584	11.569	21.766	26.865	92.769
AHDF	60.166	10.997	22.206	27.305	91.369
FVDF	57.466	10.748	22.406	27.504	77.111
ANNF	63.341	11.284	21.983	27.082	82.587
ANP-E	60.396	11.018	22.190	27.288	76.896
ANP-G	60.443	11.023	22.187	27.285	76.890
ANP-D	58.389	10.834	22.337	27.435	78.486
AD	41.434	9.126	23.826	28.925	69.482
GD-PDE	34.530	8.296	24.618	29.753	72.100
DPAF	42.873	9.244	23.678	28.813	82.814
DPAL	43.005	9.258	23.665	28.800	77.932
FDPA	44.913	9.462	23.476	28.611	84.918

TABLE 12.5

Comparison of the New Algorithms with the Techniques from Table 12.3, Using the *LENA* Color Image Corrupted by Mixed Gaussian and Impulsive Noise ($\sigma = 30$, $p = 0.12$, $p_1 = p_2 = p_3 = 0.25$)

Filter	NMSE [10^{-3}]	RMSE [dB]	SNR	PSNR [dB]	NCD [10^{-4}]
NONE	905.93	42.674	10.429	15.528	305.55
AMF	97.444	13.996	20.112	25.211	95.80
VMF	96.464	13.925	20.156	25.255	121.79
BVDF	336.46	26.006	14.731	19.829	123.93
GVDF	91.118	13.534	20.404	25.503	89.277
DDF	110.62	14.912	19.561	24.660	113.39
HDF	74.487	12.236	21.279	26.378	97.596
AHDF	68.563	11.740	21.639	26.738	96.327
FVDF	108.76	14.786	19.635	24.734	111.22
ANNF	75.652	12.332	21.212	26.310	86.836
ANP-E	90.509	13.488	20.433	25.532	97.621
ANP-G	90.523	13.489	20.432	25.531	97.603
ANP-D	74.203	12.213	21.296	26.394	85.026
AD	339.55	26.125	14.691	19.790	113.65
GD-PDE	59.371	10.924	22.264	27.363	77.510
DPAF	50.804	10.106	22.941	28.040	76.076
DPAL	49.999	10.025	23.010	28.109	72.851
FDPA	53.573	10.377	22.711	27.809	78.666

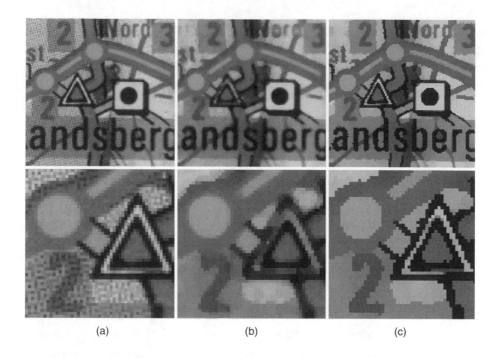

(a)   (b)   (c)

FIGURE 12.16
Comparison of efficiency of the vector median with the DPAF: (a) test image (part of a scanned map); (b) VMF (3×3 mask); (c) DPAF ($\beta = 20$, $\alpha = 1.25$, $\eta = 2$; 3 iterations).

even enhanced. The VMF gives much worse results: raster texture is still visible and image details are blurred.

Table 12.4 and Table 12.5 indicate that the new filters yield especially good results in the case of images corrupted by mixed Gaussian and impulsive noise. In addition to excellent noise attenuation properties, the new filters restore the noisy images so that they have well-preserved and even enhanced edges and corners, which make them interesting for many different computer vision applications (Figure 12.17).

The best results for the Gaussian and mixed noise attenuation for the majority of existing filters were obtained after many iterations, while for filters based on the digital paths concept the best results were achieved in the second or third iteration (see Figure 12.15).

The comparison of the new filters' efficiency with some of the standard filters is presented in Figure 12.14, where for different filters, the PSNR and NCD dependence on the amount of mixed impulsive and Gaussian noise is shown. As the intensity of the noise increases, the quantitative results obtained using the new filters become significantly better than those obtained by the standard filters (AMF, VMF, DDF).

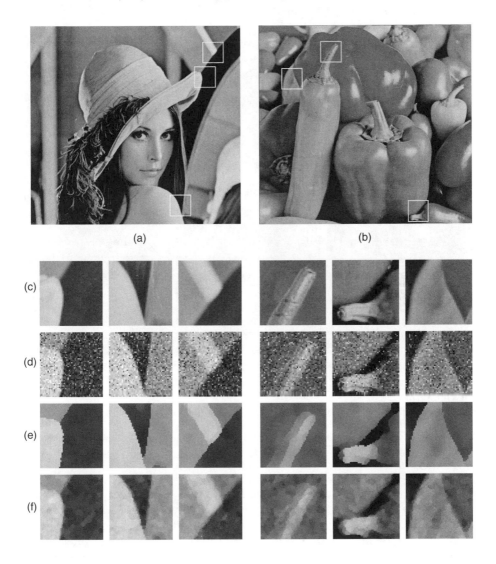

FIGURE 12.17
Color test images *LENA* (a) and *PEPPERS* (b) with depicted regions of interest (c). The chosen image regions were contaminated by mixed impulsive ($p = 0.12$, $p_1 = p_2 = p_3 = 0.3$) and Gaussian noise of $\sigma = 30$ (d) and then restored with the DPAF method (e) and VMF (f).

The simulations revealed that in the case of both Gaussian and mixed Gaussian and impulsive noise, very good results were obtained using the method GP-PDE, presented in References 79 and 80, which is based on the gradient norm described in Section 12.3.1. The visual comparison between the FDPA and the algorithm GP-PDE[79,80] is shown in Figure 12.18.

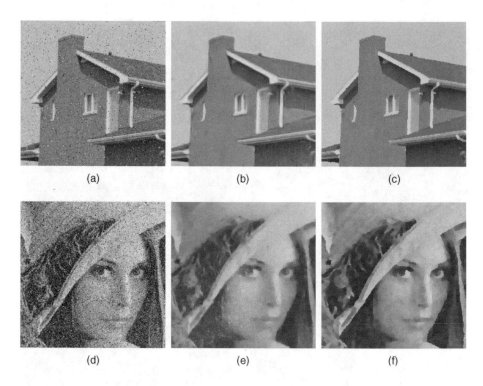

FIGURE 12.18
Comparison of the method proposed in References 79 and 80 with the new approach (DPAF): (a) test image *HOUSE* contaminated with impulsive noise ($p = 0.1$); (b) GD-PDE;[79,80] (c) DPAF; (d) test image *LENA* contaminated with mixed impulsive and Gaussian noise; (e) GD-PDE; (f) DPAF.

In conclusion, from the results listed in the tables and shown in the figures, it easily can be seen that the new filters, especially the FDPA filter, provide consistently good results. The DPAF, DPAL, and FDPA filters can be seen as universal filters, able to attenuate different types of noise while preserving image edges and corners. Simulation results show that the new class of filters yields favorable noise reduction results for various kinds of color images in comparison with the standard adaptive smoothing algorithms.

Acknowledgment

The contribution of Rachid Deriche and David Tschumperlé, who evaluated the GP-PDE algorithm[79,80] on a set of noisy images used in this chapter, is gratefully acknowledged.

References

1. G. Wyszecki and W.S. Stiles, *Color Science: Concepts and Methods, Quantitative Data and Formulae*, 2nd ed., New York: John Wiley & Sons, 1982.
2. I. Pitas, *Digital Image Processing Algorithms*, Englewood Cliffs, NJ: Prentice-Hall, 1993.
3. I. Pitas, *Digital Image Processing Algorithms and Applications*, New York: Wiley-Interscience, 2000.
4. I. Pitas and A.N. Venetsanopoulos, *Nonlinear Digital Filters: Principles and Applications*, Dordrecht: Kluwer, 1990.
5. K.N. Plataniotis and A.N. Venetsanopoulos, *Color Image Processing and Applications*, New York: Springer-Verlag, 2000.
6. S.J. Sangwine and R.E.N. Horne, Eds., *The Colour Image Processing Handbook*, London: Chapman & Hall, 1998.
7. L. Yaroslavsky and E. Murray, *Fundamentals of Digital Optics*, Boston: Birkhäuser, 1996.
8. S.K. Mitra and G. Sicuranza, Ed., *Nonlinear Image Processing (Communications, Networking and Multimedia)*, San Diego, CA: Academic Press, 2000.
9. T. Viero, K. Oistamo, and Y. Neuvo, Three-dimensional median-related filters for color image sequence filtering, *IEEE Trans. Circuits Syst. Video Technol.*, 4(2), 129–142, 1994.
10. M.I. Vardavoulia, I. Andreadis, and Ph. Tsalides, A new vector median filter for colour image processing, *Pattern Recognition Lett.*, 22, 675–689, 2001.
11. R.C. Gonzalez and R.E. Woods, *Digital Image Processing*, Reading, MA: Addison-Wesley, 1992.
12. M.J. McDonnell, Box-filtering techniques, *Comput. Graphics Image Process.*, 17, 65–70, 1981.
13. S. Haykin, *Adaptive Filter Theory*, Englewood Cliffs, NJ: Prentice-Hall, 1991.
14. K. Erler and E. Jernigan, Adaptive image restoration using recursive image filters, *IEEE Trans. Signal Process.*, 42(7), 1877–1881, July 1994.
15. R. Klette and P. Zamperoni, *Handbuch der Operatoren für die Bildverarbeitung*, Braunschweig: Vieweg Verlag, 1992.
16. A. Scher, V. Dias, and F.R. Rosenfeld, Some new image smoothing techniques, *IEEE Trans. Syst, Man Cybern.* 10, 153–158, March 1980.
17. D. Wang, A.H. Vagnucci, and C.C. Li, Gradient inverse weighted smoothing scheme and the evaluation of its performance, *Comput. Graphics Image Process.*, 15, 167–181, 1981.
18. J.S. Lee, Digital image enhancement and noise filtering by use of local statistics, *IEEE Trans. Pattern Anal. Mach. Intelligence*, 2, 165–168, 1980.
19. D. Wang and Q. Wang, A weighted averaging method for image smoothing, in *Proceedings of the 8th ICPR*, Paris, 1988, 981–983.
20. G.A. Mastin, Adaptive filters for digital image noise smoothing: an evaluation, *Comput. Vision Graphics Image Process.*, 31, 103–121, 1985.
21. J.S. Lee, Refined filtering of image noise using local statistics, *Comput. Graphics Image Process.*, 15, 380–389, 1981.
22. J.S. Lee, Digital image smoothing and the sigma filter, *Comput. Vision Graphics Image Process.*, 24, 255–269, 1983.

23. P. Saint-Marc, J.S. Chen, and G. Medioni, Adaptive smoothing: a general tool for early vision, in *Proceedings of the IEEE Conference on Computer Vision and Pattern Recognition,* San Diego, 1989, 618–624.

24. P. Saint-Marc, J.S. Chen, and G. Medioni, Adaptive smoothing: a general tool for early vision, *IEEE Trans. Pattern Anal. Mach. Intelligence,* 13(6), 514, June 1991.

25. S.M. Smith and J.M. Brady, SUSAN—a new approach to low level image processing, *Int. J. Comput. Vision,* 23(1), 45–78, 1997.

26. C. Tomasi and R. Manduchi, Bilateral filtering for gray and color image, in *Proc. of the IEEE International Conference on Computer Vision,* Bombay, India, 1998.

27. K. Chen, A feature-preserving adaptive smoothing method for early vision, Technical Report, National Laboratory of Machine Perception and Center for Information Science, Peking University, Beijing, China.

28. D. Barash, A fundamental relationship between bilateral filtering, adaptive smoothing, and the nonlinear diffusion equation, *IEEE Trans. Pattern Anal. Machine Intelligence,* 24(6), 844–847, June 2002.

29. P. Perona and J. Malik, Scale space and edge detection using anisotropic diffusion, *IEEE Trans. Pattern Anal. Mach. Intelligence,* 12, 629–639, 1990.

30. M. Spann and A. Nieminen, Adaptive Gaussian weighted filtering for image segmentation, *Pattern Recognition Lett.,* 8, 251–255, 1988.

31. S. Guillon, P. Baylou, M. Najim, and N. Keskes, Adaptive non-linear filters for 2D and 3D image enhancement, *Signal Process.,* 67, 237–254, 1998.

32. G. Lohmann, *Volumetric Image Analysis,* New York: John Wiley/Teubner, 1988.

33. L.S. Davis and A. Rosenfeld, Noise cleaning by iterated local averaging, *IEEE Trans. Syst. Man Cybern.,* SMC, 8, 705–710, 1978.

34. M. Kuwahara, K. Hachimura, S. Eiho, and M. Kinoshita, *Digital Processing of Biomedical Images,* New York: Plenum Press, 1976, 187–203.

35. F. Tomita and S. Tsuji, Extraction of multiple regions by smoothing in selected neighbourhoods, *IEEE Trans. Syst., Man Cybern.,* 7, 107–109, 1977.

36. D. de Ridder, R.P.W. Duin, P.W. Verbeek, and L.J. van Vliet, The applicability of neural networks to non-linear image processing, *Pattern Anal. Appl.,* 2, 111–128, 1999.

37. M. Nagao and T. Matsuyama, Edge preserving smoothing, *Comput. Graphics Image Process.,* 9, 394–407, 1979.

38. M. Nagao and T. Matsuyama, *A Structural Analysis of Complex Aerial Photographs,* New York: Plenum Press, 1980.

39. A. Somboonkaew, S. Chitwong, F. Cheevasuvit, K. Dejhan, and S. Mitatha, Segmentation on the edge preserving smoothing image, in *Proceedings of 20th Asian Conference on Remote Sensing,* Nov. 22–25, Hong Kong, China, 1999.

40. G.R. Arce, A generalized weighted median filter structure admitting negative weights, *IEEE Trans. Signal Process.,* 46, Dec. 1998.

41. G.R. Arce, N.C. Gallagher, Jr., and T.A. Nodes, Median filters: theory and aplications, in *Advances in Computer Vision and Image Processing,* Vol. 2, T.S. Huang, Ed., Greenwich, CT: JAI Press, 1986.

42. G.R. Arce and J. Paredes, Image enhancement with weighted medians, in *Nonlinear Image Processing,* S. Mitra and G. Sicuranza, Eds., San Diego, CA: Academic Press, 2000, 2767.

43. K.E. Barner and R.C. Hardie, Spatial rank order selection filters, in *Nonlinear Image Processing,* S.K. Mitra and G. Sicuranza, Eds., San Diego, CA: Academic Press, 2001.

44. T. Chen and H.R. Wu, Adaptive impulse detection using center-weighted median filters, *IEEE Signal Process. Lett.*, 8(1), 1–3, Jan. 2001.

45. E. Abreu and S.K. Mitra, A signal-dependent rank-ordered mean (SD-ROM) filter—a new approach for removal of impulses from highly corrupted images, *Proc. Int. Conf. Accoust. Speech Signal Process.*, 4, 2371–2374, 1995.

46. E. Abreu, Signal-dependent rank-ordered mean (SD-ROM) filter, in *Nonlinear Image Processing*, S.K. Mitra and G.L. Sicuranza, Eds., San Diego, CA: Academic Press, 2000, 111–134.

47. S.J. Ko and Y.H. Lee, Center weighted median filters and their applications to image enhancement, *IEEE Trans. Circuits Syst.*, 38, 984–993, Sept. 1991.

48. M. Schulze and J.A. Pearce, Some properties of the two-dimensional pseudo-median filter, *Proc. SPIE*, 1451, 48–57, 1991.

49. L. Alparone, M. Barni, F. Bartolini, and V. Cappellini, Adaptively weighted vector-median filters for motion-fields smoothing, *Proc. IEEE Int. Conf. Accoust., Speech Signal Processing*, May 7–10, 1996, 2267–2270.

50. D.R.K. Brownrigg, The weighted median filter, *Commun. Assoc. Comput.*, 807–818, 1984.

51. F. Catte, F. Dibos, and G. Koepfler, Image selective smoothing and edge detection by nonliner diffusion, *SIAM J. Numer. Anal.*, 29(1), 182–193, Feb. 1992.

52. F. Catte, P.-L. Lions, J.M. Morel, and T. Coll, Image selective smoothing and edge detection by nonlinear diffusion-II, *SIAM J. Numer. Anal.*, 29(3), 845–866, 1992.

53. P. Perona and J. Malik, Scale space and edge detection using anisotropic diffusion, in *Proceedings of IEEE Workshop on Computer Vision*, Miami, 1987, 16–22.

54. A. Witkin, Scale-space filtering, in *Proc. of Int. Joint Conf Artif. Intell.*, 1983, 1019–1021.

55. G. Sapiro and D.L. Ringach, Anisotropic diffusion of multivalued images with applications to color filtering, *IEEE Trans. Image Processing*, 5(11), 1582–1586, Nov. 1996.

56. G. Gerig, R. Kikinis, O. Kuebler, and F. Jolesz, Nonlinear anisotropic filtering of MRI data, *IEEE Trans. Med. Imaging*, 11(2), 221–232, June 1992.

57. S.T. Acton, Multigrid anisotropic diffusion, *IEEE Trans. Image Process.*, 7(3), 280–291, Mar. 1998.

58. S. Biswas, N.R. Pal, and S.K. Pal, Smoothing of digital images using the concept of diffusion process, *Pattern Recognition*, 29(3), 497–510, 1996.

59. J. Maeda, T. Iizawa, T. Ishizaka, C. Ishikawa, and Y. Suzuki, Segmentation of natural images using anisotropic diffusion and linking of boundary edges, *Pattern Recognition*, 31(12), 1993–1999, Dec. 1998.

60. F. Torkamani-Azar and K.E. Tait, Image recovery using the anisotropic diffusion equation, *IEEE Trans. Image Process.*, 5(11), 1573–1578, Nov. 1996.

61. Y.L. You, W. Xu, A. Tannenbaum, and M. Kaveh, Behavioral analysis of anisotropic diffusion in image processing, *IEEE Trans. Image Process.*, 5(11), 1539–1553, 1996.

62. B. Fischl and E.L. Schwartz, Learning an integral equation approximation to nonlinear anisotropic diffusion in image processing, *IEEE Trans. Pattern Anal. Machine Intelligence*, 19(4), 342–352, Apr. 1997.

63. B.B. Kimia and K. Siddiqi, Geometric heat equation and nonlinear diffusion of shapes and images, *Comput. Vision Image Understanding*, 64(3), 305–322, Nov. 1996.

64. B.B. Kimia and K. Siddiqi, Geometric heat equation and non-linear diffusion of shapes and images, in *IEEE Computer Society Conference on Computer Vision and Pattern Recognition,* 1994, 113–120.

65. J. Shah, A common framework for curve evolution, segmentation and anisotropic diffusion, in *Proc. of IEEE Conf. on Computer Vision and Pattern Recognition,* June 1996, 136–142.

66. M.J. Black, G. Sapiro, D. H. Marimont, and D. Heeger, Robust anisotropic diffusion, *IEEE Trans. Image Process.,* 7(3), Mar. 1998, 421–432.

67. B. Romeny ter Haar, *Geometry-Driven Diffusion in Computer Vision,* Boston: Kluwer, 1994.

68. G. Sapiro, *Geometric Partial Differential Equations and Image Analysis,* Cambridge, U.K.: Cambridge University Press, 2001.

69. L. Alvarez, F. Guichard, P.L. Lions, and J.M. Morel, Image selective smoothing and edge detection by nonliner diffusion, *SIAM J. Numer. Anal.,* 29(3), 845–866, 1992.

70. P. Charbonnier, G. Aubert, M. Blanc-Ferraud, and M. Barlaud, Two-deterministic half-quadratic regularization algorithms for computed imaging, in *Proc of IEEE International Conf. on Image Processing ICIP,* Austin, TX, Nov. 1994, 168–172.

71. L.I. Rudin, S. Osher, and E. Fatemi, Nonlinear total variation based noise removal algorithms, *Phys. D,* 60, 259–268, 1992.

72. G. Ramponi and C. Moloney, Smoothing speckled images using an adaptive rational operator, *IEEE Signal Process. Lett.,* 4(3), 68–71, 1997.

73. J. Weickert, *Anisotropic Diffusion in Image Processing,* Stuttgart: G.G. Teubner, 1998.

74. R. Whitaker and G. Gerig, Vector-valued diffusion, in *Geometry-Driven Diffusion in Computer Vision,* B. Romeny ter Haar, Boston: Kluwer, 1994, 93–134.

75. R.S. Lin and Y.C. Hsueh, Multichannel filtering by gradient information, *Signal Process.,* 80, 279–293, 2000.

76. P. Pujas and M.J. Aldon, Estimation of the colour image gradient with perceptual attributes, in *Proceedings of the 9th International Conference on Image Analysis and Processing,* New York: Springer-Verlag, Sept. 1997, 103–110.

77. R.A. Carmona and S. Zhong, Adaptive smoothing respecting feature directions, *IEEE Transa. Image Process.,* Special Issue, 7(3), 353–358, 1998.

78. S. DiZenzo, A note on the gradient of multi-image, *Comput. Vision Graphics Image Process.,* 33, 116–125, 1986.

79. D. Tschumperlé and R. Deriche, Constrained and unconstrained PDE's for vector image restoration, in *Proceedings of the SCIA Conference,* 2001.

80. D. Tschumperlé and R. Deriche, Diffusion PDE's on vector-valued images: local approach and geometric viewpoint, *IEEE Signal Process. Mag.,* Special Issue, Sept. 2002.

81. P. Blpmgren and T.F. Chan, Color TV: total variation methods for restoration of vector-valued images, *IEEE Trans. Image Process.,* Special Issue, 7(3), Mar. 1998, 304–309.

82. A.N. Venetsanopoulos and K.N. Plataniotis, Multichannel image processing, in *Proceedings of the IEEE Workshop on Nonlinear Signal and Image Processing,* 1995, 2–6.

83. P.E. Trahanias, I. Pitas, and A.N. Venetsanopoulos, Color image processing, in *Advances in 2D and 3D Digital Processing (Techniques and Applications)*, C.T. Leondes, Ed., San Diego, CA: Academic Press, 1994.

84. K.N. Plataniotis, D. Androutsos, and A.N. Venetsanopoulos, Multichannel filters for image processing, *Signal Process. Image Commun.*, 9(2), 143–158, 1997.

85. R. Lukac, Color image filtering by vector directional order-statistics, *Pattern Recognition Image Anal.*, 12(3), 279–285, 2002.

86. N. Nikolaidis and I. Pitas, Multichannel *L* filters based on reduced ordering, *IEEE Trans. Circuits Syst. Video Technol.*, 6(5), 470–482, Oct. 1996.

87. V. Barnett, The ordering of multivariate data, *J. R. Stat. Soc. A*, 139(3), 318–355, 1976.

88. K.N. Plataniotis and A.N. Venetsanopoulos, Vector filtering, in *The Colour Image Processing Handbook*, S.J. Sangwine and R.E.N. Horne, Eds., London: Chapman & Hall, 1998, 188–209.

89. K. Tang, J. Astola, and Y. Neuovo, Nonlinear multivariate image filtering techniques, *IEEE Trans. Image Process.*, 4(6), 788–797, June 1995.

90. R.C. Hardie and G.R. Arce, Ranking in R^p and its use in multivariate image estimation, *IEEE Trans. Circuits Syst. Video Technol.*, 1(2), 197–209, 1991.

91. I. Pitas and P. Tsakalides, Multivariate ordering in color image processing, *IEEE Trans. Circuits Syst. Video Technol.*, 1(3), 247–256, 1991.

92. J. Astola, P. Haavisto, and Y. Neuovo, Vector median filters, *IEEE Proc.*, 78, 678–689, 1990.

93. M. Barni and V. Cappellini, On the computational complexity of multivariate median filters, *Signal Process.*, 71, 45–54, 1998.

94. M. Barni, V. Cappellini, and A. Mecocci, Fast vector median filter based on Euclidean norm approximation, *IEEE Signal Process. Lett.*, 1(6), 92–94, 1994.

95. M. Gabbouj and F.A. Cheickh, Vector median–vector directional hybrid filter for colour image restoration, *Proc. EUSIPCO*, 879–881, 1996.

96. B. Smolka, M.K. Szczepanski, K.N. Plataniotis, and A.N. Venetsanopoulos, Fast modified vector median filter, in *Computer Analysis of Images and Patterns*, W. Skarbek, Ed., LNCS 2124, Berlin: Springer-Verlag, 2001, 570–580.

97. B. Smolka, M.K. Szczepanski, K.N. Plataniotis, and A.N. Venetsanopoulos, On the modified weighted vector median filter, in *Proceedings of Digital Signal Processing DSP2002*, Santorini, Greece, 2002.

98. B. Smolka, A. Chydzinski, K. Wojciechowski, K. Plataniotis, and A. Venetsanopoulos, Self-adaptive algorithm of impulsive noise reduction in color images, *Pattern Recognition*, 35, 1771–1784, 2002.

99. R. Wichman, K. Oistamo, Q. Liu, M. Grundstrom, and Y. Neuovo, Weighted vector median operation for filtering multispectral data, in *Proceedings of the Conference Visual Communications and Image Processing*, 1992, 376–383.

100. P.E. Trahanias, D. Karakos, and A.N. Venetsanopoulos, Directional processing of color images: theory and experimental results, *IEEE Trans. Image Process.*, 5(6), 868–880, 1996.

101. R. Lukac, Introducing of the weight concept to vector directional filters, *J. Elect. Eng.*, 52(3–4), 98–100, 2001.

102. R. Lukac, Adaptive impulse noise filtering by using center-weighted directional information, in *Proceedings of the CGIV 2002*, France, 2002, 86–89.

103. D. Karakos and P.E. Trahanias, Generalized multichannel image filtering structures, *IEEE Trans. Image Process.*, 6(7), 1038–1045, 1997.

104. R. Lukac, Optimised directional distance filter, *Machine Graphics Vis.*, Special Issue, 11(2/3), 311–326, 2002.

105. K.N. Plataniotis, D. Androutsos, S. Vinayagamoorthy, and A.N. Venetsanopoulos, Color image processing using adaptive multichannel filters, *IEEE Trans. Image Process.*, 6(7), 933–950, 1997.

106. K.N. Plataniotis, D. Androutsos, and A.N. Venetsanopoulos, Colour image processing using fuzzy vector directional filters, in *Proceedings of the IEEE Workshop on Nonlinear Signal and Image Processing*, Greece, 1995, 535–538.

107. S.A. Durrani, The distance-weighted k-nearest neighbor rule, *IEEE Trans. Syst. Man Cybern.*, 15, 630–636, 1977.

108. K.N. Plataniotis, D. Androutsos, and A.N. Venetsanopoulos, Adaptive multichannel filters for color image processing, *Proc. Visual Commun. Image Process. Conf. SPIE*, 2727(3), 1270–1279, 1996.

109. K.N. Plataniotis, D. Androutsos, and A.N. Venetsanopoulos, Fuzzy adaptive filters for multichannel image processing, *Signal Process. J.*, 55(1), 93–106, 1996.

110. K.N. Plataniotis, D. Androutsos, and A.N. Venetsanopoulos, Fuzzy adaptive filters for multichannel image processing, *Signal Process. J.*, 55(1), 93–106, 1996.

111. K.N. Plataniotis, D. Androutsos, S. Vinayagamoorthy, and A.N. Venetsanopoulos, An adaptive nearest neighbor multichannel filter, *IEEE Trans. Circuits Syst. Video Technol.*, 699–703, Dec. 1996.

112. K. Fukunaga, *Introduction to Statistical Pattern Recognition*, 2nd ed., London: Academic Press, 1990.

113. B. Smolka and K. Wojciechowski, Random walk approach to image enhancement, *Signal Process.*, 81(3), 465–482, 2001.

114. B. Smolka, On the new robust algorithm of noise reduction in color images, *Comput. Graphics*, 11(2/3), 311–326, 2003.

115. K.N. Plataniotis, D. Androutsos, and A.N. Venetsanopoulos, Adaptive fuzzy systems for multichannel signal processing, *Proc. IEEE*, 87(9), 1601–1622, 1999.

116. G. Borgefors, Distance transformations in arbitrary dimensions, *Comput. Vision Graphics Image Process.*, 27, 321–345, 1984.

117. G. Borgefors, Distances transformations in digital images, *Comput. Vision Graphics Image Process.*, 34, 344–371, 1986.

118. F.Y. Shih and J.J. Liu, Size-invariant four-scan Euclidean distance transformation, *Pattern Recognition*, 31(11), 1761–1766, 1998.

119. F. Preteux and N. Merlet, New concepts in mathematical morphology: the topographical distance function, *Proc. SPIE*, 1568, 66–77, 1991.

120. C.O. Kiselman, Regularity properties of distance transformations in image analysis, *Comput. Vision Image Understanding*, 64(3), 390–398, Nov. 1996.

121. N. Madras and G. Slade, *The Self-Avoiding Walk*, Boston: Birkhauser, 1993.

122. F. Spitzer, *Principles of Random Walk*, Princeton, NJ: van Nostrand, 1975.

123. V. Starovoitov, Towards a distance transform generalization, in *Proceedings of the 9th Scandinavian Conference on Image Analysis*, G. Borgefors, Ed., Uppsala, 1995, 499–506.

124. M. Schmitt, Lecture notes on geodesy and morphological measurements, in *Proceedings of the Summer School on Morphological Image and Signal Processing*, Zakopane, Poland, 1995, 36–91.

125. P.J. Toivanen, New geodesic distance transforms for gray scale images, *Pattern Recognition Lett.*, 17, 437–450, 1996.

126. O. Cuisenaire, Distance Transformations: Fast Algorithms and Applications to Medical Image Processing, Ph.D. thesis, Universite Catholique de Louvain, Oct. 1999.

127. R. Kimmel, R. Malladi, and N. Sochen, Images as embedded maps and minimal surfaces: movies, color, texture, and volumetric medical images, *Int. J. Comput. Vision*, 39(2), 111, 2000.

128. N. Sochen, R. Kimmel, and R. Malladi, A geometrical framework for low level vision, *IEEE Trans. Image Process.*, 7(3), 310, 1998.

129. B. Smolka, A. Chydzinski, K. Wojciechowski, K. Plataniotis, and A.N. Venetsanopoulos, On the reduction of impulsive noise in multichannel image processing, *Opt. Eng.*, 40(6), 902–908, 2001.

130. M. Basu and M. Su, Image smoothing with exponential functions, *Int. J. Pattern Recognition Artif. Intelligence*, 15(4), 735–752, 2001.

131. M. Szczepanski, B. Smolka, K.N. Plataniotis, and A.N. Venetsanopoulos, On the distance function approach to color image enhancement, *Discrete Appl. Math.*, in press.

132. M. Szczepanski, B. Smolka, K.N. Plataniotis, and A.N. Venetsanopoulos, On the geodesic paths approach to color image filtering, *Signal Process.*, 83, 1309–1342, 2003.

133. M.K. Szczepanski, B. Smolka, K.N. Plataniotis, and A.N. Venetsanopoulos, Enhancement of the DNA microarray chip images, in *Proceedings of Digital Signal Processing DSP 2002*, Santorini, Greece, 2002.

134. J.C. Bezdek and S.K. Pal, Eds., *Fuzzy Models for Pattern Recognition*, Pisrataway, NJ: IEEE Press, 1992.

135. S. Osher and L.I. Rudin, Feature-oriented image enhancement using shock filters, *SIAM J. Numer. Anal.*, 27(4), 919–940, Aug. 1990.

136. I. Pitas and A.N. Venetsanopoulos, Order statistics in digital image processing, *Proc. IEEE*, 80(12), 1893–1923, 1992.

137. K.N. Plataniotis, D. Androutsos, V. Sri, and A.N. Venetsanopoulos, A nearest neighbour multichannel filter, *Electron. Lett.*, 31(22), 1910–1911, 1995.

138. P.E. Trahanias, D.G. Karakos, and A.N. Venetsanopoulos, Directional processing of color images: theory and experimental results, *IEEE Trans. Image Process.*, 5(6), 868–880, 1996.

139. P.E. Trahanias and A.N. Venetsanopoulos, Vector directional filters: a new class of multichannel image processing filters, *IEEE Trans. Image Process.*, 2(4), 528–534, 1993.

13

Genetic Regulatory Networks: A Nonlinear Signal Processing Perspective

Ilya Shmulevich and Edward R. Dougherty

CONTENTS

13.1 Introduction

The term *functional genomics* refers to the study of how genes affect biological mechanisms and phenotype, in particular by applying large-scale and high-throughput experimental methods. The application of computational methods to these and other related problems is referred to as *computational genomics*. This discipline has been highly influenced by data mining, partly due to the availability of large data sets and databases. Although data mining, as a discipline, is quite broad and lies at the intersection of statistics, machine learning, pattern recognition, and artificial intelligence,[1] there are a number of challenging and important problems in computational genomics that can benefit from the application of engineering principles and methodologies, the latter being characterized by systems-level modeling and simulation.

Modern nonlinear signal processing, although encompassing many of the same subject areas, has had a different history and background, being rooted mostly in traditional signal processing. As such, the applications around which the field has developed have been of a nature substantially different from those in data mining. Whereas data-mining problems are often centered around visualization and exploratory analysis of large, high-dimensional data sets, finding patterns in data and discovering good feature sets for classification, some common tasks in signal processing include removal of interference from signals, transforming signals into more suitable representations for various purposes, and analyzing and extracting some characteristics from signals.[2]

Of importance in nonlinear signal processing in particular is the optimal design of nonlinear operators under various criteria and constraints. That is, given a "true" signal and its noise-corrupted version, the goal is to find an optimal estimator from some class of estimators (constraint), such that when it is applied to the noisy signal, some error (criterion) between its output and the true signal is minimized. Alternatively, if a representative signal is not available for training, and one is armed with only the knowledge of the noise characteristics and a class of operators, the goal is to select an optimal estimator under a different criterion, such as minimizing the variance of the noise at its output.

Although these approaches have much in common with machine learning and statistical estimation theory, the nature of the constraints and criteria, and consequently the ensuing theory and algorithms, are guided by application-specific needs, such as detail and edge preservation, robustness to outliers, and other statistical and structural constraints. At the same time, much of the theory behind nonlinear signal processing, in particular nonlinear digital filters, is tightly intertwined with dynamical systems theory, involving constructs such as finite and cellular automata. In this chapter, we consider these topics in the context of computational genomics and, in particular, models of genetic regulatory networks. We should point out that the role of nonlinear signal processing extends well beyond models and inference of genetic networks, and can also be quite useful for the analysis of gene expression data (e.g., Reference 3).

13.2 Genetic Regulatory Networks

In living organisms, genes code for proteins. These proteins are in turn used to control the regulation of other genes. Such interactions, when considered collectively, form complex gene regulatory networks. To gain an understanding of the dynamical behavior and characteristics of such complex regulatory systems, it is necessary to be able to observe them in a global, large-scale fashion. The recent development of high-throughput technologies, such as cDNA microarrays and oligonucleotide chips,[4-8] is empowering researchers in the collection of broad-scope gene information. The diagnostic potential of gene expression data has already been demonstrated. For example, cancer classification using a variety of methods has been used to exploit the class-separating power of expression data.[9-14] The next step is to dig deeper and understand the underlying mechanisms and the functions of genes in health and disease.

One approach is to model the genetic regulatory system and infer the model structure and parameters from real gene expression data. There are two main objectives. The first is to discover and understand the underlying gene regulatory mechanisms by means of inferring them from data. This generally falls within the scope of computational learning theory[15] or system identification.[16] Second, by using the inferred model, we endeavor to make useful predictions of the system under study by mathematical analysis and computer simulations. The potential clinical impact is tremendous as this type of model-based analysis not only can open a window on the physiology of an organism and disease progression, but can also eventually translate into accurate diagnosis, target identification, drug development, and treatment.

A question of immediate concern is: What class of models should be chosen such that it is compatible with the currently available data as well as with our intended goals of modeling and analysis? In other words, will we have the right type of data to infer our models and can we hope to answer our questions by working with these models? A simplified, abstract model with fewer parameters and lower complexity can succeed in capturing "high-level" phenomena and will impose fewer requirements on the data, both in terms of quality and quantity, for its inference. A fine-scale model with many parameters may be able to capture detailed "low-level" phenomena, such as kinetics of biochemical reactions, but will naturally require large amounts of data on all the relevant features.

Although there is a rather wide spectrum of approaches for modeling gene regulatory networks,[27–29] each with its own assumptions, data requirements, and goals, in this chapter, we focus on the *Boolean network* model, originally introduced by Kauffman.[17–20] Good reviews of this model can be found in References 21 through 23. Boolean networks have been one of the most intensively studied models of discrete dynamical systems, enjoying a sustained interest from both the biology and physics communities. Although structurally simple, these systems are capable of displaying a remarkably rich variety of complex behavior, having much in common with other dynamical system models. They are well suited for discovering qualitative relationships underlying genetic regulation and control, as they emphasize fundamental generic coarse-grained properties of large networks and have yielded insights into the overall behavior of large genetic networks.[23–26]

In this model, gene expression is quantized to only two levels: ON and OFF. The expression level (state) of each gene is functionally related to the expression states of some other genes, using logical rules. Dynamics are introduced by synchronous updating of all genes. Abundant biological justification for using Boolean network models for genetic network analysis is contained in Kauffman's books[21,30,31] and in the excellent reviews by Huang.[22,32] Our main focus here is to discuss the relationships between these models of genetic regulatory networks and some topics in nonlinear signal processing theory.

13.3   Boolean Networks

A Boolean network contains n elements (genes) $\{x_1, \ldots, x_n\}$. Each gene $x_i \in \{0, 1\}$ $(i = 1, \ldots, n)$ is a binary variable whose value at time $t+1$ is completely determined by the values of some other genes $x_{j_1(i)}, x_{j_2(i)}, \ldots, x_{j_{k_i}(i)}$ at time t by means of a Boolean function $f_i : \{0, 1\}^{k_i} \to \{0, 1\}$. That is, there are k_i genes assigned to gene x_i and the mapping $j_k : \{1, \ldots, n\} \to \{1, \ldots, n\}, k = 1, \ldots, k_i$ determines the "wiring" of gene x_i. Thus, we can write

$$x_i(t + 1) = f_i(x_{j_1(i)}(t), x_{j_2(i)}(t), \ldots, x_{j_{k_i}(i)}(t)). \tag{13.1}$$

In a *random Boolean network*, the functions f_i are selected randomly as are the genes that are used as its inputs. Each x_i represents the state (expression) of gene i, where $x_i = 1$ represents the fact that gene i is expressed and $x_i = 0$ means it is not expressed. A given gene transforms its inputs (regulatory factors that bind to it) into an output, which is the state or expression of the gene itself at the next time point. All genes are assumed to update synchronously in accordance with the functions assigned to them and this process is then repeated. The artificial synchrony simplifies computation while preserving the qualitative, generic properties of global network dynamics.[21,22,25] It is clear that the dynamics of the network are completely determined by Equation 13.1. Let us give an example.

Consider a Boolean network consisting of five genes $\{x_1, \ldots, x_5\}$ with the corresponding Boolean functions given by the truth tables shown in Table 13.1.

TABLE 13.1

Truth Tables of the Functions in a
Boolean Network with Five Genes

	f_1	f_2	f_3	f_4	f_5
	0	0	0	0	0
	1	1	1	0	0
	1	1	1	0	0
	1	0	0	1	0
	0	0	1	0	0
	1	1	1	1	0
	1	1	0	1	0
	1	1	1	1	1
j_1	5	3	3	3	5
j_2	2	5	1	4	4
j_3	4	4	5	4	1

Note: The indices j_1, j_2, and j_3 indicate the input connections for each of the functions.

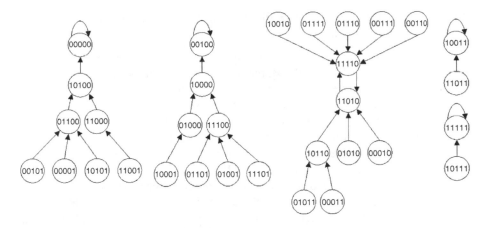

FIGURE 13.1
The state-transition diagram for the Boolean network in Table 13.1.

The *maximum connectivity*, defined as $K = \max_i k_i$, is equal to 3 in this case, although we allow some input variables to duplicate, essentially reducing the connectivity. For example, consider f_4, which is the truth table of the well-known majority function (median). We see that because $j_1(4) = 3$ and $j_2(4) = j_3(4) = 4$, $f_4(x_3, x_4, x_4) = x_4$, which is a function of only one (*essential*) variable.*

The dynamics of this Boolean network are shown in Figure 13.1. Because there are five genes, there are $2^5 = 32$ possible states that the network can be in. Each state is represented by a circle and the arrows between states show the transitions of the network according to the functions in Table 13.1. It is easy to see that because of the inherent deterministic directionality in Boolean networks as well as only a finite number of possible states, certain states will be revisited infinitely often if, depending on the initial starting state, the network happens to transition into them. Such states are called *attractors* and the states that lead into them comprise their *basins of attraction*. For example, in Figure 13.1, the state (00000) is an attractor and the seven other (transient) states that eventually lead into it are its basin of attraction.

The attractors represent the *fixed points* of the dynamical system that capture its long-term behavior. The attractors are always cyclical and may consist of more than one state. The number of transitions needed to return to a given state in an attractor is called the *cycle length*. For example, the attractor (00000) has cycle length 1 while the states (11010) and (11110) comprise an attractor with cycle length 2.

---

* The majority of x_3, x_4, and x_4 is always x_4. Other variables are called *fictitious*.

13.3.1   Attractors, Nonlinear Digital Filters, Associative Memory, and Root Signals

Real genetic regulatory networks are highly stable in the presence of perturbations of the genes. Within the Boolean network formalism, this means that when a minimal number of genes change value (say, by means of some external stimulus), the system transitions into states that reside in the same basin of attraction from which the network eventually flows back to the same attractor. Generally speaking, large basins of attraction correspond to higher stability. Such stability of networks in living organisms allows the cells to maintain their functional state within the tissue environment.[22]

Although in developmental biology, epigenetic, heritable changes in cell determination have been well established, it is now becoming evident that the same type of mechanisms may be also responsible in carcinogenesis and that gene expression patterns can be inherited without the need for mutational changes in DNA.[33] In the Boolean network framework, this can be explained by *hysteresis*, which is a change in the system's state caused by a stimulus that is not changed back when the stimulus is withdrawn.[22] Thus, if the change of some particular gene does in fact cause a transition to a different attractor, the network will often remain in the new attractor even if that gene is switched off. Thus, the attractors of a Boolean network also represent a type of memory of the dynamical system.[22]

Virtually the same type of behavior is exhibited by many nonlinear digital filters. Let us consider a popular class of nonlinear filters called *stack filters*, first introduced by Wendt et al.[34] Suppose a window of length $n = 2m + 1$ is sliding across a binary-valued one-dimensional (1D) signal of arbitrary length. At every location of the window, the contents inside the window are used as input variables to some fixed Boolean function f. That is,

$$y_i = f(x_{i-m}, \ldots, x_i, \ldots, x_{i+m}) \tag{13.2}$$

represents the output of the Boolean function corresponding to the window centered on the ith value of the input signal. The sequence of outputs y_i can be thought of as an output signal of the filter. If this Boolean function is monotone (positive), meaning that it can be written without complemented variables in its disjunctive normal form,* then the filter defined by such a Boolean function is a stack filter. It also corresponds to a neural network in which the weights of all the threshold logic gates are nonnegative. In the real-valued domain, stack filters can be defined as follows. Let $\mathbf{X} = (X_1, X_2, \ldots, X_n)$ be a real-valued vector of observations comprising the contents of the filter window. Then a stack filter $S(\mathbf{X})$ is defined as

$$S(\mathbf{X}) = \max\left\{ \min\{X_j : j \in P_1\}, \ldots, \min\{X_j : j \in P_K\}\right\},$$

---

*Thus, the disjunctive normal form contains only disjunctions and conjunctions. A monotone Boolean function can also be defined as follows: $f(x_1, \ldots, x_n)$ is called *monotone* if for any two vectors $\tilde{\alpha}, \tilde{\beta} \in \{0, 1\}^n$ such that $\alpha_i \leq \beta_i$ for every i ($1 \leq i \leq n$), we have $f(\tilde{\alpha}) \leq f(\tilde{\beta})$.

where $P_i \subseteq \{0, 1, 2, \ldots, n\}, i = 1, \ldots, K, X_0 \equiv 1$, and not all P_i are empty. The sets $P_1, \ldots, P_K$ completely define the stack filter.

It is well known that by using a property called *threshold decomposition*,[35] all statistical and deterministic properties of stack filters can be obtained by considering only the binary domain and, hence, the monotone Boolean function defining the stack filter. Clearly, the max and min operations, used above in the real-valued domain definition, are generalizations of the disjunction and conjunction operations in the binary domain.

For example, consider the well-known running median, originally introduced by Tukey (Reference 36, p. 210). If the window size is 3, that is, $\mathbf{X} = (X_1, X_2, X_3)$, then the filtering operation at each location of the signal is

$$\text{MED}(\mathbf{X}) = \max\{\min\{X_1, X_2\}, \ \min\{X_2, X_3\}, \min\{X_1, X_3\}\}.$$

The monotone Boolean function corresponding to this operation is

$$f(x_1, x_2, x_3) = x_1 x_2 \lor x_2 x_3 \lor x_1 x_3.$$

It is easy to see that for a finite-length signal,* Equation 13.2 is really a special case of Equation 13.1, where

$$f_i = f, \ k_i = n \ (i = 1, \ldots, n)$$
$$j_1(i) = i - m, \ j_2(i) = i - m + 1, \ldots, \ j_n(i) = i + m.$$

So, the stack filter is really a Boolean network with simple fixed "wiring" defined by the neighborhood structure (window). In fact, such a "sliding window" filtering process actually corresponds to a cellular automaton and can easily be extended to two- or higher-dimensional signals. Suppose also that we repeat the filtering operation such that the output of one filtering pass is used as an input to the next filtering pass with the same filter. Thus, the entire signal corresponds to a state of a Boolean network and one filtering pass corresponds to a transition from that state to the next state (i.e., input–output relationship). If the filtering process is repeated, the same cyclical phenomenon that is exhibited by Boolean networks will also occur with stack filters. That is, either the signal will converge to a *root signal* after a finite number of filtering passes or periodic behavior will ensue.[†]

Root signals represent an important concept that is used to characterize the operation of stack filters. A root signal of a given filter is a signal that is invariant to applications of that filter; i.e., the signal remains unchanged.

---

* For the case of finite-length signals, various appending strategies can be used to augment the left- and right-hand sides of the signal so that the output signal is of the same length as the input signal.

[†] It is interesting to note that similar periodic behavior exists even for some infinite networks (networks with an infinite number of nodes),[37] such as those in which every Boolean function is the majority function.

When stack filters are viewed as Boolean networks, the root signals are simply the attractors with cycle length equal to 1. Similarly, periodic behavior of some stack filters is due to the attractors with cycle length greater than 1. Root signals represent the "pass-band" characteristics of a filter, much like the frequencies passed by a linear filter, and have been studied extensively for different types of filters, such as median filters, stack filters, and morphological filters.[34,35,38–41]

Root signals also represent a type of memory of a filter. These ideas were developed in a series of papers by Yu and Coyle[42,43] on the *associative memory* of stack filters, where the set of root signals represents the filter's associative memory. This, of course, is closely related to the associative memory of neural networks. Indeed, because any positive Boolean function can be implemented as a cascade of so-called threshold logic gates,[44] a stack filter can be viewed as a neural network.[45] In fact, the monotone property of the Boolean functions ensures that each neuron makes the best possible decision regarding whether or not the input signal is greater than a given level, subject to consistency with the decision of other neurons in the network.[45]

Boolean networks are essentially deterministic finite-state machines or automata without inputs. The connection between stack filters and automata has been explored for determining the statistical behavior of recursive median filters[46,47] and stack filters.[48,49] Root signals of stack filters were also studied by using deterministic finite automata.[50] It was shown that the set of root signals for a particular stack filter constitutes a regular language, and a simple procedure for testing whether root signals of one stack filter are also root signals of another filter was developed, making it possible to test whether the root signal sets of two different stack filters are equivalent.

Finally, Boolean networks also represent an abstract model of computation, by transforming a finite configuration (input) into another configuration (output). For example, the partitioning of the state space into attractors with their respective basins of attraction is a form of classification. In Reference 42, stack filters, as simple cases of Boolean networks, were also considered to be a type of classifiers. Similar work showed that cellular automata, which are also special cases of Boolean networks, can process information[51] and are able to perform computations, such as density classification.[52,53]

13.3.2 Inference of Networks

To make progress in understanding the genetic regulation in specific organisms and develop tools for rational therapeutic intervention in diseases such as cancer, it is necessary to be able to identify the networks from real experimental data. Much recent work on Boolean networks has focused on identifying the network structure from gene expression data.[54–63] At the same time, a large body of related work in computational learning theory has addressed very similar problems, namely, learning or inferring Boolean functions from examples of their input–output behavior. A major focus in this field has been on the construction of algorithms for efficient determination of Boolean formulae from examples.[63]

While the focus in computational learning theory has mostly been on the complexity of learning and the design of efficient inference algorithms, very similar types of problems have been studied in nonlinear signal processing, specifically, in optimal filter design.[64–70] Indeed, the specific roles of computational learning[77] and pattern recognition theory[78] in nonlinear signal processing have been studied. Inference typically involves designing an estimator from some predefined class of estimators that minimizes the error of estimation among all estimators in the class. An important role in filter design is played by these predefined classes or constraints. For example, as discussed above, stack filters are represented by the class of monotone Boolean functions. Although it would seem that imposing such constraints can only result in a degradation of the performance (larger error) relative to the optimal filter with no imposed constraints, constraining may have certain advantages. These include prior knowledge of the degradation process (or in the case of gene regulatory networks, knowledge of the likely class of functions, such as canalizing functions), tractability of the filter design, and precision of the estimation procedure by which the optimal filter is estimated from observations. For example, we often know that a certain class of filters will provide a very good sub-optimal filter, while lessening the data requirements for its estimation.

To quantify the advantage (or disadvantage) of filter constraint, suppose data from a sample of size n are used to design an estimate ψ_n of the optimal filter ψ_{opt}. The error, ε_n, of ψ_n cannot be less than the error, ε_{opt}, of the optimal filter. The corresponding design cost is $\Delta_n = \varepsilon_n - \varepsilon_{opt}$, and the error of the designed filter is decomposed as $\varepsilon_n = \varepsilon_{opt} + \Delta_n$. Hence, the expected error of the designed filter is

$$E[\varepsilon_n] = \varepsilon_{opt} + E[\Delta_n].$$

The essential problem for nonlinear filtering is that satisfactory filtering often requires large windows, especially for images, and it is often impossible to obtain large enough samples to sufficiently reduce $E[\Delta_n]$. Thus, optimization is constrained to some subclass C of filters. If ψ_C is an optimal filter in C with error ε_C and design error $\Delta_{n,C}$, then $\varepsilon_C \geq \varepsilon_{opt}$ and $E[\Delta_{n,C}] \leq E[\Delta_n]$. The error of a designed constrained filter, $\psi_{n,C}$, possesses the decomposition $\varepsilon_{n,C} = \varepsilon_C + \Delta_{n,C}$. The cost of constraint is given by $\Delta_C = \varepsilon_C - \varepsilon_{opt}$. Hence, $\varepsilon_{n,C} = \varepsilon_{opt} + \Delta_C + \Delta_{n,C}$, and

$$E[\varepsilon_{n,C}] = \varepsilon_{opt} + \Delta_C + E[\Delta_{n,C}]. \tag{13.3}$$

Constraint is statistically beneficial if and only if $E[\varepsilon_{n,C}] \leq E[\varepsilon_n]$, which is true if and only if

$$\Delta_C \leq E[\Delta_n] - E[\Delta_{n,C}]. \tag{13.4}$$

The savings in design error must exceed the cost of constraint.

A fundamental problem of nonlinear digital signal processing is to find constraints for which Equation 13.4 is satisfied. The benefit of a constraint is dependent on the class of ideal and observed signals under consideration. C may be defined theoretically in accordance with knowledge of the signal degradation, such as C consisting of antiextensive filters when degradation is extensive, or it may be that experience has shown that a certain constraint works well in a given setting. There is often no fine line between these two situations. When degradation is extensive, there is no constraint, because the optimal filter lies in C and $\Delta_C = 0$. Many more interesting constraints have been considered. We first mention a few cases of binary constraints that have been studied.

Envelope constraint involves two *a priori* filters, α and β, such that $\alpha \le \beta$, and a designed filter ψ must lie in the envelope determined by α and β, meaning that $\alpha \le \psi \le \beta$.[79] C is the class of all filters ψ such that $\alpha \le \psi \le \beta$. If the envelope contains the optimal filter, then $\Delta_C = 0$; if not, then $\Delta_C > 0$ and the constraint is beneficial if and only if Equation 13.4 is satisfied.

Another way to utilize Equation 13.4 is to recognize the exponential growth of the data requirement as the number of variables grows. In image processing, it is often the case that the variables near the center of the window contribute most to the filter, whereas those on the window periphery contribute less,[80] while enormously increasing the demand for data. In this case, one can apply secondary constraints.[81] The variables near the window center are used in an unconstrained manner, whereas those at the periphery are constrained in how they contribute to the filter. The situation here is that Δ_C is not too large owing to the lesser importance of peripheral variables.

Iterative design involves a filter-decomposition constraint. A large window W is decomposed into a Minkowski sum of windows, $W = W_1 \oplus W_2 \oplus \cdots \oplus W_q$, and C contains filters of the form $\psi = \psi_q \psi_{q-1} \cdots \psi_1$, where ψ_k is defined on W_k.[82,83] Not only do iterative filters require much less sample data, they can also possess implementation advantages. Δ_C depends on the degree to which the optimal filter can be approximated by an iterative filter, not only relative to algebraic decomposition, but also relative to the action of the filter on the random signal process.

The most studied constraint in nonlinear filtering is monotonicity or increasingness, as discussed in Section 13.3.1, for which special design methods exist.[64,65,68,84,85] The precision of filter design for increasing filters has been studied in terms of their morphological bases, and the inequality in Equation 13.4 has been considered in that context.[67]

Gray-scale filters require much more data for design. A method that appears to have growing potential for gray-scale signals is aperture constraint.[86] For aperture filters, the gray scale is constrained, so that the input vector is constrained to a domain-range aperture (product window). Increasing gray-scale filters have also been considered.[87]

The purpose of constraint is to reduce design cost. This issue has been extensively studied in the theory of pattern recognition, where classifiers are typically considered to be binary operators, so that the theory can be readily applied to nonlinear filtering in the context of binary representation, as in the case of stack filters. In particular, here we consider binary-valued functions defined on d-dimensional Euclidean space.

For filter class C and classifier $\psi \in C$, the *empirical error*, $\hat{\varepsilon}_n[\psi]$, on the sample data is the fraction of times that $\psi(\mathbf{X}^i) \neq Y^i$ for $(\mathbf{X}^i, Y^i)$ in the sample. It is the error rate on the sample. The empirical-error classifier, $\hat{\psi}_{n,C}$, minimizes the empirical error. We denote its mean-absolute error by $\hat{\varepsilon}_{n,C}$. The design error is $\hat{\Delta}_{n,C} = \hat{\varepsilon}_{n,C} - \varepsilon_C$. In the decomposition of Equation 13.3, $E[\varepsilon_{n,C}]$ and $E[\Delta_{n,C}]$ are replaced by $E[\hat{\varepsilon}_{n,C}]$ and $\mathbf{E}[\hat{\Delta}_{n,C}]$, respectively.

Associated with the filter class C is its *Vapnik–Chervonenkis (VC) dimension*, V_C, whose details we leave to the literature.[88] This dimension can either be finite or infinite. If it is finite, then C is called a *VC class*. Constraining the filter class lowers the VC dimension. As a consequence of the fundamental VC theorem,[89,90] for VC classes with dimension exceeding 2, the expected value of the empirical-error design cost is given by

$$E[\hat{\Delta}_{n,C}] \leq 4\sqrt{\frac{V_C \log n + 4}{2n}}.$$

The VC dimension provides a bound in the design cost.

This bound can be applied to the design of nonlinear filters. Consider an increasing filter $\psi : \mathfrak{R}^d \rightarrow \{0, 1\}$. Such a filter possesses a minimal morphological representation in terms of erosions.[91,92] The class of all increasing filters of this kind possessing a minimal representation with m erosions has VC dimension bounded by d^m [78]. Hence,

$$E[\hat{\Delta}_{n,C}] \leq 4d^{m/2}\sqrt{\frac{\log n + 4}{2n}}.$$

The kind of issues studied in pattern recognition theory apply to genetic regulatory networks, in particular to Boolean networks.

An important role in the inference of multivariate relationships between genes was played by the *coefficient of determination* (COD), introduced by Dougherty et al.[71–73] in the context of optimal nonlinear filter design. The COD was subsequently proposed for inference of probabilistic Boolean networks as models of genetic regulatory networks.[74] The COD measures the degree to which the expression levels of an observed gene set can be used to improve the prediction of the expression of a target gene relative to the best possible prediction in the absence of observations. The method allows incorporation of knowledge of other conditions relevant to the prediction, such as the application of particular stimuli, or the presence of inactivating gene mutations, as predictive elements affecting the expression level of a given gene. Using the COD, one can find sets of genes related multivariately to a given target gene.

Let us briefly discuss the COD in the context of Boolean networks. Let x_i be a *target* gene that we wish to predict by observing some other genes $x_{i_1}, x_{i_2}, \ldots, x_{i_k}$. Also, suppose $f(x_{i_1}, x_{i_2}, \ldots, x_{i_k})$ is an optimal predictor of x_i relative to some error measure ε. For example, in the case of mean-square error (MSE) estimation, the optimal predictor is the conditional expectation of x_i given $x_{i_1}, x_{i_2}, \ldots, x_{i_k}$. Let ε_{opt} be the optimal error achieved by f. Then, the COD for x_i relative to $x_{i_1}, x_{i_2}, \ldots, x_{i_k}$ is defined as

$$\theta = \frac{\varepsilon_i - \varepsilon_{\text{opt}}}{\varepsilon_i}, \tag{13.5}$$

where ε_i is the error of the best (constant) estimate of x_i in the absence of any conditional variables. It is easily seen that the COD must be between 0 and 1 and measures the relative decrease in error from estimating x_i via f rather than by just the best constant estimate. In practice, the COD must be estimated from training data with designed approximations used in place of f. Those sets of (predictive) genes that yield the highest COD, compared to all other sets of genes, are the ones used to construct the optimal predictor of the target gene. Given limited amounts of training data, it is prudent to constrain the complexity of the predictor by limiting the number of possible predictive genes that can be used. This corresponds to limiting the connectivity K of the Boolean network. Finally, the above procedure is applied to all target genes, thus estimating all the functions in a Boolean network. The method is computationally intensive and massively parallel architectures have been employed to handle large gene sets.[75]

13.4   Concluding Remarks

The modeling and analysis of genetic regulatory networks represent an important and rapidly developing area in computational genomics research, requiring a multidisciplinary approach. In this chapter, by using Boolean networks as an illustrative example, we have discussed several intimate connections with nonlinear signal processing theories related to root signals and optimal design of nonlinear digital filters. Many methods and algorithms in nonlinear signal processing, such as the design of nonlinear filters with a specified set of root signals, can be carried over to Boolean networks and their generalizations. For example, ideas and results from mathematical morphology were recently used to characterize important mappings between so-called probabilistic Boolean networks,[76] which have been proposed as models for genetic regulatory networks. It is our belief that researchers with a background in nonlinear signal processing have the potential to make significant contributions and bring their unique perspectives to this exciting and important field.

References

1. D. Hand, H. Mannila, and P. Smyth, *Principles of Data Mining,* Cambridge, MA: MIT Press, 2001.
2. J. Astola and P. Kuosmanen, *Fundamentals of Nonlinear Digital Filtering,* Boca Raton, FL: CRC Press, 1997.
3. K.M. Bloch and G.R. Arce, Median correlation for the analysis of gene expression data, *Signal Process.,* 83(4), 811–823, 2003.
4. M. Schena, D. Shalon, R.W. Davis, and P.O. Brown, Quantitative monitoring of gene expression patterns with a complementary DNA microarray, *Science,* 270, 467–470, 1995.
5. J.E. Celis, M. Kruhøffer, I. Gromova, C. Frederiksen, M. Østergaard, T. Thykjaer, P. Gromov, J. Yu, H. Pálsdóttir, N. Magnusson, and T.F. Ørntoft, Gene expression profiling: monitoring transcription and translation products using DNA microarrays and proteomics, *FEBS Lett.,* 480(1), 2–16, 2000.
6. T.R. Hughes, M. Mao, A.R. Jones, J. Burchard, M.J. Marton, K.W. Shannon, S.M. Lefkowitz, M. Ziman, J.M. Schelter, M.R. Meyer, S. Kobayashi, C. Davis, H. Dai, Y.D. He, S.B. Stephaniants, G. Cavet, W.L. Walker, A. West, E. Coffey, D.D. Shoemaker, R. Stoughton, A. P. Blanchard, S.H. Friend, and P.S. Linsley, Expression profiling using microarrays fabricated by an ink-jet oligonucleotide synthesizer, *Nat. Biotechnol.,* 19, 342–347, 2001.
7. R.J. Lipshutz, S.P.A. Fodor, T.R. Gingeras, and D.J. Lockhart, High density synthetic oligonucleotide arrays, *Nat. Genet.,* 21, 20–24, 1999.
8. D.J. Lockhart and E.A. Winzeler, Genomics, gene expression and DNA arrays, *Nature,* 405, 827–836, 2000.
9. T.R. Golub, D.K. Slonim, P. Tamayo, C. Huard, M. Gaasenbeek, J.P. Mesirov, H. Coller, M.L. Loh, J.R. Downing, M.A. Caligiuri, C.D. Bloomfield, and E.S. Lander, Molecular classification of cancer: class discovery and class prediction by gene expression monitoring, *Science,* 286, 531–537, 1999.
10. A. Ben-Dor, L. Bruhn, N. Friedman, I. Nachman, M. Schummer, and Z. Yakhini, Tissue classification with gene expression profiles, *J. Computational Biol.,* 7, 559–583, 2000.
11. J. Khan, J.S. Wei, M. Ringner, L.H. Saal, M. Ladanyi, F. Westermann, F. Berthold, M. Schwab, C.R. Antonescu, C. Peterson, and P.S. Meltzer, Classification and diagnostic prediction of cancers using gene expression profiling and artificial neural networks, *Nat. Med.,* 7, 673–679, 2001.
12. I. Hedenfalk, D. Duggan, Y. Chen, M. Radmacher, M. Bittner, R. Simon, P. Meltzer, B. Gusterson, M. Esteller, M. Raffeld, Z. Yakhini, A. Ben-Dor, E. Dougherty, J. Kononen, L. Bubendorf, W. Fehrle, S. Pittaluga, S. Gruvverger, N. Loman, O. Johannsson, H. Olsson, B. Wifond, G. Sauter, O.P. Kallioniemi, A. Borg, and J. Trent, Gene expression profiles distinguish hereditary breast cancers, *N. Engl. J. Med.,* 34, 539–548, 2001.
13. S. Kim, E.R. Dougherty, I. Shmulevich, K.R. Hess, S.R. Hamilton, J.M. Trent, G.N. Fuller, and W. Zhang, Identification of combination gene sets for glioma classification, *Mol. Cancer Ther.,* 1, 1229–1236, 2002.
14. T. Kobayashi, M. Yamaguchi, S. Kim, J. Morikawa, S. Ogawa, S. Ueno, E. Suh, E. Dougherty, I. Shmulevich, H. Shiku, and W. Zhang, Microarray reveals

differences in both tumors and vascular specific gene expression in *de novo* CD5+ and CD5− diffuse large B-cell lymphomas, *Cancer Res.*, 63, 60–66, 2003.

15. M. Anthony and N. Biggs, *Computational Learning Theory*, Cambridge, U.K.: Cambridge University Press, 1992.

16. L. Ljung, *System Identification: Theory for the User*, Englewood Clifts, NJ: Prentice-Hall, 1999.

17. S.A. Kauffman, Metabolic stability and epigenesis in randomly constructed genetic nets, *J. Theor. Biol.*, 22, 437–467, 1969.

18. S.A. Kauffman, Homeostasis and differentiation in random genetic control networks, *Nature*, 224, 177–178, 1969.

19. K. Glass and S.A. Kauffman, The logical analysis of continuous, non-linear biochemical control networks, *J. Theor. Biol.*, 39, 103–129, 1973.

20. S.A. Kauffman, The large scale structure and dynamics of genetic control circuits: an ensemble approach, *J. Theor. Biol.*, 44, 167–190, 1974.

21. S.A. Kauffman, *The Origins of Order: Self-Organization and Selection in Evolution*, New York: Oxford University Press, 1993.

22. S. Huang, Gene expression profiling, genetic networks, and cellular states: an integrating concept for tumorigenesis and drug discovery, *J. Mol. Med.*, 77, 469–480, 1999.

23. R. Somogyi and C. Sniegoski, Modeling the complexity of gene networks: understanding multigenic and pleiotropic regulation, *Complexity*, 1, 45–63, 1996.

24. Z. Szallasi and S. Liang, Modeling the normal and neoplastic cell cycle with "realistic boolean genetic networks": their application for understanding carcinogenesis and assessing therapeutic strategies, *Pac. Symp. Biocomput.*, 3, 66–76, 1998.

25. A. Wuensche, genomic regulation modeled as a network with basins of attraction, *Pac. Symp. Biocomput.*, 3, 89–102, 1998.

26. R. Thomas, D. Thieffry, and M. Kaufman, Dynamical behavior of biological regulatory networks. I. Biological role of feedback loops and practical use of the concept of the loop-characteristic state, *Bull. Math. Biol.*, 57, 247–276, 1995.

27. P. Smolen, D. Baxter, and J. Byrne, Mathematical modeling of gene networks, *Neuron*, 26, 567–580, 2000.

28. J. Hasty, D. McMillen, F. Isaacs, and J.J. Collins, Computational studies of gene regulatory networks: *in numero* molecular biology, *Nat. Rev. Genet.*, 2, 268–279, 2001.

29. H. de Jong, Modeling and simulation of genetic regulatory systems: a literature review, *J. Computational Biol.*, 9(1), 69–103, 2002.

30. S.A. Kauffman, *At Home in the Universe*, New York: Oxford University Press, 1995.

31. S.A. Kauffman, *Investigations*, New York: Oxford University Press, 2000.

32. S. Huang, Genomics, complexity and drug discovery: insights from Boolean network models of cellular regulation, *Pharmacogenomics*, 2(3), 203–222, 2001.

33. M.C. MacLeod, A possible role in chemical carcinogenesis for epigenetic, heritable changes in gene expression, *Mol. Carcinogenesis*, 15, 241–250, 1996.

34. P. Wendt, E. Coyle, and N. Gallagher, Stack filters, *IEEE Trans. Acoust. Speech Signal Process.*, 34, 898–911, 1986.

35. J.P. Fitch, E.J. Coyle, and N. Gallagher, Median filtering by threshold decomposition, *IEEE Trans. Acoust. Speech Signal Process.*, 32(6), 1183–1188, 1984.

36. J.W. Tukey, *Exploratory Data Analysis*, Reading, MA: Addison-Wesley, 1977.

37. G. Moran, On the period-two-property of the majority operator in infinite graphs, *Trans. Am. Math. Soc.*, 347(5), 1649–1667, 1995.

38. N.C. Gallagher and G.L. Wise, A theoretical analysis of the properties of median filters, *IEEE Trans. Acoust. Speech Signal Process.*, ASSP-29(6), 1981.

39. P.T. Yu and E.J. Coyle, Convergence behavior and N-roots of stack filters, *IEEE Trans. Acoust. Speech Signal Process.*, 38(9), 1990.

40. M. Gabbouj, P.-T. Yu, and E.J. Coyle, Convergence behavior and root signal sets of stack filters, *Circuits Syst. Signal Process.*, 11(1), 1992.

41. Q. Wang, M. Gabbouj, and Y. Neuvo, Root properties of morphological filters, *Signal Process.*, 34, 131–148, 1993.

42. P.-T. Yu and E.J. Coyle, The classification and associative memory capability of stack filters, *IEEE Trans. Acoust. Speech Signal Process.*, 38(9), 1990.

43. P.-T. Yu and E.J. Coyle, On the existence and design of the best stack filter based associative memory, *IEEE Trans. Circuits Syst.*, 39(3), 171–184, 1992.

44. S. Muroga, *Threshold Logic and Its Applications*. New York: Wiley, 1971.

45. E.J. Coyle and N.C. Gallagher, Stack filters and neural networks, *IEEE Int. Symp. Circuits Syst.*, 2, 995–998, 1989.

46. G. Arce and N.C. Gallagher, Jr., Stochastic analysis for the recursive median filter process, *IEEE Trans. Inf. Theor.*, 34(4), 1988.

47. O. Yli-Harja, I. Shmulevich, J.A. Bangham, R. Harvey, S. Dasmahapatra, and S. Cox, Run-length distributions of recursive median filters using probabilistic automata, in *Proceedings of Scandinavian Conference on Image Analysis*, Kangerlussuaq, Greenland, June 7–11, 1999, 251–258,

48. I. Shmulevich, O. Yli-Harja, K. Egiazarian, and J. Astola, Output distributions of recursive stack filters, *IEEE Signal Process. Lett.*, 6(7), 175–178, July 1999.

49. P. Koivisto, O. Yli-Harja, A. Niemistö, and I. Shmulevich, Breakdown probabilities of recursive stack filters, *Signal Process.*, 81(1), 227–231, Dec. 2000.

50. I. Shmulevich, O. Yli-Harja, K. Egiazarian, and J. Astola, Root signals of stack filters and regular languages, in *Conference on Computer Science and Information Technologies*, Yerevan, Armenia, Aug. 17–23, 1999, 227–230.

51. E.F. Codd, *Cellular Automata*, New York: Academic Press, 1968.

52. M. Mitchell, J.P. Crutchfield, and P.T. Hraber, Evolving cellular automata to perform computations: mechanisms and impediments, *Phys. D*, 75, 361–391, 1994.

53. F. Jiménez Morales, J.P. Crutchfield, and M. Mitchell, Evolving two-dimensional cellular automata to perform density classification: a report on work in progress, *Parallel Comput.*, 27(5), 539–553, 2001.

54. S. Liang, S. Fuhrman, and R. Somogyi, REVEAL, a general reverse engineering algorithm for inference of genetic network architectures, *Pac. Symp. Biocomput.*, 3, 18–29, 1998.

55. T. Akutsu, S. Kuhara, O. Maruyama, and S. Miyano, Identification of gene regulatory networks by strategic gene disruptions and gene overexpressions, *Proc. 9th Annual ACM-SIAM Symposium on Discrete Algorithms (SODA'98)*, 1998, 695–702.

56. T. Akutsu, S. Miyano, and S. Kuhara, Identification of genetic networks from a small number of gene expression patterns under the Boolean network model, *Pac. Symp. Biocomput.*, 4, 17–28, 1999.

57. T. Akutsu, S. Miyano, and S. Kuhara, Inferring qualitative relations in genetic networks and metabolic pathways, *Bioinformatics*, 16, 727–734, 2000.

58. T.E. Ideker, V. Thorsson, and R.M. Karp, Discovery of regulatory interactions through perturbation: inference and experimental design, *Pac. Symp. Biocomput.*, 5, 302–313, 2000.

59. R.M. Karp, R. Stoughton, and K.Y. Yeung, Algorithms for choosing differential gene expression experiments, in *RECOMB99 (ACM)*, 1999, 208–217.

60. Y. Maki, D. Tominaga, M. Okamoto, S. Watanabe, and Y. Eguchi, Development of a system for the inference of large scale genetic networks, *Pac. Symp. Biocomput.*, 6, 446–458, 2001.

61. K. Noda, A. Shinohara, M. Takeda, S. Matsumoto, S. Miyano, and S. Kuhara, Finding genetic network from experiments by weighted network model, *Genome Informatics*, 9, 141–150, 1998.

62. I. Shmulevich, A. Saarinen, O. Yli-Harja, and J. Astola, Inference of genetic regulatory networks under the best-fit extension paradigm, in *Computational and Statistical Approaches to Genomics*, W. Zhang and I. Shmulevich, Eds., Boston: Kluwer, 2002.

63. H. Lähdesmäki, I. Shmulevich, and O. Yli-Harja, On learning gene regulatory networks under the Boolean network model, *Mach. Learning*, 52, 147–167, 2003.

64. E.J. Coyle and J.H. Lin, Stack filters and the mean absolute error criterion, *IEEE Trans. Acoust. Speech Signal Process.*, ASSP-36(8), 1244–1254, Aug. 1988.

65. E.J. Coyle, J.H. Lin, and M. Gabbouj, Optimal stack filtering and the estimation and structural approaches to image processing, *IEEE Trans. Acoust., Speech Signal Process.*, ASSP-37(12), 2037–2066, 1989.

66. R. Yang, L. Yin, M. Gabbouj, J. Astola, and Y. Neuvo, Optimal weighted median filtering under structural constraints, *IEEE Trans. Signal Process.*, 43(3), 591–604, 1995.

67. E.R. Dougherty and R.P. Loce, Precision of morphological-representation estimators for translation-invariant binary filters: increasing and nonincreasing, *Signal Process.*, 40, 129–154, 1994.

68. R.P. Loce and E.R. Dougherty, Optimal morphological restoration: the morphological filter mean-absolute-error theorem, *Vis. Commun. Image Representation*, 3(4), 1992.

69. E.R. Dougherty and J.T. Astola, *Nonlinear Filters for Image Processing*, Bellingham, WA/New York: SPIE/IEEE Press, 1999.

70. E.R. Dougherty and Y. Chen, Optimal and adaptive design of logical granulometric filters, *Adv. Imaging Electron Phys.*, 117, 1–71, 2001.

71. E.R. Dougherty, S. Kim, and Y. Chen, Coefficient of determination in nonlinear signal processing, *Signal Process.*, 80(10), 2219–2235, 2000.

72. S. Kim, E.R. Dougherty, Y. Chen, K. Sivakumar, P. Meltzer, J.M. Trent, and M. Bittner, Multivariate measurement of gene expression relationships, *Genomics*, 67, 201–209, 2000.

73. S. Kim, E.R. Dougherty, M.L. Bittner, Y. Chen, K. Sivakumar, P. Meltzer, and J.M. Trent, General nonlinear framework for the analysis of gene interaction via multivariate expression arrays, *J. Biomed. Opt.*, 5(4), 411–424, 2000.

74. I. Shmulevich, E.R. Dougherty, S. Kim, and W. Zhang, Probabilistic Boolean networks: a rule-based uncertainty model for gene regulatory networks, *Bioinformatics*, 18(2), 261–274, 2002.

75. E.B. Suh, E.R. Dougherty, S. Kim, M.L. Bittner, Y. Chen, D.E. Russ, and R. Martino, Parallel computation and visualization tools for codetermination analysis of multivariate gene-expression relations, in *Computational and Statistical*

Approaches to Genomics, W. Zhang and I. Shmulevich, Eds., Boston: Kluwer, 2002.

76. E.R. Dougherty and I. Shmulevich, Mappings between probabilistic Boolean networks, *Signal Process.,* 83(4), 799–809, 2003.

77. J. Barrera, E.R. Dougherty, and N.S. Tomita, Automatic programming of binary morphological machines by design of statistically optimal operators in the context of computational learning theory, *Electron. Imaging,* 6(1), 54–67, 1997.

78. E.R. Dougherty, and J. Barrera, Pattern recognition theory in nonlinear signal processing, *J. Math. Imaging Vision,* 16(3), 181–197, 2002.

79. J. Barrera, E.R. Dougherty, and M. Brun, Hybrid human-machine binary morphological operator design: an independent constraint approach, *Signal Process.,* 80(8), 1469–1487, 2000.

80. I. Shmulevich, V. Melnik, and K. Egiazarian, The use of sample selection probabilities for stack filter design, *IEEE Signal Process. Lett.,* 7(7), 189–192, July 2000.

81. O.V. Sarca, E.R. Dougherty, and J.T. Astola, Secondarily constrained Boolean filters, *Signal Process.,* 71(3), 247–263, 1998.

82. O.V. Sarca, E.R. Dougherty, and J.T. Astola, Two stage binary filters, *Electron Imaging,* 8(3), 219–232, 1999.

83. N.T. Hirata, E.R. Dougherty, and J. Barrera, Iterative design of morphological binary image operators, *Opt. Eng.,* 39(12), 3106–3123, 2000.

84. N. Hirata, E.R. Dougherty, and J. Barrera, A switching algorithm for design of optimal increasing binary filters over large windows, *Pattern Recognition,* 33(6), 1059–1081, 2000.

85. I. Tabus, D. Petrescu, and M. Gabbouj, A training framework for stack and Boolean filtering—fast optimal design procedures and robustness case study, *IEEE Trans. Image Process.,* 5(6), 809–826, 1996.

86. N. Hirata, E.R. Dougherty, and J. Barrera, Aperture filters, *Signal Process.,* 80(4), 697–721, 2000.

87. R.P. Loce and E.R. Dougherty, Mean-absolute-error representation and optimization of computational-morphological filters, *CVGIP: Image Understanding,* 57(1), 1995.

88. L. Devroye, L. Gyorfi, and G. Lugosi, *A Probabilistic Theory of Pattern Recognition,* New York: Springer-Verlag, 1996.

89. V. Vapnik and A. Chervonenkis, *Theory of Pattern Recognition,* Moscow: Nauka, 1974.

90. V. Vapnik and A. Chervonenkis, On the uniform convergence of relative frequencies of events to their probabilities, *Theory Probab. Its Appl.,* 16, 1971, 264–280.

91. E.R. Dougherty and D. Sinha, Computational mathematical morphology, *Signal Process.,* 38, 21–29, 1994.

92. G.J.F. Banon and J. Barrera, Decomposition of mappings between complete lattices by mathematical morphology. I. General lattices, *Signal Process.,* 30(3), 299–327, 1993.

Index

A

Absorbing states, 396
Acoustic echo, 223–224
Acoustic echo cancelers
  experimental results for, 244–251
  low-complexity nonlinear filters for,
      244–252
  nonlinear adaptive
    adaptivity of, 225–226
    algorithms for, 236–243
    commercial uses of, 226
    efficiency of, 252
    linear filters, 230–231
    low-complexity, 225–236
    memoryless preprocessor, 230–231
    multi-memory-decomposition
        structure, 231–232
    parallel-cascade structure filters,
        232–233
    polynomial filters, 227–230
    Volterra filters, 227–230, 234–235
  overview of, 223–225
Active bubble, 390
Adaptive filters
  data model, 2–3
  energy conservation in, 1–2
  error nonlinearities
    correlated regression data, 27–29
    description of, 23–24
    independent regressors, 26
    long filter approximation, 29–32
    variance relation for, 24–25
    white regression data, 26–27
  fourth-order moment approximation, 22
  long filter approximation, 23
  mean-square behavior, 6–9
  mean-square stability, 9–10
  nearest neighbor, 467
  nearest neighbor multichannel filter, 468
  NLMS algorithm, 15–17, 237–239, 245
  overview of, 1–2
  RLS algorithm, 17–19
  small-step-size approximation, 12–15

steady-state performance, 10–11
weighted variance relation, 4–6
Adaptive noise reduction filtering, 448–453
Additive white Gaussian noise, 267
Affine filters
  applications of, 70–91
  center, 56–57, 81
  definition of, 55, 69
  description of, 38
  image deblocking, 80–83
  inverse synthetic aperture radar image
      filtering, 78–80
  linear, 75
  median, 55–56, 70
  multiresolution signal representations,
      71–77
  nonlinear, 75
  optimization of, 57–59
  robust frequency-selective filtering, 71
  time-frequency cross-term filtering, 83–91
Affine projection algorithms
  description of, 225, 239–243
  experimental results for, 244–246
Aggregate processes, 198
Algorithms
  affine projection, *see* Affine projection
      algorithms
  constant modulus, 187–188, 266
  expectation maximization
    description of, 338–339
    horizontal scanning, 339–340
    initialization of, 339–341
    vertical counting, 340–341
  fast recursive least squares, 237
  FS-CMA, 268
  HMTseg, 355–356
  Lloyd, 409
  LMF, 27
  LMS, 26–27, 237–239
  Newton's, 409
  NLMS, 15–17, 237–239, 245
  nonlinear acoustic echo cancelers,
      236–243
  RLS, 17–19